Statistics in the Social and Behavioral Sciences Series

Multiple Correspondence Analysis
and Related Methods

Chapman & Hall/CRC

Statistics in the Social and Behavioral Sciences Series

Aims and scope

Large and complex datasets are becoming prevalent in the social and behavioral sciences and statistical methods are crucial for the analysis and interpretation of such data. This series aims to capture new developments in statistical methodology with particular relevance to applications in the social and behavioral sciences. It seeks to promote appropriate use of statistical methods in these applied sciences by publishing a broad range of reference works, textbooks and handbooks.

The scope of the series is wide, including applications of statistical methodology in sociology, psychology, economics, education, marketing research, political science, criminology, public policy, demography, survey methodology and official statistics. The titles included in the series are designed to appeal to applied statisticians, as well as students, researchers and practitioners from the above disciplines. The inclusion of real examples and case studies is therefore essential.

Proposals for the series should be submitted directly to:

Chapman & Hall/CRC
Taylor and Francis Group
Informa
24-25 Blades Court
Deodar Road
London SW15 2NU, UK

Statistics in the Social and Behavioral Sciences Series

Multiple Correspondence Analysis and Related Methods

Edited by

Michael Greenacre and Jörg Blasius

Chapman & Hall/CRC
Taylor & Francis Group
Boca Raton London New York

Chapman & Hall/CRC is an imprint of the
Taylor & Francis Group, an informa business

Chapman & Hall/CRC
Taylor & Francis Group
6000 Broken Sound Parkway NW, Suite 300
Boca Raton, FL 33487-2742

Printed in the United States of America on acid-free paper
10 9 8 7 6 5 4 3 2 1

International Standard Book Number-10: 1-58488-628-5 (Hardcover)
International Standard Book Number-13: 978-1-58488-628-0 (Hardcover)

Visit the Taylor & Francis Web site at
http://www.taylorandfrancis.com

and the CRC Press Web site at
http://www.crcpress.com

In fond memory of Ruben Gabriel

Kuno Ruben Gabriel, 1929–2003

Preface

This book is aimed at a wide spectrum of researchers and students who are concerned with the analysis of tabular data, chiefly data measured on categorical scales. In other words, all social scientists who work with empirical data will benefit from this book, as well as most environmental scientists, market researchers, psychologists, and archaeologists, to name but a few, where applications of correspondence analysis already abound.

The idea for this book grew out of the international conference on correspondence analysis and related methods (CARME 2003), held at the Universitat Pompeu Fabra in Barcelona from 29 June to 2 July 2003. A total of 88 scientific papers were delivered at the conference, attended by 175 researchers from 18 countries, testifying to the ever-increasing interest in this multipurpose, multicultural, and multidisciplinary statistical method. The extension of correspondence analysis to more than two variables, called multiple correspondence analysis (MCA), and various methods related to it were so prominent at this meeting that we decided to embark on this book project. The idea was to gather experts in the field and to assemble and edit a single text that encompassed the subject, especially the different approaches taken by researchers from different statistical "schools."

For the record, this is the third time we have embarked on a project like this. The first book, *Correspondence Analysis in the Social Sciences* (Greenacre and Blasius 1994), was edited after the first conference organized in Cologne in 1991 and contained various methodological and applied chapters, the latter written mostly by sociologists, in an attempt to show the usefulness of correspondence analysis in exploring social science data. The second book, *Visualization of Categorical Data* (Blasius and Greenacre 1998), which was edited after the second conference organized in Cologne in 1995, broadened the content to all methods that have as their goal the graphical display of categorical data. Once again, both statisticians and social science researchers contributed to the book, which has been as successful as the first one in communicating a new field of multidisciplinary research to a wider audience. The present book, *Multiple Correspondence Analysis and Related Methods*, carries on this tradition, giving a state-of-the-art description of this new field of research in a self-contained textbook format.

A total of 40 authors have contributed to this book, and the editing process was by no means a simple task. As in the two previous books, our objective has been to produce a unified text, with unified presentation and notation. Cross-referencing between chapters was introduced, and a common reference list and index was established. In addition, we have included several introductory chapters so that readers with little experience in the field can be gently introduced to the subject. In our selection of chapters, we tried to be inclusive as well as exhaustive—inclusive of the different cultural origins of the subject's development and exhaustive of the methodological and applications fields, covering the whole subject and a wide variety of application areas. Another goal was to make the book as practically oriented as possible and to include details about software and computational aspects; most chapters have a "software note" at the end, and there is an appendix at the end of the book that summarizes the computational steps of the basic method and some related ones.

Before we move on to the acknowledgments, we feel a note of sadness that our dear friend, Ruben Gabriel, so prominent in this area of methodology, passed away a month before the CARME 2003 conference. Ruben was to have been one of the keynote speakers and a contributor to this book, so it was fitting that we agreed to dedicate this book to his memory and include him in this way in our project. To quote from the obituary written by Jack Hall, Ruben's colleague at the Department of Statistics, University of Rochester:

> Ruben Gabriel was born in Germany. His family fled to France in the 1930's and then to Palestine where he was raised on a Kibbutz. He studied at the London School of Economics and at the Hebrew University of Jerusalem, earning a PhD in Demography in 1957. He was on the faculty of the Department of Statistics at Hebrew University for many years, including a term as chair, and joined the University of Rochester as Professor of Statistics in 1975, serving as chair from 1981 to 1989 and retiring in 1997. While at the University, he also served as Professor of Biostatistics in the Medical Center, collaborating on medical research with faculty in many departments.
>
> Ruben had a distinguished statistical research career, with 150 scientific publications, and was honored as a Fellow of the American Statistical

Association and of the Institute of Mathematical Statistics and an elected member of the International Statistical Institute. Of special note was his "biplot," a graphical data analytic tool to assist in the understanding of the structure of an array of data—say of several variables on each of several units (e.g., persons)—now widely used in data analysis in many fields.

The CARME 2003 conference and this book would not have been possible without the generous support of the Fundación BBVA in Madrid. In the first instance, we would like to thank this organization and especially its director, Rafael Pardo (who also contributes to the book), for their interest in fostering research on correspondence analysis, both on the theoretical and applied levels. We also thank the other sponsors of the conference: the Spanish Ministry of Science and Technology, grant MCYT2096; IdesCAT (the Catalan Statistical Institute); DURSI (the Department of Universities, Research and Information Society of the Catalan government); and the Faculty of Economic Sciences of the Universitat Pompeu Fabra.

Then, to all our authors: you have been very patient with us, and we hope that we have been patient with you! It was a long process, but we hope that you will appreciate the fruits of your labors and share together in the success of this venture. Our respective institutions—the Department of Economics and Business at Pompeu Fabra University, Barcelona, and the Seminar für Soziologie at the University of Bonn—have given us the academic freedom to dedicate many hours to this task that is of our own choosing, and we thank them for that. We also thank Martin Blankenstein (University of Bonn) for preparing the reference list of this book and Andreas Mühlichen (University of Bonn) for assistance in preparing a large number of figures. For their work on the CARME 2003 conference, we acknowledge the assistance of Anna Cuxart, Clara Riba, Frederic Udina, Robert Diez, and Friederika Priemer, and we thank all our colleagues and family for providing heartfelt support.

Finally, we thank Chapman & Hall and editor Rob Calver in London for supporting the publication of this book, and Mimi Williams, project editor, Florida, for handling the complex task of producing this multi-authored book in such a cooperative and friendly way.

Michael Greenacre and Jörg Blasius

Barcelona and Bonn

About the authors

Elena Abascal Fernández is a professor of statistics at the Public University of Navarra, Pamplona, Spain. Her interests are mainly in multivariate data analysis, descriptive factorial analysis, and classification techniques and their application to marketing research and to the analysis of survey data.

Address: Departamento de Estadística e Investigación Operativa, Universidad Pública de Navarra, Campus de Arrosadía, E-31006 Pamplona, Spain.

Email: eabascal@unavarra.es.

Carolyn J. Anderson is an associate professor in the Departments of Educational Psychology, Psychology, and Statistics at the University of Illinois at Urbana-Champaign. Her major interests are in the analysis of multivariate categorical data and latent variable models.

Address: University of Illinois, 1310 South Sixth Street, MC-708, Champaign, IL, 61820.

Email: cja@uiuc.edu.

Web address: http://www.psych.uiuc.edu/~canderso.

Matthias C. Angermeyer is professor and, since 1995, head of the Department of Psychiatry, University of Leipzig. His main research interests are epidemiology of mental disorders, public attitudes toward people with mental illness, and evaluation of mental health services.

Address: University of Leipzig, Department of Psychiatry, Johannisallee 20/1, 04317 Leipzig, Germany.

Mónica Bécue-Bertaut is a professor in the Department of Statistics and Operational Research, Universitat Politècnica de Catalunya, Spain. Her interests are primarily in textual statistical methods and text mining tools and their application to open-ended responses, especially to multilingual responses. She is coauthor (with Ludovic Lebart and André Salem) of the book *Análisis Estadístico de Textos.*

Address: Departament d'Estadística i Investigació Operativa, Universitat Politècnica de Catalunya, Edifici FME, c/ Pau Gargallo, 5, E-08028 Barcelona, Spain.

Email: Monica.Becue@upc.es.

Antonio Blázquez-Zaballos is an assistant professor in the Departamento de Estadística, Universidad de Salamanca, Spain. His main research interest is in multivariate data analysis, particularly in the generalizations of biplots to mixed data types. Recent work includes contributions to the detection of genotype-environment interaction.

Address: Departamento de Estadística, Universidad de Salamanca, C/ Espejo s/n, 37007, Salamanca, Spain.

E-mail: abz@usal.es.

Jörg Blasius is a professor of sociology at the Institute for Political Science and Sociology, Bonn, Germany. His research interests are mainly in exploratory data analysis, data collection methods, sociology of lifestyles, and urban sociology. Together with Michael Greenacre, he has coedited two books on correspondence analysis and visualizing of categorical data.

Address: University of Bonn, Institute for Political Science and Sociology, Seminar of Sociology, Lennéstr. 27, 53113 Bonn, Germany.

Email: jblasius@uni-bonn.de.

Web address: http://www.soziologie.uni-bonn.de/blasius1.htm.

Stéphanie Bougeard is a researcher in the veterinary epidemiological team of the French Agency for Food Safety (AFSSA). Her research interests are in factorial methods for qualitative data. This work is a part of her Ph.D. thesis.

Address: Department of Epidemiology, French Agency for Food Safety (AFSSA), Les Croix, BP53, F-22 440 Ploufragan, France.

Email: s.bougeard@afssa.fr.

Henri Caussinus is an emeritus professor, Laboratory of Statistics and Probabilities, Université Paul Sabatier, Toulouse, France. One of his main interests is the use of probabilistic models for exploratory data analysis, with special emphasis on graphical displays and on the choice of their relevant dimension.

Address: Laboratoire de Statistique et Probabilités, Université Paul Sabatier, 118 route de Narbonne, F-31062 Toulouse Cedex 04, France.

Email: Caussinus@cict.fr.

Jan de Leeuw is Distinguished Professor and Chair of Statistics at the University of California at Los Angeles. His interests are mainly in multivariate analysis, optimization, and computational statistics.

Address: Department of Statistics, University of California at Los Angeles, Box 951552, Los Angeles, CA 90095-1664.

E-mail: deleeuw@stat.ucla.edu.

Web address: http://gifi.stat.ucla.edu.

M. Purificación Galindo Villardón is a professor and head in the Departamento de Estadística, Universidad de Salamanca, Spain. Her main research interest is in multivariate data analysis, particularly the application of biplot and correspondence analysis to clinical and environmental data. Recent work includes contributions to three-way nonsymmetrical correspondence analysis.

Address: Departamento de Estadística, Universidad de Salamanca, C/ Espejo s/n, 37007, Salamanca, Spain.

E-mail: pgalindo@usal.es.

Ignacio García Lautre is an associate professor of statistics at the Public University of Navarra, Pamplona, Spain. His interests are mainly in multivariate data analysis, descriptive factorial analysis, classification techniques, and their application to economic data and, more generally, to the analysis of survey data.

Address: Departamento de Estadística e Investigación Operativa, Universidad Pública de Navarra, Campus de Arrosadía, E-31006 Pamplona, Spain.

Email: nacho@unavarra.es.

Beatriz Goitisolo is a professor of statistics and econometrics at the University of Basque Country, Bilbao, Spain. Her interests are mainly in multivariate data analysis in general and in particular in the analysis of several contingency tables and their applications to social sciences.

Address: Departamento de Economía Aplicada III (Econometría y Estadística), Facultad de Ciencias Económicas y Empresariales, Avda Lehendakari Agirre 83, E-48015 Bilbao, Spain.

Email: Beatriz. Goitisolo@ehu.es.

John C. Gower, formerly head of the Biomathematics Division, Rothamsted Experimental Station, is currently a professor of statistics at the Open University, U.K. His research interests are mainly in exploratory multivariate data analysis and visualization, especially biplot methodology, Procrustes analysis, multidimensional scaling, analysis of asymmetry, and classification methods.

Address: Department of Statistics, Walton Hall, The Open University, Milton Keynes, MK7 6AA, U.K.
Email: j.c.gower@open.ac.uk.
Web address: http://statistics.open.ac.uk/staff/jg1.html.

Michael Greenacre is a professor of statistics in the Department of Economics and Business, Pompeu Fabra University in Barcelona. His research interests are mainly in exploratory analysis and visualization of large data sets, and applications of multivariate analysis in sociology and marine ecology. He has published two books on correspondence analysis and co-edited two more with Jörg Blasius.

Address: Departament d'Economia i Empresa, Universitat Pompeu Fabra, Ramon Trias Fargas 25-27, E-08005 Barcelona, Spain.
Email: michael@upf.es.
Web address: http://www.econ.upf.es/~michael.

Patrick J.F. Groenen is a professor of statistics at the Econometric Institute, Erasmus University Rotterdam, the Netherlands. His research interests are exploratory multivariate analysis, optimization, visualization, and multidimensional scaling. Together with Ingwer Borg, he has written a textbook on multidimensional scaling.

Address: Econometric Institute, Erasmus University Rotterdam, P.O. Box 1738 DR Rotterdam, the Netherlands.
E-mail: groenen@few.eur.nl.

Mohamed Hanafi is a researcher at the École Nationale des Ingénieurs des Industries Agricoles et Alimentaires (ENITIAA). His interests are primarily in multivariate analysis with applications in sensory analysis and chemometrics.

Address: ENITIAA-INRA Unité de Sensométrie et Chimiométrie, Rue de la Géraudière, BP 82225, F-44322 Nantes Cedex 03, France.
Email: hanafi@enitiaa-nantes.fr.

Willem J. Heiser is a professor and head of the Section of Psychometrics and Research Methodology, Department of Psychology, at Leiden University. His interests are multivariate analysis techniques, multidimensional scaling, optimal scaling, classification methods, and the history of statistics. He has been the president of the Psychometric Society, 2003–2004 and is the present editor of the *Journal of Classification*.

Address: Wassenaarseweg 52, P.O. Box 9555, 2300 RB Leiden, The Netherlands.
Email: heiser@fsw.leidenuniv.nl.

Heungsun Hwang is an assistant professor of marketing, HEC Montréal, Montréal, Canada. His recent interests include generalizations of growth curve models and correspondence analysis to capture subject heterogeneity.

Address: Department of Marketing, HEC Montréal, 3000 Chemin de la Côte Ste Catherine, Montréal, Québec, H3T 2A7, Canada.

Email: heungsun.hwang@hec.ca.

Alex J. Koning is an assistant professor at the Econometric Institute, Erasmus University Rotterdam, the Netherlands. His research interests include goodness-of-fit tests, statistical quality control, and nonparametric statistics.

Address: Econometric Institute, Erasmus University Rotterdam, P.O. Box 1738 DR Rotterdam, the Netherlands.

E-mail: koning@few.eur.nl.

Pieter M. Kroonenberg occupies the chair "Multivariate Analysis, in particular of three-way data" in the Department of Education and Child Studies, Leiden University, the Netherlands. His major interest is in three-mode analysis in all its facets, but he is also interested in other multivariate data-analytic methods and applying such methods to data from different fields. He is presently completing a book on the practice of three-mode analysis.

Address: Department of Education, Leiden University, Wassenaarseweg 52, 2333 AK Leiden, the Netherlands.

Email: kroonenb@fsw.leidenuniv.nl.

Web address: http://three-mode.leidenuniv.nl.

M. Isabel Landaluce Calvo is a professor of statistics at the University of Burgos, Burgos, Spain. Her interests are mainly in multivariate data analysis, descriptive factorial analysis, and classification techniques and their application to marketing research and to the analysis of survey data.

Address: Departamento de Economía Aplicada, Universidad de Burgos, Plaza Infanta Doña Elena s/n, E-09001 Burgos, Spain.

Email: iland@ubu.es.

Ludovic Lebart is a senior researcher at the Centre National de la Recherche Scientifique and a professor at the Ecole Nationale Supérieure des Télécommunications in Paris. His research interests are the exploratory analysis of qualitative and textual data. He has coauthored several books on descriptive multivariate statistics, survey methodology, and exploratory analysis of textual data.

Address: Ecole Nationale Supérieure des Télécommunications, 46 rue Barrault, 75013 Paris, France.

Email: lebart@enst.fr.

Web address: http://www.lebart.org.

Herbert Matschinger is, since 1996, a senior researcher in the Department of Psychiatry, University of Leipzig. Previously, he was a researcher in the Department of Psychiatric Sociology at the Central Institute of Mental Health in Mannheim. His main research interests are generalized canonical analysis, mixture modeling, IRT modeling, and analysis of repeated measurements.

Address: University of Leipzig, Department of Psychiatry, Johannisallee 20/1, 04317 Leipzig, Germany.

Email: math@medizin.uni-leipzig.de.

Oleg Nenadić is a research assistant at the Institute for Statistics and Econometrics, University of Göttingen. His interests are mainly in computational statistics, multivariate analysis, and visualization.

Address: Georg-August-Universität Göttingen, Institut für Statistik und Ökonometrie, Platz der Göttinger Sieben 5, 37075 Göttingen, Germany.

Email: onenadi@uni-goettingen.de.

Web address: http:// www.statoek.wiso.uni-goettingen.de/.

Ndèye Niang is an assistant professor of statistics at the Institut d'Informatique d'Entreprise, the engineering school in computer science of CNAM in Paris and a member of the data analysis group of CEDRIC, the computer science research team of CNAM. Her interests are in multivariate analysis and applications to quality control.

Address: Chaire de Statistique Appliquée, CNAM, 292 rue Saint Martin, F-75141 Paris Cedex 03, France.

Email: niang@cnam.fr.

Shizuhiko Nishisato is a professor emeritus, University of Toronto, Canada. He is a former President of the Psychometric Society, a former editor of *Psychometrika*, and a fellow of the American Statistical Association. He coined the term "dual scaling," the subject of his lifelong work.

Address: OISE/University of Toronto, 252 Bloor Street West, Toronto, Ontario, M5S 1V6, Canada.

Email: snishisato@oise.utoronto.ca.

Hicham Noçairi is a Ph.D. student at Ecole Nationale des Ingénieurs des Industries Agricoles et Alimentaires (ENITIAA) in Nantes, France. His interests are primarily in discrimination methods in presence of multicollinearity among predictors, with applications in chemometrics.

Address: ENITIAA-INRA Unité de Sensométrie et Chimiométrie, Rue de la Géraudière, BP 82225, F-44322 Nantes Cedex 03, France.

Email: nocairi@enitiaa-nantes.fr.

Jérôme Pagès is a professor and head of the Laboratory of Applied Mathematics at Agrocampus Rennes in Rennes, France. His interest is primarily in exploratory data analysis, especially in the treatment of multiple tables. In collaboration with Brigitte Escofier, he published a book about simple and multiple factor analysis, which is a classic one in France.

Address: Agrocampus Rennes, 65 rue de Saint-Brieuc, CS 84215, F-35042 Rennes Cedex, France.

Email: jerome.pages@agrocampus-rennes.fr.

Rafael Pardo is director of the Fundación BBVA in Madrid and professor of research at the CSIC (Spanish National Council for Scientific Research). His current areas of research are scientific and environmental culture in late modern societies, social capital, and research methodology.

Address: Fundacion BBVA, Paseo de Recoletos 10, E-28001 Madrid, Spain.

Email: rpardoa@fbbva.es.

El-Mostafa Qannari is a professor at École Nationale des Ingénieurs des Industries Agricoles et Alimentaires (ENITIAA) in Nantes, France. He is also head of a research unit affiliated with INRA (Institut de Recherche Agronomique). His interests are primarily in multivariate analysis with applications in sensory analysis and chemometrics.

Address: ENITIAA-INRA Unité de Sensométrie et Chimiométrie, Rue de la Géraudière, BP 82225, F-44322 Nantes Cedex 03, France.

Email: qannari@enitiaa-nantes.fr.

Henry Rouanet is a guest researcher at the Centre de Recherche Informatique de Paris 5 (CRIP5), Université René Descartes, Paris. His main interests are analysis of variance and Bayesian inference. He has coauthored several books about statistics and geometric data analysis.

Address: UFR Math-Info, Université René Descartes, 45 rue des Saints-Pères, F-75270 Paris Cedex 06, France.
Email: rouanet@math-info.univ-paris5.fr.
Web address: http://www.math-info.univ-paris5.fr/~rouanet.

Anne Ruiz-Gazen is an associate professor, GREMAQ, Université des Sciences Sociales and Laboratory of Statistics and Probabilities, Université Paul Sabatier, Toulouse, France. Her main research interests are robust statistics and multivariate data analysis.
Address: GREMAQ, Université Toulouse I, 21 allée de Brienne, F-31000 Toulouse, France.
Email: ruiz@cict.fr.

Gilbert Saporta is a professor and head of the chair of Applied Statistics at CNAM-Paris (Conservatoire National des Arts et Métiers). He is responsible of the data analysis group of CEDRIC, the computer science research team of CNAM. His interests are in applied multivariate analysis, scoring techniques, and time-dependent data. He has been president of SFdS, the French statistical society.
Address: Chaire de Statistique Appliquée, CNAM, 292 rue Saint Martin, F-75141 Paris Cedex 03, France.
Email: saporta@cnam.fr.
Web address: http://cedric.cnam.fr/~saporta.

Yoshio Takane is a professor of psychology at McGill University, Montréal, Canada. He is a past president of the Psychometric Society. His recent interests are primarily in the development of methods for structured analysis of multivariate data, and artificial neural network simulations.
Address: Department of Psychology, McGill University, 1205 Dr. Penfield Avenue, Montréal, Québec, H3A 1B1, Canada.
Email: takane@ takane2.psych.mcgill.ca.
Web address: http://takane.brinkster.net/Yoshio/.

Victor Thiessen is a professor in the Department of Sociology and Social Anthropology, Dalhousie University in Halifax, Nova Scotia. He has published articles on survey methodology as well as on the educational and occupational aspirations of youth. His current investigations focus on (a) the various pathways along which young Canadians navigate their way from schooling to employment and (b) the effects of information and communication technologies in schools and at home on educational attainments.

Address: Department. of Sociology and Social Anthropology, Dalhousie University, Halifax, NS, B3H 4P9, Canada.
Email: victor. thiessen@dal.ca.

Anna Torres-Lacomba is an associate professor of marketing research at the Universidad Carlos III, Madrid, Spain. Her interests are mainly in multivariate data analysis applied to marketing problems, specially the use of correspondence analysis for measuring preferences and brand image.

Address: Departament d'Economia i Empresa, Pompeu Fabra University, Ramon Trias Fargas 25–27, E-08005 Barcelona, Spain.
Email: anna.torres@upf.edu

Wijbrandt van Schuur is an associate professor in the Sociology Department of the University of Groningen and a member of the sociological graduate school ICS. His research interests are the development and application of measurement models, especially nonparametric IRT models such as the Mokken models, unidimensional unfolding, and the circumplex. His substantive interests are in the areas of political and cultural sociology.

Address: Department of Sociology, University of Groningen, Grote Rozenstraat 31, NL-9712 TG Groningen, the Netherlands.
Email: h.van.schuur@ppsw.rug.nl.

José L. Vicente-Villardón is a professor in the Departamento de Estadística, Universidad de Salamanca, Spain. His main research interest is in multivariate data analysis, especially biplot, related methods, and integration of several data matrices. Recent work is related to biplots based on generalized linear models and three-way generalizations of canonical correspondence analysis.

Address: Departamento de Estadística, Universidad de Salamanca, C/ Espejo s/n, 37007, Salamanca, Spain.
Email: villardon@usal.es.

Matthijs J. Warrens is a Ph.D. student in the Section of Psychometrics and Research Methodology, Department of Psychology, Leiden University. His interests are methods for optimal scaling and item-response theory.

Address: Wassenaarseweg 52, P.O. Box 9555, 2300 RB Leiden, the Netherlands.
Email: Warrens@fsw.leidenuniv.nl.

Amaya Zárraga is a professor of statistics and data analysis at the University of Basque Country, Bilbao, Spain. Her interests are mainly in multivariate data analysis and in particular the analysis of several contingency tables and their applications to social sciences.

Address: Departamento de Economía Aplicada III (Econometría y Estadística), Facultad de Ciencias Económicas y Empresariales, Avda Lehendakari Agirre 83, E-48015 Bilbao, Spain.

Email: amaya.zarraga@ehu.es.

Table of Contents

SECTION I

Introduction

CHAPTER 1

Correspondence Analysis and Related Methods in Practice

Jörg Blasius and Michael Greenacre

CONTENTS

1.1 Introduction

Correspondence analysis as we know it today has come a long way in the 30 years since the publication of Benzécri's seminal work, *Analyse des Données* (Benzécri et al. 1973) and, shortly thereafter, Hill's paper on applied statistics, "Correspondence analysis: a neglected multivariate method" (Hill 1974), which drew the English-speaking world's attention to the existence of this technique. In a bibliography of publications on correspondence analysis, Beh (2004) documents an almost exponential increase in articles on the subject (Figure 1.1). However, this bibliography focused mostly on methodological journals, and it does not include the explosion of applications of correspondence analysis in fields as diverse as archaeology, linguistics, marketing research, psychology, sociology, education, geology, ecology, and medicine—indeed, in all the physical, social, human, and biological sciences.

Correspondence analysis (CA) is an exploratory multivariate technique for the graphical and numerical analysis of almost any data matrix with nonnegative entries, but it principally involves tables of frequencies or counts. It can be extended to analyze presence/absence data, rankings and preferences, paired comparison data, multiresponse tables, multiway tables, and square transition tables, among others. Because it is oriented toward categorical data, it can be used to analyze almost any type of tabular data after suitable data transformation, or recoding, as exemplified by a recent book by Murtagh (2005).

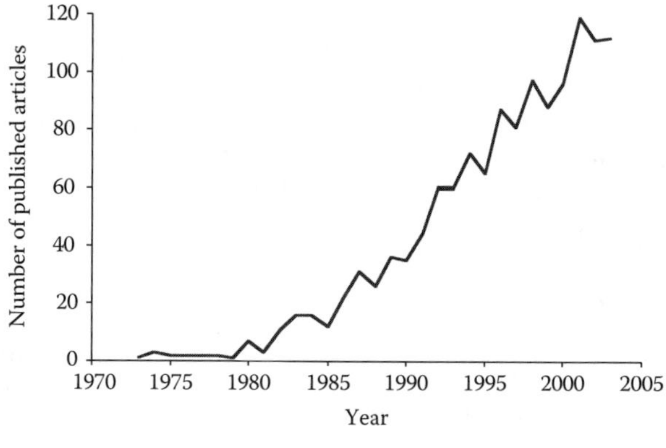

Figure 1.1 Growth in publications on correspondence analysis in selected journals. (Based on papers in 66 statistical journals and 22 books, thus representing a lower bound on the number of articles published. From Beh 2004).

There are several different ways of defining and thinking about CA, which is evident in the rediscovery of the method by so many different authors in the last century. We like to think of CA as a type of principal component analysis (PCA) of categorical data, where we consider the geometric definition of PCA rather than its statistical one. Similar to PCA, the rows or columns of a data matrix are assumed to be points in a high-dimensional Euclidean space, and the method aims to redefine the dimensions of the space so that the principal dimensions capture the most variance possible, allowing for lower-dimensional descriptions of the data. The fact that CA analyzes categorical data rather than metric data opens up the PCA style of data analysis to a world of new possibilities. Categorical data abound in almost all areas of research, especially in the social sciences, where most kinds of survey data are collected by means of nominally or ordinally scaled categorical items.

The history of correspondence analysis can be traced back to Hirschfeld (1935) (later changing his name to Hartley), who gave an algebraic formulation of the correlation between rows and columns of a contingency table. Fisher (1940) used the same ideas in the framework of discriminant analysis and is also regarded as one of the "founding fathers" of the technique. Independent from this approach, Guttman (1941) developed a method for constructing scales for categorical data, where he treated the general case for more than two qualitative variables. Since we are concentrating mostly on the multiple form of CA in this book, Louis Guttman should be credited as being the originator of the ideas behind present-day multiple correspondence analysis (MCA); he even used the term "principal components" and "chi-square distance" in his description of the method (Guttman 1950a,b). In the early 1950s Hayashi (1950, 1952, 1954) built on Guttman's ideas to create a method that he called "quantification of qualitative data," and was later followed in this tradition by the "dual scaling" ideas of Nishisato (1980, 1994). Apart from these brief historical remarks, we direct the interested reader to various texts where the history of CA is described, notably de Leeuw (1973), Nishisato (1980), Greenacre (1984), and Gifi (1990).

Our personal approach, due to our belief that the geometric approach has the most benefits, is to follow in great part the ideas of Jean-Paul Benzécri. Benzécri, a mathematician and linguist, developed CA and MCA in France in the 1960s and 1970s as part of a philosophy that placed the data firmly at the center of attention of the researcher. According to Benzécri, the data are king, not the model one might want to propose for them: one of his famous principles states that "the model should follow the data, not the inverse." He gathered around him an influential team in France who made a large contribution to the early development of MCA—notably, Ludovic Lebart and Brigitte Escofier, to mention but two of the original key coworkers. Benzécri's original ideas are worthy of consideration because they do represent an extreme, counterbalancing the excessive attention paid to confirmatory modeling in statistics. Pretty much at the same time, and finding the right balance between Benzécri's ideas and statistical practice, one of the most important data analysis schools was developing in the Netherlands, inspired by Jan de Leeuw in Leiden and including Willem Heiser, Jacqueline Meulman, and Pieter Kroonenberg, to name but a few. De Leeuw engendered a style and culture of research on MCA and related methods that remain the strongest and most important at the present time (for an excellent review, see Michailidis and de Leeuw 1998). This group is best known for their collectively authored book under the nom de plume of Gifi, published internationally in 1990, but existing in other editions published in Leiden since 1981. In their approach, MCA (called homogeneity analysis) is the central analytical tool that is used to embed categorical data, through optimal scaling of the categories, into the interval-scale-based world of classical multivariate statistical analysis.

In the English-speaking world, interest in CA accelerated with the publication of textbooks by Lebart et al. and Greenacre, both published in 1984. In the late 1980s, several CA procedures were included in the leading statistical software packages of the time, notably SPSS, BMDP, and SAS. At the same time, the number of applications significantly increased, in the social sciences especially, influenced by the work of the French sociologist Pierre Bourdieu (see Rouanet et al. 2000), which has been translated into many languages. With these applications and with the further developments of the method, the number of yearly publications in this area increased steeply. In his overview of publications in the field of CA/MCA, Beh (2004) reports just one publication for 1973, 16 for 1983, 60 for 1993, and 112 for 2003 (the last year of his report), as shown in Figure 1.1.

In this book we have 23 chapters on the topic of MCA, starting from the basics and leading into state-of-the-art chapters on methodology, each with applications to data. There are data from social science research (Chapters 1, 2, 3, 7, 8, 22, and the Appendix), education (Chapters 4 and 5), health (Chapter 6), item responses in psychology (Chapters 9 and 10), marketing data and product preference (Chapters 11 and 19), health sciences (Chapter 12), food preferences (Chapter 13), urban research (Chapter 14), elections in political science (Chapter 15), credit scoring (Chapter 16), epidemiology of animal health (Chapter 17), animal classification (Chapter 18), international comparisons (Chapter 20), psychological experiments (Chapter 21), and microarray studies in genetics (Chapter 23).

In this first and introductory chapter, we will start with a simple example of CA applied to a small two-way contingency table, explain the basics of the method, and then introduce MCA and related methods, ending up with a short overview of the rest of the book's contents.

1.2 A simple example

Correspondence analysis is an exploratory method based on well-known geometrical paradigms. To provide an initial illustration of the method, we use data from the most recently available survey from the International Social Survey Program (ISSP 2003). The original respondent-level data are available at the "Zentralarchiv für Empirische Sozialforschung" (Central Archive for Empirical Social Research) in Cologne, Germany (http://www.gesis.de). We start with a simple cross-tabulation of how respondents reacted to the statement, "When my country does well in international sports, it makes me proud to be {Country Nationality}." We chose respondents from five specific countries—U.K., U.S., Russia, Spain, and France—because their respective cities (London, New York, Moscow, Madrid, and Paris) were involved in the final decision for the summer Olympics in 2012. Respondents could give one of five possible responses to the above question—(1) "agree strongly," (2) "agree," (3) "neither agree nor disagree," (4) "disagree," (5) "disagree strongly"—as well as various "nonresponses." Table 1.1 shows the frequencies, and Table 1.2 the (column) percentages

Table 1.1 Frequencies of the cross-table "international sports × country."

	U.K.	U.S.	Russia	Spain	France	Total
Agree strongly	230	400	1010	201	365	2206
Agree	329	471	530	639	478	2447
Neither nor	177	237	141	208	305	1068
Disagree	34	28	21	72	50	205
Disagree strongly	6	12	11	14	97	140
Total	776	1148	1713	1134	1295	6066

Table 1.2 Column percentages of the cross-table "international sports × country."

	U.K.	U.S.	Russia	Spain	France	Average
Agree strongly	29.6	34.8	59.0	17.7	28.2	36.4
Agree	42.4	41.0	30.9	56.4	36.9	40.3
Neither nor	22.8	20.6	8.2	18.3	23.6	17.6
Disagree	4.4	2.4	1.2	6.5	3.9	3.4
Disagree strongly	0.8	1.1	0.6	1.2	7.5	2.3
Total	100.0	100.0	100.0	100.0	100.0	100.0

for each country. To keep our explanation simple at this stage and to avoid discussion about the handling of missing data, we have restricted attention to those respondents with no missing data for this question as well as for the other variables used in extended analyses of these data later in this chapter. These additional variables are on other aspects of national identity as well as several demographic variables: sex, age, marital status, and education. The topic of missing data is treated in Chapters 8 and 12 of this book.

Table 1.1 shows sample sizes ranging from 776 for U.K. to 1713 for Russia. In all countries, most people either "agree strongly" or "agree" with the statement on international sports. However, there are some differences between the countries: whereas in Russia most people "agree strongly," in Spain there are fewer than 20% giving this response. On the other hand, in France there is the largest share of respondents (7.5%) who "disagree strongly" with the statement. Calculating the chi-square statistic for testing independence on this table produces a value of $\chi^2 = 879.3$ (16 degrees of freedom), which is highly significant (P-value close to 0); Cramer's V measure of association is $V = 0.190$.

Analyzing Table 1.1 by CA gives a map of the pattern of association between countries and response categories, shown in Figure 1.2. This graphical representation, called the *symmetric map*, plots the *principal coordinates* of the rows and columns (to be explained more fully below). This two-dimensional map is not an exact representation of the data because it would require four dimensions to represent this 5 × 5 table perfectly. (We say the *dimensionality* of the table is four.) The two-dimensional map in Figure 1.2 accounts for 95.6% of the total "variance" in the table, where the measure of variance is closely related to the chi-square statistic. The objective of CA is to represent the maximum possible variance in a map of few dimensions, usually two dimensions. In this case there is only a small (4.4%) proportion of variance that is

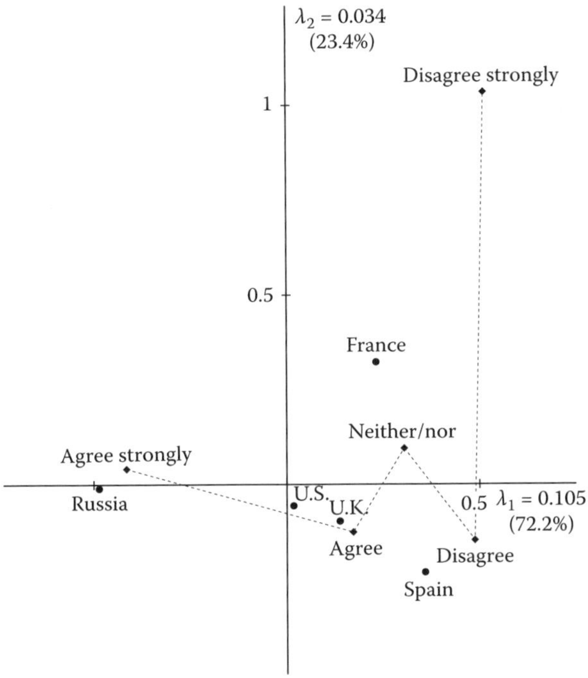

Figure 1.2 Symmetric correspondence analysis map of Table 1.1.

not represented here and that is effectively discarded because it is unlikely to be of interest. This strategy is identical in spirit to the coefficient of determination in linear regression, where we say that the predictors in a regression model explain a certain percentage of variance, with the remainder unexplained and relegated to "residual" or "error" variance. Each orthogonal axis in CA accounts for a separate part of variance, similar to uncorrelated predictors in a regression model: in Figure 1.2 the first axis explains 72.2%, the second an additional 23.4%.

Interpretation of the map consists in inspecting how the categories lie relative to one another and how the countries are spread out relative to the categories. The first (horizontal) dimension reflects a clear subdivision of the responses toward "international sports," with the category "agree strongly" on the left and the other four categories on the right. Furthermore, all categories retain their original order along the first dimension, although the intercategory distances are different: for example, "disagree" and "disagree strongly" are very close to each other along this dimension, while "agree strongly" is relatively far from "agree" (projecting the categories perpendicularly onto the horizontal axis). The first dimension can be interpreted as "level of pride toward achievements in international sport," especially focused on the contrast between "strong agreement" and the other response categories. As for the countries, we see Russia on the left, opposing the other countries on the right; thus of these five nations, the Russians feel most proud when Russia is doing well in international sports. At the opposite right-hand side of this axis, we see that the French and Spanish are the least proud of the five nations in this respect, but there is also a big difference between these two along the second (vertical) dimension.

The second dimension mainly reflects the outlying position of "disagree strongly" as well as France compared with the other categories and countries. As we already noted by inspecting Table 1.2, 7.5% of the French chose this category compared with approximately 1% respondents from the other countries. This contrast is so strong that it accounts for most of the 23.4% of the variance along this second dimension.

In Table 1.2 the U.S. and U.K. have very similar response patterns, which are not much different from the overall, or average, pattern. Geometrically, this is depicted by these two countries lying close to each other, toward the origin of the map. The average response pattern is given in the last column of Table 1.2, which is the percentage responses for the whole data set.

All features in the map are relative in the sense that we can see that the Spanish are less proud of sporting achievement than the British, for example, but the map does not tell us how much less. The only point that is represented perfectly in the map is the center, or origin, which coincides with the average for the data set at hand, so we can judge in the map how the countries deviate from the average. To get an idea of absolute scale, an alternative way of representing the table is the so-called *asymmetric map*, where one set of points, usually the "describing variable" (the rows in this case), is depicted in *standard coordinates* (Figure 1.3), and the other set, the "variable being described" (the columns here), is depicted in principal coordinates as before (Greenacre and Blasius, 1994: Preface). In this map the country points, still in principal coordinates, are in the same positions as in Figure 1.2, but the category points are now reference points in the space depicting a 100% response in each of the five respective categories (notice the change in scale of Figure 1.3 compared with Figure 1.2). Now the "strong agreement" point, for example, represents a fictitious country where all respondents "strongly agreed." Thus we can see that Russia, for example, deviates from the average toward strong agreement, but now we can also judge how far away Russia is from being a country with 100% strong agreement.

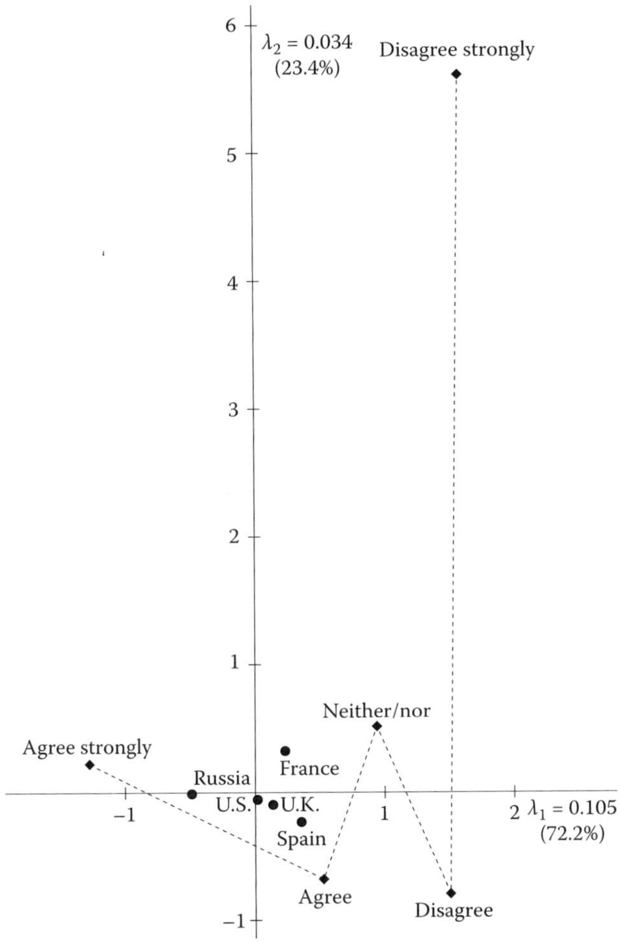

Figure 1.3 Asymmetric correspondence analysis map of Table 1.1.

Because the overall level of strong disagreement is low, this category is very far away from the countries in Figure 1.3, but France is closer to the strong disagreement "pole" than the others. Although enlightening as to the absolute level of variation existing among the countries, the asymmetric map is generally not preferred; the outlying positions of the set of points in standard coordinates (the response categories in Figure 1.3) force the other set of points (the countries) into a bunch at the center of the map. In the symmetric map of Figure 1.2, the category points—represented in principal coordinates—can be seen to lie in positions along the axes, which are a scaled-down version of those in Figure 1.3. This fact, further explained in the next sections, underpins our interpretation of the symmetric map, which is the preferred map in CA.

1.3 Basic method

There are several different, but mathematically equivalent, ways to define CA. Because our approach is chiefly graphical and in the French tradition of Benzécri et al. (1973), we see CA as an adaptation to categorical data of PCA, which is a method for identifying dimensions explaining maximum variance in metric data. Both methods are based on decompositions of centered and normalized matrices, using either the eigenvalue-eigenvector decomposition of a square symmetric matrix or the singular-value decomposition (SVD) of a rectangular matrix. As in Greenacre (1984), we present the theory in terms of the SVD, which is an approach that is better equipped to show the relationship between row and column solutions. Similar to PCA, CA provides eigenvalues that are squared singular values (called principal inertias in CA), percentages of explained variance (percentages of inertia), factor loadings (correlations with principal axes), and communalities (percentages of explained inertia for individual rows or columns). In PCA, visualizations of the results are also made, but they are less common than in the CA/MCA framework, where the map is the central output for the interpretation.

For the presentation of the basic CA algorithm, we use a simple cross-table, or contingency table, of two variables with I rows and J columns, denoted by $\mathbf{N}$, with elements n_{ij}. As a first step, the *correspondence matrix* $\mathbf{P}$ is calculated with elements $p_{ij} = n_{ij}/n$, where n is the grand total of $\mathbf{N}$, the sample size in this case. CA analyzes the matrix $\mathbf{P}$ and is not concerned with the sample size n unless aspects of statistical inference such as confidence intervals are of interest (see Chapter 7). Corresponding to each element p_{ij} of $\mathbf{P}$ is a row sum $p_{i\cdot}$ ($= n_{i\cdot}/n$) and column sum $p_{\cdot j}$ ($= n_{\cdot j}/n$), denoted by r_i and c_j respectively. These marginal relative frequencies, called *masses*, play dual roles in CA, serving to center and to normalize the correspondence matrix.

Under the null hypothesis of independence, the expected values of the relative frequencies p_{ij} are the products $r_i c_j$ of the masses. Centering involves calculating differences $(p_{ij} - r_i c_j)$ between observed and expected relative frequencies, and normalization involves dividing these differences by the square roots of $r_i c_j$, leading to a matrix of standardized residuals $s_{ij} = (p_{ij} - r_i c_j) / \sqrt{r_i c_j}$. In matrix notation this is written as:

$$\mathbf{S} = \mathbf{D}_r^{-1/2}(\mathbf{P} - \mathbf{r}\mathbf{c}^\mathsf{T})\mathbf{D}_c^{-1/2} \tag{1.1}$$

where $\mathbf{r}$ and $\mathbf{c}$ are vectors of row and column masses, and $\mathbf{D}_r$ and $\mathbf{D}_c$ are diagonal matrices with these masses on the respective diagonals. The sum of squared elements of the matrix of standardized residuals, $\sum_i \sum_j s_{ij}^2 = \text{trace}(\mathbf{S}\mathbf{S}^\mathsf{T})$, is called the *total inertia* and is the amount that quantifies the total variance in the cross-table. Because the standardized residuals in $\mathbf{S}$ resemble those in the calculation of the chi-square statistic, χ^2, apart from the division by n to convert original frequencies to relative ones, we have the following simple relationship:

$$\text{total inertia} = \chi^2/n \tag{1.2}$$

The association structure in the matrix $\mathbf{S}$ is revealed using the SVD:

$$\mathbf{S} = \mathbf{U}\mathbf{\Sigma}\mathbf{V}^\mathsf{T} \tag{1.3}$$

where $\mathbf{\Sigma}$ is the diagonal matrix with singular values in descending order: $\sigma_1 \geq \sigma_2 \geq \ldots \geq \sigma_S > 0$, where S is the rank of matrix $\mathbf{S}$. The columns of $\mathbf{U}$, called *left singular vectors*, and those of $\mathbf{V}$, the *right singular vectors*, are orthonormal: $\mathbf{U}^\mathsf{T}\mathbf{U} = \mathbf{V}^\mathsf{T}\mathbf{V} = \mathbf{I}$. The connection between the SVD and the eigenvalue decomposition can be seen in the following:

$$\mathbf{S}^\mathsf{T}\mathbf{S} = \mathbf{V}\mathbf{\Sigma}\mathbf{U}^\mathsf{T}\mathbf{U}\mathbf{\Sigma}\mathbf{V}^\mathsf{T} = \mathbf{V}\mathbf{\Sigma}^2\mathbf{V}^\mathsf{T} = \mathbf{V}\mathbf{\Lambda}\mathbf{V}^\mathsf{T}$$

$$\mathbf{S}\mathbf{S}^\mathsf{T} = \mathbf{U}\mathbf{\Sigma}\mathbf{V}^\mathsf{T}\mathbf{V}\mathbf{\Sigma}\mathbf{U}^\mathsf{T} = \mathbf{U}\mathbf{\Sigma}^2\mathbf{U}^\mathsf{T} = \mathbf{U}\mathbf{\Lambda}\mathbf{U}^\mathsf{T}$$

showing that the right singular vectors of $\mathbf{S}$ correspond to the eigenvectors of $\mathbf{S}^\mathsf{T}\mathbf{S}$, the left singular vectors correspond to the eigenvectors of $\mathbf{S}\mathbf{S}^\mathsf{T}$, and the squared singular values σ^2 in $\mathbf{\Sigma}^2$ correspond to the eigenvalues λ of $\mathbf{S}^\mathsf{T}\mathbf{S}$ or $\mathbf{S}\mathbf{S}^\mathsf{T}$, where $\mathbf{\Lambda}$ is the diagonal matrix of eigenvalues. Within the context of CA, these eigenvalues are termed *principal inertias*, and their sum $\sum_s \lambda_s$ is equal to the total inertia since $\text{trace}(\mathbf{S}\mathbf{S}^\mathsf{T}) = \text{trace}(\mathbf{S}^\mathsf{T}\mathbf{S}) = \text{trace}(\mathbf{\Sigma}^2) = \text{trace}(\mathbf{\Lambda})$.

The SVD provides all the results we need to make CA maps. The principal and standard coordinates can be calculated for the row and column categories:

$$\text{principal coordinates of rows:} \quad \mathbf{F} = \mathbf{D}_r^{-1/2}\,\mathbf{U}\boldsymbol{\Sigma} \tag{1.4}$$

$$\text{standard coordinates of rows:} \quad \mathbf{A} = \mathbf{D}_r^{-1/2}\,\mathbf{U} \tag{1.5}$$

$$\text{principal coordinates of columns:} \quad \mathbf{G} = \mathbf{D}_c^{-1/2}\,\mathbf{V}\boldsymbol{\Sigma} \tag{1.6}$$

$$\text{standard coordinates of columns:} \quad \mathbf{B} = \mathbf{D}_c^{-1/2}\,\mathbf{V} \tag{1.7}$$

For a two-dimensional map we would use either (a) the first two columns of the coordinates matrices **F** and **G** for the symmetric map, (b) **A** and **G** for the asymmetric map of the columns (called "column principal" in SPSS, for example), or (c) **F** and **B** for the asymmetric map of the rows ("row principal"). The proportion of inertia explained would be $(\sigma_1^2 + \sigma_2^2)/\sum_s \sigma_s^2$, i.e., $(\lambda_1 + \lambda_2)/\sum_s \lambda_s$. For further details about the computation of CA, see Blasius and Greenacre (1994).

1.4 Concepts of correspondence analysis

In this section we summarize the main concepts underlying CA, mostly geometric concepts. Each concept will be illustrated briefly in the context of the CA of the "international sports" example of Table 1.1 and Figure 1.2 and Figure 1.3.

1.4.1 Profiles, average profiles, and masses

As mentioned previously, CA is based on relative values; the sample size is not important for the construction of the map. The data table can be expressed as proportions (or percentages) relative to the row or column margins. Table 1.2 illustrates the latter possibility, showing for each country the percentage responses across the categories of the variable "international sport": for example, "U.K.: agree strongly" is 230/776 = 0.296 (or 29.6%). The columns containing the relative frequencies for the single countries are called *profiles*, in this case *column profiles*. Furthermore, we calculate average profiles as the row or column margins relative to the grand total: for example, the profile value for "All countries: agree strongly" is 2206/6066 = 0.364 (or 36.4%). This *average column profile* is given in the last column of Table 1.2. In the

CA map, the average column profile is represented at the origin where the axes cross.

Table 1.2 shows that Russia has a profile value in the category "agree strongly" clearly above average (0.590 > 0.364), whereas the respective value of Spain is clearly below average (0.177 < 0.364). Under the condition that the two countries are well represented in the map, which is very likely, as 95.6% of the variation is explained by the first two dimensions, Russia should be associated strongly with "agree strongly," whereas Spain should be just the opposite. That this is true has already been seen in Figure 1.3. Comparing the column profiles with the average column profile as well as with one another gives a first understanding of which columns (countries) should be located close to one another and which should be separated.

Table 1.2 shows the column profiles and the average column profile. In the same way, we could calculate the *row profiles*, expressing the frequencies in each row of Table 1.1 relative to their corresponding row total. Further, we can compute the *average row profile*, i.e., the column sums (or column totals, see the last row in Table 1.1) divided by the sample size. We could then compare the elements of the row profiles with their corresponding elements in the average row profile, which gives a first understanding of which row categories are in the same part of the map and which are relatively distinct from one another. In the map of the row profiles, the origin again reflects the position of the average row profile.

The distinction between the presentation of the rows, i.e., discussing the differences between the row profiles compared with the average row profile, and the presentation of the column profiles, i.e., discussing the differences between the column profiles compared with the average column profile, is also reflected in the two possibilities of visualizing the data using asymmetric maps. Figure 1.3 shows the column profiles in a map spanned by the rows. The categories of the rows are expressed in terms of "artificial countries," for example, the position of "agree strongly" reflects a hypothetical "country" containing only respondents all strongly agreeing with the statement. We also could calculate an asymmetric map in which the profiles of the rows are visualized in a space spanned by the columns. Because the differences between the countries are of central interest, this possibility of visualization is less meaningful. Finally, and most often done, row and column profiles are shown together (symmetric map). However, there is a close relation between standard and principal coordinates because there are only scale-factor differences between them along the axes. The standard coordinates can be obtained by dividing the principal coordinates by the singular values (i.e., by the square roots of the principal inertias), for example for the rows: $\mathbf{A} = \mathbf{F\Sigma}^{-1}$ (see Equation 1.4 to Equation 1.7).

In addition to being elements of the average profile, the marginal values described previously (denoted by r_i and c_j in Section 1.3) are also used as weights in CA to give more or less importance to the display of the individual row and column profiles. For this reason they are called masses, and their dual role in the analysis will become apparent when we explain the concepts of total variance and distance in the next two sections.

1.4.2 *Chi-square statistic and total inertia*

A common way of describing the relationships in contingency tables is the chi-square statistic, which tests for significant associations between rows and columns. The chi-square statistic is defined as the sum of squared deviations between observed and expected frequencies, divided by the expected frequencies, where the expected frequencies are those calculated under the independence model:

$$\chi^2 = \sum_i \sum_j \frac{(n_{ij} - \hat{n}_{ij})^2}{\hat{n}_{ij}} \quad \text{where } \hat{n}_{ij} = n_{i\cdot} \times n_{\cdot j} / n, \{i = 1, 2, \ldots, I; j = 1, 2, \ldots, J\}$$

Repeating this calculation for relative frequencies p_{ij}, we obtain the chi-square statistic divided by the grand total n of the table:

$$\frac{\chi^2}{n} = \sum_i \sum_j \frac{(p_{ij} - \hat{p}_{ij})^2}{\hat{p}_{ij}}, \quad \text{where } \hat{p}_{ij} = r_i \times c_j \tag{1.8}$$

This is exactly the total inertia defined in Section 1.3, the sum of squared standardized residuals in the matrix $\mathbf{S}$ of Equation 1.1. Calculating the chi-square value for Table 1.1 gives $\chi^2 = 879.3$ and a total inertia of $\chi^2/n = 0.145$. Comparing the total inertia with other solutions taken from literature, the value is relatively high, which means that there is a relatively large amount of variation in the data or, equivalently, there are relatively large differences between the countries in terms of their national pride concerning international sports.

Table 1.3 Chi-square components of Table 1.1.

	U.K.	U.S.	Russia	Spain	France	Total
Agree strongly	9.66	0.73	240.46	108.36	23.83	383.04
Agree	0.81	0.13	37.52	72.05	3.77	114.28
Neither nor	11.93	6.02	85.52	0.35	26.00	129.82
Disagree	2.31	3.00	23.51	29.59	0.89	59.30
Disagree strongly	7.92	7.93	20.60	5.66	150.70	192.81
Total	32.63	17.81	407.61	216.01	205.19	879.25

Note: The contributions to total inertia are these values divided by $n = 6066$.

Table 1.3 shows the chi-square components for each cell, showing how much each column (country), each row (response category for "international sports"), and each cell (country × response category) contribute to the deviation from independence. The larger the deviation, the larger the contribution to chi-square (equivalently, to total inertia), and the larger will be the influence on the geometric orientation of the axes in CA. The largest deviations from independence are due to Russia, with a contribution of 46.4% (407.61/879.25). Within "Russia" the response "agree strongly" has a chi-square component of 240.46, i.e., 59.0% of the inertia of Russia and 27.3% of the total inertia. This explains why the first and most important principal axis in the CA of Table 1.1 showed Russia and "agree strongly" clearly separated from other rows and columns. In contrast, U.K. and especially the U.S. have very little contribution to total inertia (3.7% and 2.0%, respectively), i.e., they are close to the average. A comparison of the three column profiles of U.K., U.S., and Russia with the average column profile (Table 1.2) supports these findings: the differences between the respective profile elements are large for Russia and small for U.K. and U.S. Furthermore, the contributions to total inertia are also reflected in the map: U.K. and U.S. are relatively close to the origin, and Russia is relatively far away.

With respect to the response categories, the strongest impacts on total inertia are from "agree strongly" and "disagree strongly": both are relatively far away from the origin. However, although "agree strongly" has a higher contribution to total inertia (383.04/879.25 = 0.436 or 43.6%) than "disagree strongly" (21.9%), the profile of the latter is farther away from the origin (see Figure 1.2). This is due to the different masses of the two response categories, as will be explained in more detail in Section 1.4.5.

1.4.3 Chi-square distances

In the CA of a contingency table such as Table 1.1, distances between any pair of row profiles or between any pair of column profiles become apparent. From Equation 1.8, we can rewrite the total inertia as

$$\frac{\chi^2}{n} = \sum_i \sum_j c_j \frac{(p_{ij}/c_j - r_i)^2}{r_i}$$

where p_{ij}/c_j is an element of the jth column profile: $p_{ij}/c_j = n_{ij}/n_{\cdot j}$; and r_i is the corresponding element of the average column profile: $r_i = n_{i\cdot}/n$. This shows that the total inertia can be interpreted geometrically as the *weighted sum of squared distances* of the column profiles to the average profile: the weight of the column profile is its mass c_j, and the squared distance is a Euclidean-type distance where each squared difference is divided by the corresponding average value r_i. For this reason this distance is known as the chi-square (χ^2) distance. For example, from Table 1.2, the χ^2 distance between U.K. and the "average country" is:

$$d_{\text{UK,O}} = \sqrt{\frac{(.296-.364)^2}{.364} + \frac{(.424-.403)^2}{.403} + \frac{(.228-.176)^2}{.176} + \frac{(.044-.034)^2}{.034} + \frac{(.008-.023)^2}{.023}}$$

$$= 0.205$$

In a similar fashion, the χ^2 distance between U.K. and the U.S. is:

$$d_{\text{UK,US}} = \sqrt{\frac{(.296-.348)^2}{.364} + \frac{(.424-.410)^2}{.403} + \frac{(.228-.206)^2}{.176} + \frac{(.044-.024)^2}{.034} + \frac{(.008-.011)^2}{.023}}$$

$$= 0.151$$

In Figure 1.2 we can see that the positions of the origin, U.K., and the U.S., closely agree with these interpoint χ^2 distances. Similar distance calculations can be made between the row profiles and their average and between pairs of row profiles.

In the CA solution, the χ^2 distances between profiles are visualized as ordinary Euclidean distances. For example, the (squared) χ^2 distance between two columns j and j' is exactly equal to the (squared) Euclidean distance $\sum_s (g_{js} - g_{j's})^2$ between the points in principal coordinates in the full S-dimensional space.

1.4.4 *Reduction of dimensionality*

As in PCA and in other data reduction techniques, as few dimensions as possible are used for interpretation. Due to limitations in the graphical display, in CA/MCA usually only planar maps are used, showing pairs of dimensions at a time, although three-dimensional graphical displays are becoming easier to use. (See, for example, Rovan 1994, and the RGL package in the R language presented in the computational appendix of this book.) The determination of the number of dimensions to be interpreted can be performed in various ways, similar to PCA: (a) consider all those with eigenvalues that explain more than average inertia, (b) examine a "scree plot" of the eigenvalues to identify the "elbow" in the descending sequence, or (c) use the application-based method of including all dimensions that have a coherent substantive interpretation (see also Blasius 1994). The issue of eigenvalues and their percentages of explained inertia is more problematic in MCA, a subject that will be treated in detail in Chapter 2.

1.4.5 *Contributions to inertia*

The principal coordinate positions of the row and column points, relative to their respective averages, are given by Equation 1.4 and Equation 1.6, respectively. Because chi-square distances are equal to distances between the points represented in principal coordinates in the full space, an alternative way to calculate total inertia and the inertial contributions of Section 1.4.2 is as a weighted sum of squares of principal coordinates. For example, multiplying the mass of the ith row (r_i) by the squared coordinate of the ith row on the sth dimension (f_{is}^2) gives the amount of inertia of row i on axis s. The sum of these inertias over all rows and over all dimensions, i.e., $\sum_i \sum_s r_i f_{is}^2$, gives the total inertia. The same holds for the columns, and we have the equality:

$$\text{total inertia} = \sum_i \sum_s r_i f_{is}^2 = \sum_j \sum_s c_j g_{js}^2 \tag{1.9}$$

Summing over points on single dimensions s gives the principal inertias λ_s: $\sum_i r_i f_{is}^2 = \sum_j c_j g_{js}^2 = \lambda_s$, $s = 1, 2, \ldots, S$, again recovering the total inertia as $\sum_s \lambda_s$. Summing over dimensions for a single point gives the inertia of the corresponding row or columns. For example, the inertia of the *i*th row is $r_i \sum_s f_{is}^2$. Using these decompositions of inertia in terms of points and dimensions, we can compute:

1. The contribution from each row or each column to total inertia. This is the amount of variance each row or column contributes to the geometric model as a whole (as described in Section 1.4.2 in terms of chi-square, or inertia components).
2. Same as item 1, but with respect to single dimensions.
3. The contribution of each dimension to total inertia, i.e., the explained variance of each dimension.
4. The contribution of each dimension to the inertia of a row or column, i.e., the explained variance of each dimension to a point. The square roots of these values are often called *factor loadings* because they are also correlation coefficients between the point and the dimension.
5. The amount of explained variance of the first S^* dimensions to each row or to each column. These coefficients are called *qualities* in CA, known as *communalities* in PCA.

Together with the eigenvalues (principal inertias), these five coefficients constitute the standard numerical output in CA, as given by such statistical packages as SPSS and XLSTAT.

Mass plays a crucial role in the inertia contributions of each row and each column, and it must be considered when interpreting a CA map or, even earlier, when starting to perform a CA. In the given example, the contribution of "agree strongly" to total inertia is about twice that of "disagree strongly" (Table 1.3), but the mass of "agree strongly" is about 15 times higher than the mass of "disagree strongly" (Table 1.2), resulting in a shorter distance to the origin. In general, categories with low frequencies (i.e., with low masses) tend to be outlying because their distribution is often quite different from the average. In such cases these categories can contribute quite highly to the total inertia. For example, the fifth category "disagree strongly" has a low mass (r_5 = 0.023; see Table 1.2) but has a relatively high share of the total inertia (150.70/879.25 = 0.171; see Table 1.3) owing to its high profile value on France (97/140 = 0.693; see Table 1.1). For this reason, categories with very low relative frequencies should be carefully monitored in CA, and if they contribute too much to the solution, they should be combined with other categories in a substantively meaningful way.

1.4.6 Reconstruction of the data

In log-linear analysis, contingency tables can be reconstructed by means of interaction effects of different order. In CA, this reconstruction is performed by means of the margins and the bilinear decomposition inherent in the SVD. A common feature of both techniques is that the sparsest model is chosen. In log-linear analysis, this is the model with the fewest interaction effects; in CA, it is the model with the fewest dimensions. (A detailed description of the relationship between these models is given by van der Heijden et al. 1989, 1994; see also Goodman 1991.) Exact data reconstruction in CA can be obtained using the row and column principal coordinates and singular values on all dimensions:

$$\mathbf{P} = \mathbf{r}\mathbf{c}^{\mathsf{T}} + \mathbf{D}_r\mathbf{F}\mathbf{\Sigma}^{-1}\mathbf{G}^{\mathsf{T}}\mathbf{D}_c \tag{1.10}$$

where $\mathbf{r}\mathbf{c}^{\mathsf{T}}$ is the matrix of expected relative frequencies under independence. Various alternative forms of this reconstruction formula are possible, for example in terms of row standard and column principal coordinates:

$$\mathbf{P} = \mathbf{r}\mathbf{c}^{\mathsf{T}} + \mathbf{D}_r\mathbf{A}\mathbf{G}^{\mathsf{T}}\mathbf{D}_c$$

since $\mathbf{A} = \mathbf{F}\mathbf{\Sigma}^{-1}$ (see Equation 1.4 and Equation 1.5).

In CA maps, where a reduced number S^* of dimensions are used (setting $\sigma_{S^*+1} = \sigma_{S^*+2} = \ldots = \sigma_S = 0$ in Equation 1.10), the data reconstruction is not exact, approximating the data as well as the percentage of inertia accounted for by the solution. The reconstruction formula shows how CA can be considered as a model for the table, with parameters fitted by weighted least squares.

1.5 Stacked tables

In the case of simple CA, described previously, the frequencies of a single contingency table are used as input information. In this section, we describe how CA can be used to visualize several contingency tables at a time.

As a first example of a stacked table, the variable of interest, "country," is cross-tabulated by several variables describing the countries, in this case several variables on national identity; the cross-tables are stacked one on top of each other, i.e., row-wise. One of the most famous applications of CA to such a table is given by Bourdieu (1979). In his book *La Distinction*, he describes the French population, differentiated by classes of occupation, using a large set of lifestyle indicators (for example, preferences in arts and music); see Blasius and Winkler (1989).

Table 1.4 Possibilities of stacked tables.

NI1 × Country	NI1 × Sex	NI1 × Mar.Status	NI1 × Edu.Level	NI1 × Age
NI2 × Country	NI2 × Sex	NI2 × Mar.Status	NI2 × Edu.Level	NI2 × Age
NI3 × Country	NI3 × Sex	NI3 × Mar.Status	NI3 × Edu.Level	NI3 × Age
NI4 × Country	NI4 × Sex	NI4 × Mar.Status	NI4 × Edu.Level	NI4 × Age
NI5 × Country	NI5 × Sex	NI5 × Mar.Status	NI5 × Edu.Level	NI5 × Age
NI6 × Country	NI6 × Sex	NI6 × Mar.Status	NI6 × Edu.Level	NI6 × Age
NI7 × Country	NI7 × Sex	NI7 × Mar.Status	NI7 × Edu.Level	NI7 × Age

We could also extend the set of tables columnwise by adding cross-tables of "international sports" with variables such as sex, marital status, educational level, and age group. In this case, one could describe which of the sociodemographic characteristics are most important to explain the responses toward "international sports." Table 1.4 shows the possible combinations of stacking tables for seven indicators on national identity (NI1 to NI7) with country and four sociodemographic indicators.

The simplest case is the CA on a single table, for example, NI1 × Country, which has been shown previously in Figure 1.2 and Figure 1.3. The next possibility is to add other indicators of national identity, as shown in the first column of Table 1.4. Another possibility is the columnwise stacking, as shown in the first row in Table 1.4. The third and most complex possibility is to analyze all 35 cross-tables as a single matrix input to CA. Before we analyze this complex table, we show the CA solution of the stacked table with country as column variable and seven statements toward national identity as row variables. These statements are (in the order as given in the questionnaire, the first letter indicating the variables in the forthcoming figures):

a. I would rather be a citizen of {Country} than of any other country in the world.
b. There are some things about {Country} today that make me feel ashamed of {Country}.
c. The world would be a better place if people from other countries were more like the {Country nationality}.
d. Generally speaking, {Country} is a better country than most other countries.
e. People should support their country even if the country is in the wrong.
f. When my country does well in international sports, it makes me proud to be {Country nationality}.
g. I am often less proud of {Country} than I would like to be.

All variables have five categories as previously: (1) "agree strongly," (2) "agree," (3) "neither nor," (4) "disagree," and (5) "disagree strongly" (the number indicating the category in the figures).

Compared with the CA on the simple table where the two-dimensional solution accounts for 95.6% of the variation, the two-dimensional solution of the stacked table accounts for only 72.0%, even though the dimensionality of the stacked table is also four. In Figure 1.4, four of the seven "agree strongly" responses (questions b, e, f, and g) are located on the negative side of the first dimension; the remaining three (questions a, c, and d) are on the negative part of dimension 2. For the countries, Russia is on the same side of dimension 1 as strong agreements to "there are some things about Russia today that make me feel ashamed," "people should support their country even if the country in the wrong," "... international sports ...," and "I'm often less

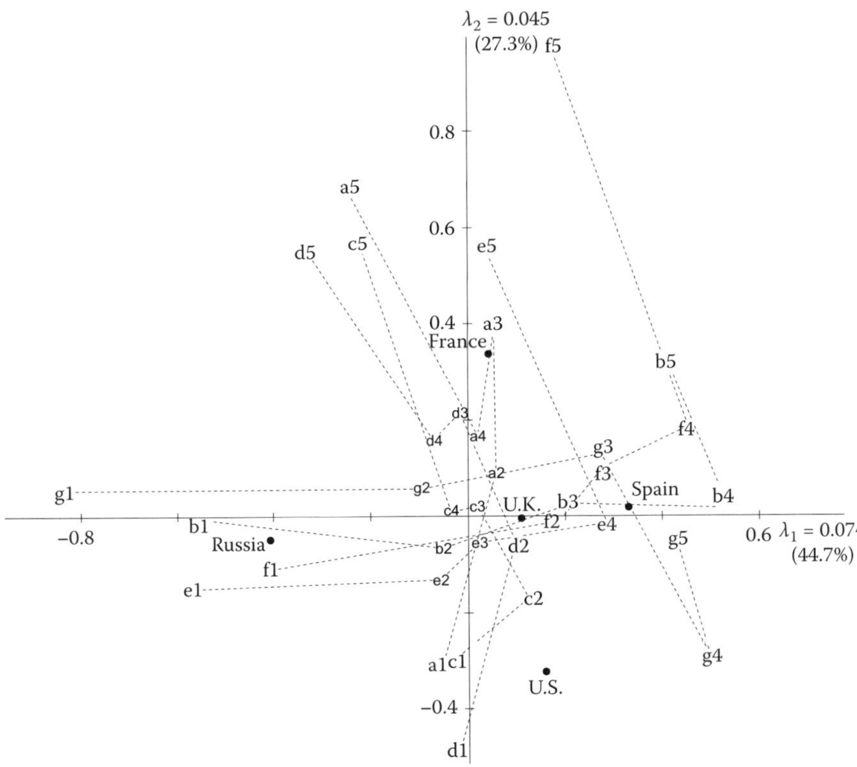

Figure 1.4 Symmetric CA map for stacked table of "national identity" indicators by "country."

proud of Russia than I would like to be," which suggests a mixture of pride toward achievement in international sports and a critical reflection toward their own country and their own attitudes. In contrast, the respondents from the U.S. agreed relatively strongly to statements that reflect national pride: "I would rather be a citizen of the U.S. than of any other country in the world," "the world would be a better place if people from other countries were more like Americans," and "generally speaking, the U.S. is a better country than most other countries."

The first dimension mirrors mainly a contrast between agreement and disagreement toward the statements b, e, f, and g, whereby the "disagree strongly" responses for e and f are located on the positive side of dimension 2. The second factor reflects the contrast between a strong agreement toward statements that suggest a kind of superiority of the home country and a strong disagreement toward most of the other statements on national identity (including the critical ones). It can be concluded that, of the five nations analyzed here, Americans are most proud of their country, whereas the French are the least proud. The British are close to the center, i.e., close to average. The Spanish are closest to some disagreement on national identity, but they relatively often avoid the choice of the strongest disagreement.

The first column of Table 1.5 shows the decomposition of inertia over the seven subtables of the stacked table. Total inertia of the stacked table is 0.1646, which can be shown to be the average of the inertias of

Table 1.5 Decomposition of inertias ($N = 6066$) for all cross-tables.

	Country	Sex	Marital Status	Educational Level	Age Groups	Average
Citizen of country	0.1690	0.0003	0.0258	0.0279	0.0544	0.0555
Feel ashamed of country	0.1898	0.0068	0.0096	0.0306	0.0098	0.0493
World would be better	0.0978	0.0029	0.0131	0.0502	0.0291	0.0386
Country is better than others	0.1289	0.0029	0.0093	0.0188	0.0204	0.0361
People should support country	0.1754	0.0009	0.0165	0.0377	0.0268	0.0515
Well in international sport	0.1450	0.0001	0.0114	0.0295	0.0108	0.0394
I am often less proud of country	0.2465	0.0031	0.0145	0.0155	0.0127	0.0585
Average	0.1646	0.0024	0.0143	0.0300	0.0234	0.0469

the seven subtables: $(0.1690 + 0.1898 + \cdots + 0.2465)/7 = 0.1646$. The largest differences between the countries are in the responses toward statement g, "I am often less proud ...," which corresponds to the subtable making the highest contribution to this average.

We now analyze the complete data matrix, extending the previous data set columnwise by stacking the sets of cross-tables with a number of sociodemographic characteristics. The final table is a supermatrix containing seven variables stacked row-wise and five variables stacked columnwise. The seven variables on national identity have been cross-tabulated with the following variables (with abbreviations as used in Figure 1.5):

Country: U.K., U.S., Russia, Spain, France — as already done
Sex: male (m), female (f)
Marital status: married (mar), widowed (wid), divorced (div), separated (sep), single (sin)
Education: no formal education (E1), lowest formal education (E2), above lowest formal education (E3), higher secondary (E4), above secondary (E5), university (E6)
Age groups: till 25 (A1), 26 to 35 (A2), 36 to 45 (A3), 46 to 55 (A4), 56 to 65 (A5), 66 and older (A6).

In total there are 24 sociodemographic categories, including the countries, so the resulting table of frequencies is 35×24. The CA map is given in Figure 1.5.

The dimensionality of the solution of the stacked table is determined by the minimum of the I rows and J columns minus the respective number of variables, i.e., $\min(I - Q_r; J - Q_c) = \min(35 - 7; 24 - 5) = 19$, where Q_r = number of row variables and Q_c = number of column variables. Total inertia of this supertable is 0.0469, which is again the average of the inertias of the 35 subtables (Table 1.5) (see Greenacre 1994). The first dimension explains 37.2% of the total variation, the second another 31.2%. Because the solution has 19 dimensions, 68.4% explanatory power for the first two dimensions is reasonable and is quite close to the previous solution. However, total inertia is—compared with the previous solution—relatively low because there are many cross-tables in the supermatrix with low levels of association.

Comparing the solutions of the row-wise stacked table with "country" as column variable and the supertable containing 35 cross-tables shows that, on the level of the responses toward national identity, there are almost no differences; the general structure of the response categories is the same. The same holds for the countries; they keep

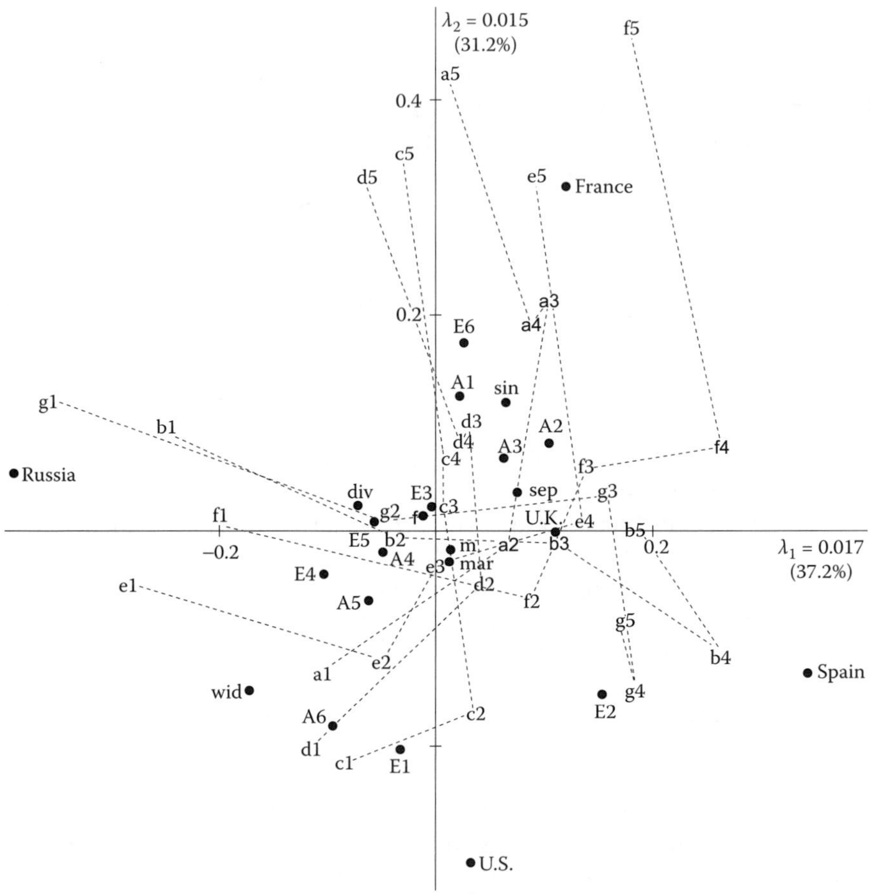

Figure 1.5 Symmetric CA map based on 35 tables, stacked row-wise and column-wise.

their relative positions in the two-dimensional space. In addition to the previous solution, one only can add some findings for the additional sociodemographic variables considered in Figure 1.5. For example, respondents 66 years and older (A6) and widowed persons (wid) are relatively often proud to live in the country where they live. Young people (A1), singles (sin), as well as respondents with a university degree (E6) have relatively often a low national identity. Furthermore, attitudes toward national identity seem to be uncorrelated with sex: there are almost no differences between males and females (compare their location close to the origin in Figure 1.5).

As already noted, in the complex table total inertia is the average value of the 35 subtables. Table 1.5 shows the inertia of all 35 tables as well as the average inertias of the sociodemographic characteristics and of the seven national identity indicators. It can be seen that the highest inertias belong to the cross-tabulations with country, i.e., the most variation in the data is caused by country differences. Further, there are almost no sex differences for the seven items on national identity, although there are some findings that might be worthwhile to report. For example, the association between "sex" and "international sport" is much smaller than the association between "sex" and "feel ashamed of country."

1.6 Multiple correspondence analysis

In the previous examples we analyzed the relation between two variables or between two different sets of variables. In this section, we are interested in the relationships within a set of variables, for example, the interrelationships between the statements on national identity. Thus, for example, we could find out if there is an association between a "strong agreement toward international sports" and a "strong agreement toward people should support their country." In the previous analysis of stacked tables, we could only see whether these categories had the same association with sociodemographic variables.

This new case, which is reminiscent of principal component analysis, involves all the cross-tables of a set of variables, such as the national identity indicators, with themselves. Assembling all these cross-tables into a square supermatrix of cross-tables, we obtain what is known in CA literature as the *Burt matrix*, which we denote by **C**. Alternatively, a data structure known as the *indicator matrix* can be constructed based on the original data. The indicator matrix, denoted by **Z**, is a respondents-by-categories table with as many rows as respondents (6066 in our example) and as many columns as response categories (35 for the seven national identity indicators). The elements of **Z** are zeros apart from ones in the positions to indicate the categories of response of each respondent (**Z** is often called a matrix of dummy variables). The Burt matrix is related quite simply to the indicator matrix as follows: $\mathbf{C} = \mathbf{Z}^{\mathsf{T}}\mathbf{Z}$. If the usual CA algorithm is applied to an indicator matrix or to a Burt matrix, the method is called *multiple correspondence analysis* (MCA). In MCA there is no distinction between describing variables and variables to be described, as is the case in simple CA of single or stacked tables. In MCA all variables have the same status. The relationship between the analyses of **C** and **Z** in MCA is discussed in depth in Chapter 2. In the following, we illustrate the method by analyzing the 6066×35 indicator matrix **Z** that codes the responses to the seven national identity questions. The graphical solution is given in Figure 1.6.

Inspecting Figure 1.6, we see that the first dimension contrasts the strong agreements and the strong disagreements (positive part) from the middle categories (negative part). With two exceptions (statements b and g), the second dimension contrasts the positive statements from the negative ones. Therefore, it can be seen as an overall dimension toward national identity, with a relatively high national identity in the positive part and a relatively low national identity in the negative part. The variables a, c, d, e, and f form a horseshoe, a typical structure we usually find in ordered categorical data (for more details, see Chapters 2 and 4). However, there are two points to be mentioned. First, neither item b, "there are some things ...," nor item g, "I am often less proud...," fulfill this structure, and maybe even worse, the most-opposite categories "b1" and "b5" as well as "g1" and "g5" are close to each other. One reason for this finding might be that a significant number of respondents misunderstood the direction of the question, which would result in such a structure (Blasius and Thiessen 2001b). Another reason is that two dimensions are not sufficient to mirror the structure of these two variables adequately. In a higher-dimensional solution "b1" and "b5" as well as g1 and g5 might be far away from each other. Second, the horseshoe belongs to the first dimension, i.e., the second dimension is more important for substantive interpretation. This might be caused by the joint analysis of the data from five countries, where respondents in different countries may understand the single questions (slightly) differently (see Chapter 20).

Notice that we have not indicated the percentages of inertia explained in Figure 1.6. There are several methods of scaling the solution in MCA that change the fit (see Chapters 2 and 3), but the overall structure of the variable categories in the space remains the same and, hence, the substantive interpretation too. Using the indicator matrix as input, the measure of fit in terms of explained inertia is heavily underestimated. Various solutions to this problem are proposed in Chapter 2 (see Section 2.3.4).

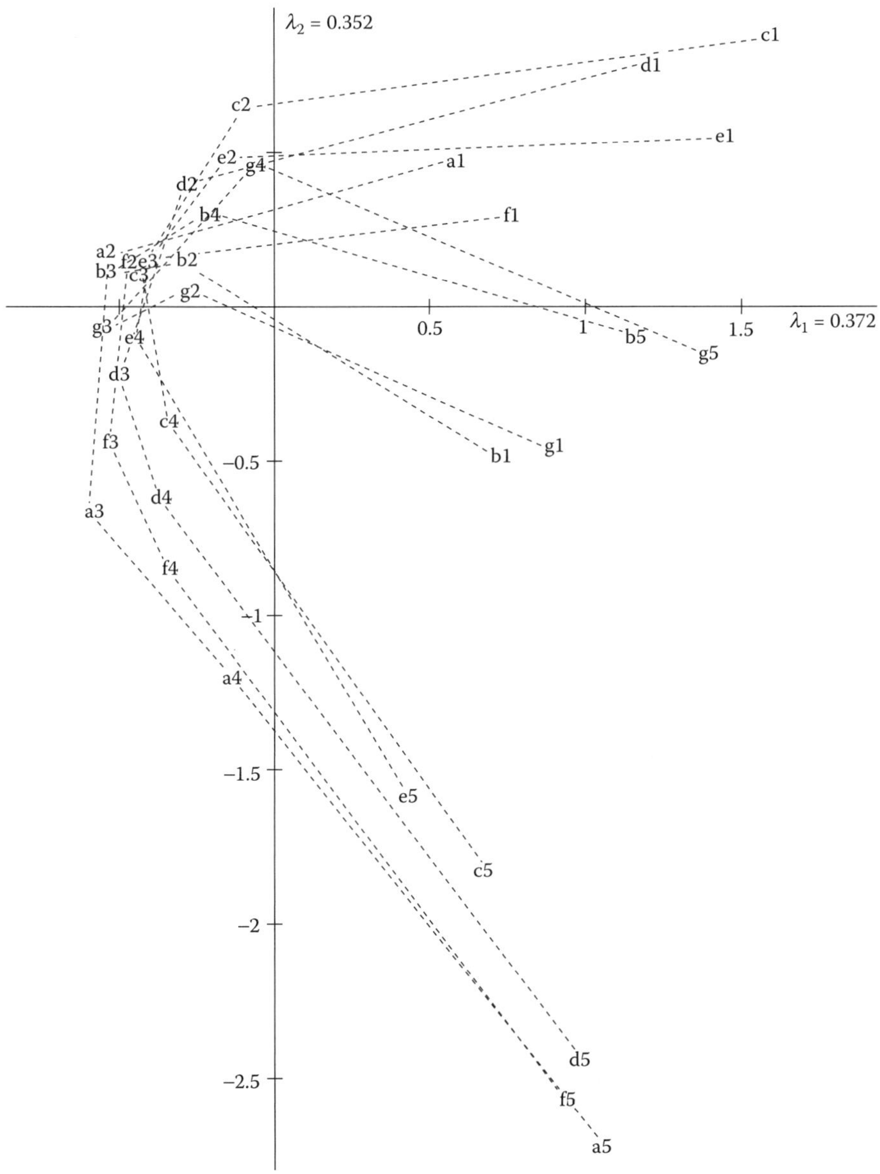

Figure 1.6 MCA map of indicators on national identity.

1.7 Categorical principal component analysis

In the previous sections we discussed CA and its extension to stacked tables and MCA. There are several situations in which the variable categories are ordered, for example, ordered from "agree strongly" to "disagree strongly," as in our example. Where MCA does not impose any constraints on the data, principal component analysis (PCA) has linear constraints, i.e., it assumes that the categories are ordered and that the distances between the categories are constant. Categorical principal component analysis (CatPCA), also known as nonlinear PCA (NLPCA), can be understood as a technique intermediate between (linear) PCA and (nonlinear) MCA. In its most utilized form, CatPCA takes into account the ordering of the categories but allows the intervals between categories to vary. CatPCA is the PCA of a data matrix of categorical data where the category values (1 to 5 in the example on national identity) are replaced by optimal scale values on each dimension. (For more details, see Chapter 4 as well as Gifi 1990 and Heiser and Meulman 1994.) The optimal scaling process allows order constraints to be imposed so that ordered categorical variables get increasing, or at least nondecreasing, quantifications in the low-dimensional solution space (usually two dimensional). When the ordering of the categories in CatPCA is not consistent with the implied ordering, this manifests itself in the form of tied optimal quantifications for two or more subsequent categories. Unlike classical PCA and unlike MCA, the number S^* of dimensions required must be specified in advance because the solutions are not nested.

The results of a CatPCA are mapped in the form of straight lines through the origin for the respective variables, with the response categories indicated on each vector (see Figure 1.7). The first dimension of the CatPCA to the seven indicators on national identity accounts for 34.6% of the variation, and the second accounts for another 23.3%; thus 57.9% of the variation is explained with the first two dimensions. The axes are labeled on the "agree strongly" responses, with the ticks showing the categories. Figure 1.7 shows that the distances between the successive categories are different. With respect to questions b and g, the last two categories ("disagree" and "disagree strongly") are tied (shown by an empty circle). In all questions, the largest difference is between "agree strongly" and "agree." As already shown in the MCA solution (Figure 1.6), questions b, "there are some things about ...," and g, "I am often less proud ...," seem to measure something different than the other five questions, appearing almost uncorrelated with them. Furthermore, questions a, c, and d, all three measuring pride toward country as part of

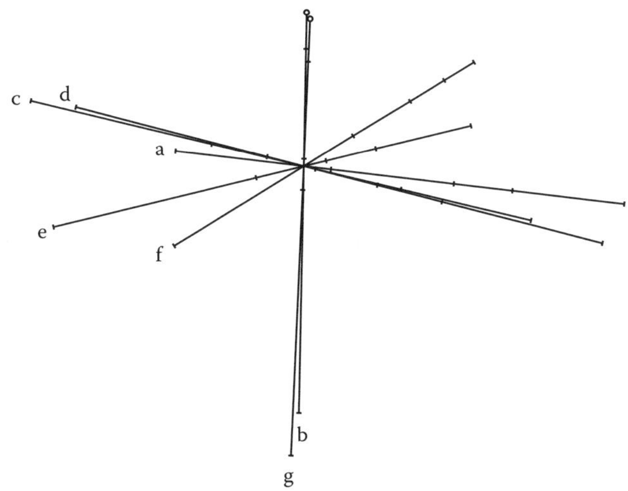

Figure 1.7 CatPCA map of indicators on national identity.

one's national identity, are very close to each other. More details on the theory and practice of CatPCA are given in Chapter 4.

1.8 Active and supplementary variables

For simple CA, MCA, and CatPCA, it is possible to project categories of additional variables on an already existing solution configuration. These additional variables are called supplementary variables, sometimes also referred to as "illustrative" or "passive" as opposed to the "active" variables of the analysis that determine the solution space. Supplementary variables have no influence on the geometric orientation of the axes; rather, they support and complement the interpretation of the configuration of active variable categories. One can think of supplementary points as additional points in the row or column profile spaces; these points have zero mass and thus play no role in the analysis apart from interpreting their positions.

To display supplementary points, we use the so-called *transition formulas* that relate row and column solutions, which are derived from the linear relationships between left and right singular vectors in an SVD. For example, to obtain the principal coordinates **F** of the row points from the principal coordinates **G** of the column points, we have from Equation 1.1, Equation 1.3, Equation 1.4, and Equation 1.6 the following relationship:

$$\mathbf{G} = \mathbf{D}_c^{-1}\mathbf{P}^\mathsf{T}\mathbf{F}\boldsymbol{\Sigma}^{-1} \tag{1.11}$$

The matrix $\mathbf{D}_c^{-1}\mathbf{P}^\mathsf{T}$ contains the column profiles of Table 1.2 (but written here as relative frequencies as row vectors), while $\mathbf{F}\boldsymbol{\Sigma}^{-1}$ is the matrix of row standard coordinates, denoted by $\mathbf{A}$ in Equation 1.5. This shows the barycentric relationship between rows and columns in CA: the column points (in principal coordinates) are weighted averages of the row points (in standard coordinates). This formula allows any additional column profile to be displayed on the map by computing its position as a weighted average of the row (standard) coordinates using its profile elements as weights. A similar transition formula can be obtained that allows supplementary row points to be displayed.

1.9 Multiway data

In CA of stacked tables and MCA, only two-way interaction effects are taken into account in determining the solution. These methods are often described as being *joint bivariate* in this sense. In some situations, however, higher-order interactions need to be taken into account. A common approach to analyzing multiway data is to code two or more variables interactively, which we illustrate in the context of the national identity data. We saw in Section 1.5 that there was little difference between average male and female attitudes. But there could exist some larger male–female differences in certain countries that are masked by the comparison of all males with all females. To visualize male–female differences for each country in terms of their attitudes to national identity—in other words a three-way interaction—we would create a new "interactive" variable called "country-gender" with 10 categories (5 countries × 2 genders) and then use this variable instead of the separate country and gender variables. Examples of interactive coding are given by Greenacre (1993a), who investigates whether the choice of a car depends on income group in combination with age group, as well as Carlier and Kroonenberg (1998), who examine the connection of region and profession at different points in time on the basis of data from official statistics. In Chapter 21 a data set with a four-way structure is analyzed, and two variables are interactively coded so that the data structure becomes three-way.

1.10 Content of the book

The aim of this book is to present introductory and state-of-the-art material on MCA and related techniques, both from a methodological and an applications perspective. Information on this topic is currently scattered, not all of it is in English, and the information has not been unified notationally or directly compared. Some literature is very statistical, while other works are only applications-oriented and lack important details on the methodology and general rules of interpretation. This volume is a compromise between statistical methodology and applications and is designed to explain the methodology to social scientists and other researchers in an intuitive, example-based way. At the same time, it provides statisticians and other methodologists with the theoretical background for understanding MCA and methods related to it. The book is subdivided into five parts: Introduction, Multiple Correspondence Analysis, Analysis of Sets of Tables, MCA and Classification, and Related Methods. Here we give a brief overview of these parts and the chapters comprising them.

1.10.1 Introduction

The first four chapters of the book contain the main concepts and the theoretical background of CA, MCA, and some of the related methods discussed and applied in the subsequent chapters. In these four chapters we provide an overview of these techniques and show how they are related to one another. In this first chapter we have already given an overview of CA of a single table and of stacked tables between two sets of variables, and a brief introduction to MCA. We have introduced the basic geometric concepts of profiles, masses, inertia, chi-square distances, and the reduction of dimensionality, all of which are important for the understanding of CA and the interpretation of CA maps. We also focused specifically on the measure of variance in the tables, i.e., the total inertia, and we showed how, in the multiple case, total inertia is equal to the average of the inertias of the tables constituting the stacked matrix. This is an aspect that is useful in the understanding of MCA.

In Chapter 2, Michael Greenacre discusses in detail some ways to generalize simple CA to MCA. He starts with the case of canonical correlation analysis, which leads to a solution that maximizes the correlation between the row variable and the column variable of a two-way contingency table. The extension to the multivariate case involves applying a more general definition of correlation among sets of variables, leading to a version of MCA also known as homogeneity analysis. The geometric version of this is shown to reduce to the CA of the data coded as an indicator matrix or a Burt matrix. In both cases, a large amount of the total variance is induced purely by the data coding; hence, Greenacre proposes an adjusted version of MCA as the method of choice, where the coordinates have been rescaled to better estimate the fit of the solution. Joint correspondence analysis is also treated here, in which the effects of the main diagonal blocks in the Burt tables are excluded. The methodology is illustrated extensively using empirical data on attitudes toward science from the International Social Survey Program (ISSP).

Chapter 3, written by John Gower, aims to show similarities (and dissimilarities) between a number of techniques that are all concerned with two-way arrays. Two-way arrays can be of many different types: for example, tables of values on a single variable observed on a two-way classification, two-way contingency tables, square correlation matrices based on metric data, indicator matrices, and Burt tables. All techniques to analyze two-way arrays have in common either a decomposition in terms of simple structures such as main effects plus interactions or the singular-value decomposition. Gower discusses and compares principal component analysis, correspondence analysis, and multiple correspondence analysis. He also compares the fit measures for each of these techniques and the effect of scaling of the axes on the final solutions. For his empirical example he uses data from the ISSP on national identity.

Chapter 4, written by Jan de Leeuw, is a state-of-the-art description of nonlinear principal component analysis (NLPCA), also known as categorical principal component analysis. De Leeuw starts with an explanation of PCA and extends it to the nonlinear case, providing algorithmic details that are needed to understand the background of the method. Furthermore, he shows the relation with MCA (or homogeneity analysis, in the Dutch terminology) as well as with multiple regression, and an alternative way of performing NLPCA using a logistic approach. He demonstrates the methodology using mainly a Dutch data set on primary schoolchildren.

1.10.2 Multiple correspondence analysis

The next seven chapters are on multiple correspondence analysis, showing this methodology from several different points of view. Chapter 5 by Henry Rouanet gives the French view of PCA and MCA, especially the role played by Jean-Paul Benzécri in the methodological development of these methods, and goes on to cite the influence of Pierre Bourdieu on their application. This approach to PCA and MCA, known as *analyse des données* in French, is called "geometric data analysis" (LeRoux and Rouanet 2004a). He starts with an explanation of PCA and extends it to MCA, where he gives the basic rules of interpretation. As examples, he uses two data sets, one from the judgment of basketball experts on high-level potential players and one from the Education Program for Gifted Youth.

In Chapter 6, Shizuhiko Nishisato discusses different aspects of the correlational structure of multichoice data from the view of dual scaling. Dual scaling of multivariate categorical data leads to the same solution as MCA, but it is part of a general framework of data scaling used in many different contexts. This chapter shows how the method can capture both linear and nonlinear associations between the variables. Further, Nishisato gives an overview on forced classification of dual scaling, which can be understood as a procedure for discriminant analysis for categorical data. As an empirical example, he uses a small data set from a health survey.

Chapter 7, written by Ludovic Lebart, is dedicated to validation techniques in MCA. One of the criticisms of MCA is that it does not involve techniques for statistical inference. However, there are several methods to validate the findings statistically, two of which are discussed in this chapter. The first is based on external validation, which involves external data, usually included as supplementary or passive variables, leading to cross-validation of the results. The other possibility is based on internal validation, using resampling techniques such as the bootstrap and other Monte Carlo methods. Lebart illustrates the different techniques using a British data set in which the respondents were asked about their standard of living and expectations for the future.

In Chapter 8, Michael Greenacre and Rafael Pardo discuss the application of subset correspondence analysis to the case of MCA. The idea here is to concentrate on some response categories of the variables only, excluding others from the solution. For example, missing responses on several questions can be analyzed alone, or substantive responses excluding missing values can be analyzed and mapped. In the former case, patterns of missing values can be explored on their own, focusing on their relationships with sociodemographic characteristics. In the latter case, the advantage of this method would be to keep all information that is available in the study and not lose any respondent data, as happens when applying listwise deletion of cases.

The authors demonstrate their methodology using ISSP data about attitudes toward women in the labor force.

In Chapter 9, Matthijs Warrens and Willem Heiser discuss the scaling of unidimensional models with MCA. The objective of this chapter is to determine what information on the parameters can be obtained from the application of MCA when exact unidimensional models are used as gauges, or benchmarks. The authors discuss eight possible models—where each model is either deterministic or probabilistic, dichotomous or polytomous, and monotonic or unimodal— and how these models are related to MCA. In the literature, these models are known under names such as Guttman and Rasch scales. The authors show the structure of these models and how they are generated, as well as the MCA solutions of simulated item-response data.

Chapter 10, written by Wijbrandt van Schuur and Jörg Blasius, has a similar purpose as Chapter 9 by Warrens and Heiser. Different item-response data—for example, dominance and cumulative data, proximity or unfolding data—give certain graphical patterns when mapped by MCA. This allows the authors to differentiate between unfolding and Rasch data, for example, on the basis of the results of an MCA. After discussing some of the typical patterns that these models provide, the authors apply MCA to two data sets that are labeled as dominance and unfolding data. Whereas data on religious beliefs from the Dutch World Value Survey form a dominance structure, as expected, there is no evidence that the unfolding data form an "unfolding structure."

In Chapter 11, Yoshio Takane and Heungsun Hwang discuss the topic of regularization in the MCA context. Regularization can be considered as an important and general way to supplement insufficient data by prior knowledge or to incorporate certain desirable properties in the estimates of parameters in the model. Because MCA does not always provide estimates that are on average closest to the population parameters, the authors propose an alternative estimation procedure for MCA, called regularized MCA. Using two small data sets taken from the literature, the authors compare the regularized solution with the ones obtained by MCA.

1.10.3 Analysis of sets of tables

Chapters 12 through 15 deal with different sets of data to be analyzed simultaneously. In Chapter 12, written by Herbert Matschinger and

Matthias Angermeyer, the first set contains the substantive responses to a set of items, while the second set contains the nonsubstantive "don't know" responses to the same items. Because listwise deletion would reduce the sample size dramatically, the authors discuss different methods that retain the cases that have missing values. A solution with high stability is obtained when applying a generalized canonical approach called OVERALS, a method implemented in the SPSS module "Categories." The stability of the solutions are demonstrated using a bootstrap procedure. The authors demonstrate their approach on ten items on attitudes toward psychotropic drugs.

Chapter 13, written by Jérôme Pagès and Mónica Bécue-Bertaut, is dedicated to multiple factor analysis for contingency tables (MFACT). The extension of simple CA to stacked tables is straightforward when the data come from the same source, for example, from the same study (as is the case in the examples given in Section 1.5). In this case the tables have the same margins and thus the same origins and no between-tables variance. However, there are many situations in which the data come from different sources, for example from different studies within the same country or from studies in different countries. With the application of MFACT it is possible, for example, to exclude external country-differences, such as different margins, from the solution. As an empirical application, the authors use textual data on food preferences from three countries.

Chapter 14, written by Amaya Zárraga and Beatriz Goitisolo, presents an approach called simultaneous analysis (SA) for the study of several contingency tables with different margins. Their approach bears a strong resemblance to MFACT but differs in the way the objects forming the common margin of the tables are weighted (for example, the rows if the tables are concatenated columnwise). In MFACT a single weight is used based on the overall margin across all the tables, whereas in SA the individual weights from each table are taken into account. Thus, these two approaches would be equivalent in the particular case when the row margins are equal or proportional. The method is demonstrated by an example from urban research, where the data are the distribution of employed and unemployed males and females in different regions of Spain.

Chapter 15, written by Elena Abascal, Ignacio García Lautre, and Isabel Landaluce, deals with the application of multiple factor analysis (MFA) to a combination of metric and categorical data. Their aim is to analyze election votes from five parties in 50 Spanish provinces. The problem is that two of the five parties have many structural zeros in the data matrix because they have no candidates in several provinces.

In effect, the authors have to combine metric data, percentages of votes, for three parties and a combination of metric data and structural zeros for the other two parties. The structural zeros cannot be set to real zeros (this would be wrong because, if there were candidates, these candidates would be chosen by a significant number of voters); thus these two variables must be treated as categorical. Applying MFA to this mixture of metric and categorical data gives an appropriate global insight into the votes in the Spanish provinces.

1.10.4 Multiple correspondence analysis and classification

Chapters 16 through 18 compare MCA with different classification approaches such as discriminant analysis and logistic regression. Chapter 16, written by Gilbert Saporta and Ndèye Niang, uses the respondent scores from MCA (also called factor scores) to discriminate among different groups. The chapter starts with a brief description of linear methods for discrimination: Fischer's linear discriminant analysis and logistic regression. These linear methods are then extended to include categorical variables into the models, the categorical variables being scaled with help of MCA. As an empirical example, the authors use credit-scoring data from automobile insurers in Belgium to discriminate between two groups of clients: those with one or more claims and those without a claim.

Chapter 17, written by Stéphanie Bougeard, Mohamed Hanafi, Hicham Noçairi, and El-Mostafa Qannari, deals with a similar problem of classification but in a biological context. In this case the authors have a dependent variable with three categories and a large set of categorically scaled exploratory variables. Their approach, called multiblock canonical correlation analysis, is a blend of a single analysis of multivariate regression and generalized canonical correlation analysis. They deal with animal health data (a rabbit disease) and attempt to differentiate between affected and unaffected farms, with an intermediate group in between.

Chapter 18, written by Henri Caussinus and Anne Ruiz-Gazen, focuses on a projection-pursuit approach for categorical data. Considering a typical cases-by-variables data table in the columns, exploratory projection pursuit aims to find low-dimensional projections displaying interesting features in the structure of the case distribution, for example, outliers or a partition into groups. The authors start with the investigation of the properties in metric data, where they rely on a mixture model; thereby, the "noninteresting noise" is a normal distribution, while the "nonnormal" mixing distribution is the structure of interest. The authors then extend their method to indicator matrices and discuss the application of different metrics to elicit different types of structure in the data. As an empirical example, they use measurements on skulls from wolves and dogs.

1.10.5 Related methods

The last part of the book (Chapters 19 through 23) contains methodologies and applications of methods that are closely related to MCA. Chapter 19, written by Anna Torres-Lacomba, deals with the correspondence analysis of a stacked table and its relationship to categorical conjoint measurement. Conjoint analysis is a popular technique in marketing research to predict what products or services people will prefer, assessing in the process the weight consumers give to various attributes that describe the products. The author is interested in conjoint measurement, with product preference expressed on a categorical scale. She shows that correspondence analysis can be used to analyze this kind of data and that the graphical descriptions provide a useful understanding of the results. As an empirical application, she uses data from marketing research on perfumes.

Chapter 20, written by Jörg Blasius and Victor Thiessen, deals with the problem of comparing cross-national data. Such data are meaningful only if there is a common understanding of the questions in all countries and if there is an acceptable level of data quality. The authors employ a three-step approach to assess the comparability and quality of the data: classical PCA, CatPCA, and biplots. When the results of PCA are compared with those of CatPCA, large divergences are a first indicator of the existence of measurement problems. These are then explored systematically using the biplot methodology. The proposed approach can be used as a streaming technique to explore the quality and the structure of the data before applying confirmatory models. As an empirical example, the authors use data from 24 countries of the ISSP on dual- and single-earner households.

Chapter 21, written by Pieter Kroonenberg and Carolyn Anderson, deals with three-way contingency tables. In the modeling approach, such data are often analyzed using different kinds of log-linear models or, more recently, Goodman's multiplicative models, to detect effects and interaction effects in the data. The authors start with a brief introduction into these kinds of models and then move to the additive models and to three-way correspondence analysis. They discuss the measuring, then the modeling, and finally the plotting of global and marginal dependence. As an empirical example, the authors use experimental data where a subject had to classify different types of signals that appeared on a computer screen.

Chapter 22, written by Patrick Groenen and Alex Koning, presents a new way to visualize interaction effects in analysis of variance. The main effects depend on the number of categories, but each interaction effect is characterized by the product of the categories for each variable, and there is an ever-increasing number of combinations of categories as the number of variables increases. The authors describe an interaction-decomposition model that allows the visualization of two-way interactions; they also demonstrate that the interpretation of the interaction plot is similar to that in MCA. As an empirical example, the authors use data on holiday spending.

In the final Chapter 23, José L. Vicente-Villardón, Purificación Galindo-Villardón, and Antonio Blázquez-Zaballos discuss logistic biplots in which a logistic response model can be visualized in the form of a linear biplot. The geometry of the biplot is such that the coordinates of individuals and variables are calculated to have logistic responses along the latent dimensions. Although the method is based on a nonlinear response, the representation is linear, and the directions of the variable vectors on the biplot show the directions of increasing logit values. After giving the technical background of the technique, the authors illustrate their method using microarray gene-expression data.

The book is completed by an appendix, written by Oleg Nenadić and Michael Greenacre, on the computation of MCA. Using data from the ISSP on attitudes toward science (the same data as in Chapter 2), the authors give a step-by-step description of the MCA algorithm, including the adjustments of inertia, joint correspondence analysis, supplementary variables, and subset analysis. The chapter includes all program code in the R language so that readers can repeat these calculations for themselves using the R package, which is freely downloadable at www.r-project.org. This program code, as well as various data sets used in the chapters of this book, are available in electronic form on the Web site of the CARME network: www.carme-n.org.

CHAPTER 2

From Simple to Multiple Correspondence Analysis

Michael Greenacre

CONTENTS

2.1 Introduction

Simple correspondence analysis (CA) is primarily applicable to a two-way contingency table, leading to a map that visualizes the association between two categorical variables. Multiple correspondence analysis (MCA) tackles the more general problem of associations among a set of more than two categorical variables. We shall see that the generalization to more than two variables is neither obvious nor well defined. In other areas of multivariate analysis, such as regression and log-linear modeling, the situation is less complicated: for example, the transition from the regression of a response variable on a single predictor variable to the case of several predictors is quite straightforward. The main problem we face here is that the notion of association between two categorical variables is a complex concept. There are several ways to generalize this concept to more than two variables.

Of the many different ways that exist to define MCA, we shall consider two approaches: first, the definition which is perhaps the easiest to understand, namely that of correlation between sets of variables, known as canonical correlation, and second, the geometric approach, which is directly linked to data visualization and which has many similarities to Pearson-style principal component analysis. In the explanation of each approach, we will consider the case of two variables and then describe possible generalizations to more than two variables.

As an illustration of the theory, we shall use a data set from the International Social Survey Program on environment (ISSP 1993), looking specifically at questions on attitudes toward science. The survey questions that we consider are as follows:

> How much do you agree or disagree with each of these statements?
>
> A. We believe too often in science, and not enough in feelings and faith.
> B. Overall, modern science does more harm than good.
> C. Any change humans cause in nature — no matter how scientific — is likely to make things worse.
> D. Modern science will solve our environmental problems with little change to our way of life.

Each question has five possible response categories:

1. Agree strongly
2. Agree
3. Neither agree nor disagree
4. Disagree
5. Disagree strongly

To avoid the issue of cross-cultural differences, we use data for the West German sample only (the ISSP surveys still distinguish between former West and East Germany). We shall also show how to relate the

MCA results to the external demographic variables of sex, age, and education, which were also coded as categorical variables as follows:

Sex: male, female
Age (six groups): 16–24, 25–34, 35–44, 45–54, 55–64, 65 and older
Education (six groups): primary incomplete, primary completed, secondary incomplete, secondary completed, tertiary incomplete, tertiary completed

A listwise deletion of respondents with missing data has been performed because we do not want to deal with the further complicating issue of missing data here (see Chapter 8). This reduces the original West German sample by about 14% to leave $n = 871$ respondents with complete data, which form the data set used in this chapter.

In the appendix of this book, most of the numerical results in this chapter are given along with the code in the R language to perform the computations (R Development Core Team 2005).

2.2 Canonical correlation analysis

2.2.1 Two variables

We start by considering just the first two variables, A (concerning belief in science) and B (concerning harm caused by science), both worded unfavorably toward science, so that disagreement indicates a favorable attitude toward science. Because there are only two variables, all 871 responses to these questions can be coded, with no loss of information, in the form of a cross-tabulation, given in Table 2.1. The correlational approach investigates how to measure the association between these two categorical variables. Several measures of association already exist for categorical data, some of which depend on whether the variables are measured on a nominal or ordinal scale, or whether we are trying to predict one variable from the other. In the following, we shall focus our interest on the classical product-moment correlation coefficient applicable to metric data and on the *quantification* of the categories, i.e., how to achieve numerical values for the response categories to calculate a correlation coefficient between the variables. Because the categories are ordered, a simple way out would be to use the existing values 1 to 5, as they are coded in the data file, thereby assuming that there is an equal interval difference between adjacent points on each scale.

Table 2.1 Cross-tabulation of 871 West German respondents with respect to two questions on attitudes to science.

We believe too often in science, not enough in feelings & faith	Overall, modern science does more harm than good					
	B1 agree strongly	B1 agree	B3 neither/nor	B4 disagree	B5 disagree strongly	SUM
A1-agree strongly	27	28	30	22	12	119
A2-agree	38	74	84	96	30	322
A3-neither/nor	3	48	63	73	17	204
A4-disagree	3	21	23	79	52	178
A5-disagree strongly	0	3	5	11	29	48
SUM	71	174	205	281	140	871

But note that such a choice would be incorrect if one of the variables were nominal, for example "province of residence" or "religious denomination."

There are two ways to calculate the correlation coefficient: one is from the original respondent-level data, which are the 871 pairs of responses to the two questions; the other, more compact, approach is directly from Table 2.1, since this table gives the frequencies of occurrence of all pairs of categories. Suppose that the responses to questions A and B are coded in the indicator matrices $\mathbf{Z}_1$ and $\mathbf{Z}_2$, respectively, whose columns are zero-one dummy variables: that is $\mathbf{Z}_1$ and $\mathbf{Z}_2$ are both 871×5 matrices. Then Table 2.1 is the cross-product $\mathbf{Z}_1^{\mathsf{T}}\mathbf{Z}_2$ of the two indicator matrices. Furthermore, suppose that the proposed scale values for the categories of the two variables are contained in the vectors $\mathbf{s}_1$ and $\mathbf{s}_2$, so that the individual quantified responses are in the vectors $\mathbf{Z}_1\mathbf{s}_1$ and $\mathbf{Z}_2\mathbf{s}_2$. To simplify the notation greatly, it is convenient to consider the quantified responses as initially mean-centered, $\mathbf{1}^{\mathsf{T}}\mathbf{Z}_1\mathbf{s}_1 = \mathbf{1}^{\mathsf{T}}\mathbf{Z}_2\mathbf{s}_2 = 0$, so that the covariance s_{12} between the two variables and their variances s_1^2 and s_2^2 can be written as:

$$s_{12} = (1/n)\mathbf{s}_1^{\mathsf{T}}\mathbf{Z}_1^{\mathsf{T}}\mathbf{Z}_2\mathbf{s}_2 = \mathbf{s}_1^{\mathsf{T}}\mathbf{P}_{12}\mathbf{s}_2$$

$$s_1^2 = (1/n)\mathbf{s}_1^{\mathsf{T}}\mathbf{Z}_1^{\mathsf{T}}\mathbf{Z}_1\mathbf{s}_1 = \mathbf{s}_1^{\mathsf{T}}\mathbf{D}_1\mathbf{s}_1 \quad \text{and} \quad s_2^2 = (1/n)\mathbf{s}_2^{\mathsf{T}}\mathbf{Z}_2^{\mathsf{T}}\mathbf{Z}_2\mathbf{s}_2 = \mathbf{s}_2^{\mathsf{T}}\mathbf{D}_2\mathbf{s}_2$$

where $\mathbf{P}_{12} = (1/n)\mathbf{Z}_1^{\mathsf{T}}\mathbf{Z}_2$ is called the *correspondence matrix*, containing the relative frequencies, i.e., Table 2.1 divided by its grand total of

n = 871. $\mathbf{D}_1$ and $\mathbf{D}_2$ are diagonal matrices of the marginal relative frequencies, or *masses*, of the two variables. (In Chapter 1 these are denoted by $\mathbf{D}_r$ and $\mathbf{D}_c$ for "rows" and "columns"; in this chapter we use indices 1 and 2, since we are going to extend the concepts to more than two variables.)

Using the above notation the correlation can be written as:

$$r = \frac{s_{12}}{s_1 s_2} = \frac{\mathbf{s}_1^\mathsf{T}\mathbf{P}_{12}\mathbf{s}_2}{\sqrt{\mathbf{s}_1^\mathsf{T}\mathbf{D}_1\mathbf{s}_1\mathbf{s}_2^\mathsf{T}\mathbf{D}_2\mathbf{s}_2}} \tag{2.1}$$

which can be calculated directly from Table 2.1 and its margins. Because this calculation involves some important concepts in CA, we shall go through it in detail, using the values 1 to 5 for the categories of each variable.

- From Table 2.1 we have the values of the marginal relative frequencies (masses) for the categories of the two variables:

$$(1/n)\mathbf{1}^\mathsf{T}\mathbf{Z}_1 = (1/871)[\ 119 \quad 322 \quad 204 \quad 178 \quad 48\ ]$$

$$= [\ 0.137 \quad 0.370 \quad 0.234 \quad 0.204 \quad 0.055\ ]$$

$$(1/n)\mathbf{1}^\mathsf{T}\mathbf{Z}_2 = (1/871)[\ 71 \quad 174 \quad 205 \quad 281 \quad 140\ ]$$

$$= [\ 0.082 \quad 0.200 \quad 0.235 \quad 0.323 \quad 0.161\ ]$$

- Assuming the equal interval scales 1, 2, 3, 4, 5 for the two variables, their averages are

$$(0.137 \times 1) + (0.370 \times 2) + \cdots + (0.055 \times 5) = 2.672$$

$$(0.082 \times 1) + (0.200 \times 2) + \cdots + (0.161 \times 5) = 3.281$$

and the centered vectors $\mathbf{s}_1$ and $\mathbf{s}_2$ are

$$\mathbf{s}_1 = \begin{bmatrix} -1.672 \\ -0.672 \\ +0.328 \\ +1.328 \\ +2.328 \end{bmatrix} \qquad \mathbf{s}_2 = \begin{bmatrix} -2.281 \\ -1.281 \\ -0.281 \\ +0.719 \\ +1.719 \end{bmatrix}$$

- The *correspondence matrix* is the matrix of relative frequencies (we only give some elements of the matrix):

$$\mathbf{P}_{12} = (1/871)\begin{bmatrix} 27 & \cdots & 12 \\ 38 & \cdots & 30 \\ \vdots & \ddots & \vdots \\ 0 & \cdots & 29 \end{bmatrix} = \begin{bmatrix} 0.03100 & \cdots & 0.01378 \\ 0.04363 & \cdots & 0.03444 \\ \vdots & \ddots & \vdots \\ 0.00000 & \cdots & 0.03330 \end{bmatrix}$$

  and the diagonal matrices of masses, $\mathbf{D}_1$ and $\mathbf{D}_2$, contain the marginal relative frequencies (masses) computed above.
- Hence, the covariance, variances and correlation are

$$\begin{aligned} s_{12} = \mathbf{s}_1^{\mathsf{T}}\mathbf{P}_{12}\,\mathbf{s}_2 &= (0.03100 \times -1.672 \times -2.281) + \cdots + \\ &\quad (0.03330 \times 2.328 \times 1.719) \\ &= 0.4988 \end{aligned}$$

$$\begin{aligned} s_1^2 = \mathbf{s}_1^{\mathsf{T}}\mathbf{D}_1\mathbf{s}_1 &= 0.137 \times (-1.672)^2 + \cdots + 0.055 \times (2.328)^2 \\ &= 1.233 \end{aligned}$$

$$\begin{aligned} s_2^2 = \mathbf{s}_2^{\mathsf{T}}\mathbf{D}_2\mathbf{s}_2 &= 0.082 \times (-2.281)^2 + \cdots + 0.161 \times (1.719)^2 \\ &= 1.412 \end{aligned}$$

$$r = \frac{s_{12}}{s_1 s_2} = \frac{0.4988}{\sqrt{1.233 \times 1.412}} = 0.3780$$

All of the above calculations clearly depend on the equal-interval scale values in $\mathbf{s}_1$ and $\mathbf{s}_2$ assumed at the start. We now consider these scale values as unknowns to be determined, and we pose the following question: what scale values for $\mathbf{s}_1$ and $\mathbf{s}_2$ will give the highest correlation (Equation 2.1) between the two variables? This is exactly the problem of *canonical correlation* between the five dummy variables in $\mathbf{Z}_1$ and the five dummy variables in $\mathbf{Z}_2$. Because the correlation remains the same if any linear transformations of $\mathbf{s}_1$ and $\mathbf{s}_2$ are made, we need to introduce *identification conditions* that fix the scale of $\mathbf{s}_1$ and $\mathbf{s}_2$ in order to find the optimal solution. The usual identification conditions are that the two variables are standardized, i.e., that the means are zero, as previously: $(1/n)\mathbf{1}^{\mathsf{T}}\mathbf{Z}_1\mathbf{s}_1 = (1/n)\mathbf{1}^{\mathsf{T}}\mathbf{Z}_2\mathbf{s}_2 = 0$ and, furthermore, that the variances are 1: $\mathbf{s}_1^{\mathsf{T}}\mathbf{D}_1\mathbf{s}_1 = \mathbf{s}_2^{\mathsf{T}}\mathbf{D}_2\mathbf{s}_2 = 1$. Under these conditions, we now show that the optimal solution coincides exactly with the so-called *standard coordinates* of the response categories on the first principal dimension of a simple CA of the original cross-tabulation (see Chapter 1).

Consider the singular-value decomposition (SVD) of the following normalized matrix:

$$\mathbf{D}_1^{-1/2}\mathbf{P}_{12}\mathbf{D}_2^{-1/2} = \mathbf{U}\mathbf{\Sigma}\mathbf{V}^\mathsf{T} \quad \text{where} \quad \mathbf{U}^\mathsf{T}\mathbf{U} = \mathbf{V}^\mathsf{T}\mathbf{V} = \mathbf{I} \tag{2.2}$$

where $\mathbf{\Sigma}$ is the diagonal matrix of singular values, and $\mathbf{U}$ and $\mathbf{V}$ are the matrices of left and right singular vectors as columns. Then, writing Equation 2.2 for one pair of left and right vectors, $\mathbf{u}$ and $\mathbf{v}$, corresponding to a singular value σ, we have, after multiplying on the left by $\mathbf{u}^\mathsf{T}$ and on the right by $\mathbf{v}$, and using the orthogonality of the singular vectors:

$$\mathbf{u}^\mathsf{T}\mathbf{D}_1^{-1/2}\mathbf{P}_{12}\mathbf{D}_2^{-1/2}\mathbf{v} = \sigma$$

So if we let $\mathbf{s}_1 = \mathbf{D}_1^{-1/2}\mathbf{u}$ and $\mathbf{s}_2 = \mathbf{D}_2^{-1/2}\mathbf{v}$, then $\mathbf{s}_1^\mathsf{T}\mathbf{P}_{12}\mathbf{s}_2 = \sigma$, which is the formula for the covariance. Furthermore, the identification conditions $\mathbf{s}_1^\mathsf{T}\mathbf{D}_1\mathbf{s}_1 = \mathbf{s}_2^\mathsf{T}\mathbf{D}_2\mathbf{s}_2 = 1$ are satisfied, since the singular vectors have length 1: $\mathbf{u}^\mathsf{T}\mathbf{u} = \mathbf{v}^\mathsf{T}\mathbf{v} = 1$, so it appears that the correlation is given by the singular value σ. However, the centering conditions have not been imposed, and these can be introduced by first centering the matrix to be decomposed as follows, subtracting the product of the row and column margins from each element of the correspondence matrix:

$$\mathbf{D}_1^{-1/2}(\mathbf{P}_{12} - \mathbf{P}_{12}\mathbf{1}\mathbf{1}^\mathsf{T}\mathbf{P}_{12}^\mathsf{T})\mathbf{D}_2^{-1/2} = \mathbf{U}\mathbf{\Sigma}\mathbf{V}^\mathsf{T} \tag{2.3}$$

where $\mathbf{P}_{12}\mathbf{1}$ is the (column) vector of row margins of $\mathbf{P}_{12}$, i.e., the row masses (denoted by $\mathbf{r}$ in Chapter 1), and $\mathbf{1}^\mathsf{T}\mathbf{P}_{12}^\mathsf{T}$ is the (row) vector of column margins, the column masses (denoted by $\mathbf{c}^\mathsf{T}$). In the parlance of CA, this is known as "removing the trivial solution" because the uncentered matrix (Equation 2.2) has a trivial maximal solution with a singular value of 1 for $\mathbf{s}_1$ and $\mathbf{s}_2$ equal to $\mathbf{1}$ (thus, $\mathbf{U}$, $\mathbf{V}$, and $\mathbf{\Sigma}$ in Equation 2.2 all have this one extra trivial singular component, which is eliminated by the centering in Equation 2.3).

We thus have the following result: each singular value is a correlation between variables A and B, based on the scale values from the transformed singular vectors $\mathbf{s}_1$ and $\mathbf{s}_2$, and so the maximum correlation is attained for the first (i.e., the largest) singular value of Equation 2.3 or, equivalently, the second largest singular value of the uncentered matrix

(Equation 2.2). The solutions $\mathbf{s}_1$ and $\mathbf{s}_2$ are exactly the vectors of standard coordinates in CA on the first principal axis. The largest singular value σ_1 of Equation 2.3, also called the first canonical correlation, is equal to 0.4106 in our example, compared with the value of 0.3780 obtained with the equal-interval (1 to 5) scales. The scale values are:

$$\mathbf{s}_1^{\mathsf{T}} = [-1.017 \quad -0.560 \quad -0.248 \quad 1.239 \quad 2.741]$$

$$\mathbf{s}_2^{\mathsf{T}} = [-1.571 \quad -0.667 \quad -0.606 \quad -0.293 \quad 1.926]$$

These scale values are standardized, but since any linear transformation leaves the correlation unchanged, it is convenient to transform them so that the endpoints also have values 1 and 5, with a range of 4, in order to compare with the previous equal-interval scales. For example, for the first variable, the range of values is 2.741 – (–1.017) = 3.758, so in order to make the range exactly four units, we should multiply all the values by 4/3.758 = 1.064, in which case the lowest value is now $-1.017 \times 1.064 = -1.083$. Then the addition of 2.083 to all the values will bring the scale to have lowest and highest values equal to 1 and 5, respectively. This procedure gives the following rescaled values for the two sets of response categories:

$$\text{rescaled row values} = [1 \quad 1.486 \quad 1.818 \quad 3.402 \quad 5]$$

$$\text{rescaled column values} = [1 \quad 2.034 \quad 2.103 \quad 3.132 \quad 5]$$

The scale points do emerge in their expected order in both cases, but it is interesting to study their relative spacings. Compared with the equal-interval values considered previously, these rescaled values show that the categories "disagree" and "disagree strongly" for question A are further spaced out, with relatively small differences between scale values assigned to the categories "strongly agree," "agree," and "neither/nor." For question B the difference between "disagree" and "disagree strongly" is even larger, almost two full units. For both questions the neutral "neither/nor" category is not in the center of the scale but close to the agreement category.

Before moving onto the case of several variables, we remark that, in the above, only one set of scale values has been derived for each variable, corresponding to the first singular value σ_1. Further sets of scale values can be determined in a stepwise manner by maximizing the correlation between another pair of subject scores based on different scale values, say $\tilde{\mathbf{s}}_1$ and $\tilde{\mathbf{s}}_2$, where the subject scores are uncorrelated with those already obtained, i.e., $\tilde{\mathbf{s}}_1^{\mathsf{T}}\mathbf{D}_1\mathbf{s}_1 = \tilde{\mathbf{s}}_2^{\mathsf{T}}\mathbf{D}_2\mathbf{s}_2 = 0$. The solution is given by the second set of singular vectors of Equation 2.3, transformed as before to standard coordinates, corresponding to the second singular value, σ_2, which is the second canonical correlation. For a table of order $I \times J$, this process can be continued to obtain a total of $\min\{I-1, J-1\}$ canonical correlations and associated scale values: in our 5×5 example, four sets of scale values and canonical correlations can be calculated. The canonical correlations are the square roots of the *principal inertias* usually reported on the axes of the map (see Chapter 1 and Section 2.3 below).

2.2.2 *Several variables*

To make the transition to the case of several variables, notice that the problem is almost identical if we reformulate it as maximizing the correlation between the two variables and their average (or their sum). In general, for two variables z_1 and z_2 with correlation ρ, the correlation between either of them and their average $\frac{1}{2}(z_1 + z_2)$ (or their sum $z_1 + z_2$) is equal to $\sqrt{(1+\rho)/2}$, so that maximizing ρ is equivalent to maximizing the correlation between the variables and their average (or sum). The only real difference is the value of the maximum found: in the latter formulation this will be $\sqrt{(1+\rho)/2}$ and not the value of ρ itself. The average of two categorical variables leads us to consider the matrix of the two indicator matrices $[\mathbf{Z}_1 \quad \mathbf{Z}_2]$, where the average of the two quantifications of the variables, based on $\mathbf{s}_1$ and $\mathbf{s}_2$, respectively, is equal to

$$\frac{1}{2}\left(\mathbf{Z}_1\mathbf{s}_1 + \mathbf{Z}_2\mathbf{s}_2\right) = \frac{1}{2}[\mathbf{Z}_1 \quad \mathbf{Z}_2]\begin{bmatrix}\mathbf{s}_1 \\ \mathbf{s}_2\end{bmatrix}$$

Consider now what happens when we apply the standard CA algorithm to the superindicator matrix $\mathbf{Z} = [\mathbf{Z}_1 \ \mathbf{Z}_2]$. Because $\mathbf{Z}$ has total sum $2n$, with each of the n rows summing to a constant 2 and column sums equal to the marginal frequencies of each variable, the correspondence matrix is $[1/(2n)]\mathbf{Z}$, the row mass matrix is $(1/n)\mathbf{I}$, and the column mass matrix is $\mathbf{D} = \frac{1}{2}\mathrm{diag}(\mathbf{D}_1,\mathbf{D}_2)$, where $\mathrm{diag}(\mathbf{D}_1,\mathbf{D}_2)$ is the diagonal matrix formed by the two diagonal matrices $\mathbf{D}_1$ and $\mathbf{D}_2$ defined. Hence, the SVD to compute the CA solution of $\mathbf{Z}$ is (in its uncentered form, see Equation 2.2):

$$\sqrt{n}\,\frac{\mathbf{Z}}{2n}\mathbf{D}^{-1/2} = \mathbf{U}\Gamma\mathbf{V}^{\mathsf{T}} \qquad \text{where} \quad \mathbf{U}^{\mathsf{T}}\mathbf{U} = \mathbf{V}^{\mathsf{T}}\mathbf{V} = \mathbf{I}$$

which—in one of its symmetric eigenvalue formulations—can be written as:

$$\left(\sqrt{n}\,\frac{\mathbf{Z}}{2n}\mathbf{D}^{-1/2}\right)^{\mathsf{T}}\left(\sqrt{n}\,\frac{\mathbf{Z}}{2n}\mathbf{D}^{-1/2}\right)=\frac{1}{4n}\mathbf{D}^{-1/2}\mathbf{Z}^{\mathsf{T}}\mathbf{Z}\mathbf{D}^{-1/2}=\mathbf{V}\boldsymbol{\Gamma}^2\mathbf{V}^{\mathsf{T}}$$

$$\text{where } \mathbf{V}^{\mathsf{T}}\mathbf{V}=\mathbf{I}$$

that is,

$$\frac{1}{4n}\mathbf{D}^{-1/2}\mathbf{C}\mathbf{D}^{-1/2}=\mathbf{V}\boldsymbol{\Gamma}^2\mathbf{V}^{\mathsf{T}}=\mathbf{V}\boldsymbol{\Lambda}\mathbf{V}^{\mathsf{T}} \tag{2.4}$$

where $\mathbf{C} = \mathbf{Z}^{\mathsf{T}}\mathbf{Z}$ and $\boldsymbol{\Lambda} = \boldsymbol{\Gamma}^2$. The matrix $\mathbf{C}$, called the *Burt matrix*, is an important data structure in MCA: it is the matrix of all two-way cross-tabulations of the categorical variables, which in the present case of two categorical variables can be written as:

$$\mathbf{C}=\begin{bmatrix}\mathbf{Z}_1^{\mathsf{T}}\mathbf{Z}_1 & \mathbf{Z}_1^{\mathsf{T}}\mathbf{Z}_2\\ \mathbf{Z}_2^{\mathsf{T}}\mathbf{Z}_1 & \mathbf{Z}_2^{\mathsf{T}}\mathbf{Z}_2\end{bmatrix}=n\begin{bmatrix}\mathbf{D}_1 & \mathbf{P}_{12}\\ \mathbf{P}_{21} & \mathbf{D}_2\end{bmatrix}$$

We use the notation $\boldsymbol{\Lambda} = \boldsymbol{\Gamma}^2$, that is, the squares γ^2 of the singular values, or principal inertias of $\mathbf{Z}$ that appear on the diagonal of $\boldsymbol{\Gamma}^2$, are denoted by λ on the diagonal of $\boldsymbol{\Lambda}$. Writing Equation 2.4 for a single eigenvector $\mathbf{v}$, partitioned into two subvectors $\mathbf{v}_1$ and $\mathbf{v}_2$ (one corresponding to the rows of the original table, the other to the columns) and multiplying as before on the left by $\mathbf{v}^{\mathsf{T}}$ and on the right by $\mathbf{v}$, defining $\mathbf{s} = \mathbf{D}^{-1/2}\mathbf{v}$ similarly partitioned into $\mathbf{s}_1$ and $\mathbf{s}_2$, we obtain the eigenequation:

$$\frac{1}{4}\begin{bmatrix}\mathbf{s}_1\\ \mathbf{s}_2\end{bmatrix}^{\mathsf{T}}\begin{bmatrix}\mathbf{D}_1 & \mathbf{P}_{12}\\ \mathbf{P}_{21} & \mathbf{D}_2\end{bmatrix}\begin{bmatrix}\mathbf{s}_1\\ \mathbf{s}_2\end{bmatrix}=\gamma^2=\lambda$$

that is,

$$\tfrac{1}{4}(\mathbf{s}_1^{\mathsf{T}}\mathbf{D}_1\mathbf{s}_1+\mathbf{s}_1^{\mathsf{T}}\mathbf{P}_{12}\mathbf{s}_2+\mathbf{s}_2^{\mathsf{T}}\mathbf{P}_{21}\mathbf{s}_1+\mathbf{s}_2^{\mathsf{T}}\mathbf{D}_2\mathbf{s}_2)=\gamma^2=\lambda \tag{2.5}$$

The maximum value of Equation 2.5, given by the largest nontrivial eigenvalue $\lambda_1 = \gamma_1^2$, coincides with the solution of simple CA of the single two-way table with correspondence matrix $\mathbf{P}_{12}$, except that its maximum is now equal to $\tfrac{1}{4}(1 + \sigma_1 + \sigma_1 + 1) = \tfrac{1}{2}(1 + \sigma_1)$, where σ_1 is the maximized canonical correlation in simple CA. According to our previous remarks, $\tfrac{1}{2}(1 + \sigma_1)$ is exactly the square of the correlation between either of the two (quantified) categorical variables and their average (or sum). Hence $\mathbf{s}_1$ and $\mathbf{s}_2$ derived above are identical to the $\mathbf{s}_1$ and $\mathbf{s}_2$ of simple CA, and what we have derived are the standard coordinates of the columns of the indicator matrix $\mathbf{Z}$. Notice that the eigenvalue λ_1 above is also the singular value of $\mathbf{C}$ because $\mathbf{C}$ is symmetric: in the language of geometric CA (see Chapter 1 and Section 2.3 below), λ_1 is the square root of the principal inertia of the Burt matrix $\mathbf{C}$.

The alert reader will have noticed that in Equation 2.4 the identification condition on $\mathbf{s}$ implied by the standardization of $\mathbf{v}$ in the SVD is that its weighted sum of squares is equal to 1, that is, $\frac{1}{2}(\mathbf{s}_1^\mathsf{T}\mathbf{D}_1\mathbf{s}_1 + \mathbf{s}_2^\mathsf{T}\mathbf{D}_2\mathbf{s}_2) = \mathbf{s}^\mathsf{T}\mathbf{D}\mathbf{s} = 1$, and not that the subvectors $\mathbf{s}_1$ and $\mathbf{s}_2$ are individually normalized to be 1. It can be shown, however, that if $\mathbf{s}_1$ and $\mathbf{s}_2$ constitute a solution corresponding to an eigenvalue $\frac{1}{2}(1 + \sigma)$, then $\mathbf{s}_1$ and $-\mathbf{s}_2$ constitute another solution corresponding to the eigenvalue $\frac{1}{2}(1 - \sigma)$ (see, for example, Greenacre 1984: section 5.1). The orthogonality of these eigenvectors, $\mathbf{s}_1^\mathsf{T}\mathbf{D}_1\mathbf{s}_1 - \mathbf{s}_2^\mathsf{T}\mathbf{D}_2\mathbf{s}_2 = 0$, together with the overall normalization constraint, imply the individual normalizations $\mathbf{s}_1^\mathsf{T}\mathbf{D}_1\mathbf{s}_1 = \mathbf{s}_2^\mathsf{T}\mathbf{D}_2\mathbf{s}_2 = 1$. As far as the individual centering constraints are concerned, these are automatic, since each set of dummy variables (columns of $\mathbf{Z}_1$ and $\mathbf{Z}_2$) has the same sum, equal to $\mathbf{1}$, the vector of ones, so that each set has the same centroid, equal to the overall centroid $(1/n)\mathbf{1}$.

The scene is now set for one possible generalization of CA to the multivariable case, where there are Q categorical variables, coded in indicator matrices $\mathbf{Z}_1, \mathbf{Z}_2, \ldots, \mathbf{Z}_Q$. The problem can be defined as finding a set of scale values $\mathbf{s}_1, \mathbf{s}_2, \ldots, \mathbf{s}_Q$ for the variables so that an overall measure of correlation is maximized. To generalize the two-variable case, the measure of choice is the sum of squared correlations of the individual scores $\mathbf{Z}_1\mathbf{s}_1, \mathbf{Z}_2\mathbf{s}_2, \ldots, \mathbf{Z}_Q\mathbf{s}_Q$ with the summated score $\mathbf{Z}\mathbf{s}$, where $\mathbf{Z}$ and $\mathbf{s}$ are the concatenations of the $\mathbf{Z}_q$'s and $\mathbf{s}_q$'s, respectively. We specify an overall identification constraint $\mathbf{s}^\mathsf{T}\mathbf{D}\mathbf{s} = 1$, where $\mathbf{D} = (1/Q)\mathrm{diag}(\mathbf{D}_1, \mathbf{D}_2, \ldots, \mathbf{D}_Q)$. This overall constraint does not imply that individual variances $\mathbf{s}_q^\mathsf{T}\mathbf{D}_q\mathbf{s}_q$ will be 1 in the final solution—in contrast to the case $Q = 2$ described in the previous paragraph.

Again there are two ways to achieve the solution, one way by performing a CA of the superindicator matrix $\mathbf{Z} = [\mathbf{Z}_1\ \mathbf{Z}_2, \ldots, \mathbf{Z}_Q]$, alternatively a CA of the Burt matrix $\mathbf{C}$, which is now a block matrix with Q blocks row-wise and columnwise. We denote the number of categories for the qth categorical variable by J_q and let $J = \sum_q J_q$ be the total number of categories. Then $\mathbf{Z}$ is of order $n \times J$ and $\mathbf{C}$ is of order $J \times J$. Since $\mathbf{Z}$ has total sum nQ, with row sums equal to a constant Q and column sums equal to the marginal frequencies of each variable, the correspondence matrix is $(1/Qn)\mathbf{Z}$, the row mass matrix is $(1/n)\mathbf{I}$, and the column mass matrix is $\mathbf{D}$. Hence, the SVD to compute the CA solution of $\mathbf{Z}$ is (in its uncentered form, see Equation 2.2):

$$\sqrt{n}\,\frac{\mathbf{Z}}{Qn}\mathbf{D}^{-1/2} = \mathbf{U}\Gamma\mathbf{V}^{\mathsf{T}} \quad \text{where } \mathbf{U}^{\mathsf{T}}\mathbf{U} = \mathbf{V}^{\mathsf{T}}\mathbf{V} = \mathbf{I} \tag{2.6}$$

To eliminate the trivial solution, the matrix to be decomposed is (see Equation 2.3):

$$\sqrt{n}\left(\frac{\mathbf{Z}}{Qn} - \frac{1}{n}\mathbf{1}\mathbf{1}^{\mathsf{T}}\mathbf{D}\right)\mathbf{D}^{-1/2}$$

where $(1/n)\mathbf{1}$ is the vector of row masses and $\mathbf{1}^{\mathsf{T}}\mathbf{D}$ is the vector of column masses of the indicator matrix (denoted by $\mathbf{c}^{\mathsf{T}}$ in simple CA). The SVD for the CA of the Burt matrix $\mathbf{C}$ (uncentered) is

$$\mathbf{D}^{-1/2}\frac{\mathbf{C}}{Q^2 n}\mathbf{D}^{-1/2} = \mathbf{V}\Gamma^2\mathbf{V}^{\mathsf{T}} = \mathbf{V}\Lambda\mathbf{V}^{\mathsf{T}} \quad \text{where } \mathbf{V}^{\mathsf{T}}\mathbf{V} = \mathbf{I} \tag{2.7}$$

and $\mathbf{C} = \mathbf{Z}^{\mathsf{T}}\mathbf{Z}$. Once again, the centered form of the matrix on the left-hand side of Equation 2.7 removes the trivial solution in the form of the expected relative frequencies:

$$\mathbf{D}^{-1/2}\left(\frac{\mathbf{C}}{Q^2 n} - \mathbf{D}\mathbf{1}\mathbf{1}^{\mathsf{T}}\mathbf{D}\right)\mathbf{D}^{-1/2}$$

The right-hand singular vectors, which give us the scale values for the Q variables, are identical in the two problems. The maximum value of the average squared correlation is given by the square of the first singular value in the (centered) analysis of $\mathbf{Z}$, that is, the first singular value in the (centered) analysis of $\mathbf{C}$. Notice that the singular values λ in the analysis of $\mathbf{C}$ are also eigenvalues, since the matrix being decomposed is positive definite symmetric. The standard coordinates $\mathbf{x}$ that provide the scale values, partitioned into $\mathbf{x}_1, \mathbf{x}_2, \ldots, \mathbf{x}_Q$ for the Q variables, are given by the usual transformation of the singular vectors:

$$\mathbf{x} = \mathbf{D}^{-1/2}\mathbf{v}$$

where $\mathbf{v}$ is the first right-hand singular vector, that is, the first column of $\mathbf{V}$. Only the principal coordinates are slightly different in the two problems, since the singular values differ.

We now apply the above theory to the four variables described in Section 2.2.1. In terms of our notation: $Q = 4$, $J_q = 5$ for all q, $J = 20$, $\mathbf{Z}$

is 871×20, and **C** is 20×20. Some rows of the original data matrix and the indicator matrix are given in Table A.1 and Table A.2 of the computational appendix at the end of this book, which details all of the computing steps to obtain the solution in this example. The full Burt matrix is reproduced in Table 2.2. In Table 2.3 we reproduce the standard coordinates for the first and second optimal solutions, along with their corresponding correlation measures. In addition, the squared correlation of each quantified variable with the total score is given, showing that the correlation measure is equal to their average. The first and optimal set of scale values, with an average squared correlation of 0.457, is monotonically increasing for questions A, B, and C, but question D has a quite different pattern, with the extreme poles opposing the intermediate categories. This is evidence that there is a possible problem with the responses to question D, which was worded in the reverse sense compared with the other questions. The second set of scale values captures an axis of "polarization," where all four questions have the pattern of the extreme categories opposing the intermediate ones, and here question D fits in more with the others. This interpretation is supported by the squared correlations, which show a low value for question D in the first solution. MCA thus effectively acts as an item analysis, and this result shows us that question D has degraded the reliability of the total score based on the second optimal solution and should preferably be removed.

To clarify the link between MCA and reliability theory, consider the Q variables as items measuring an underlying construct. Using the average squared correlation of 0.457, that is, 0.676 in the square root, as a measure of the reliability is an overestimate because even for random data we would find positive correlation between items and their sum (in fact, the average squared correlation between Q uncorrelated items and their sum is equal to $1/Q$). Cronbach's alpha is a measure of reliability that compensates for this and is classically defined as:

$$\alpha = \left(\frac{Q}{Q-1}\right)\left(1 - \frac{\sum_q s_q^2}{s^2}\right) \tag{2.8}$$

where s_q^2 is the variance of the qth item score, and s^2 is the variance of the summated score. In MCA the sum of item score variances ($\sum_q s_q^2$) is equal to $\mathbf{a}^\mathsf{T}\mathbf{D}\mathbf{a}$, which from the identification conditions described above is a fixed value, equal to Q. The variance of the summated score (s^2) is equal to Q^2 times the variance of the average score **z**, that is, Q^2

Table 2.2 Data on attitudes to science and the environment, showing the complete Burt matrix of all pairwise cross-tables of the four variables.

	Variable A					Variable B					Variable C					Variable D				
	A1	A2	A3	A4	A5	B1	B2	B3	B4	B5	C1	C2	C3	C4	C5	D1	D2	D3	D4	D5
A1	119	0	0	0	0	27	28	30	22	12	49	40	18	7	5	15	25	17	34	28
A2	0	322	0	0	0	38	74	84	96	30	67	142	60	41	12	22	102	76	68	54
A3	0	0	204	0	0	3	48	63	73	17	18	75	70	34	7	10	44	68	58	24
A4	0	0	0	178	0	3	21	23	79	52	16	50	40	56	16	9	52	28	54	35
A5	0	0	0	0	48	0	3	5	11	29	2	9	9	16	12	4	9	13	12	10
B1	27	38	3	3	0	71	0	0	0	0	43	19	4	3	2	9	17	10	10	25
B2	28	74	48	21	3	0	174	0	0	0	36	88	34	15	1	16	51	42	45	20
B3	30	84	63	23	5	0	0	205	0	0	37	90	57	19	2	10	53	63	51	28
B4	22	96	73	79	11	0	0	0	281	0	27	88	75	74	17	6	66	70	92	47
B5	12	30	17	52	29	0	0	0	0	140	9	31	27	43	30	19	45	17	28	31
C1	49	67	18	16	2	43	36	37	27	9	152	0	0	0	0	25	24	15	38	50
C2	40	142	75	50	9	19	88	90	88	31	0	316	0	0	0	15	97	67	89	48
C3	18	60	70	40	9	4	34	57	75	27	0	0	197	0	0	5	51	83	41	17
C4	7	41	34	56	16	3	15	19	74	43	0	0	0	154	0	6	44	30	51	23
C5	5	12	7	16	12	2	1	2	17	30	0	0	0	0	52	9	16	7	7	13
D1	15	22	10	9	4	9	16	10	6	19	25	15	5	6	9	60	0	0	0	0
D2	25	102	44	52	9	17	51	53	66	45	24	97	51	44	16	0	232	0	0	0
D3	17	76	68	28	13	10	42	63	70	17	15	67	83	30	7	0	0	202	0	0
D4	34	68	58	54	12	10	45	51	92	28	38	89	41	51	7	0	0	0	226	0
D5	28	54	24	35	10	25	20	28	47	31	50	48	17	23	13	0	0	0	0	151

Note: Table 2.1 is the A × B block of this matrix.

Table 2.3 Results of CA of 871 × 20 indicator matrix **Z** (see Table A.2 in the appendix) or, equivalently, of Burt matrix **C** in Table 2.2, showing standard coordinates (scale values) for the four variables on the first two dimensions of the solution (F1 and F2).

	F1	F2
A1	−1.837	0.727
A2	−0.546	−0.284
A3	0.447	−1.199
A4	1.166	0.737
A5	1.995	2.470
sq. corr.	0.510	0.382
B1	−2.924	1.370
B2	−0.642	−0.667
B3	−0.346	−0.964
B4	0.714	−0.280
B5	1.354	2.108
sq. corr	0.579	0.517
C1	−2.158	0.909
C2	−0.247	−0.592
C3	0.619	−1.044
C4	1.349	0.635
C5	1.468	3.017
sq. corr.	0.627	0.488
D1	−1.204	1.822
D2	0.221	−0.007
D3	0.385	−1.159
D4	0.222	−0.211
D5	−0.708	1.152
sq. corr.	0.113	0.337
rho	0.457	0.431
Cronbach's alpha	0.605	0.560

Note: *sq. corr.* is the squared correlation of the quantified variable with the total score; *rho* is the corresponding singular value of **C**, i.e., the squared singular value (or principal inertia) of **Z**, which is the arithmetic average of the four corresponding squared correlations; *Cronbach's alpha* is the measure of reliability discussed in Section 2.2.2.

times the λ that we are maximizing. Hence, we can write the maximum value of Equation 2.8 as:

$$\alpha = \left(\frac{Q}{Q-1}\right)\left(1 - \frac{Q}{Q^2\lambda}\right) = \left(\frac{Q}{Q-1}\right)\left(1 - \frac{1}{Q\lambda}\right) \tag{2.9}$$

so that maximum λ (the first singular value of $\mathbf{C}$ in Equation 2.7, which is also an eigenvalue as we have said previously) corresponds to maximum reliability. Hence, the maximum value of Cronbach's alpha for the first two solutions is, respectively (see Table 2.3):

$$\alpha_1 = \frac{4}{3}\left(1 - \frac{1}{4\times 0.457}\right) = 0.605 \quad \text{and} \quad \alpha_2 = \frac{4}{3}\left(1 - \frac{1}{4\times 0.431}\right) = 0.560$$

If question D is removed, as would be suggested by its low item correlation with the total, a recomputation of the solution gives a much higher value, 0.602, of the maximum average squared correlation, and an increase in Cronbach's alpha to 0.669. (We do not report the complete results here.)

Table 2.4 shows all the squared intercorrelations as well as the variances and covariances of the four quantified questions, according to the first optimal solution. This table also demonstrates empirically that the optimal λ can be computed either as (a) the variance of the total score, or (b) the average of the four squared correlations of the respective questions with the total, or (c) the average of all the elements of the full variance–covariance matrix between the four questions.

2.2.3 *Homogeneity analysis*

An alternative but equivalent definition of the correlational definition of MCA is based on Guttman's criterion of "internal consistency" (see, for example, Nishisato 1994). The idea is to look for scale values $\mathbf{s}_1, \mathbf{s}_2, \ldots, \mathbf{s}_Q$ that give individual scores $\mathbf{Z}_1\mathbf{s}_1, \mathbf{Z}_2\mathbf{s}_2, \ldots, \mathbf{Z}_Q\mathbf{s}_Q$ that are as close to one another (i.e., *homogeneous*) as possible. Lack of closeness can be measured by the sum of squares of the differences of each individual's Q scale values from the corresponding mean score in the vector $(1/Q)(\mathbf{Z}_1\mathbf{s}_1 + \mathbf{Z}_2\mathbf{s}_2 + \cdots + \mathbf{Z}_Q\mathbf{s}_Q) = (1/Q)\mathbf{Z}\mathbf{s}$, which we shall again denote by $\mathbf{z}$. The overall objective is thus to

Table 2.4 Squared intercorrelations as well as variances and covariances of the four quantified questions according to the first optimal solution.

	A	B	C	D	Squared correlation with total
A	1.1151	*0.1396*	*0.1270*	*0.0059*	**0.5100**
B	0.4440	1.2666	*0.1868*	*0.0059*	**0.5793**
C	0.4406	0.5697	1.3716	*0.0480*	**0.6273**
D	0.0403	0.0369	0.1274	0.2467	**0.1129**
Ave. covariance	0.5100	0.5793	0.6273	0.1129	**0.4574**

Note: The squared correlations between the four variables A, B, C, and D are quantified by their scale values on the first dimension (upper right-hand-side triangle of table, in italics) as well as their squared correlations with the total score (right-hand column; cf. *sq.corr.* in column F1 of Table 2.3). The variances (diagonal of table) and covariances (lower left-hand-side triangle of table) are also quantified, with the average covariance of each variable with itself and the others shown in the last row in boldface (e.g., 0.5793 = (0.4440 + 1.2666 + 0.5697 + 0.0369)/4). Note that these average covariances are identical to the squared correlations with the total. Hence, the variance of the average score (the quantity maximized by MCA, underlined) is both (a) the average of the four squared correlations of the question scores with the total score and (b) the average of the four average covariances; in other words, it is the average of the full 4 × 4 variance–covariance matrix. Note further the sum of the variances of the four variables, 1.1151 + 1.2666 + 1.3716 + 0.2467 = 4, which is the identification condition on the scale values. (To calculate variances and covariances, divide by n = 871, not n – 1.)

minimize, in this case, the following function of **s**, which is the average of the Q squared differences for each individual, averaged in turn over all n individuals:

$$\frac{1}{nQ}\left[(\mathbf{Z}_1\mathbf{s}_1-\mathbf{z})^{\mathsf{T}}(\mathbf{Z}_1\mathbf{s}_1-\mathbf{z})+(\mathbf{Z}_2\mathbf{s}_2-\mathbf{z})^{\mathsf{T}}(\mathbf{Z}_2\mathbf{s}_2-\mathbf{z})+\cdots+(\mathbf{Z}_Q\mathbf{s}_Q-\mathbf{z})^{\mathsf{T}}(\mathbf{Z}_Q\mathbf{s}_Q-\mathbf{z})\right] \tag{2.10}$$

This approach is known as *homogeneity analysis* (Gifi 1990), and the objective function (Equation 2.10) is called the loss function. Here "loss" refers to loss of homogeneity, since perfect homogeneity would be when all the differences $\mathbf{Z}_q\mathbf{s}_q - \mathbf{z}$ are zero. Once more an identification condition on **s** is required, otherwise the trivial solution when all elements of **s** are constant will be found, giving a loss of zero. With the same quadratic constraint $\mathbf{s}^\mathsf{T}\mathbf{D}\mathbf{s} = 1$ as previously, it can be shown that minimum loss is achieved by the same optimal scale values described above, and the value of the minimum is equal to 1 minus the value of the corresponding largest eigenvalue of the superindicator matrix $\mathbf{Z}$. In our example, the successively maximized eigenvalues of 0.457 and 0.431 (see Table 2.4) correspond to minimum losses of 0.543 and 0.569, respectively.

2.3 Geometric approach

The geometric approach to CA, introduced in Chapter 1, turns out to be slightly more problematic to generalize to the multivariable case. A lot of the controversy about CA stems from this difficulty, and here we shall clarify the issues involved in MCA as a graphical method as well as propose a specific version of MCA that acceptably addresses these problems. We shall approach the geometry from both the chi-square distance scaling perspective and the biplot perspective.

Figure 2.1 shows the usual CA map of the contingency table in Table 2.1. The map is established using the theory described in Chapter 1 and Section 2.2.1, namely the SVD of the matrix of standardized residuals, followed by the calculation of the principal coordinates to represent the points in a map. The principal coordinates are the standard coordinates multiplied by the respective singular values (see Chapter 1). Since standard coordinates have unit normalization, principal coordinates are normalized to have (weighted) sum of squares equal to the respective squared singular value of the associated solution.

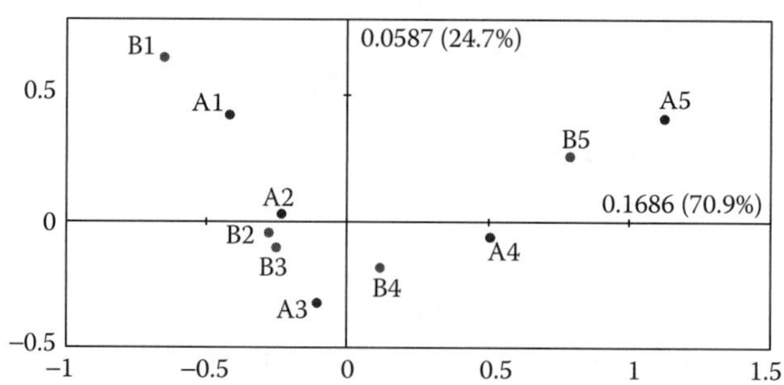

Figure 2.1 Symmetric CA map of Table 2.1. Percentage of inertia displayed in the map is 95.6%. Categories for questions A and B are from 1 (agree strongly) to 5 (disagree strongly).

The squared singular value σ^2 is called the *principal inertia*, corresponding to a principal axis, or dimension (see Chapter 1). Figure 2.1 shows the points represented by their principal coordinates calculated for the first two principal axes.

2.3.1 Chi-square distance scaling

Simple CA is justified mainly by its use of the chi-square (χ^2) distance as a measure of dissimilarity between row profiles and between column profiles of a two-way table. In Figure 2.1, where both rows and columns are displayed in principal coordinates, distances between row points optimally approximate the χ^2 distances between row profiles, and the distances between column points optimally approximate the χ^2 distances between column profiles. Recall from Chapter 1 that the squared χ^2 distance between the row profiles, for example, has this form:

$$d^2(i,i') = \sum_{j=1}^{J} \left(\frac{p_{ij}}{r_i} - \frac{p_{i'j}}{r_{i'}} \right)^2 \Bigg/ c_j \qquad (2.11)$$

so that every jth squared difference between the profile elements is weighted inversely by the column margin c_j.

MCA is the application of CA to either the superindicator matrix $\mathbf{Z}$ or the Burt matrix $\mathbf{C}$. While the χ^2 distance makes sense for a two-way contingency table, it has less justification when applied to the rows and to the columns of the superindicator matrix or the Burt matrix. As an illustration of this problem, consider the same four-variable example on attitudes to science in the environmental context. As shown in Equation 2.11, the χ^2 distances between rows and between columns are calculated between their profiles: in the case of distances between row profiles, the column masses of the correspondence matrix are used inversely as weights in the calculation of distance. The row profiles of the superindicator matrix $\mathbf{Z} = [\mathbf{Z}_1 \;\; \mathbf{Z}_2 \;\; \mathbf{Z}_3 \;\; \mathbf{Z}_4]$ are vectors with elements equal to zero apart from four values of 1/4 in the positions of the categories selected by the corresponding case. When calculating the distance between two rows, differences between coincident zero values and coincident values of 1/4 are zero, thus making no contribution to the distance measure, and so it is only differences between noncoincident categories that count in the distance function. These nonzero squared differences (each equal to 1/16 in this case), arising from disagreements between respondents, are then weighted by the inverses of the corresponding column masses, proportional to the respective categories' marginal frequencies, and added up to give the χ^2 distances. For the rows, this distance measure appears fairly reasonable, and the weighting is in accordance with the χ^2 concept that the contribution of categories with low frequency needs to be boosted because their variance is inherently lower. However, Gower (Chapter 3, this volume) prefers an unweighted version of this distance measure.

The situation for the column profiles of **Z**, however, is quite different and difficult, if not impossible, to justify. Here we make the distinction between calculating (a) distances between two categories of the same variable and (b) distances between two categories of different variables. Let us denote the relative frequency of the jth column category by c_j (that is, for a particular variable, the quantities c_j sum to 1). As shown by Greenacre (1989), the squared χ^2 distances between two column categories of a superindicator matrix are, in the two cases:

1. $1/c_j + 1/c_{j'}$ between categories j and j' of the same variable q
2. $1/c_j + 1/c_{j'} - 2p_{jj'}/(c_j c_{j'})$ between categories j and j' of different variables q and q'

where $p_{jj'}$ is the relative frequency of occurrence of categories j and j' (in fact, the above formulas are the same, since the frequency of co-occurrence of categories j and j' of the same variable is zero). The former "within-variable" distance makes little sense, since it depends only on the marginal frequencies, no matter what the relationship with the other variables is. The latter "between-variable" distance has at least a slight justification in that the distance decreases as association between categories j and j' increases, but again the dominant role played by the marginal frequencies is hard to defend.

The situation improves if we consider intercategory distances calculated on the Burt matrix rather than the indicator matrix. Because the Burt matrix is symmetric, it makes no difference whether we calculate the χ^2 distances between rows or between columns. The squared distance between categories can be described verbally as follows:

1. Between categories j and j' of the same variable q: This within-variable squared distance is the average of the $(Q - 1)$ squared χ^2 distances between categories j and j' calculated in the cross-tabulations of variable q with all the other variables $q' \neq q$, but also including an unnecessary term from the cross-tabulation of q with itself. (This term involves the distance

between two unit profiles in a submatrix on the diagonal of **C** and is thus a large component of the overall distance, tending to inflate the distance.)

2. Between categories j and j' of different variables q and q': This between-variable squared distance is an average of $(Q - 2)$ squared χ^2 distances between profiles of categories j and j' across variables not equal to q or q', but including two additional terms that can also be considered unnecessary. (These measure distances between a profile and a unit profile again on the diagonal of **C**, again tending to inflate the between-category distance.)

In spite of the above theoretical difficulties to justify the full-space chi-square geometry, MCA as regularly applied — that is, the CA of **Z** or of **C** — successfully recovers interesting patterns of association between the variables. It seems that the low-dimensional projections of the points are more valid than their full-dimensional counterparts, which is a paradox from the multidimensional scaling viewpoint. Another worrying aspect is the inflation of the total inertias of **Z** and of **C**, which leads to all percentages of inertia on the principal axes being artificially low. This inflation can also be understood by considering the calculation of total inertia for the Burt matrix **C** and the high contributions made by the diagonal matrices on its block diagonal. It is clear that MCA of a two-variable data set will not give the same results as a CA; the standard coordinates will be the same, but the principal inertias (and hence the principal coordinates) and their percentages of inertia will be different. In Section 2.3.3 we define another variant of MCA, called joint correspondence analysis, that resolves all these issues to a certain extent. We shall also show that a simple adjustment of the scale in the MCA solution dramatically improves the fit from a multidimensional scaling viewpoint.

2.3.2 Biplot

The biplot is concerned with data reconstruction in a joint map of the rows and columns, rather than distance reconstruction. In the simple case of a two-way table, we can think of reconstructing different variants of the table depending on the way we think of the table: either as a set of rows, or a set of columns, or just a two-way table of entries where rows and columns are symmetric entities (see Chapter 3). As an illustration of this approach we consider a two-way table as a set of rows. For example, Table 2.5a shows the row profiles of the two-way

Table 2.5 (a) Row profiles of Table 2.1, including average row profile. (b) Approximate row profiles estimated from biplot of Figure 2.2 (the average profile is always represented exactly by the origin of the map).

	(a) Original profiles					
	B1	B2	B3	B4	B5	Sum
A1	0.227	0.235	0.252	0.185	0.101	1
A2	0.118	0.230	0.261	0.298	0.093	1
A3	0.115	0.235	0.309	0.358	0.083	1
A4	0.017	0.118	0.129	0.444	0.292	1
A5	0.000	0.063	0.104	0.229	0.604	1
Average	0.075	0.176	0.211	0.303	0.235	1
	(b) Estimated profiles					
	B1	B2	B3	B4	B5	Sum
A1	0.226	0.239	0.253	0.181	0.102	1
A2	0.117	0.229	0.265	0.294	0.094	1
A3	0.024	0.226	0.283	0.393	0.074	1
A4	0.002	0.135	0.169	0.387	0.307	1
A5	0.026	0.034	0.034	0.329	0.578	1
Average	0.075	0.176	0.211	0.303	0.235	1

Note: The difference between the two tables is the error of biplot approximation, measured as 100 − 95.6% = 4.4% of the total inertia of the table.

table in Table 2.1, that is, conditional on each response category of question A, the proportions of respondents falling into the response categories of question B. The biplot can be thought of as a way to reconstruct these row profiles in a map. Greenacre and Hastie (1987) and Greenacre (1993a) show how the asymmetric map of CA, with row points in principal coordinates and column points in standard coordinates, is a biplot of these profiles. The direction vector defined by each column point, called a *biplot axis*, can be calibrated in profile units, and the approximate value of the profile can be read off the map by simply projecting the row points onto the column axis (Figure 2.2). The success of the reconstruction of the data from the biplot in this way is measured by the percentage of inertia explained by the map: in this case it is 95.6%, so the reconstruction has an error of only 4.4%. Table 2.5b reports the estimated values from the biplot of Figure 2.2, testifying to the high accuracy of the data reconstruction.

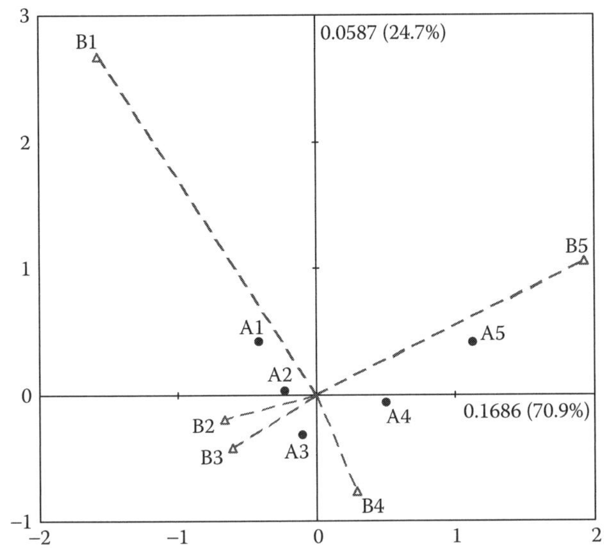

Figure 2.2 Asymmetric CA map of Table 2.1 showing biplot axes. The categories of the row points A1 to A5 can be projected onto each biplot axis to read off approximations of corresponding profile values. The accuracy of this approximation is as good as the percentage of inertia displayed, which is 95.6%; hence it is excellent.

Interpreting Figure 2.2 we can see, for example, a direction opposing the categories B2 and B3 pointing bottom left and category B5 top right. If we project A1, A2, and A3 onto this diagonal "dimension," it is clear that they project more or less at the same position, showing that their profile values on B2, B3, and B5 are similar, with profile values on B2 and B3 above average and those on B5 below average (the origin of the biplot always represents the average profile values exactly). This deduction from the map can be confirmed in Table 2.5a, and it is only for categories A4 and A5 that there are distinct changes along this direction, increasing in percentage response to B5 and decreasing on B2 and B3. Likewise, with respect to the other diagonal "dimension" from top left to bottom right, which opposes B1 and B4, we see that A3, A4, and A5 project at the same positions and thus are estimated to have similar profile values on B1 and B4. This can be mostly verified in Table 2.5a, the only exception being the profile of A5 on B4, which has an observed frequency much lower than the corresponding values for A3 and A4. This error of approximation would be part of the 4.4% unexplained inertia.

Thinking about the joint map in this way sheds light on the problematic aspects of the CA of the indicator matrix **Z** or the Burt matrix **C**. In the case of **Z**, it is futile to expect a good approximation of a matrix of zeros and ones in a two-dimensional map of points. Another measure of quality is needed; for example, one could deduce from a joint map the most likely set of responses of each case (row) and then count how many of these are correct predictions (see Gower 1993; Greenacre 1994). The situation is similar for the Burt matrix **C**: any attempt to approximate the diagonal matrices down the diagonal of the Burt matrix is clearly in conflict with the approximation of the more interesting and relevant contingency tables in the rest of the matrix. In both cases the percentages of inertia will be artificially low because of the structurally high-dimensional nature of the matrices being analyzed.

Figure 2.3 shows the CA of the Burt matrix of Table 2.3, which represents only 35.1% of the total inertia; yet its interpretation is clear: we can see the same pattern of association for questions A and B already seen in Figure 2.1, along with a similar pattern of association with question C. But the response categories for question D are not

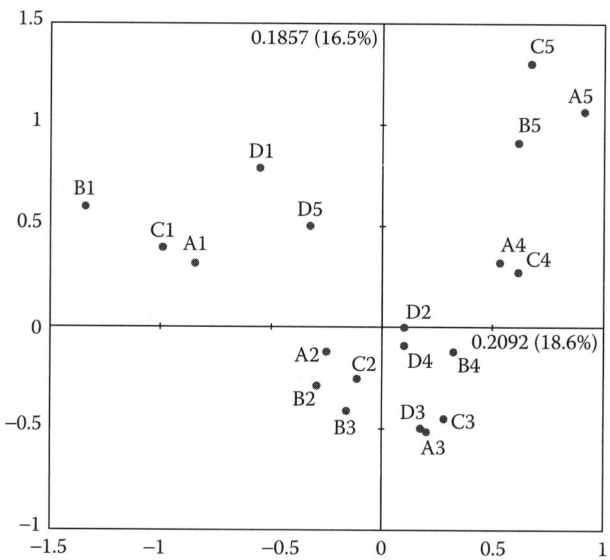

Figure 2.3 Multiple correspondence analysis map of the Burt matrix of Table 2.3. The percentage of explained inertia is 35.1%.

at all in line with the other three. The categories D1 and D5 of strong agreement and strong disagreement lie within the arch formed by the other questions, quite close together even though they are at opposite ends of the scale. This shows clearly the incompatibility of this question with the others.

Notice how different the scale of Figure 2.3 is compared with Figure 2.1, and how the points have been pulled outward in the analysis of the Burt matrix. Most of the problem of low percentages of inertia is due to this scale change, and this can be rectified by a simple scale adjustment of the solution. This is best explained after an account of joint correspondence analysis.

2.3.3 Joint correspondence analysis

As we have just seen, when applying CA to the Burt matrix, the diagonal submatrices on the "diagonal" of the block matrix **C** inflate both the chi-square distances between profiles and the total inertia by artificial amounts. In an attempt to generalize simple CA more naturally to more than two categorical variables, joint correspondence analysis (JCA) accounts for the variation in the "off-diagonal" tables of **C** only, ignoring the matrices on the block diagonal. Hence, in the two-variable case ($Q = 2$) when there is only one off-diagonal table, JCA is identical in all respects to simple CA (which is not the case for MCA of **Z** or **C**, which give different principal inertias).

The solution can no longer be obtained by a single application of the SVD, and various algorithms have been proposed by Greenacre (1988), Boik (1996), and Tateneni and Browne (2000). For example, Greenacre (1988) describes an alternating least-squares algorithm that treats the matrices on the block diagonal as missing values. The algorithm proceeds in the following steps:

1. Perform MCA by applying CA to the Burt matrix **C**, and choose the dimensionality S^* of the solution (e.g., $S^* = 2$ is the most typical).
2. Optionally perform an adjustment of the solution along each of the S^* dimensions to improve the approximation to the off-diagonal block matrices (see Section 2.3.4 below).
3. From the resulting map, reconstruct the values in the diagonal blocks of **C** in the same way as the data were reconstructed in the biplot, using the reconstruction formula (see appendix, Equation A.7). Replace the original values in the diagonal blocks by these estimates, calling this the modified Burt matrix $\mathbf{C}^*$.

4. Perform another CA on the resultant matrix **C*** with modified block diagonal.
5. Repeat steps 3 and 4, that is, substitute the diagonal blocks from the reconstructed values in the new solution, performing again the CA to obtain another solution, and so on, until the process converges. Convergence can be measured by the maximum absolute difference between the values substituted into the diagonal blocks at the present iteration and their corresponding values substituted during the previous iteration.

Figure 2.4 shows the results of a two-dimensional JCA applied to the Burt matrix of Table 2.3, where we have intentionally left the scale exactly as in Figure 2.3. Comparing these two figures we can see the high degree of similarity in the pattern of the response categories, but mainly a change in the scale, with the JCA map being reduced in scale on both axes, but especially on the second. Most of the properties of simple CA carry over to JCA, most importantly the reconstruction of profiles with respect to biplot axes (Greenacre 1993a: chapter 16).

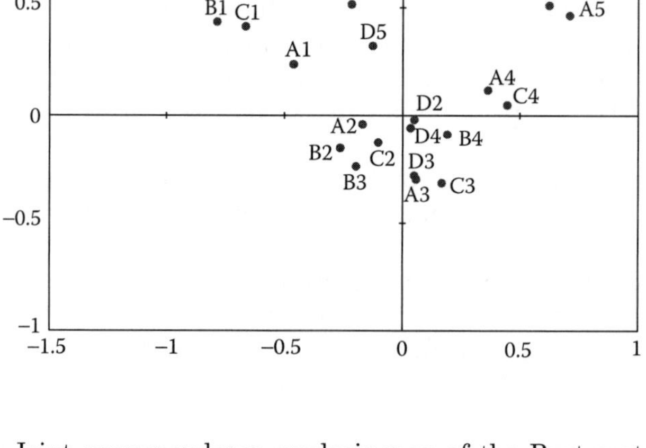

Figure 2.4 Joint correspondence analysis map of the Burt matrix of Table 2.3. The percentage of explained inertia is 85.7%.

Compared with regular CA and MCA, there are two aspects to remember in computing the percentage of inertia explained by the map. First, the percentage has to be calculated for the two dimensions of the solution together, not separately, since the dimensions in JCA are not nested. Second, in the final solution (the CA of the modified Burt matrix at the final iteration), the usual way of calculating the proportion of explained inertia involves the ratio between the sum of the first two principal inertias and the total inertia, but both the numerator and denominator of this sum include an amount due to the modified diagonal blocks, which are fitted exactly by the solution. This amount, which can be calculated in various ways (see the computational appendix, Section A.5), needs to be discounted from both the numerator and denominator to obtain the percentage of (off-diagonal) inertia explained. In this example, the percentage of inertia accounted for by the JCA map is 85.7%, much higher than the 35.1% explained in the MCA map based on the Burt matrix. The value of 85.7% appropriately measures the success of approximating the off-diagonal blocks relative to the total inertia of these blocks only, unaffected by the diagonal blocks. This would be the quality of the map considered as an MCA biplot as well: that is, express all six off-diagonal blocks as profiles (rows or columns profiles, in upper or lower triangle of the Burt matrix), then the quality of reconstructing these profiles as described in Section 2.3.2 would be 85.7%, and the error of reconstruction, or residual, would be 14.3%.

2.3.4 Adjustment of the inertias in MCA

Since the main difference between MCA and JCA in Figure 2.3 and Figure 2.4 is change in scale, it is possible to remedy partially the percentage-of-inertia problem in a regular MCA by a compromise between the MCA solution and the JCA objective by using simple scale readjustments of the MCA solution. In this approach the total inertia is measured (as in JCA) by the average inertia of all off-diagonal blocks of **C**, calculated either directly from the tables themselves or by adjusting the total inertia of **C** by removing the fixed contributions of the diagonal blocks as follows:

$$\text{average off-diagonal inertia} = \frac{Q}{Q-1}\left(\text{inertia}(\mathbf{C}) - \frac{J-Q}{Q^2}\right) \qquad (2.12)$$

Parts of inertia are then calculated from the principal inertias λ_s^2 of **C** (or from the principal inertias λ_s of **Z**) as follows: for each $\lambda_s \geq 1/Q$, calculate the adjusted inertias:

$$\lambda_s^{\text{adj}} = \left(\frac{Q}{Q-1}\right)^2 \left(\lambda_s - \frac{1}{Q}\right)^2 \tag{2.13}$$

and then express these as percentages of Equation 2.12. Although these percentages underestimate those of a JCA, they dramatically improve the results of an MCA and are recommended in all applications of MCA. A further property of the adjusted principal inertias λ_s^{adj} is that they are identical to the principal inertias σ_s^2 of simple CA in the case of two categorical variables, where $Q = 2$: $\lambda_s^{\text{adj}} = 4(\lambda_s - 1/2)^2$, since we have shown earlier in Section 2.2.2 the relationship $\lambda_s = 1/2(1 + \sigma_s)$.

In our example the total inertia of **C** is equal to 1.1138, and the first seven principal inertias are such that $\lambda_s \geq 1/Q$, that is, $\lambda_s^2 \geq 1/Q^2 = 1/16$. The average off-diagonal inertia is equal to 0.17024, as shown in Table 2.6 along with the different possibilities for inertias and percentages of inertia. Thus what appears to be a percentage explained in two dimensions of 22.2% (= 11.4 + 10.8) in the analysis of the indicator matrix **Z**, or 35.1% (= 18.6 + 16.5) in the analysis of the Burt matrix C, is shown to have a lower bound of 79.1% (= 44.9 + 34.2) when the principal inertias are adjusted. Compared with the adjusted MCA solution, the JCA solution for this example (Figure 2.4) gives an additional benefit of 6.4 percentage points in the explained inertia, with a percentage explained of 85.7%. We stress again that in JCA the solutions are not nested, so the percentages are reported for the whole solution (in this case, a two-dimensional one), not for individual dimensions.

We propose the adjusted solution as the one to be routinely reported; not only does it considerably improve the measure of fit, but it also removes the inconsistency about which of the two matrices to analyze, indicator or Burt. The adjusted solution is given in Figure 2.5 and has the same standard coordinates as Figure 2.3, but it uses the adjusted principal inertias to calculate the principal coordinates, leading to the improved quality of display. Again we have left the scale identical to Figure 2.3 and Figure 2.4 for purposes of comparison.

Benzécri (1979) has proposed the same adjusted inertias (Equation 2.13), but expresses them as percentages of their own sum over the dimensions s for which $\lambda_s \geq 1/Q$ (see an example of this in Chapter 5). This approach goes to the opposite extreme of giving an overly optimistic expression of explained inertia, since it explains 100% in the space of the dimensions for which $\lambda_s \geq 1/Q$ (there are six dimensions

Table 2.6 Eigenvalues (principal inertias) of indicator matrix **Z** and Burt matrix **C**, their percentages of inertia, the adjusted inertias, and their lower-bound estimates of the percentages of explained inertia for the off-diagonal tables of the Burt matrix.

Dimension	Eigenvalue of **Z**	Percentage Explained	Eigenvalue of **C**	Percentage Explained	Adjusted Eigenvalue	Percentage Explained
1	0.4574	11.4	0.2092	18.6	0.07646	44.9
2	0.4310	10.8	0.1857	16.5	0.05822	34.2
3	0.3219	8.0	0.1036	9.2	0.00920	5.4
4	0.3065	7.7	0.0939	8.3	0.00567	3.3
5	0.2757	6.9	0.0760	6.7	0.00117	0.7
6	0.2519	6.3	0.0635	5.6	0.00001	0.0

Note: The average off-diagonal inertia on which these latter percentages are based is equal to $\frac{4}{3}(1.12768 - \frac{16}{16}) = 0.17024$, where 1.12768 is the total inertia of the Burt matrix (i.e., the average of the inertias of all its submatrices, including those on the block diagonal). Note that the sum of explained inertia over these six dimensions must be less than 100% in the case of the adjusted eigenvalues because the percentages are lower-bound estimates and, in any case, the six-dimensional MCA solution does not fully explain all pairwise cross-tabulations.

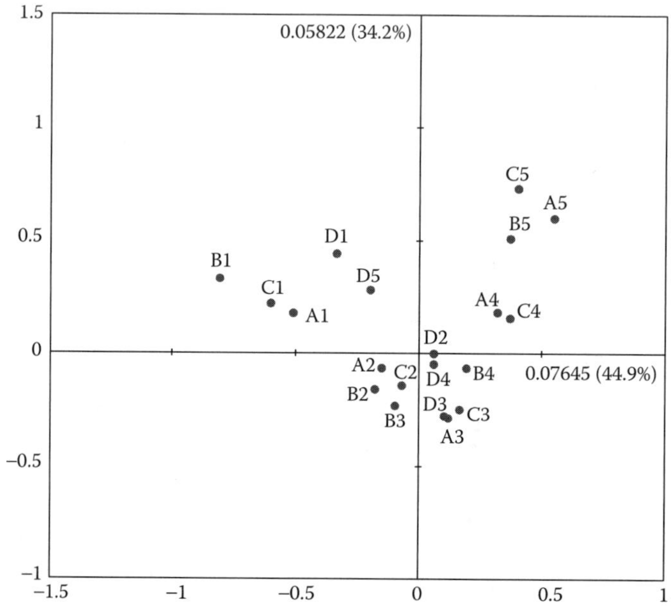

Figure 2.5 Multiple correspondence map of Table 2.3 with adjustment of the principal inertias (and thus the scale of the principal coordinates) along each dimension. The percentage of explained inertia is at least 79.1%. Note the change in scale compared with Figure 2.3.

in our example, as seen in Table 2.6) when in fact the data are not exactly reconstructed in the map.

2.4 Supplementary points

Up to now we have described three different ways to perform MCA:

Variant 1: CA of the indicator matrix
Variant 2: CA of the Burt matrix
Variant 3: An adjusted version of variants 1 or 2 that unifies and rectifies the scaling issue, giving a single solution irrespective of which matrix is considered, with highly improved measures of explained inertia

In all these variations, the standard coordinates of category points remain the same; only the principal inertias change. Furthermore, we have introduced an alternative method, JCA, that has a different solution from the above variants and that is analogous to least-squares factor analysis in that it concentrates on between-variable associations only. In all of these cases it is possible to display supplementary points in the usual way to enrich the interpretation (see Chapter 5, where this aspect is discussed in depth for the MCA case). Here we define a way of displaying supplementary points that does not depend on the variant of the method used. In our illustration, we will consider three supplementary demographic variables: gender, age and education (full category descriptions are given in Section 2.1).

To motivate our approach we consider the case of the indicator matrix, where supplementary categories can be thought of as either row or column points. For example, male and female categories can be added as two supplementary-column dummy variables or as two supplementary rows containing the frequencies for males and for females across the response categories. These two alternatives are equivalent up to scale factors, as we now explain. To position a supplementary column category using the so-called transition, or barycentric, relationship between rows and columns (see, for example, Greenacre 1984 and Chapter 1 of this volume), we have to consider all the respondent points (rows) in standard coordinate positions in the map. Then any column category, active or supplementary, is situated (in principal coordinates) at the average of the respondents who fall into that category. Alternatively, to position a supplementary row, for example "male," we have to consider all the active column categories in standard coordinate positions; then the "male" row point will be at the weighted average of column points, using the profile of "male" across the active columns. Remember that the "male" frequency profile sums to 1 across the Q questions, so its position is an average of averages; for each question, the group "males" has an average position according to male frequencies across the categories of that particular question, and the final position of "male" is a simple average of these averages. We can show that the position of a supplementary point as a row is the same as the supplementary column dummy, but it is shrunk on each dimension by the corresponding singular value, that is, by the same scale factor that links principal to standard coordinates on each dimension (see Greenacre 1984: chapter 5.1 for a proof of this result). Thus a simple way to unify the representation of supplementary points in all situations would be to think of supplementary categories always as the averages of the principal coordinate positions of respondents, in which case both approaches will give exactly the same results.

Our proposal is thus the following: using the principal coordinates of respondent points, calculate average positions for supplementary categories, for example, the average position for male points, female points. Since it is only for the indicator matrix (variant 1 listed above) that we (automatically) have respondent points in the CA, we need to make precise what we mean by respondent points in the case of the Burt matrix and the adjusted analysis (variants 2 and 3 above, respectively). In these last cases, respondent points are displayed, at least theoretically, as supplementary points, that is, as averages of the column categories, in standard coordinates, according to their respective response patterns. Because in all three variants of MCA these standard coordinates are identical, respondent points will have exactly the same positions in all three cases. Thus when we average their positions according to a supplementary variable, showing for example average male and average female points, the results will also be identical. But notice that, to obtain these average positions, we do not actually have to calculate all the original respondent points. The calculations can be made much more efficiently, thanks to transition relationships, by simply adding cross-tabulations as supplementary rows or columns. The following summarizes the calculations in each case, assuming that a supplementary variable is coded in indicator form as $\mathbf{Z}_s$, so that $\mathbf{Z}_s^\mathsf{T}\mathbf{Z}$ denotes the concatenated set of cross-tabulations of the supplementary variable with the Q active variables:

1. In the case of the indicator matrix $\mathbf{Z}$, we would already have the principal coordinates of the respondents, so we can either make the calculation of averages directly or add as supplementary row points the cross-tabulations $\mathbf{Z}_s^\mathsf{T}\mathbf{Z}$ of the supplementary variable with the active variables.
2. In the case of the Burt matrix $\mathbf{C}$, we do not need to calculate the positions of respondents (if required for other reasons, this would be done by adding $\mathbf{Z}$ as supplementary rows to the Burt matrix $\mathbf{C}$). Instead, we can simply add the cross-tabulations $\mathbf{Z}_s^\mathsf{T}\mathbf{Z}$ as supplementary rows (or $\mathbf{Z}^\mathsf{T}\mathbf{Z}_s$ as supplementary columns), which leads to the same positions for the supplementary categories as in variant 1.
3. In the case of the adjusted analysis, we do as in variant 2, since it is only *a posteriori* that we adjust the eigenvalues, and this adjustment affects only the positions of the principal coordinates of the active category points, not the supplementary categories that, we repeat, are defined as averages of the respondent points;
4. In the case of JCA, it is again a simple addition of supplementary rows or columns, as in variants 2 and 3, to the modified Burt matrix $\mathbf{C}^*$ at the final iteration of the algorithm.

Table 2.7 shows the cross-tabulations $\mathbf{Z}_s^\mathsf{T}\mathbf{Z}$, and Figure 2.6 shows the positions of the supplementary points in the adjusted MCA (variant 3,

Table 2.7 Supplementary rows added to the indicator matrix **Z**, Burt matrix **C**, or modified Burt matrix **C*** to represent the supplementary variable sex, age, and education. These are the cross-tabulations of the three supplementary variables with four active variables A, B, C, and D.

Supplementary	Variable A					Variable B					Variable C					Variable D				
Variables	A1	A2	A3	A4	A5	B1	B2	B3	B4	B5	C1	C2	C3	C4	C5	D1	D2	D3	D4	D5
Sex																				
Male	43	144	120	92	28	25	75	104	146	77	58	157	105	82	25	25	136	92	101	73
Female	76	178	84	86	20	46	99	101	135	63	94	159	92	72	27	35	96	110	125	78
Age [a]																				
age1	9	34	17	25	6	5	21	16	33	16	13	32	22	19	5	7	26	19	30	9
age2	33	68	45	51	13	15	30	52	76	37	35	60	57	43	15	10	46	48	56	50
age3	18	63	40	26	11	15	27	31	62	23	24	61	35	25	13	12	39	42	35	30
age4	19	45	47	27	8	13	29	34	41	29	24	57	33	26	6	10	39	38	38	21
age5	18	43	33	25	5	10	26	35	37	16	21	45	26	24	8	9	48	23	28	16
age6	22	69	22	24	5	13	41	37	32	19	35	61	24	17	5	12	34	32	39	25
Education [b]																				
edu1	7	15	7	8	1	6	10	11	7	4	7	8	11	12	0	5	9	12	8	4
edu2	59	155	84	68	12	34	93	95	112	44	79	156	73	54	16	28	110	90	95	55
edu3	29	84	65	54	10	19	47	55	82	39	34	90	68	36	14	11	69	58	63	41
edu4	11	27	18	26	12	6	12	18	37	21	18	24	18	28	6	7	13	23	29	22
edu5	5	20	11	8	5	4	5	11	16	13	6	14	14	9	6	4	12	7	14	12
edu6	8	21	19	14	8	2	7	15	27	19	8	24	13	15	10	5	19	12	17	17

[a] age1: 16–24 years; age2: 25–34 years; age3: 35–44 years; age4: 45–54 years; age5: 55–64 years; age6: 65 years and older.

[b] edu1: no or some primary education; edu2: primary education completed; edu3: some secondary education; edu4: secondary education completed; edu5: some tertiary education; edu6: tertiary education completed.

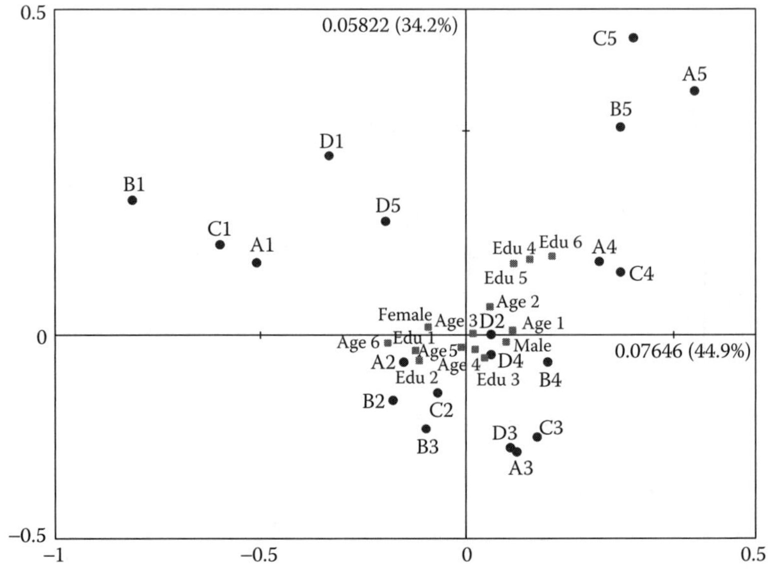

Figure 2.6 Adjusted MCA solution, showing positions of supplementary categories of sex, age, and education.

that is, the supplementary points are superimposed on Figure 2.5). Here we can see that the age groups and education groups show a horizontal trend, with younger respondents on the right moving over to older respondents on the left, and lower education groups on the left moving over to higher-education groups on the right. In addition, we can see that the three upper education groups (from secondary education completed upward) separate out at the right toward the strong disagreement poles of the questions, indicating that they are particularly strongly in favor of science. We also find the average male point on the right-hand side and average female point on the left-hand side. Remember that these supplementary variables have been added separately to the map and not in combination, that is, the fact that male is on the right and higher-education groups are on the right does not imply that it is only higher-educated males that are more favorable toward science. To see the positions of higher-educated females, for example, interactive coding of the demographic variables would have to be performed. This can be done, for example, by coding the six education groups for males and for females separately, giving 12 combinations of gender and education, each represented as a separate supplementary point.

2.5 Discussion and conclusions

We have shown that extending CA of two variables to the case of several variables is not a simple issue, especially in the geometric case. As explained in Chapter 1, simple CA is typically applied to situations where two different types of variables are cross-tabulated, for example, country of residence and a response to a survey question. MCA is applied to one set of variables, preferably all with the same response scales, that revolve around a particular issue, and where we are interested in the association patterns among the variables. Putting this in another way, in simple CA we are interested in associations between two variables or between two sets of variables, while in MCA we are interested in associations within a set of variables.

Although the geometric concepts of simple CA do not carry over easily to the multiple-variable case, adjustment of the principal inertias and alternative methods, such as JCA, partially rectify the situation. Because MCA has attractive properties of optimality of scale values (thanks to achieving maximum intercorrelation and thus maximum reliability in terms of Cronbach's alpha), the compromise offered by the adjusted MCA solution is the most sensible one and the one that we recommend. The adjustment, described in Section 2.3.4, is easy to calculate and simply changes the scale on each dimension of the map to best approximate the two tables of association between pairs of variables, leaving everything else in the solution intact. Thanks to this adjustment we obtain estimates of the explained inertias that are much closer to the true values than the pessimistic values obtained in MCA. For this reason we propose the adjusted MCA solution as the one to be used as a matter of course in all (graphical) MCA applications. The adjusted solution also has the nice property that it is identical to simple CA of a cross-tabulation in the case of two variables. JCA also perfectly reproduces simple CA in the two-variable case, since it is also focused exclusively on the single off-diagonal cross-tabulation. JCA has the additional advantage in the multivariable case of optimizing the fit to all of the off-diagonal cross-tabulations of interest.

Categorical supplementary points can be added to any of the variants of MCA, as well as JCA, as averages of respondents that fall into the corresponding categories. This amounts to simply adding the cross-tabulations of the supplementary variables with the active variables as supplementary rows or columns. The JCA and adjusted MCA solutions have the advantage that the active points are reduced in scale compared with the solutions for the indicator and Burt matrices, thus leading to a greater relative dispersion of the supplementary points in the joint map of the active and supplementary points.

We have not dealt with the important topic of imposing constraints on the MCA solution, such as linear or order constraints. Chapter 4 gives a comprehensive and up-to-date treatment of these issues.

Software note

The analyses of this chapter were performed using XLSTAT (www.xlstat.com), distributed by Addinsoft, and the R functions for CA, MCA, and JCA, written by Oleg Nenadić. Both implementations, described in the appendix of this book, include the adjustment of inertias described in this chapter.

CHAPTER 3

Divided by a Common Language: Analyzing and Visualizing Two-Way Arrays

John C. Gower

CONTENTS

3.1 Introduction: two-way tables and data matrices

According to George Bernard Shaw, the U.S. and the U.K. are two nations divided by a common language. I sometimes think that the range of data-analytical methods concerned with approximating two-way arrays is divided by the common language of algebraic canonical forms, especially spectral decompositions and singular-value decompositions. This common mathematical foundation may be a powerful basis for writing unified software, but it can obfuscate the statistical interpretations appropriate to the different methods. In the following, I shall be concerned more with statistical differences than with computational similarities.

I use "two-way array" as a neutral term to include methods that are concerned with two-way tables as well as those concerned with data matrices. The distinction is fundamental. A two-way table is concerned with a single variable classified by I rows and J columns of equal status, e.g., a table of average incomes for different age groups (the rows) in different cities (columns). With rows and columns of different status, a data matrix is multivariate, with rows corresponding to samples or cases or other observational units, and columns corresponding to variables, for example, a study of European countries (rows) for which various economic indicators (columns, e.g., inflation rate, gross domestic product) have been measured. A further fundamental consideration is whether the variables are quantitative (e.g., inflation rate) or categorical (e.g., whether a country has the euro or not), or a mixture of the two types. The concepts of quantitative and categorical variables can be further refined, for example, to quantitative ratio and interval scales or to ordered and nominal categorical variables.

Notice that a two-way table that classifies a single variable may be considered as a data matrix with IJ rows and three variables (the columns), two referring to the categorical row- and column-classifying variables (for example, age group and city) and a third to the classified variable itself (for example, average income). Even this simple situation includes important variants, such as when the third variable is categorical, such as an ordinal indicator of "quality of life" classified by age group and city, or when the two-way table is a contingency table where there is no third variable, the entries being counts n_{ij} of co-occurrence of the categories (i, j) of the two classifying variables. In this last case, there are two forms of data matrix: in the first there are IJ rows with the entry n_{ij} in a third column corresponding to the entries i and j in the first two columns, while in the second there are n rows and only two columns, the entries i and j occurring as n_{ij} identical, but not necessarily contiguous, rows (this latter form is sometimes termed the response pattern matrix; see, for example, Chapters 1, 2, and 6). Our previous example of average income for age groups and cities is also an important special case, since the quantitative variable "income" is effectively replicated n_{ij} times within each cell of the two-way table. Not only the mean, but also other summary measures such as the maximum, minimum, median, and variance could be of interest, as well as the associated counts n_{ij} themselves. In controlled experimental situations, n_{ij} is often a constant, and then a table of associated counts has little interest, but in sample surveys, associated contingency tables contain valuable response information.

To express whatever information there may be in an array, statisticians have developed two approaches: (a) to approximate them by parsimonious parametric models and (b) to give a visual approximation, usually in two dimensions. The two approaches are linked, because a two-dimensional approximation implies an underlying parsimonious model, though the converse is not necessarily true; for example, simple additive models lack simple visualizations.

A fundamental statistical consideration is whether one or more of the variables may be considered as dependent on explanatory variables. Usually, the variable classified in a two-way table can be regarded as dependent on the classifying variables. To anticipate a little, the log-linear analysis of a contingency table recognizes that the counts in the body of the table depend on the classifying variables, and this structure persists, less overtly, in correspondence analysis (Section 3.3.2). However, principal component analysis and multiple correspondence analysis do not distinguish between independent and response variables. Indeed, many of the methods for analyzing data matrices do not distinguish between dependent and explanatory variables, although the whole area typified by graphical modeling, path analysis, and covariance structures is rooted in the complexities of interdependencies.

Two-way tables and data matrices are not the only arrays with which we are concerned. Correlation, Burt, and dissimilarity matrices can be derived from data matrices. Sometimes their approximation is of primary interest; at other times they occur as a secondary computational step in some wider analysis. In all cases, visualizations based on eigenvectors are valuable for summarizing results, but there are issues about how these vectors should be scaled and interpreted. Most of these displays are varieties of biplot (Gower and Hand 1996), where the "bi" refers to the row and column entities and not to the usual bidimensionality of the visualizations. In this chapter I shall try to disentangle the differences and similarities between the different data types, methodologies, scalings, and interpretations. We have stressed the importance of distinguishing (a) two-way tables from data matrices and (b) quantitative from categorical variables, giving a total of four

Table 3.1 Some well-known methods of analysis classified by type of two-way array (rows) and type of variable (columns).

	Quantitative	Qualitative
Two-Way Table	Biadditive models (BM)	Correspondence analysis (CA)
Data Matrix	Principal component analysis (PCA)	Multiple correspondence analysis (MCA)

major types of data. Table 3.1 lists some of the better-known methods of analysis associated with this classification.

This book is mainly concerned with categorical variables. However, it is simpler to begin with the similar methods appropriate for quantitative variables: the biadditive (bilinear) models that can be fitted to a two-way quantitative table and the PCA of a data matrix of quantitative variables.

3.2 Quantitative variables

3.2.1 Biadditive models

Here, we consider a two-way table $\mathbf{X}$ of quantitative values classified by I rows and J columns. The simplest biadditive model is

$$\mathbf{X} = m\mathbf{1}\mathbf{1}^{\mathsf{T}} + \mathbf{a}\mathbf{1}^{\mathsf{T}} + \mathbf{1}\mathbf{b}^{\mathsf{T}} + \mathbf{c}\mathbf{d}^{\mathsf{T}} + \text{residual} \tag{3.1}$$

where $\mathbf{1}$ denotes a vector of ones of appropriate order. Hence, we are concerned with approximating $\mathbf{X}$ by the special rank-3 matrix $(m\mathbf{1} + \mathbf{a}, \mathbf{1}, \mathbf{c})(\mathbf{1}, \mathbf{b}, \mathbf{d})^{\mathsf{T}}$. In Equation 3.1, $\mathbf{a}$ and $\mathbf{b}$ represent additive effects, and the product of $\mathbf{c}$ and $\mathbf{d}$ represents a multiplicative, or biadditive, interaction effect. Often, the terms "linear" and "bilinear" are used, but these properly refer to variables measured on continuous scales rather than the discrete sets of constants denoted by $\mathbf{a}$, $\mathbf{b}$, $\mathbf{c}$, and $\mathbf{d}$. Notice that, as with linear models, values of $\mathbf{X}$ are regarded as those of a response variable dependent on explanatory variables m, $\mathbf{a}$, $\mathbf{b}$, $\mathbf{c}$, and $\mathbf{d}$. As usual, the model is not fully identified, since constants can be added to any of the parameters $\mathbf{a}$, $\mathbf{b}$, $\mathbf{c}$, or $\mathbf{d}$ to define a new parameterization of $\mathbf{X}$ with the same form as Equation 3.1. A convenient unique solution is given by estimating the main effects in the absence of the interaction term, as $a_i = x_{i\cdot} - x_{\cdot\cdot}$, $b_j = x_{\cdot j} - x_{\cdot\cdot}$, and then

c and **d** are obtained by appealing to the Eckart–Young theorem (see Section 3.4.1 for a summary) using the dominant pair of singular vectors of the matrix

$$\mathbf{Y} = \mathbf{X} - \mathbf{a}\mathbf{1}^{\mathsf{T}} - \mathbf{1}\mathbf{b}^{\mathsf{T}} - x_{..}\mathbf{1}\mathbf{1}^{\mathsf{T}} \tag{3.2}$$

which represents the residual after fitting the main effects. To be more precise, when the rank of $\mathbf{Y} = R$ (usually, $R = \min(I - 1, J - 1)$), we have R bilinear terms in the singular-value decomposition (SVD):

$$\mathbf{Y} = \mathbf{U}\boldsymbol{\Sigma}\mathbf{V}^{\mathsf{T}} = \sigma_1\mathbf{u}_1\mathbf{v}_1^{\mathsf{T}} + \sigma_2\mathbf{u}_2\mathbf{v}_2^{\mathsf{T}} + \cdots + \sigma_R\mathbf{u}_R\mathbf{v}_R^{\mathsf{T}} \tag{3.3}$$

and we select the first of these for the interaction term $\mathbf{c}\mathbf{d}^{\mathsf{T}}$. Thanks to the Eckart–Young theorem, the first term of Equation 3.3 provides the best (in terms of least squares) rank-1 matrix approximation to **Y**. Further bilinear terms accounting for the interaction can be obtained from the subdominant singular vectors of **Y**, with the first r terms giving the best rank-r fit. Thus the following rank-$(r + 2)$ model for the original table is fitted:

$$\mathbf{X} = m\mathbf{1}\mathbf{1}^{\mathsf{T}} + \mathbf{a}\mathbf{1}^{\mathsf{T}} + \mathbf{1}\mathbf{b}^{\mathsf{T}} + \sum_{s=1}^{r}\mathbf{c}_s\mathbf{d}_s^{\mathsf{T}} + \text{residual} \tag{3.4}$$

Notice that $\mathbf{1}^{\mathsf{T}}\mathbf{a} = 0$ and $\mathbf{1}^{\mathsf{T}}\mathbf{b} = 0$, so Equation 3.2 implies that $\mathbf{1}^{\mathsf{T}}\mathbf{Y} = \mathbf{0}$ and $\mathbf{Y}\mathbf{1} = \mathbf{0}$, so that $\mathbf{1}^{\mathsf{T}}\mathbf{c}_s = 0$ and $\mathbf{1}^{\mathsf{T}}\mathbf{d}_s = 0$ for $s = 1,2,\ldots, r$. In the SVD (Equation 3.3), both the $\mathbf{u}_s$'s and $\mathbf{v}_s$'s form orthonormal sets of vectors, and each $\mathbf{c}_s\mathbf{d}_s^{\mathsf{T}}$ in Equation 3.3 is estimated by the corresponding $\sigma_s\mathbf{u}_s\mathbf{v}_s^{\mathsf{T}}$. So, the scalings of $\mathbf{c}_s$ and $\mathbf{d}_s$ are not fully determined because we could set $\mathbf{c}_s = \mu\mathbf{u}_s$ and $\mathbf{d}_s = \mu^{-1}\sigma_s\mathbf{v}_s$. Three possible choices of μ are $\mu = 1$, $\mu = \sqrt{\sigma_s}$ and $\mu = \sigma_s$. The important thing is that all choices of μ, **a**, **b**, **c**, and **d** give the same approximation to **X** but give different visualizations that may inadvertently influence interpretations.

When $r = 2$, we can obtain a two-dimensional biplot visualization of the interaction by plotting $(\mathbf{c}_1, \mathbf{c}_2)$ to give coordinates for the I row points R_i and $(\mathbf{d}_1, \mathbf{d}_2)$ for the J column points C_j. Both row points and column points will be centered at the origin O of the biplot. Interpretation is via the inner product (or scalar product) of each pair of row and column vectors OR_i and OC_j, respectively, which approximates the corresponding element y_{ij} of the interaction **Y** (for the visualization of interaction effects, see Chapter 22). Although main effects have been removed, information on them can also be included in the biplot (Gower 1990).

We can write the fitted model as

$$\hat{\mathbf{X}} = \frac{1}{IJ}(\mathbf{1}^\mathsf{T}\mathbf{X}\mathbf{1})\mathbf{1}\mathbf{1}^\mathsf{T} + \frac{1}{J}\mathbf{X}\mathbf{1}\mathbf{1}^\mathsf{T} + \frac{1}{I}\mathbf{1}\mathbf{1}^\mathsf{T}\mathbf{X} + \sum_{s=1}^{r}\mathbf{c}_s\mathbf{d}_s^\mathsf{T}$$

from which it is clear that the interaction matrix (Equation 3.2) can be expressed as the double-centering of $\mathbf{X}$: $\mathbf{Y} = (\mathbf{I} - \frac{1}{I}\mathbf{1}\mathbf{1}^\mathsf{T})\mathbf{X}(\mathbf{I} - \frac{1}{J}\mathbf{1}\mathbf{1}^\mathsf{T})$. The matrices $(\mathbf{I} - \frac{1}{I}\mathbf{1}\mathbf{1}^\mathsf{T})$ and $(\mathbf{I} - \frac{1}{J}\mathbf{1}\mathbf{1}^\mathsf{T})$ are centering matrices: the former (which premultiplies $\mathbf{X}$) centers the columns with respect to column averages, and the latter (which postmultiplies $\mathbf{X}$) centers with respect to row averages. Hence, the model approximation to the interaction is

$$\mathbf{Y} = \left(\mathbf{I} - \frac{1}{I}\mathbf{1}\mathbf{1}^\mathsf{T}\right)\mathbf{X}\left(\mathbf{I} - \frac{1}{J}\mathbf{1}\mathbf{1}^\mathsf{T}\right) \approx \sum_{s=1}^{r}\mathbf{c}_s\mathbf{d}_s^\mathsf{T} \tag{3.5}$$

where $\approx$ denotes least-squares approximation. This can be compared with the similar result for correspondence analysis, given later in Equation 3.12.

3.2.2 *Principal component analysis*

In PCA it is customary to think of the $I \times J$ data matrix $\mathbf{X}$ as representing the coordinates of I points referred to J-dimensional Cartesian coordinates. Karl Pearson (1901) was concerned with the "lines and planes of closest fit" to the resulting cloud of points. In effect, he was looking for an r-dimensional, or rank-r fit, to $\mathbf{X}$. The relationship with the Eckart–Young theorem is clear. The only difference is that planes of closest fit necessarily pass through the centroid of the cloud, so we want an approximation to the deviations of $\mathbf{X}$ from their column means rather than to $\mathbf{X}$ itself. Thus, PCA is concerned with the singular-value decomposition of $(\mathbf{I} - \frac{1}{I}\mathbf{1}\mathbf{1}^\mathsf{T})\mathbf{X}$, where premultiplication by $(\mathbf{I} - \frac{1}{I}\mathbf{1}\mathbf{1}^\mathsf{T})$ centers the variables (columns) with respect to their averages. Hence, in terms of approximating the data, Equation 3.5 is replaced by

$$\left(\mathbf{I} - \frac{1}{I}\mathbf{1}\mathbf{1}^\mathsf{T}\right)\mathbf{X} \approx \sum_{s=1}^{r}\mathbf{c}_s\mathbf{d}_s^\mathsf{T} \tag{3.6}$$

Although the computations associated with the models presented in Equation 3.5 and Equation 3.6 are very similar, differing mainly in the way $\mathbf{X}$ is centered, there are two major statistical differences. First, with PCA there is no notion of dependent variable. Second, in PCA the variables may be in different measurement units, and changing a measurement scale (e.g., replacing inches with centimeters) will change $\mathbf{X}$ in a way that leads to different singular-value decompositions (SVDs) that are not simply related. Since it is desirable to have results that are invariant to the vagaries of the measurement units used, it is usual in PCA to scale each column of $\mathbf{X}$ so that the results are independent of measurement units. Usually, and most simply, this scaling is achieved by normalization: that is each column (variable) of the centered $\mathbf{X}$ is divided by its standard deviation, so that each column has unit variance, but this is not the only possible method of scaling. This invariance difficulty is not a problem with biadditive models, where the entries of $\mathbf{X}$ are all concerned with measurements of a single variable.

With the asymmetric nature of the PCA setup, where the variables correspond to coordinate axes, we have a different kind of biplot than that of the biadditive model. The I units are represented by points P_i whose coordinates are given by the rows of $\mathbf{XV} = \mathbf{U\Sigma}$, i.e., the choice $\mu = \sigma_s$ mentioned earlier for the biadditive model biplot. The variables are represented by J points V_j, whose coordinates are given by the first r elements in the jth row of $\mathbf{V}$, and are usually drawn as vectors OV_j from the origin O to the points V_j. These r-dimensional vectors are approximations to the J-dimensional Cartesian coordinate axes. Gower and Hand (1996) explain how the directions indicated by these vectors, called *biplot axes*, can be marked with scales just like any other coordinate axes, a process known as *calibration*. Then the inner-product approximations $\hat{x}_{ij}$ given by the SVD of the multiplicative part of Equation 3.6 can be obtained, just as with conventional coordinate axes, by projecting the ith point P_i onto the jth biplot axis and reading off the value on the calibrated scale. The same device could be used to evaluate the inner products of biadditive biplots, by arbitrarily selecting either the rows or the columns to play the role of coordinate axes.

3.2.3 Computation using singular-value decomposition or eigendecomposition

Computations are no problem because software for calculating the SVD $\mathbf{Y} = \mathbf{U\Sigma V}^{\mathrm{T}}$ is freely available, for example, in the R package (R Development Core Team 2005). As an alternative, it is always possible to calculate an SVD from the eigendecomposition (or spectral decomposition) of the symmetric positive semidefinite matrix $\mathbf{Y}^\mathsf{T}\mathbf{Y} = \mathbf{V}\boldsymbol{\Sigma}^2\mathbf{V}^\mathsf{T}$, which yields $\mathbf{V}$ and $\boldsymbol{\Sigma}$, and then obtain $\mathbf{U} = \mathbf{Y}\mathbf{V}\boldsymbol{\Sigma}^{-1}$. However, we now encounter another statistical issue and one of the confusions to which we referred at the start. When $\mathbf{Y}$ is a centered and normalized data matrix, then $\mathbf{Y}^\mathsf{T}\mathbf{Y} = \mathbf{R}$, say, is a correlation matrix. Sometimes PCA is seen as approximating $\mathbf{R}$, but this is the wrong way to think about it. Indeed when PCA is interpreted as fitting lines and planes of closest fit, it takes no account of possible correlations. This is most easily appreciated by recalling that in the J-dimensional coordinate representation, the square of the distance $d_{ii'}$ between rows i and i' of $\mathbf{X}$ is given by Pythagoras's formula

$$d_{ii'}^2 = \sum_{j=1}^{J}(x_{ij} - x_{i'j})^2$$

where, manifestly, the variables are treated independently. It is true that we can derive the correlation matrix $\mathbf{R}$ as a computational step in evaluating $\mathbf{V}$, $\boldsymbol{\Sigma}$, and then $\mathbf{U}$, but that does not imply that we are analyzing the correlations themselves; fundamentally, we are approximating $\mathbf{Y}$. In Section 3.3.3, we shall see that a similar issue applies to interpretations of MCA.

It is, rather, in factor analysis that the approximation of $\mathbf{R}$ is of primary importance. Then the SVD of $\mathbf{R}$ (which coincides with its eigendecomposition because $\mathbf{R}$ is symmetric) gives the simplest r-dimensional approximation. Hotelling (1933), in an influential paper, proposed this as a solution to the factor analysis problem and, unfortunately, named it "principal component analysis," identifying the eigenvectors as factor loadings and thus forever clouding the distinction between factor analysis and the earlier definition of PCA due to Pearson. Factor analysis is concerned with identifying factors and hence with the so-called reification of loadings, that is, rotating solutions to generate factors that hopefully have substantive interpretations, an operation that has little, or no, relevance for PCA per se. One method of factor analysis does indeed involve direct approximation to $\mathbf{R}$ but excludes the unit diagonal elements or, at least, introduces further parameters, called specific factors, to account for the diagonal. Thus, rather than minimizing $\sum_j \sum_{j'} (r_{jj'} - \hat{r}_{jj'})^2$, which is given by the spectral decomposition of $\mathbf{R}$, the diagonal is excluded by minimizing $\sum\sum_{j<j'} (r_{jj'} - \hat{r}_{jj'})^2$. This requires an iterative algorithm that introduces additional complications if the fitted $\hat{\mathbf{R}}$ (with unit diagonal) is required to be positive semidefinite, as

is a correlation matrix itself. There are further difficulties with the spectral-decomposition solution, for not only does it approximate the uninteresting unit diagonal elements, but it approximates each off-diagonal element twice, once as $r_{jj'}$ and once as $r_{j'j}$. Thus, the diagonal elements get half the weight of the off-diagonal elements. Although this has the desirable effect of down-weighting the diagonal, zero diagonal weights would be preferable. Bailey and Gower (1990) explore some of the algebraic consequences of giving the diagonal weights intermediate between zero and unity. A similar distinction between MCA and joint correspondence analysis (JCA) is discussed in Section 3.3.3.

By defining a distance $d_{jj'}^2 = 1 - r_{jj'}$ or $d_{jj'}^2 = 1 - r_{jj'}^2$, multidimensional scaling methods give approximations to **R** that avoid the difficulties of approximating the diagonal, which is automatically zero, as it should be for distances. Indeed, classical scaling/principal coordinate analysis requires only the eigendecomposition of the double-centered form of **R**, $(\mathbf{I} - \frac{1}{J}\mathbf{1}\mathbf{1}^\mathsf{T})\mathbf{R}(\mathbf{I} - \frac{1}{J}\mathbf{1}\mathbf{1}^\mathsf{T})$. When $d_{jj'}^2 = 1 - r_{jj'}^2$, **R** is replaced by a matrix of squared correlations (Hill 1969). This result indicates a relationship between the spectral decomposition of **R** itself and its double-centered forms. However, any form of metric or nonmetric multidimensional scaling could be used.

Because **R** is symmetric, we have $\mathbf{R} = \mathbf{U}\boldsymbol{\Sigma}^2\mathbf{U}^\mathsf{T}$, so the row and column vectors are the same and methods for approximating **R** give biplots with only one set of J points. In the simplest case, these coordinates are given by the first r columns of $\mathbf{U}\boldsymbol{\Sigma}$, the fitted correlations being estimated by the scalar product. Because we know that the diagonal elements are units, in the exact representation the J points must lie on a unit hypersphere. In the approximation in, say, two dimensions, ideally the distribution should be circular, and then departures from circularity highlight where the approximation is good or bad. Alternatively, we can plot $\mathbf{U}\boldsymbol{\Sigma}^2$ as the points with calibrated biplot axes whose directions are given by **U** to give a simple way of evaluating all scalar products. Similar biplots can be based on covariance matrices (Underhill 1990), where approximations to the diagonal variances are of interest but the unit hypersphere property is not available for exploitation.

3.3 Categorical variables

We now turn to the cases where **X** is formed of categorical variables. We shall see that all the issues discussed above for quantitative variables are relevant for categorical variables too.

3.3.1 Tables of categorical variables

The simplest situation corresponding to a quantitative variable classified by two categorical variables is when, for each cell, a single observation is made of a third categorical variable with L levels. For example, we might classify which of L political parties is represented in each of I years and J districts. One method of analysis is to find optimal scale values $\mathbf{s} = (s_1, s_2, \dots, s_L)$ for the parties. By an optimal scale, we mean scale values that lead to the best fit in a least-squares estimation of a model such as Equation 3.1, where the matrix $\mathbf{X}$ contains the (unknown, to be estimated) scale values corresponding to the categories of the "response" variable. Of course, such scale values can be linearly transformed with no material effect: thus, we seek a zero-centered and normalized scale such that $\mathbf{s}^\mathsf{T}\mathbf{s} = 1$, for example, or as is more common, $\mathbf{s}^\mathsf{T}\mathbf{L}\mathbf{s} = 1$, where $\mathbf{L}$ is the diagonal matrix of frequencies (or relative frequencies) of the L categories.

Rather than analyze the simple model obtained from Equation 3.1 by deleting the multiplicative term, it is just as easy to analyze the general linear (regression) model $\mathbf{Zs} = \mathbf{X}$, where $\mathbf{Z}$ is an indicator matrix with L columns and N rows (for a two-way table $N = IJ$), and $\mathbf{X}$ is an appropriately constructed design matrix. Thus, as described in Section 3.1, $\mathbf{X}$ is now in the form of a data matrix that, at its simplest, gives row/column information. Each row of $\mathbf{Z}$ refers to one of the observations, and each column refers to one of the L categories. Thus, the elements of $\mathbf{Z}$ are zero, except for a single unit in each row, indicating the category observed in that row. Hence, the row totals are 1 ($\mathbf{Z1} = \mathbf{1}$), and the column totals are the frequencies of the L categories (thus $\mathbf{L} = \mathbf{Z}^\mathsf{T}\mathbf{Z}$ and $\mathbf{1}^\mathsf{T}\mathbf{Z} = \mathbf{1}^\mathsf{T}\mathbf{L}$). $\mathbf{Zs}$ gives the scores associated with each row, leading to a total sum of squares of $\mathbf{s}^\mathsf{T}\mathbf{Z}^\mathsf{T}\mathbf{Zs} = \mathbf{s}^\mathsf{T}\mathbf{Ls}$. The fitted sum of squares after least-squares estimation is given by $\mathbf{s}^\mathsf{T}\mathbf{Z}^\mathsf{T}\mathbf{X}(\mathbf{X}^\mathsf{T}\mathbf{X})^{-1}\mathbf{X}^\mathsf{T}\mathbf{Zs}$. The best model fit, i.e., minimum residual sum of squares, occurs for the eigenvector $\mathbf{s}$ corresponding to the maximum eigenvalue λ of

$$\mathbf{Z}^\mathsf{T}\mathbf{X}(\mathbf{X}^\mathsf{T}\mathbf{X})^{-1}\mathbf{X}^\mathsf{T}\mathbf{Zs} = \lambda\mathbf{Ls} \tag{3.7}$$

To identify the solution, the scores $\mathbf{s}$ can be constrained to satisfy $\mathbf{s}^\mathsf{T}\mathbf{Ls} = 1$. Note that this constraint is very weak; any other quadratic constraint could be used without any substantive effect.

In deriving Equation 3.7, we have not worked in deviations from the means. To do so requires replacing $\mathbf{X}$ by $(\mathbf{I} - \frac{1}{N}\mathbf{11}^\mathsf{T})\mathbf{X}$, as is usual in multiple regression. Then, Equation 3.7 has a zero eigenvalue corresponding to the vector **1**, and it follows that $\mathbf{1}^T\mathbf{Ls} = 0$, (i.e., $\mathbf{1}^T\mathbf{Zs} = 0$) for any other eigenvector **s**. Thus, the scored values **Zs** are automatically centered, and it is not necessary also to replace **Z** by $(\mathbf{I} - \frac{1}{N}\mathbf{1}\mathbf{1}^T)\mathbf{Z}$. These results are interesting because:

1. Ordinary linear models require only matrix inversion. The estimation of optimal scores requires the additional eigenvalue calculation (Equation 3.7).
2. We have seen that extending the additive model to the biadditive model (Equation 3.1) also requires an eigenvalue calculation in the form of the SVD of the residual matrix (Equation 3.2). This implies that to fit the biadditive model with optimal scores requires two eigenvalue calculations. Computation is more difficult, but an iterative algorithmic solution can be constructed.
3. The linear optimal-scores problem is the simplest example of where working in deviations from the mean interacts with constraints. Some constraints are weak identification constraints (e.g., $\mathbf{s}^T\mathbf{Ls} = 1$), and some are not imposed but arise as a consequence of the solution (e.g., $\mathbf{1}^T\mathbf{Zs} = 0$). In general, substantive strong constraints, such as that all elements of **s** must be nonnegative, are not required in the material covered by this chapter, but the distinction between weak identification constraints and strong constraints can cause confusion (see Gower 1998 for a discussion).

In these methods, once the optimal scores are estimated, then **Zs** can be treated as a set of quantitative observations, and the previously discussed biplot displays for biadditive models remain available.

When, rather than a single observation, we have n_{ij} observations of a categorical variable within the (i,j)th cell of a two-way table, the mean is not available as it is for quantitative variables, but we can adopt some other summary statistic such as the most frequent category and proceed as above. Then the optimal-scores approach can be applied using all the data, where **Z** is defined as before but with a number of rows larger than IJ, and **s** is determined from a variant of the eigenequation (Equation 3.7).

It is interesting to look a little more closely at the structure of **X**. For a two-way table, we can write $\mathbf{X} = (\mathbf{Z}_2, \mathbf{Z}_3)$ where $\mathbf{Z}_2$ and $\mathbf{Z}_3$ are indicator matrices for the categorical variables classifying the rows and columns. We reserve $\mathbf{Z}_1$ for the dependent categorical variable

$\mathbf{Z}$, defined above. We can construct $\mathbf{C}$, the cross-product matrix for $(\mathbf{Z}_1, \mathbf{Z}_2, \mathbf{Z}_3)$:

$$\mathbf{C} = \begin{pmatrix} \mathbf{Z}_1^\mathsf{T}\mathbf{Z}_1 & \mathbf{Z}_1^\mathsf{T}\mathbf{Z}_2 & \mathbf{Z}_1^\mathsf{T}\mathbf{Z}_3 \\ \mathbf{Z}_2^\mathsf{T}\mathbf{Z}_1 & \mathbf{Z}_2^\mathsf{T}\mathbf{Z}_2 & \mathbf{Z}_2^\mathsf{T}\mathbf{Z}_3 \\ \mathbf{Z}_3^\mathsf{T}\mathbf{Z}_1 & \mathbf{Z}_3^\mathsf{T}\mathbf{Z}_2 & \mathbf{Z}_3^\mathsf{T}\mathbf{Z}_3 \end{pmatrix} \tag{3.8}$$

$\mathbf{C}$ is the so-called Burt matrix for the three categorical variables. Its qth diagonal block is a diagonal matrix $\mathbf{Z}_q^\mathsf{T}\mathbf{Z}_q$ giving the category frequencies for the qth categorical variable; the (q,q')th block is the contingency table $\mathbf{Z}_q^\mathsf{T}\mathbf{Z}_{q'}$ for the qth and q'th variables. The Burt matrix is of primary interest in MCA (see Section 3.3.3) for any number of categorical variables, but for the moment we are interested only in the three-variable case where the first variable is treated as a dependent variable for which we are estimating optimal scores $\mathbf{s}$. It can be shown that if we invert $\mathbf{C}$, then the first block-diagonal term delivers the left-hand side of Equation 3.7, from which the optimal scores derive.

If there is no third categorical variable and we are interested only in the counts n_{ij} in each cell of the table, we are moving toward the realm of simple CA, but for completeness we first mention possible ways to model these counts. In the present case of a two-way contingency table, one way of proceeding is to treat the expected counts as some function (the link function) of an additive model, as on the right-hand side of Equation 3.1. The appeal to expectation implies a specification of an appropriate distribution, such as Poisson, for counts. This is the approach of generalized linear models (GLMs) (McCullagh and Nelder 1989). GLMs are not restricted to two-way tables and so can handle multifactor additive interaction terms; an important special case is the class of log-linear models. GLMs can be further generalized, to include multiplicative bilinear terms, as in Equation 3.3 (e.g., van Eeuwijk 1995; de Falguerolles and Francis 1992; see also Chapter 21 of this volume).

3.3.2 Correspondence analysis

Simple correspondence analysis (CA) offers another way of approximating a contingency table $\mathbf{N}$. The method is rooted in the association between the row and column factors of the two-way table. Suppose the marginal relative frequencies of $\mathbf{N}$, called row and column *masses* in CA, are denoted by $\mathbf{r}$ and $\mathbf{c}$, that is, $\mathbf{r} = (1/n)\mathbf{N1}$ and $\mathbf{c} = (1/n)\mathbf{N}^\mathsf{T}\mathbf{1}$, where $n = \mathbf{1}^\mathsf{T}\mathbf{N1}$ is the grand total of $\mathbf{N}$. Then the matrix of expected values under the hypothesis of no association between rows and columns is equal to $n\mathbf{rc}^\mathsf{T}$, with elements nr_ic_j. Similarly to the way that biadditive models express interaction as departures from the main effects, so CA is concerned with departures from independence as given by the elements of the matrix:

$$\mathbf{N} - n\mathbf{rc}^\mathsf{T}$$

As with the biadditive model, the SVD is used to give a least-squares approximation to the "interaction." In my opinion, how this is done in CA is not straightforward. This is because at least two ways of measuring departures from independence are at issue. In addition to $\mathbf{N} - n\mathbf{rc}^\mathsf{T}$, we can consider a contribution to what is called the *total inertia* in CA, that is, Pearson's chi-square divided by n (see Chapter 1). The square root of this contribution is

$$\frac{1}{\sqrt{n}} \frac{\left(n_{ij} - nr_ic_j\right)}{\sqrt{nr_ic_j}}$$

or in matrix terms:

$$(1/n)\mathbf{D}_r^{-1/2}(\mathbf{N} - n\mathbf{rc}^\mathsf{T})\mathbf{D}_c^{-1/2} = \mathbf{D}_r^{-1/2}(\mathbf{P} - \mathbf{rc}^\mathsf{T})\mathbf{D}_c^{-1/2} \tag{3.9}$$

where $\mathbf{P} = (1/n)\mathbf{N}$ is the table of relative frequencies, termed the *correspondence table.* Since the overall sample size n is not of interest in CA, interest is focused on $\mathbf{P}$ rather than the original contingency table $\mathbf{N}$. In Equation 3.9, $\mathbf{D}_r$ and $\mathbf{D}_c$ are diagonal matrices of the so-called row and column *masses*, so that $\mathbf{D}_r\mathbf{1} = \mathbf{r}$ and $\mathbf{D}_c\mathbf{1} = \mathbf{c}$, showing that $\mathbf{D}_r^{1/2}\mathbf{1}$ and $\mathbf{D}_c^{1/2}\mathbf{1}$ are unit vectors. Note that $(1/n)\mathbf{D}_r^{-1/2}\mathbf{ND}_c^{-1/2}(\mathbf{D}_c^{1/2}\mathbf{1}) = \mathbf{D}_r^{1/2}\mathbf{1}$ and $(1/n)(\mathbf{1}^\mathsf{T}\mathbf{D}_r^{1/2})\mathbf{D}_r^{-1/2}\mathbf{ND}_c^{-1/2} = \mathbf{1}^\mathsf{T}\mathbf{D}_c^{1/2}$, so these unit vectors correspond to a unit singular value of $(1/n)\mathbf{D}_r^{-1/2}\mathbf{ND}_c^{-1/2}$, giving the SVD:

$$\mathbf{D}_r^{-1/2}\mathbf{PD}_c^{-1/2} = \mathbf{D}_r^{1/2}\mathbf{11}^\mathsf{T}\mathbf{D}_c^{1/2} + \sum_{s=1}^{r} \sigma_s \mathbf{u}_s \mathbf{v}_s^\mathsf{T}$$

or, equivalently:

$$\mathbf{D}_r^{-1/2}(\mathbf{P} - \mathbf{rc}^\mathsf{T})\,\mathbf{D}_c^{-1/2} = \sum_{s=1}^{r} \sigma_s \mathbf{u}_s \mathbf{v}_s^\mathsf{T} \tag{3.10}$$

The usual orthogonality relationships of SVD show that $\mathbf{1}^\mathsf{T}\mathbf{D}_c^{1/2}\mathbf{u}_s = \mathbf{1}^\mathsf{T}\mathbf{D}_c^{1/2}\mathbf{v}_s = 0$, $s = 1,2,\ldots,r$, giving the origin as a weighted centroid. Notice that with CA, unlike biadditive models, we need not first calculate the "residuals," in this case, $\mathbf{P} - \mathbf{rc}^\mathsf{T}$ but just the SVD of $\mathbf{D}_r^{-1/2}\mathbf{P}\mathbf{D}_c^{-1/2}$, obtaining the "trivial" solution $\mathbf{D}_r^{1/2}\mathbf{1}\mathbf{1}^\mathsf{T}\mathbf{D}_c^{1/2} = \mathbf{D}_r^{-1/2}\mathbf{rc}^\mathsf{T}\mathbf{D}_c^{-1/2}$ as the first bilinear term of the decomposition, corresponding to a singular value of 1; hence, the bilinear terms from second onward would be those required. Equation 3.10 shows that the approximation to the Pearson chi-squared contributions can be visualized as a biplot of $\sigma_s^\alpha\mathbf{u}_s, \sigma_s^\beta\mathbf{v}_s$ $(s = 1,2,\ldots,r)$ for $\alpha + \beta = 1$ (we return below to choices of α and β). This straightforward biplot of the standardized residuals in Equation 3.9 is not in the standard form of CA described below.

One of the alternative ways of interpreting the standardized residuals that can lead us to CA is to rewrite the SVD (Equation 3.10) as:

$$\mathbf{D}_r^{1/2}(\mathbf{D}_r^{-1}\mathbf{P}\mathbf{D}_c^{-1} - \mathbf{1}\mathbf{1}^\mathsf{T})\,\mathbf{D}_c^{1/2} = \sum_{s=1}^{r}\sigma_s\mathbf{u}_s\mathbf{v}_s^\mathsf{T} \tag{3.11}$$

which can be compared with the interaction term (Equation 3.5) for the biadditive model. The elements of $\mathbf{D}_r^{-1}\mathbf{P}\mathbf{D}_c^{-1}$ are known as *contingency ratios* and give the ratios of the observed frequencies to those expected under independence. Thus, rather than as a function of the Pearson contributions, Equation 3.11 can be interpreted as a *weighted* least-squares approximation to the departures of the contingency ratios from unity. Although Equation 3.10 and Equation 3.11 are algebraically equivalent, Equation 3.10 gives an unweighted approximation to the Pearson contributions, while Equation 3.11 gives a weighted approximation to the contingency ratios $p_{ij}/(r_i c_j)$. In the latter case we plot $\sigma_s^\alpha\mathbf{D}_r^{-1/2}\mathbf{u}_s, \sigma_s^\beta\mathbf{D}_c^{-1/2}\mathbf{v}_s (s = 1,2,\ldots,r)$, which is the standard CA representation. The vectors $\mathbf{a}_s = \mathbf{D}_r^{-1/2}\mathbf{u}_s$ and $\mathbf{b}_s = \mathbf{D}_c^{-1/2}\mathbf{v}_s$ are normalized so that $\mathbf{a}_s^\mathsf{T}\mathbf{D}_r\mathbf{a}_s = \mathbf{b}_s^\mathsf{T}\mathbf{D}_c\mathbf{b}_s = 1$, and are thus called *standard coordinates* (see Chapter 1). A result that encapsulates much of the above is that:

$$(\mathbf{I} - \mathbf{1}\mathbf{r}^\mathsf{T})\mathbf{D}_r^{-1}\mathbf{P}\mathbf{D}_c^{-1}(\mathbf{I} - \mathbf{c}\mathbf{1}^\mathsf{T}) = \mathbf{A}\mathbf{\Sigma}\mathbf{B}^\mathsf{T} \tag{3.12}$$

where $\mathbf{A}^\mathsf{T}\mathbf{D}_r\mathbf{A} = \mathbf{B}^\mathsf{T}\mathbf{D}_c\mathbf{B} = \mathbf{I}$, relating a double-centered version of the contingency ratios to a weighted SVD of $\mathbf{P} - \mathbf{rc}^\mathsf{T}$ (compare with Equation 3.5 for biadditive models).

When α or β is 1, the resulting coordinates $\sigma_s^\alpha\mathbf{D}_r^{-1/2}\mathbf{u}_s = \sigma_s^\alpha\mathbf{a}_s$ and $\sigma_s^\alpha\mathbf{D}_r^{-1/2}\mathbf{u}_s = \sigma_s^\alpha\mathbf{b}_s$ are called *principal coordinates* because they give approximate interpoint distances between rows or between columns with respect to principal axes. This leads us to give an alternative derivation of CA (and the most popular one) in terms of chi-squared distances. The squared chi-squared distance between the ith and i'th rows of the contingency table $\mathbf{N}$ is:

$$\sum_{j=1}^{J} \frac{1}{c_j} \left(\frac{p_{ij}}{r_i} - \frac{p_{i'j}}{r_{i'}} \right)^2 \tag{3.13}$$

so this distance is zero when the ith and i'th rows of $\mathbf{N}$ are in the same proportions with their respective row totals, that is, when their *profiles* (vectors of relative frequencies) are the same. Coordinates that generate the distances (Equation 3.13) are given by the rows of $\mathbf{Y} = \mathbf{D}_r^{-1}\,\mathbf{P}\mathbf{D}_c^{-1/2}$. From Equation 3.10,

$$\mathbf{Y} = \mathbf{D}_r^{-1/2}(\mathbf{D}_r^{1/2}\mathbf{1}\mathbf{1}^{\mathsf{T}}\mathbf{D}_c^{1/2} + \mathbf{U\Sigma V}^{\mathsf{T}}) = \mathbf{1}\mathbf{1}^{\mathsf{T}}\mathbf{D}_c^{1/2} + \mathbf{D}_r^{-1/2}\mathbf{U\Sigma V}^{\mathsf{T}}$$

The first term on the right-hand side is constant for every row of $\mathbf{Y}$ and, thus, represents a translation that has no effect on distances. The second term includes the orthogonal matrix $\mathbf{V}$, which also has no effect on distances. Thus, we see that the rows of $\mathbf{D}_r^{-1/2}\,\mathbf{U\Sigma} = \mathbf{A\Sigma}$ generate the chi-squared distances, the first r columns of which are precisely the coordinates obtained by setting $\alpha = 1$, as in the biplot derived from Equation 3.11. CA can thus be interpreted as a weighted version of classical scaling of these chi-square distances between row profiles, where each row is weighted proportionally to the row masses (see Greenacre 1984). Similarly, concentrating rather on the chi-squared distances between column profiles (columns of $\mathbf{N}$ divided by their column totals), we can interpret CA as the weighted classical scaling of these distances, with the columns weighted by the column masses, yielding the principal coordinates of the columns. CA has several well-known properties that are useful for interpretation, such as the principal of distributional equivalence and expressions for the row points as weighted averages of the column points (see, for example, Greenacre 1984).

To obtain our final biplots (usually in two dimensions, $r = 2$) we must choose both α and β. When $\alpha = 1$ and $\beta = 0$ or $\alpha = 0$ and $\beta = 1$ for the so-called *asymmetric maps*, or $\alpha = 1$ and $\beta = 1$ giving the *symmetric map*, we have seen that the contingency ratio approximations based on Equation 3.11 generate approximate row and column chi-squared distances. The inner-product also approximates the contingency ratios, except when $\alpha = \beta = 1$, which conflicts with the inner-product interpretation of a biplot. These same settings can also be used with Pearson contributions based on Equation 3.10, with chi-squared distance replaced by, possibly, a less interesting distance.

The biplot with $\alpha = 1/2$ and $\beta = 1/2$ respects the inner-product and gives a symmetric treatment of the variables, but it is seldom used. This biplot does not have the attractive distance properties and other advantages of CA based on Equation 3.11 described above, but one could use it to make a biplot of the standardized Pearson residuals in Equation 3.10, that is, by using the singular vectors $\mathbf{u}_s$ and $\mathbf{v}_s$ in place of $\mathbf{a}_s$ and $\mathbf{b}_s$. if only scalar products were of interest.

3.3.3 Multiple correspondence analysis and related methods

When the data matrix comprises Q categorical variables, we can write them as a concatenation of indicator matrices: $\mathbf{Z} = [\mathbf{Z}_1\ \mathbf{Z}_2\ \mathbf{Z}_3\ \dots\ \mathbf{Z}_Q]$, where $\mathbf{Z}_q$ is the indicator matrix for the qth categorical variable, with J_q columns corresponding to its J_q categories. Hence, $\mathbf{Z}$ has $J = \sum_{q=1}^{Q} J_q$ columns and $I = n$ rows, where n is the sum of any cross-tabulation $\mathbf{N}_{qq'} = \mathbf{Z}_q^\mathsf{T}\mathbf{Z}_{q'}$. The Burt matrix $\mathbf{C}$ for these Q variables is now a $Q \times Q$ block matrix:

$$\mathbf{C} = \mathbf{Z}^\mathsf{T}\mathbf{Z} = \begin{bmatrix} \mathbf{Z}_1^\mathsf{T}\mathbf{Z}_1 & \mathbf{Z}_1^\mathsf{T}\mathbf{Z}_2 & \cdots & \mathbf{Z}_1^\mathsf{T}\mathbf{Z}_Q \\ \mathbf{Z}_2^\mathsf{T}\mathbf{Z}_1 & \mathbf{Z}_2^\mathsf{T}\mathbf{Z}_2 & \cdots & \mathbf{Z}_2^\mathsf{T}\mathbf{Z}_Q \\ \vdots & \vdots & \ddots & \vdots \\ \mathbf{Z}_Q^\mathsf{T}\mathbf{Z}_1 & \mathbf{Z}_Q^\mathsf{T}\mathbf{Z}_2 & \cdots & \mathbf{Z}_Q^\mathsf{T}\mathbf{Z}_Q \end{bmatrix} = \begin{bmatrix} \mathbf{L}_1 & \mathbf{N}_{12} & \cdots & \mathbf{N}_{1Q} \\ \mathbf{N}_{21} & \mathbf{L}_2 & \cdots & \mathbf{N}_{2Q} \\ \vdots & \vdots & \ddots & \vdots \\ \mathbf{N}_{Q1} & \mathbf{N}_{Q2} & \cdots & \mathbf{L}_Q \end{bmatrix} \tag{3.14}$$

Several methods for analyzing $\mathbf{Z}$ or $\mathbf{C}$ have been proposed and are briefly discussed below, where I shall try to unscramble their intricate interrelationships. As before, we use the notation $\mathbf{L}$ for the diagonal matrix of marginal frequencies, which in this multivariable case means that $\mathbf{L}$ is the $J \times J$ matrix with submatrices $\mathbf{L}_1, \mathbf{L}_2, \dots, \mathbf{L}_Q$ down the diagonal. (Thus, to link up with the notation of previous chapters, the diagonal matrix of masses in the MCA of $\mathbf{C}$, denoted in Chapter 2 by $\mathbf{D}$, is $\mathbf{D} = (1/nQ)\mathbf{L}$.)

The first method of analysis is nonlinear principal component analysis (NLPCA, also known as categorical PCA, or CatPCA, described in detail in Chapter 4). In NLPCA, we try to find scale values for the category levels that give a "best" (in a sense to be defined) PCA. Thus, we define scale values $\mathbf{s}_1, \mathbf{s}_2, \dots, \mathbf{s}_Q$ gathered together in a vector $\mathbf{s}$ with J elements, and then perform a PCA on the corresponding quantified data matrix: $[\mathbf{Z}_1\mathbf{s}_1\mathbf{Z}_2\mathbf{s}_2\ \dots\ \mathbf{Z}_Q\mathbf{s}_Q]$. Writing $\lambda_1, \lambda_2, \dots, \lambda_Q$ for the resultant eigenvalues, we can seek an r-dimensional solution that maximizes $\lambda_1 + \lambda_2 + \cdots + \lambda_r$. However, the scores may be made arbitrarily large, so we maximize the ratio: $\sum_{s=1}^{r} \lambda_s / \sum_{s=1}^{Q} \lambda_s$. As before, any multiple of $\mathbf{s}$ will give the same maximum, so for identification we must normalize $\mathbf{s}$ in some way. It is now not sufficient to make the sum of squares $\mathbf{s}^\mathsf{T}\mathbf{s}$ or weighted sum of squares $\mathbf{s}^\mathsf{T}\mathbf{L}\mathbf{s}$ a constant, for that would give trivial exact fits in r dimensions by choosing $\mathbf{s}_1, \mathbf{s}_2, \mathbf{s}_3, \ldots, \mathbf{s}_r$ arbitrarily and $\mathbf{s}_q = 0$, for all $q > r$. Identification conditions are needed for each subset $\mathbf{s}_q$ of scale values individually, usually by centering and normalization as follows: $\mathbf{1}^\mathsf{T}\mathbf{L}_q\mathbf{s}_q = 0$ and $\mathbf{s}_q^\mathsf{T}\mathbf{L}_q\mathbf{s}_q = 1$, $q = 1, \ldots, Q$. This implies that the sums-of-squares-and-products matrix is a correlation matrix, which is, of course, a common normalization for PCA. Note that a typical element of the sum-of-squares-and-products matrix is $\mathbf{s}_q^\mathsf{T}\mathbf{N}_{qq'}\mathbf{s}_{q'}$. NLPCA, possibly with order constraints on the scores, can be computed by an alternating least-squares algorithm (see Gifi 1990). One important difference between NLPCA and PCA is that, unlike PCA, solutions for increasing settings of r in NLPCA are not nested; each solution has to be computed *ab initio*. Once the scores $\mathbf{s}$ have been computed, the subsequent analysis is identical to PCA, with all its associated visualizations.

MCA can be derived in several ways. One way is to perform a CA of $\mathbf{Z}$, regarded as a two-way contingency table. Thus, in terms of our previous notation for simple CA, we have that $\mathbf{N} = \mathbf{Z}$, $\mathbf{D}_r = (1/n)\mathbf{I}$ and $\mathbf{D}_c = (1/nQ)\mathbf{L}$. Although the CA approximation of a contingency table $\mathbf{N}$ weighted by row and column margins is justifiable, it is less easy to see why a similar approximation to an indicator matrix $\mathbf{Z}$, with its zero/one entries, has any intrinsic interest. Rather, in the spirit of PCA, it would be better to approximate $\mathbf{Z}$ by another indicator matrix of lower rank. This requires a definition of what is meant by the rank of a matrix of categorical, rather than quantitative, variables, together with a categorical variable form of the Eckart–Young theorem. Gower (2002) suggests how this might be done, but the practical details remain to be developed. As far as any distance interpretation is concerned, note that because the row sums of $\mathbf{Z}$ are constant, equal to the number of variables Q, the chi-squared distance (Equation 3.12) between rows i and i' for MCA of $\mathbf{Z}$ simplifies to

$$\frac{1}{Q^2} \sum_{j=1}^{J} \frac{1}{c_j} (z_{ij} - z_{i'j})^2 \tag{3.15}$$

which is proportional to a Pythagorean distance weighted inversely by the category-level frequencies. Effectively, Equation 3.15 generates a coefficient where rare category levels have greater weight than common ones. Because **Z** is an indicator matrix with zero/one elements, the elements of Equation 3.15 are nonzero only when the *i*th and *i'*th cases have mismatching category levels, so that the inter-row chi-squared distance for MCA can be interpreted as a weighted mismatching dissimilarity coefficient. Gower and Hand (1996) suggest that sometimes it might be better to ignore the weights in Equation 3.15 to define what they term the "extended matching coefficient" that gives equal unit weights to all category levels. In its simplest case, when all categories have two levels, the extended matching coefficient coincides with the simple matching coefficient. Whatever dissimilarity is chosen, including chi-squared distance, it can be approximated by any form of multidimensional scaling (MDS) to give a map of n points corresponding to the cases labeling the n rows of **Z**. If one finds centroid properties useful, each category level can be represented by a point at the centroid or weighted centroid of all those having that category level. In methods proposed by Gower and Hand (1996), all category levels are represented by regions. If the use of row-wise chi-squared distance with MCA is somewhat questionable, there seems to be absolutely no substantive interest in columnwise chi-squared distances derived from **Z**, as pointed out by Greenacre (1989).

An alternative derivation of MCA, like NLPCA, seeks optimal scores, but now optimality is defined in a different way. We consider the vector **Zs** giving the total scores for the individuals (rows) of the data matrix. We choose the scores **s** to maximize the ratio of the sum of squares of the row scores to the total sum of squares. That is, we maximize:

$$\frac{\mathbf{s}^\mathsf{T}\mathbf{Z}^\mathsf{T}\mathbf{Z}\mathbf{s}}{\mathbf{s}^\mathsf{T}\mathbf{L}\mathbf{s}} = \frac{\mathbf{s}^\mathsf{T}\mathbf{C}\mathbf{s}}{\mathbf{s}^\mathsf{T}\mathbf{L}\mathbf{s}} \tag{3.16}$$

This requires the solution of the two-sided eigenvalue problem:

$$\mathbf{C}\mathbf{s} = \lambda\mathbf{L}\mathbf{s} \tag{3.17}$$

The eigenequation, Equation 3.17, has the "trivial" solution $\mathbf{s} = \mathbf{1}$, $\lambda = Q$, so we choose the eigenvector **s** corresponding to the next largest eigenvalue. Ignoring the trivial solution is equivalent to working in terms of deviations from the general mean. Because Equation 3.16 is a ratio, the scaling of the eigenvectors is arbitrary and does not affect the value of the eigenvalues. I prefer to normalize so that $\mathbf{s}^\mathsf{T}\mathbf{L}\mathbf{s} = Q$, which is consistent with the NLPCA choice $\mathbf{s}_q^\mathsf{T}\mathbf{L}_q\mathbf{s}_q = 1$ $(q = 1, 2, \ldots, Q)$, although it does not imply the separate normalizations except, interestingly, for the important special case when $Q = 2$ (see Chapter 2, Section 2.2.2). To be entirely consistent with MCA, we would normalize so that $\mathbf{s}^\mathsf{T}\mathbf{L}\mathbf{s} = nQ$, that is, $\mathbf{s}^\mathsf{T}\mathbf{D}\mathbf{s} = 1$ in the notation of Chapter 2, but the difference has no substantive effect. Maximizing the sum of squares of the row scores is the same as minimizing the sum of squares of the scores within rows, so we can interpret the method as seeking homogeneous within-rows scores, and hence, in this derivation, MCA is often termed homogeneity analysis (Gifi 1990). As is usual with eigenvalue formulations, nested solutions in r dimensions are available and can be plotted in r-dimensional diagrams, but it is not obvious how the remaining dimensions should be interpreted in the optimal-scores context (this point is discussed at length in Chapter 4). One interpretation is that they give better approximations to the chi-squared distances (Equation 3.14), but we have already seen that alternative distances may be preferred.

NLPCA and homogeneity analysis both operate on similar matrices: for NLPCA, on a matrix with elements $\{ \mathbf{s}_q^\mathsf{T}\mathbf{N}_{qq'}\mathbf{s}_{q'} \}$, and for homogeneity analysis, on the Burt matrix with typical block-matrix elements $\{ \mathbf{N}_{qq'} \}$. Both methods are based on similar but different eigenvector decompositions. There is an even closer relationship between NLPCA and MCA/homogeneity analysis. Suppose $\mathbf{s}_1, \mathbf{s}_2, \ldots, \mathbf{s}_Q$ are the optimal scores for a one-dimensional NLPCA solution ($r = 1$); then we can form $\zeta_q = \nu_{1q}\mathbf{s}_q$ ($q = 1, 2, \ldots, Q$), where ν_{1q} is the qth term of $\mathbf{v}_1$, the first eigenvector of the PCA of the optimally scored variables. It can be shown that ζ_q gives optimal scores for the qth variable in a homogeneity analysis. A similar result holds for the $r = Q - 1$ dimensional NLPCA solution and the homogeneity-analysis optimum. These equivalences do not hold for intermediate values of r. A fuller analysis of the relationships between NLPCA and MCA is given by van Rijkevorsel and de Leeuw (1988, p. 11), Gower (1998) and in Chapter 4 of this book.

The Burt matrix is prominent in MCA. Its (q, q')th off-diagonal block is the two-way contingency table formed from the qth and q'th categorical variables. Equation 3.17 can be written as:

$$\mathbf{L}^{-1/2}\mathbf{C}\mathbf{L}^{-1/2}(\mathbf{L}^{1/2}\mathbf{s}) = \lambda\mathbf{L}^{1/2}\mathbf{s} \tag{3.18}$$

The solution of Equation 3.18 requires the eigenvectors of the normalized Burt matrix $\mathbf{L}^{-1/2}\mathbf{C}\mathbf{L}^{-1/2}$, which has unit diagonal block-matrices and off-diagonal blocks $\{(1/n)\mathbf{D}_q^{-1/2}\mathbf{N}_{qq'}\mathbf{D}_{q'}^{-1/2}\}$, where $\mathbf{D}_q = (1/n)\mathbf{L}_q$, precisely of the form (Equation 3.10) required by CA, apart from the absence here of the centering, which merely involves calculating the trivial solution, as explained previously. With CA in mind, we may want to approximate $\mathbf{L}^{-1/2}\mathbf{C}\mathbf{L}^{-1/2}$ to give a simultaneous approximation of all the two-way contingency tables. Thus, the approximations of $\mathbf{Z}\mathbf{L}^{-1/2}$ and of $\mathbf{L}^{-1/2}\mathbf{C}\mathbf{L}^{-1/2}$ give rise to the same eigenvalue equations. This is an exact analogue of the ambivalence in PCA, discussed in Section 3.2.3, as to whether we are interested in approximating $\mathbf{X}$ or a correlation matrix derived from $\mathbf{X}$. The approximation of $\mathbf{L}^{-1/2}\mathbf{B}\mathbf{L}^{-1/2}$ gives a simultaneous analysis of all the contingency tables, which seems to be of greater interest than the approximation of $\mathbf{Z}\mathbf{L}^{-1/2}$. As with PCA, the unit diagonal blocks of $\mathbf{L}^{-1/2}\mathbf{B}\mathbf{L}^{-1/2}$ can be excluded from the approximation, thus leading to joint correspondence analysis, or JCA (Greenacre 1988 and Chapter 2, Section 2.3.3). The measures of goodness of fit associated with all these variants are discussed in Section 3.4.

3.4 Fit and scaling

3.4.1 *Basic results*

To establish notation, we begin with some well-known results concerning an $I \times J$ matrix $\mathbf{X}$ of rank R (usually $R = \min\{I, J\}$). For convenience, we assume that $I > J$, if necessary, when $\mathbf{X}$ is a two-way table by permuting its rows and columns; for a data matrix there are nearly always more cases than variables. All least-squares fits with which we are concerned are based on the SVD:

$$\mathbf{X} = \mathbf{U}\boldsymbol{\Sigma}\mathbf{V}^{\mathsf{T}} \tag{3.19}$$

where $\mathbf{U}$ and $\mathbf{V}$ are matrices with orthogonal columns called *singular vectors* ($\mathbf{U}^{\mathsf{T}}\mathbf{U} = \mathbf{V}^{\mathsf{T}}\mathbf{V} = \mathbf{I}$) and $\boldsymbol{\Sigma}$ is zero apart from diagonal positive *singular values* arranged in descending order: $\sigma_1 \geq \sigma_2 \geq \ldots \geq \sigma_R > 0$. The Eckart–Young theorem (1936) says that the rank-r matrix $\hat{\mathbf{X}}$ that minimizes the residual sum of squares $\|\mathbf{X} - \hat{\mathbf{X}}\|^2$ is given by $\hat{\mathbf{X}} = \mathbf{U}\boldsymbol{\Sigma}_{(r)}\mathbf{V}^{\mathsf{T}}$, differing from the SVD of $\mathbf{X}$ only in setting $\sigma_s = 0$ for $s > r$. Furthermore, because $\hat{\mathbf{X}}^{\mathsf{T}}(\mathbf{X} - \hat{\mathbf{X}}) = \mathbf{0}$, the total sum of squares can be partitioned into orthogonal components to give an analysis of variance:

$$\begin{aligned}\|\mathbf{X}\|^2 = \operatorname{trace}(\mathbf{X}^{\mathsf{T}}\mathbf{X}) &= \|\hat{\mathbf{X}}\|^2 + \|\mathbf{X} - \hat{\mathbf{X}}\|^2 \\ &= (\sigma_1^2 + \sigma_2^2 + \cdots + \sigma_r^2) + (\sigma_{r+1}^2 + \sigma_{r+2}^2 + \cdots + \sigma_R^2)\end{aligned} \tag{3.20}$$

where the two terms in parentheses are attributable to the fit and residual, respectively. The quality of the fit is usually expressed as the ratio of the fitted sum of squares to the total sum of squares:

$$\text{fit}(\mathbf{X}) = \frac{\sigma_1^2 + \sigma_2^2 + \cdots + \sigma_r^2}{\sigma_1^2 + \sigma_2^2 + \cdots + \sigma_R^2} = \frac{\Sigma_r^2}{\Sigma_R^2} \tag{3.21}$$

where the right-hand side establishes an obvious notation used in the following, where Σ (not bold) is distinguished from singular-value matrices $\boldsymbol{\Sigma}$ (boldface). The denominator, $\|\mathbf{X}\|^2$, of Equation 3.21 often simplifies in special cases. From the SVD, we have the spectral decomposition $\mathbf{X}^\mathsf{T}\mathbf{X} = \mathbf{V}\boldsymbol{\Sigma}^2\mathbf{V}^\mathsf{T}$ with eigenvalues $\boldsymbol{\Lambda} = \boldsymbol{\Sigma}^2$ and corresponding eigenvectors in the columns of $\mathbf{V}$. It follows that Equation 3.21 can be rewritten as:

$$\text{fit}(\mathbf{X}) = \Lambda_r / \Lambda_R \tag{3.22}$$

Spectral decomposition can aid computation, but our concern remains with fits to $\mathbf{X}$ and not to $\mathbf{X}^\mathsf{T}\mathbf{X}$.

3.4.2 Fits for quantitative data

When I is the number of cases and J the number of variables, then $(\mathbf{I} - \frac{1}{I}\mathbf{1}\mathbf{1}^\mathsf{T})\mathbf{X}$ gives deviations from the means to which the Eckart–Young theorem can be applied directly. Then, Equation 3.21 gives the measure of fit for PCA (and also for NLPCA and similarly for the residual matrix [Equation 3.5] of the biadditive model). If, in addition, the columns are normalized, then $\mathbf{X}^\mathsf{T}(\mathbf{I} - \frac{1}{I}\mathbf{1}\mathbf{1}^\mathsf{T})\mathbf{X}$ becomes a correlation matrix $\mathbf{R}$, with trace $(\mathbf{R}) = J$, and the PCA fit (Equation 3.22) simplifies to

$$\text{fit}(\mathbf{X}) = \Lambda_r / J \tag{3.23}$$

However, with a least-squares fit to the correlation matrix itself, the Eckart–Young theorem is based on the SVD of the correlation matrix and *its* singular values are the eigenvalues of $\mathbf{R}$, giving:

$$\text{fit}(\mathbf{R}) = \Lambda_r^2 / \Lambda_R^2 = \Sigma_r^4 / \Sigma_R^4 \tag{3.24}$$

Section 3.2.3 pointed out that fits to **R** give disparate weights to the diagonal and off-diagonal terms. The simplest way of fitting a correlation matrix while ignoring the diagonal is to use an algorithm that replaces the unit diagonal of **R** by a diagonal matrix **Δ** chosen so that the rank-r component of the SVD of $\mathbf{R} - \mathbf{I} + \mathbf{\Delta}$ is a symmetric matrix **H** with diag(**H**) = **Δ**. Then Equation 3.20 becomes

$$\|\mathbf{R} - \mathbf{I} + \mathbf{\Delta}\|^2 = \|\mathbf{\Delta}\|^2 + 2\sum\sum_{j<j'} r_{jj'}^2$$
$$= \|\mathbf{H}\|^2 + \|(\mathbf{R} - \mathbf{I} + \mathbf{\Delta}) - \mathbf{H}\|^2$$

where $\|\mathbf{H}\|^2 = \|\mathbf{\Delta}\|^2 + 2\sum\sum_{j<j'} h_{jj'}^2$ and $(\mathbf{R} - \mathbf{I} + \mathbf{\Delta}) - \mathbf{H}$ has a zero diagonal. The term $\|\mathbf{\Delta}\|^2$ cancels out to give an orthogonal analysis of variance:

$$\sum\sum_{j<j'} r_{jj'}^2 = \sum\sum_{j<j'} h_{jj'}^2 + \sum\sum_{j<j'} (r_{jj'} - h_{jj'})^2 \tag{3.25}$$

with fit $\sum\sum_{j<j'} h_{jj'}^2 / \sum\sum_{j<j'} r_{jj'}^2$. Excluding the effects of the unwanted diagonal in Equation 3.24 gives:

$$\text{fit}(\mathbf{R} - \mathbf{I} + \mathbf{\Delta}) = (\Lambda_r^2 - \|\mathbf{\Delta}\|^2)/(\Lambda_R^2 - \|\mathbf{\Delta}\|^2) \tag{3.26}$$

where now the eigenvalues are those of $\mathbf{R} - \mathbf{I} + \mathbf{\Delta}$.

The results for Equation 3.22, Equation 3.24, and Equation 3.26 are summarized in the first three rows of Table 3.2.

3.4.3 *Fit for categorical variables*

Similar results apply to two-way CA, based on Equation 3.10—the SVD of $\mathbf{D}_r^{-1/2}[(1/n)\mathbf{N} - \mathbf{r}\mathbf{c}^\mathsf{T}]\mathbf{D}_c^{-1/2}$—with fit defined as in Equation 3.22. For a Q-variable indicator matrix **Z**, the MCA solution is based on the SVD of the correspondence matrix ("**N**" divided by its total "n") equal to $(1/nQ)\mathbf{Z}$, $\mathbf{D}_r = (1/n)\mathbf{I}$, and $\mathbf{D}_c = (1/nQ)\mathbf{L}$. This gives the CA of the matrix $(1/Q^{1/2})(\mathbf{I} - 1/n\mathbf{1}\mathbf{1}^\mathsf{T})\mathbf{Z}\mathbf{L}^{-1/2}$, with fit again given by Equation 3.22 and of the derived Burt matrix by Equation 3.24. With a total of J categories, the sum of the squared singular values, called the *total inertia* in MCA, is equal to $(J - Q)/Q$, which can replace J in Equation 3.23.

One might ask how the MCA of a two-variable indicator matrix **Z** relates to the CA of the generated two-way contingency table ($Q = 2$). It can be shown (see, e.g., Greenacre 1984: chapter 5; Gower and Hand

Table 3.2 Measures of fit for some of the methods discussed.

Analysis Method	Measure of Fit	Associated SVD
PCA	$\Sigma_r^2/\Sigma_R^2 = \Lambda_r/\Lambda_R$	$\text{SVD}((\mathbf{I}-\frac{1}{I}\mathbf{1}\mathbf{1}^\mathsf{T})\mathbf{X})$
R (correlation matrix)	Λ_r^2/Λ_R^2	$\text{SVD}(\mathbf{R})$
R (without diagonal)	$(\Lambda_r^2 - \lVert\Delta\rVert)/(\Lambda_R^2 - \lVert\Delta\rVert)$	$\text{SVD}(\mathbf{R}+\boldsymbol{\Delta}-\mathbf{I})$
CA	$\Sigma_r^2/\Sigma_R^2 = \Lambda_r/\Lambda_R$	$\text{SVD}(\mathbf{D}_r^{-1/2}(\mathbf{P}-\mathbf{r}\mathbf{c}^\mathsf{T})\mathbf{D}_c^{-1/2})$
MCA (two variables, indicator matrix)	$(1+\Sigma)_r/(I+J-2)$	$\text{SVD}((1/Q^{1/2})(\mathbf{I}-\frac{1}{n}\mathbf{1}\mathbf{1}^\mathsf{T})\mathbf{Z}\mathbf{L}^{-1/2})$
MCA (two variables, Burt matrix)	$(1+\Sigma)_r^2/(I+J-2+2\Sigma_R^2)$	$\text{SVD}(\frac{1}{2}\mathbf{L}^{-1/2}\mathbf{Z}^\mathsf{T}(\mathbf{I}-\frac{1}{n}\mathbf{1}\mathbf{1}^\mathsf{T})\mathbf{Z}\mathbf{L}^{-1/2})$

Note: See text for notation. Note that the singular values in each row are those of the matrices in the final column, with agreement in some rows. For example, the values σ_k in the various Σ are the same in the last three rows. See the text for a full explanation and description of the notation.

1996: Chapter 10; and Chapter 2 in this volume) that the fit in r dimensions is

$$\text{fit}(\mathbf{Z}) = (1+\Sigma)_r \big/ (I+J-2) \tag{3.27}$$

where the singular values $\boldsymbol{\Sigma}$ are those of Equation 3.10.

We can also form the normalized Burt matrix, corrected for deviations from the means, $(1/2)\mathbf{L}^{-1/2}\mathbf{Z}^\mathsf{T}(\mathbf{I} - 1/n\mathbf{1}\mathbf{1}^\mathsf{T})\mathbf{Z}\mathbf{L}^{-1/2}$, which has singular values that are the squares of those of the indicator matrix, and the measure of fit is given by the fourth powers of singular values, as in Equation 3.24:

$$\text{fit}(\mathbf{C}) = \left(1+\Sigma\right)_r^2 \Big/ \left(I+J-2+2\Sigma_R^2\right) \tag{3.28}$$

Thus, when $Q = 2$, the measures of fit associated with the three approximations—(a) classical CA of the two-way table $\mathbf{N}$, (b) CA of the indicator matrix $\mathbf{Z}$, and (c) CA of the Burt matrix $\mathbf{C}$—all differ, but their solutions are all functions of the same singular values and singular vectors. Because each is approximating a different, albeit related, matrix, it is not valid to compare the fits. The results of Equation 3.27 and Equation 3.28 are summarized in the last two rows of Table 3.2.

Analogous to excluding the unit diagonal of a correlation matrix, we can exclude the diagonal blocks from our measures of fit. This gives $\mathbf{U}(\mathbf{I}_{(r)} + \mathbf{\Sigma}_{(r)})\mathbf{V}^{\mathsf{T}}$ as the r-dimensional fit to Equation 3.10, with the following nonorthogonal analysis of variance:

$$
\begin{array}{rc}
\text{Fitted sum of squares:} & \left(1+\Sigma\right)_r^2 \\
\text{Residual sum of squares:} & r+\Sigma_R^2-\Sigma_r^2 \\
\text{Total sum of squares:} & \Sigma_R^2
\end{array}
$$

By omitting the diagonal blocks from the Burt matrix, the fitted sum of squares has increased as a proportion of the total sum of squares, but it would be misleading to represent the fit in ratio form (Equation 3.21), which depends crucially on orthogonality. In fact, the residual sum of squares is greater than for the classical solution, which is easily recovered by accepting the vectors and adjusting their scaling. When $Q > 2$, a straightforward regression method can be used to adjust the singular values to give an improved fit to the off-diagonal blocks; details are given by Greenacre (1988) and Gower and Hand (1996: chapter 10). When $Q = 2$, this process recovers classical CA with its orthogonal analysis of variance, but for $Q > 2$, the resulting analysis of variance remains nonorthogonal. However, by replacing the unit blocks by the new fitted values and iterating, much in the manner described above for ignoring the diagonal of a correlation matrix, we arrive at joint correspondence analysis, or JCA (Greenacre 1988; see also Chapter 2, Section 2.3). At convergence, the diagonal blocks of the reconstituted matrix are the same as those for the fitted values, and the analysis of variance is orthogonal, as in Equation 3.25. Then, once again, it is valid to express the fit as a proportion of the total sum of squares (Equation 3.20). Gower and Hand (1996: p. 207) give an example comparing orthogonal and nonorthogonal fits to the Burt matrix. To sum up; be careful with

1. Comparing different least-squares criteria (e.g., $\|\mathbf{X} - \hat{\mathbf{X}}\|^2$ with $\|(\mathbf{X}^{\mathsf{T}}\mathbf{X}) - (\hat{\mathbf{X}}^{\mathsf{T}}\hat{\mathbf{X}})\|^2$)
2. The effects of diagonal blocks in the correlation and Burt matrices
3. Nonorthogonal analyses of variance (i.e., fit + residual $\neq$ total sum of squares)

3.4.4 Scaling of axes

Section 3.3.2 touched on the different scalings that can be given to singular vectors used for plotting correspondence analysis or indeed any SVD associated with a two-way array. Inevitably, different scalings affect visualizations, but recent results suggest not seriously.

We have seen that we can plot the vectors from an SVD of a matrix **X** as two sets of points with coordinates in r dimensions, $\sigma_s^{\alpha}\mathbf{u}_s, \sigma_s^{\beta}\mathbf{v}_s$ ($s = 1,2,\ldots, r$), and have used the settings $\alpha = \beta = 1/2$; $\alpha = 1$, $\beta = 0$; $\alpha = 0$, $\beta = 1$; and $\alpha = 1$, $\beta = 1$. Only when $\alpha + \beta = 1$ are the plots consistent with the SVD of **X**, and hence only then is an inner-product interpretation fully admissible. Yet it has long been observed that even a choice of scaling where $\alpha + \beta \neq 1$ usually gives visually similar maps and hence similar interpretations. In the following discussion, we try to quantify this empirical observation.

If $\hat{\mathbf{X}} = \mathbf{U}_{(r)}\mathbf{V}^{\mathsf{T}}$ is the r-dimensional Eckart–Young approximation to **X**, we are interested in how closely approximations of the set of approximations in the "γ-family" $\hat{\mathbf{X}}_{(\gamma)}=\mathbf{U}\boldsymbol{\Sigma}_{(r)}^{\gamma}\mathbf{V}^{\mathsf{T}}$, where $\gamma = \alpha + \beta$, might agree. Using a simple measure of agreement between fits for different values of γ, Gabriel (2002a) and Gower (2004) show that when $r = 2$ or $r = 3$, the agreements within the range $\frac{1}{2} < \gamma < 2$ are astonishingly good. In Section 3.4.3 we considered fits of the form $\mathbf{U}(\mathbf{I}_{(r)} + \boldsymbol{\Sigma}_{(r)})\mathbf{V}^{\mathsf{T}}$, which are in the "U/V family," i.e., with the same singular vectors but singular values not in the γ-family; for the most part, fits remain good but sometimes can be much worse than for the γ-family.

These results are encouraging, showing that most visualizations associated with the Eckart–Young approximation are not sensitive to reasonable scalings of axes. This explains why interpretations of the symmetric map in CA (where $\alpha = \beta = 1$) are not invalid to interpret as approximate biplots when—strictly speaking—only the asymmetric plots (where $\alpha = 1$, $\beta = 0$ or $\alpha = 0$, $\beta = 1$) would be true biplots giving optimal fit to the original data.

It should be noted in the visualization of categorical data, however, that the map with coordinates $\sigma_s^{\alpha}\mathbf{D}_r^{-1/2}\mathbf{u}_s, \sigma_s^{\beta}\mathbf{D}_c^{-1/2}\mathbf{v}_s (s = 1,2,\ldots,r)$ of CA, based on chi-square geometry, has neither a **U/V** nor γ-family relationship with the coordinates $\sigma_s^{\alpha}\mathbf{u}_s, \sigma_s^{\beta}\mathbf{v}_s (s = 1,2,\ldots, r)$ based on a strict least-squares approximation to the standardized residuals (Equation 3.10).

As examples of the alternative visualizations possible in the case of simple CA, we present in Figure 3.1, Figure 3.2, and Figure 3.3 a series of two-dimensional maps ($k = 1,2$) of the same data presented in Chapter 1, Table 1.1. Figure 3.1 is a biplot of the data using coordinates $\sigma_s^{\alpha}\mathbf{u}_s, \sigma_s^{\beta}\mathbf{v}_s$ and the scaling $\alpha = \beta = 1/2$, representing the standardized

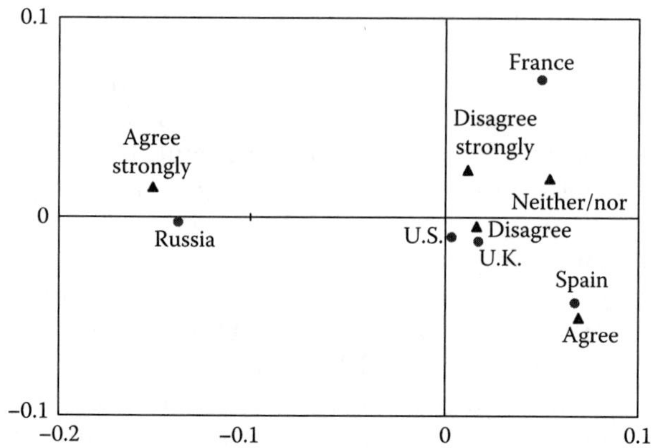

Figure 3.1 Biplot of the data in Table 3 using coordinates $\sigma_s^\alpha \mathbf{u}_s, \sigma_s^\beta \mathbf{v}_s$ and scaling $\alpha = \beta = 1/2$.

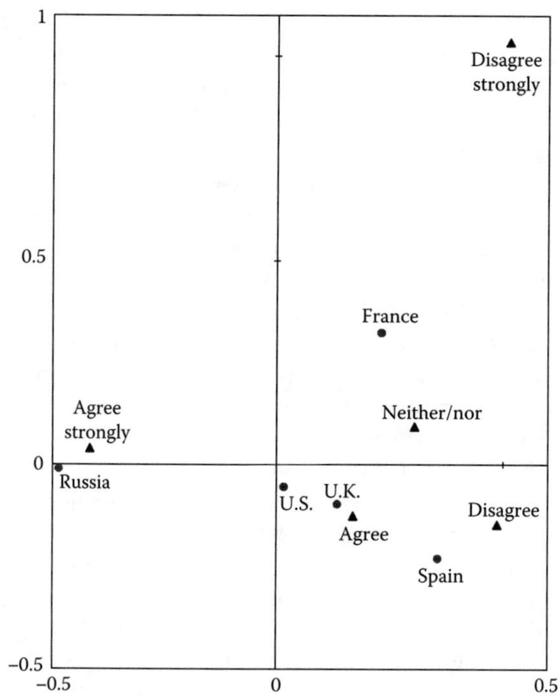

Figure 3.2 Symmetric CA map (identical to Figure 1.1 of Chapter 1) using $\sigma_s^\alpha \mathbf{D}_r^{-1/2}\mathbf{u}_s, \sigma_s^\beta \mathbf{D}_c^{-1/2}\mathbf{v}_s$ with principal coordinate scaling $\alpha = \beta = 1$, allowing interpretation in terms of chi-square distances between rows and between columns.

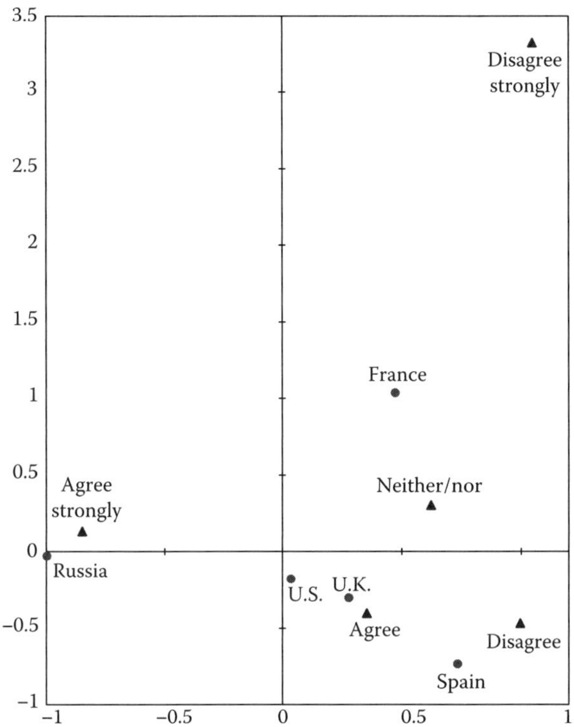

Figure 3.3 MCA map based on the Burt matrix using principal coordinates $\frac{1}{2}(1+\sigma_s)^\alpha \mathbf{D}_r^{-1/2}\mathbf{u}_s$, $\frac{1}{2}(1+\sigma_s)^\beta \mathbf{D}_c^{-1/2}\mathbf{v}_s$.

residuals as scalar products, with no nice distance interpretation between rows or between columns. Figure 3.2 is a symmetric CA map (identical to Figure 1.1 of Chapter 1) using $\sigma_s^\alpha \mathbf{D}_r^{-1/2}\mathbf{u}_s, \sigma_s^\beta \mathbf{D}_c^{-1/2}\mathbf{v}_s$ with principal coordinate scaling $\alpha = \beta = 1$, representing chi-square distances between rows and between columns. Figure 3.3 is an MCA map based on the Burt matrix using coordinates $\frac{1}{2}(1+\sigma_s)^\alpha \mathbf{D}_r^{-1/2}\mathbf{u}_s$, $\frac{1}{2}(1+\sigma_s)\mathbf{D}_c^{-1/2}\mathbf{v}_s$, which would be the principal coordinates of the rows or columns of the Burt matrix. The fits are identical in Figure 3.1 and Figure 3.2 (95.6%). The fit in Figure 3.3 is much less (38.1%) because of the unnecessary fitting of the diagonal blocks of the Burt matrix, but it is identical to Figure 3.2 as far as the interpretation along individual axes is concerned, since the only difference compared with Figure 3.2 is a scale change along each axis.

3.5 Discussion and conclusion

The common language underlying everything listed in Table 3.1, and much more, is evident, but whether this helps with understanding is questionable. Indeed, it obscures the major classifications of the table that distinguish (a) two-way tables from multivariate data matrices and (b) quantitative from categorical variables.

With quantitative variables, there is a clear statistical difference between fitting a biadditive model to a two-way table and the PCA of a data matrix. I think that it is not helpful that the SVD-based methodologies have very considerable overlap. Also unhelpful is the entanglement of PCA with approximating a correlation matrix and with factor analysis.

With categorical variables, similar considerations apply, but there are additional complexities, for example the distinction between the models expressed by Equation 3.10 and Equation 3.11. Both rely on the same SVD, but there are major differences in the coordinates and in their interpretation. I find Equation 3.10 entirely clear as an analysis of departures from independence that is expressible in terms of approximations to Pearson's chi-square (or inertia) contributions. It seems that applications to categorical data rarely use this direct approximation but prefer the CA Equation 3.11 or Equation 3.12, which leads to chi-square distance approximations. When not committed to chi-square distance, one can always appeal to the optimal-scores approach that leads to the canonical correlation derivation of Equation 3.11, as described in Chapter 2, Section 2.2. The trouble with optimal scores is that usually (but not for NLPCA) they give primarily a one-dimensional solution. To proceed to further dimensions, one has either to condition subsequent dimensions on the earlier dimensions or to embed the problem in a higher-dimensional space (see Gower et al. [2006] for an exposition). Higher-dimensional spaces have an associated inner product and hence a distance, leading back to chi-square distance. At least Equation 3.11 allows the simultaneous representation of rows and columns to give approximations to the two sets of chi-square distances and an inner product that approximates the contingency ratios. Nevertheless, I prefer the simplicity of Equation 3.10.

A multitude of problems surface with MCA. Analogous to the relationship between approximating a data matrix and a derived correlation matrix, we have the relationship between approximating an indicator matrix and the derived Burt matrix. It has already been suggested that the CA of an indicator matrix $\mathbf{Z}$ by evaluating the SVD of $\mathbf{ZL}^{-1/2}$ is suspect because chi-squared distance derived from an indicator matrix is questionable (see Chapter 2, Section 2.3.4, where this problem is addressed and scale adjustments of the solution are proposed to remedy the problem). Other distances defined on binary or multilevel categorical variables are easier to justify and can be easily analyzed and visualized by any multidimensional scaling method. The homogeneity analysis optimal-score approach fully justifies one-dimensional representations of the CA derivation but, as before, leads back to chi-squared distance to justify higher-dimensional solutions. NLPCA combines an optimal-score methodology with a simple distance and might often be preferred.

The Burt matrix is the analogue of the correlation matrix but, rather than having a simple unit diagonal, has diagonal blocks of unit matrices with many zero values. Attempts to approximate these will only degrade the solution, and I think that JCA offers the only sensible way of approximating the contingency tables inherent in a Burt matrix.

Section 3.4 mentioned the links between the CA of **Z** and the MCA of the resulting Burt matrix **C**, in the case $Q = 2$; there are no similar relationships for quantitative variables. The links are algebraically fascinating and serve as a pedagogic tool for illustrating some of the pitfalls encountered with measures of goodness of fit. Apart from that, they are best disregarded. In my opinion, despite the names, there is little statistical connection between simple CA and MCA.

These remarks notwithstanding, all the methods discussed in this chapter and elsewhere in this book are useful, as is easily demonstrated by examining the applications reported. However, to get the most out of them, it is not sufficient just to be able to operate a favorite software package. One also needs to have a deeper understanding of the methods and of possible difficulties.

CHAPTER 4

Nonlinear Principal Component Analysis and Related Techniques

Jan de Leeuw

CONTENTS

4.1 Introduction

Principal component analysis (PCA) is a multivariate data analysis technique used for many different purposes and in many different contexts. PCA is the basis for low-rank least-squares approximation of a data matrix, for finding linear combinations with maximum or minimum variance, for fitting bilinear biplot models, for computing factor-analysis approximations, and for studying regression with errors in variables. It is closely related to simple correspondence analysis (CA) and multiple correspondence analysis (MCA), which are discussed in Chapters 1 and 2 of this book, respectively.

PCA is used wherever large and complicated multivariate data sets have to be reduced to a simpler form. We find PCA in microarray analysis, medical imaging, educational and psychological testing, survey analysis, large-scale time series analysis, atmospheric sciences, high-energy physics, astronomy, and so on. Jolliffe (2002) provides a comprehensive overview of the theory and applications of classical PCA.

4.2 Linear PCA

Suppose we have measurement of n objects or individuals on m variables, collected in an $n \times m$ matrix $\mathbf{X} = \{x_{ij}\}$. We want to have an approximate representation of this matrix in p-dimensional Euclidean space. There are many seemingly different, but mathematically equivalent, ways to define PCA. We shall not dwell on each and every one of them, but we consider the one most relevant for the nonlinear generalizations of PCA we want to discuss.

Our definition of PCA is based on approximating the elements of the data matrix $\mathbf{X}$ by the inner products of vectors in p-dimensional R^p. We want to find n vectors $\mathbf{a}_i$ corresponding with the objects and m vectors $\mathbf{b}_j$ corresponding with the variables such that $x_{ij} \approx \mathbf{a}_i^{\mathsf{T}}\mathbf{b}_j$. The elements of the $n \times p$ matrix $\mathbf{A}$ are called *component scores*, while those of the $m \times p$ matrix $\mathbf{B}$ are *component loadings*.

We measure degree of approximation by using the least-squares loss function

$$\sigma(\mathbf{A},\mathbf{B}) = \sum_{i=1}^{n}\sum_{j=1}^{m}(x_{ij} - \mathbf{a}_i^{\mathsf{T}}\mathbf{b}_j)^2 \tag{4.1}$$

PCA is defined as finding the scores $\mathbf{A}$ and the loadings $\mathbf{B}$ that minimize this loss function. Another way of formulating the same problem is that we want to find p new unobserved variables, collected in the columns of $\mathbf{A}$, such that the observed variables can be approximated well by linear combinations of these unobserved variables.

As originally shown by Householder and Young (1938), the solution to this problem can be found by first computing the singular-value decomposition (SVD) $\mathbf{X} = \mathbf{K\Lambda L}^{\mathsf{T}}$, where $\mathbf{\Lambda}$ is a diagonal matrix and $\mathbf{K}^{\mathsf{T}}\mathbf{K} = \mathbf{L}^{\mathsf{T}}\mathbf{L} = \mathbf{I}$. The general solution can be established by truncating the SVD by keeping only the largest p singular values $\mathbf{\Lambda}_p$ and corresponding singular vectors $\mathbf{K}_p$ and $\mathbf{L}_p$, and then setting $\hat{\mathbf{A}} = \mathbf{K}_p\mathbf{\Lambda}_p^{1/2}\mathbf{S}$ and $\hat{\mathbf{B}} = \mathbf{L}_p\mathbf{\Lambda}_p^{1/2}\mathbf{T}$, where $\mathbf{S}$ and $\mathbf{T}$ are any two nonsingular matrices of order p satisfying $\mathbf{ST}^{\mathsf{T}} = \mathbf{I}$. The minimum value of the loss function is equal to

$$\sigma(\hat{\mathbf{A}},\hat{\mathbf{B}}) = \sum_{s=p+1}^{m} \lambda_s^2(\mathbf{X}) \tag{4.2}$$

where the $\lambda_s(\mathbf{X})$ are the ordered singular values of $\mathbf{X}$ (so that the λ_s^2 are the ordered eigenvalues of both $\mathbf{X}^{\mathsf{T}}\mathbf{X}$ and $\mathbf{XX}^{\mathsf{T}}$).

We illustrate this with an example, similar to the box problem in Thurstone (1947: 140). We use 20 rectangles and describe them in terms of seven variables (base, height, diagonal, area, circumference, ratio of base to height, and ratio of height to base). The data matrix, in which base and height are uncorrelated, is given in Table 4.1. The PCA model

Table 4.1 Rectangles.

Base	Height	Diag.	Area	Circumf.	Base/Height	Height/Base
1	1	1.41	1	4	1.00	1.00
2	2	2.82	4	8	1.00	1.00
3	3	4.24	9	12	1.00	1.00
4	4	5.66	16	16	1.00	1.00
5	5	7.07	25	20	1.00	1.00
6	6	8.49	36	24	1.00	1.00
7	7	9.90	49	28	1.00	1.00
8	8	11.31	64	32	1.00	1.00
9	9	12.73	81	36	1.00	1.00
10	10	14.14	100	40	1.00	1.00
11	10	14.87	110	42	1.10	0.91
12	9	15.00	108	42	1.33	0.75
13	8	15.26	104	42	1.63	0.62
14	7	15.65	98	42	2.00	0.50
15	6	16.16	90	42	2.50	0.40
16	5	16.76	80	42	3.20	0.31
17	4	17.46	68	42	4.25	0.23
18	3	18.24	54	42	6.00	0.17
19	2	19.10	38	42	9.50	0.11
20	1	20.02	20	42	20.00	0.05

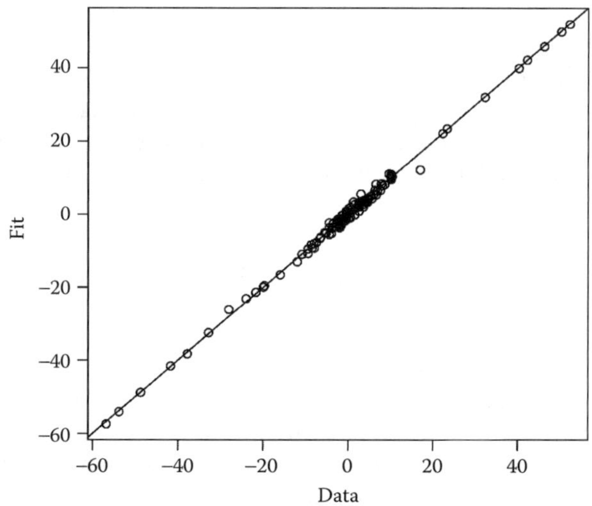

Figure 4.1 PCA fit for rectangles.

fits excellently in two dimensions (99.6% of the sum of squares is "explained"). A plot of the data and the fitted values is in Figure 4.1.

The representation in Figure 4.2 nicely reproduces the V shape of the base–height plot. In Figure 4.2 we have followed the biplot conventions from Gower and Hand (1996), in which loadings are plotted as directions on which we can project the scores. We see, for example, that the last ten rectangles have the same projection on the circumference direction, and that the base/height and height/base directions are very similar because these two variables have a high negative correlation of –0.74.

4.3 Least-squares nonlinear PCA

4.3.1 Introduction

When we talk about nonlinear PCA (NLPCA) in this chapter, we have a specific form of nonlinearity in mind. PCA is a *linear* technique, in the sense that observed variables are approximated by linear combinations of principal components. It can also be a *bilinear* technique, in the sense that elements of the data matrix are approximated by inner products, which are bilinear functions of component scores and component loadings. The nonlinearities in the forms of PCA that we discuss are introduced as nonlinear transformations of the variables, and we still preserve the basic (bi)linearity of the technique. We do not discuss

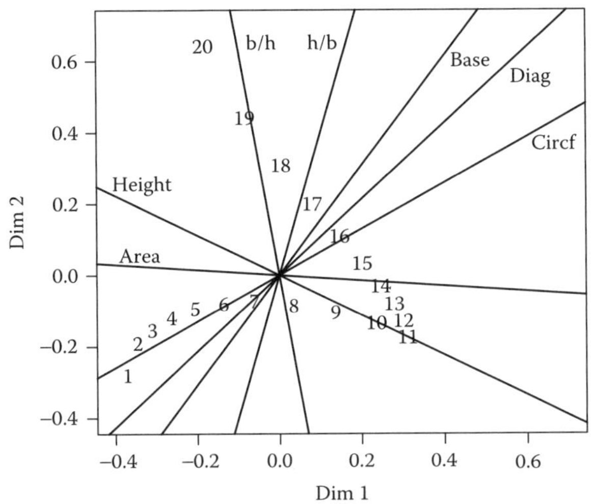

Figure 4.2 PCA solution for rectangles.

more-complicated techniques in which the observed variables are approximated by nonlinear functions of the principal components.

NLPCA can be used, for instance, if we do not have actual numerical values as our data but each variable merely ranks the objects. In other examples, similar to MCA, variables are categorical and partition the objects into a finite number of sets or categories. Binary variables (true/false, yes/no, agree/disagree, and so on) are a very common special case of both ordinal and categorical variables. And in yet other examples, variables may have numerical values, but we want to allow for the possibility of computing transformations to improve the fit of the bilinear model.

Observe that multivariate data matrices in most cases have a property called column conditionality. This means that it makes sense to compare observations within a single column, or variable, but it does not make sense to compare objects from different columns. Each variable orders or measures or classifies the objects into ranges of values specific to the variable, and those ranges may not be comparable. For preference rankings, for instance, the individuals in the experiment order the stimuli, and comparisons are only possible within an individual. This means that, for preference rankings, the individuals are actually the variables ranking the objects. This concept of conditionality is closely related to the classical psychometric distinction between Q and R techniques (Stephenson 1953).

We have seen in the previous section that we evaluate fit of PCA in p dimensions by computing the sum of squares of the residual singular values of $\mathbf{X}$ (or the sum of the residual eigenvectors of the product moment matrix $\mathbf{X}^\top\mathbf{X}$). This makes it natural to look for transformations or quantifications of the variables that minimize the same criterion. Thus, we do not minimize loss merely over component scores $\mathbf{A}$ and component loadings $\mathbf{B}$, but also over the admissible transformations of the columns of $\mathbf{X}$. The loss function becomes

$$\sigma(\mathbf{A},\mathbf{B},\mathbf{X}) = \sum_{i=1}^{n}\sum_{j=1}^{m}\left(x_{ij} - \mathbf{a}_i^\top \mathbf{b}_j\right)^2 \tag{4.3}$$

and we minimize, in addition, over $\mathbf{x}_j \in X_j$, where $X_j \subseteq \mathcal{R}^n$ are the admissible transformations for variable j. By using Equation 4.2, this is the same as finding

$$\min_{\mathbf{x}_j \in X_j} \sum_{s=p+1}^{m} \lambda_s(\mathbf{X}) \tag{4.4}$$

This form of NLPCA, in the special case of monotone transformations, has been proposed by, among others, Lingoes and Guttman (1967), Kruskal and Shepard (1974), and Roskam (1968).

The notion of admissible transformation needs some additional discussion. We have already mentioned the class of monotone transformations as an important example. But other examples can also be covered. We could, for instance, allow low-order polynomial transformations for all or some of the variables. Or, combining the two ideas, monotone polynomials. We could also look for convex or concave transformations, increasing or not. Or we could look for low-order splines on a given knot sequence, which again may or may not be restricted to be monotone. For categorical variables with a small number of categories we may simply allow the class of all possible transformations, which is also known as the class of *quantifications*, in which category labels are replaced by real numbers, as in MCA. NLPCA has been extended to these wider classes of admissible transformations by Young et al. (1978) and Gifi (1990).

All special cases of transformations mentioned so far are covered by the general restriction that the transformed variable must be in a convex cone $\mathcal{R}$ in $\mathcal{R}^n$. Convex cones are defined by the conditions that $\mathbf{x} \in \mathcal{K}$ implies $\alpha\mathbf{x} \in \mathcal{K}$ for all real $\alpha \geq 0$ and $\mathbf{x} \in \mathcal{K}$ and $\mathbf{y} \in \mathcal{K}$ implies $\mathbf{x} + \mathbf{y} \in \mathcal{K}$. It is easy to see that all classes of transformations discussed above are indeed convex cones. In fact some of them, such as the low-order polynomials and splines, are linear subspaces, which are special cones for which $\mathbf{x} \in \mathcal{K}$ implies $\alpha\mathbf{x} \in \mathcal{K}$ for all real α.

It is also clear that if a transformation $\mathbf{x}$ is in one of the cones mentioned above, then a positive linear function $\alpha\mathbf{x} + \beta$ with $\alpha \geq 0$ is in the cone as well. As a consequence of this we need to normalize our transformations, both to identify them and to prevent the trivial solution in which all transformations are identically set to zero. Another way of saying this is that we redefine our cones to consist only of centered vectors, and we want all transformations $\mathbf{x}$ to be on the unit sphere. Thus, the sets of admissible transformations X_j are of the form $\mathcal{K}_j \cap S$, where $\mathcal{K}_j$ is a convex cone of centered vectors.

The use of normalizations implies that the product moment matrix $\mathbf{X}^T\mathbf{X}$ is actually the *correlation matrix* of the variables. Thus, the optimization problem for NLPCA in p dimensions is to find admissible transformations of the variables in such a way that the sum of the $n - p$ smallest eigenvalues of the correlation matrix is minimized or, equivalently, such that the sum of the p largest eigenvalues is maximized. We write our NLPCA problem in the final form as

$$\max_{x_j \in \mathcal{K}_j \cap S} \sum_{s=1}^{p} \lambda_s(\mathbf{R}(\mathbf{X})) \tag{4.5}$$

where $\mathbf{R}(\mathbf{X})$ is the correlation matrix of the transformed variables in $\mathbf{X}$. This seems a natural and straightforward way to generalize PCA. Allowing for nonlinear transformations of the variables makes it possible to concentrate more variation in the first few principal components. Instead of looking at high-dimensional projections, we can look at low-dimensional projections together with plots of the nonlinear transformations that we compute (de Leeuw and Meulman 1986).

4.3.2 Aspects

Instead of tackling the optimization problem (Equation 4.5) directly, as is done in most earlier publications, we embed it in a much larger family of problems for which we construct a general algorithm. Let us look at Equation 4.5 in which we maximize any convex function ϕ of the correlation matrix $\mathbf{R}(\mathbf{X})$—not just the sum of the p largest eigenvalues. We call any convex real-valued function defined on the space of correlation matrices an *aspect* of the correlation matrix (de Leeuw 1988, 1990).

Of course, we first have to show that, indeed, the sum of the p largest eigenvalues is a convex function of the correlation matrix. For this we use the very useful lemma that if $f(\mathbf{x},\mathbf{y})$ is convex in $\mathbf{x}$ for every $\mathbf{y}$, then $g(\mathbf{x}) = \max_{\mathbf{y}} f(\mathbf{x},\mathbf{y})$ is also convex in $\mathbf{x}$. The sum of the p largest eigenvalues of a matrix $\mathbf{R}$ is the maximum of tr $\mathbf{L}^T\mathbf{RL}$ over all $m \times p$ matrices $\mathbf{L}$ with $\mathbf{L}^T\mathbf{L} = \mathbf{I}$. Thus, the aspect is the pointwise maximum of a family of functions that are linear, and thus convex, in $\mathbf{R}$, and the lemma applies.

We take the opportunity to give some additional examples of convex aspects that illustrate the considerable generality of our approach. A very simple aspect is the sum of the correlation coefficients. It does not use eigenvalues to measure how closely variables are related, but it does measure the strength of the overall relationships. Related aspects are the sum of even powers of the correlation coefficients, or the sum of odd powers of the absolute values of the correlation coefficients. Observe that the sum of squares of the correlation coefficients is actually equal to the sum of squares of the eigenvalues of the correlation matrix. Because the sum of the eigenvalues is a constant, maximizing the sum of squares is the same as maximizing the variance of the eigenvalues. This aspect gives another way to concentrate as much of the variation as possible in the first few principal components.

4.3.3 Algorithm

The algorithm we propose is based on the general *principle of majorization*. Majorization methods are discussed extensively in de Leeuw (1994), Heiser (1995), Lange et al. (2000), and de Leeuw and Michailidis (in press). We give only a very brief introduction.

In a majorization algorithm the goal is to minimize a general real-valued function $g(\mathbf{x})$ over $\mathbf{x} \in X$, with. Of course, maximization of g is the same as minimization of $-g$, so the introduction below also applies to maximization problems.

Majorization requires us to construct a function $f(\mathbf{x},\mathbf{y})$, defined on $X \times X$, that satisfies

$$f(\mathbf{x},\mathbf{y}) \geq g(\mathbf{x}) \quad \text{for all } \mathbf{x},\mathbf{y} \in X \tag{4.6a}$$

$$f(\mathbf{x},\mathbf{x}) = g(\mathbf{x}) \quad \text{for all } \mathbf{x} \in X \tag{4.6b}$$

Thus, for a fixed $\mathbf{y}$, $f(\mathbf{x},\mathbf{y})$ is above $g(\mathbf{x})$, and it touches $g(\mathbf{x})$ at the point $(\mathbf{y},g(\mathbf{y}))$. We say that f majorizes g.

Majorizing functions are used to construct the following iterative algorithm for minimizing $g(\mathbf{x})$. Suppose we are at step k.

Step 1 Given a value $\mathbf{x}^{(k)}$, construct a majorizing function $f(\mathbf{x},\mathbf{x}^{(k)})$.
Step 2 Set $\mathbf{x}^{(k+1)} = \underset{\mathbf{x}\in X}{\operatorname{argmin}} f(\mathbf{x},\mathbf{x}^{(k)})$.
Step 3 If $\| \mathbf{x}^{(k+1)} - \mathbf{x}^{(k)} \| < \varepsilon$ for some predetermined $\varepsilon > 0$, stop; else go to Step 1.

Now $g(\mathbf{x}^{(k+1)}) \leq f(\mathbf{x}^{(k+1)},\mathbf{x}^{(k)})$ because of majorization, and $f(\mathbf{x}^{(k+1)},\mathbf{x}^{(k)}) < f(\mathbf{x}^{(k)},\mathbf{x}^{(k)})$ because $\mathbf{x}^{(k+1)}$ minimizes the majorization function. But $f(\mathbf{x}^{(k)},\mathbf{x}^{(k)}) = g(\mathbf{x}^{(k)})$ because the majorization function touches at the current point. Thus, we have the *sandwich inequality,* which says

$$g(\mathbf{x}^{(k+1)}) \leq f(\mathbf{x}^{(k+1)},\mathbf{x}^{(k)}) < f(\mathbf{x}^{(k)},\mathbf{x}^{(k)}) = g(\mathbf{x}^{(k)})$$

and a majorization step consequently always decreases the loss function. For this algorithm to be of practical use, the majorizing function f needs to be easy to maximize, otherwise nothing substantial is gained by following this route.

We now apply majorization theory to maximizing our convex aspect, ϕ. Because we are maximizing, we need to find a minorization function. The convexity of the aspect, and the fact that a convex function is always above its tangents, gives the inequality

$$\phi(\mathbf{R}(\mathbf{X})) \geq \phi(\mathbf{R}(\mathbf{Y})) + \sum_{1\leq i\neq j\leq n}\sum \left.\frac{\partial\phi}{\partial r_{ij}}\right|_{\mathbf{R}=\mathbf{R}(\mathbf{Y})} \left(\mathbf{x}_i^{\mathsf{T}}\mathbf{x}_j - \mathbf{y}_i^{\mathsf{T}}\mathbf{y}_j\right) \tag{4.7}$$

for all matrices $\mathbf{X}$ and $\mathbf{Y}$ of normalized admissible transformations. The normalization ensures that the diagonal terms in the double sum on the right disappear.

Each step in the majorization algorithm requires us to maximize the right-hand side of Equation 4.7. We do this by *block relaxation,* that is, by maximizing over one transformation at a time, keeping the other transformations fixed at their current values (de Leeuw 1994). Thus in each iteration we solve m of these optimal scaling problems, transforming or quantifying each of the variables in turn.

By separating out the part of Equation 4.7 that depends only on $\mathbf{x}_j$, we find that each optimal scaling problem amounts to solving a least-squares problem of the form

$$\min_{\mathbf{x}_j\in\mathcal{K}_j\cap S} \left(\mathbf{x}_j - \tilde{\mathbf{x}}_j^{(k)}\right)^{\mathsf{T}}\left(\mathbf{x}_j - \tilde{\mathbf{x}}_j^{(k)}\right) \tag{4.8}$$

Here $\tilde{\mathbf{x}}_j^{(k)}$ is the current *target*, defined by

$$\tilde{\mathbf{x}}_j^{(k)} = \sum_{\ell<j} g_{j\ell}^{(k,j)} \mathbf{x}_\ell^{(k+1)} + \sum_{\ell<j} g_{j\ell}^{(k,j)} \mathbf{x}_\ell^{(k)}$$

and the matrices $\mathbf{H}^{(k,j)}$ are the partial derivatives, evaluated while updating variable j in iteration k. Thus,

$$h_{j\ell}^{(k,j)} = \left.\frac{\partial \phi}{\partial r_{j\ell}}\right|_{\mathbf{R}=\mathbf{R}(\mathbf{x}_1^{(k+1)},\ldots,\mathbf{x}_{j-1}^{(k+1)},\mathbf{x}_j^{(k)}\ldots,\mathbf{x}_m^{(k)})}$$

The formula looks complicated, but the only thing it does is keep track of the iteration indices. If we have an expression for the partial derivatives and a way to solve the least-squares problem in Equation 4.8, then we have a simple and general way to maximize the corresponding aspect. From the software point of view, we can write a high-level algorithm that uses as arguments subroutines to compute aspects and their partial derivatives. Thus, with relatively little extra work, users can plug in their own aspects.

If the aspect we use is the sum of the correlation coefficients, then all elements of $\mathbf{H}^{(k,j)}$ are equal to +1, and thus, the target is just the sum of all variables (except for the one we are updating). If the aspect is a single correlation coefficient in the matrix, say $\mathbf{r}_{j\ell}$, then the target when updating $\mathbf{x}_j$ will be $\mathbf{x}_\ell$ and vice versa. In the general case, we have to recompute the correlations and the partials after updating each variable. This can be expensive computationally. If our aspect is the classical NLPCA sum of the p largest eigenvalues, for instance, then

$$\frac{\partial \phi}{\partial \mathbf{R}} = \mathbf{L}\mathbf{L}^{\mathsf{T}}$$

with $\mathbf{L}$ the normalized eigenvectors corresponding with the p largest eigenvalues of $\mathbf{R}$. Computing the partials means solving an eigenvalue problem. De Leeuw (1990) discusses some (minor) variations of the algorithm that allow for updating all variables before recomputing the correlations and the partial derivatives.

It is also shown in de Leeuw (1990) that Equation 4.8 can be minimized by first projecting on the cone, thus ignoring the normalization constraint, and then normalizing afterward. Generally, such cone projection problems are simple to solve. In the categorical case, for instance, we merely have to compute category averages. In the monotone case, we must perform a monotone regression to project the target on the cone (de Leeuw 2005). In the polynomial case, we must solve a polynomial regression problem.

4.3.4 *Relation with multiple correspondence analysis*

MCA is a special case of our general aspect approach. It corresponds with maximizing the largest eigenvalue of the correlation matrix (and with the case in which all variables are categorical). As shown in Chapter 2, MCA solves the generalized eigenproblem for the Burt matrix. This corresponds with finding the stationary values of the ratio

$$\lambda(\mathbf{a}) = \frac{\sum_{j=1}^{m}\sum_{\ell=1}^{m}\mathbf{a}_j^{\mathsf{T}}\mathbf{C}_{j\ell}\mathbf{a}_\ell}{m\sum_{j=1}^{m}\mathbf{a}_j^{\mathsf{T}}\mathbf{C}_{jj}\mathbf{a}_j}$$

Change variables by letting $\mathbf{a}_j = \nu_j\mathbf{y}_j$, where $\mathbf{y}_j^{\mathsf{T}}\,\mathbf{C}_{jj}\mathbf{y}_j = 1$. Then

$$\lambda(\mathbf{v},\mathbf{y}) = \frac{\mathbf{v}^{\mathsf{T}}\mathbf{R}(\mathbf{y})\mathbf{v}}{m\mathbf{v}^{\mathsf{T}}\mathbf{v}}$$

where $\mathbf{R}(\mathbf{y})$ is the correlation matrix induced by the quantifications in $\mathbf{a}$. It follows that

$$\max_{\mathbf{y}}\max_{\mathbf{v}}\lambda(\mathbf{v},\mathbf{y}) = \max_{\mathbf{y}}\lambda_{max}(\mathbf{R}(\mathbf{y}))$$

which is what we wanted to show.

Thus, the dominant MCA solution gives us the quantifications maximizing the largest eigenvalue aspect. And the largest eigenvalue of the induced correlation matrix is the largest eigenvalue of the MCA problem. But what about the remaining MCA solutions? They provide additional solutions of the stationary equations for maximizing the largest eigenvalue aspect, corresponding with other nonglobal minima, local maxima, and saddle points. As was pointed out very early on by Guttman (1941) the first MCA solution should be distinguished clearly from the others, because the others correspond with suboptimal solutions of the stationary equations. In fact, each MCA eigenvector has its own associated induced correlation matrix. And each MCA eigenvalue is an eigenvalue (and not necessarily the largest one) of the correlation matrix induced by the corresponding MCA eigenvector.

It goes without saying that simple CA is the special case in which we have only two variables, and both are categorical. The correlation matrix has only one nonconstant element, and all reasonable aspects will be monotone functions of that single correlation coefficient. Maximizing the aspect will give us the maximum correlation coefficient, and the CA solutions will be the transformations solving the stationary equations of the maximum correlation problem.

4.3.5 Relation with multiple regression

Multiple regression and PCA are quite different techniques, but nevertheless there are some important relationships. Consider the PCA problem of maximizing the sum of the $m - 1$ largest eigenvalues of the correlation matrix. This is the same, of course, as minimizing the smallest eigenvalue, and thus, it can be interpreted as looking for a singularity in the transformed data matrix. This form of regression analysis dates back to Pearson (1901). It is a form of regression analysis, except that in the usual regression analysis we single out one variable as the outcome and define the rest as the predictors, and we measure singularity by finding out whether and how far the outcome variable is in the space spanned by the predictors.

More precisely, the squared multiple correlation coefficient of variable j with the remaining $m - 1$ variables can be written as

$$\phi(\mathbf{R}(\mathbf{X})) = \max_{\mathbf{b}}(1 - \mathbf{b}^\top\mathbf{R}\mathbf{b})$$

where the vector $\mathbf{b}$ is restricted to have $b_j = 1$. By the lemma we used previously, this is a convex function of $\mathbf{R}$, which can be maximized by our majorization algorithm. The partials are simply

$$\frac{\partial\phi}{\partial\mathbf{R}} = -\mathbf{b}\mathbf{b}^\top$$

This aspect can be easily extended to the sum of all m squared multiple correlation coefficients of each variable with all others, which has been discussed in the context of factor analysis by Guttman (1953) and others.

So far, we have written down our theory for the case in which we are maximizing a convex aspect. As we noted previously, the same results apply for minimizing a concave aspect. Some aspects are more naturally discussed in this form. Consider, for example, the determinant of the correlation matrix. Minimizing the determinant can also be thought of as looking for a singularity, i.e., as yet another way of approaching regression. The representation

$$\log \| \mathbf{R} \| = \min_{\Gamma \gtrsim 0} \log \| \boldsymbol{\Gamma} \| + \operatorname{tr} \boldsymbol{\Gamma}^{-1}\mathbf{R} - m,$$

where $\boldsymbol{\Gamma} \geq 0$ means we require $\boldsymbol{\Gamma}$ to be positive semidefinite, shows that the logarithm of the determinant is a concave function of the correlation matrix. Also

$$\frac{\partial \phi}{\partial \mathbf{R}} = \mathbf{R}^{-1}$$

which means that the target for updating a variable is its *image*, in the sense of Guttman (1953), the least-squares prediction of the variable from all others. Minimizing the determinant can be done by sequentially projecting images on cones of admissible transformations.

4.3.6 *Bilinearizability*

There is more that can be said about the relationship between MCA and maximizing the correlation aspects of NLPCA. Most of the theory we discuss here is taken from de Leeuw (1982), Bekker and de Leeuw (1988), de Leeuw (1988), and de Leeuw et al. (1999).

Let us start by looking at the condition of *bilinearizability* of regressions. This means that we can find transformation of the variables (in our class of admissible transformations) such that all bivariate regressions are exactly linear. In the case of m categorical variables with Burt table $\mathbf{C}$, this means that the system of bilinearizability equations

$$\mathbf{C}_{j\ell}\mathbf{y}_{\ell} = r_{j\ell}\mathbf{C}_{jj}\mathbf{y}_{j} \tag{4.9}$$

has a solution, normalized by $\mathbf{y}_j^T \mathbf{C}_{jj}\mathbf{y}_j = 1$ for all j. The corresponding induced correlation matrix $\mathbf{R}(\mathbf{y})$ has m eigenvalues λ_s and m corresponding normalized eigenvectors $\mathbf{v}_s$. We can now define the m vectors $\mathbf{a}_{js} = v_{js}\mathbf{y}_j$, and we find

$$\sum_{l=1}^{m} \mathbf{C}_{j\ell}\mathbf{a}_{\ell s} = \sum_{l=1}^{m} \mathbf{C}_{j\ell}\mathbf{y}_{\ell}v_{\ell s} = \mathbf{C}_{jj}\mathbf{y}_{j}\sum_{l=1}^{m} r_{j\ell}v_{\ell s} = \lambda_s v_{js}\mathbf{C}_{jj}\mathbf{y}_{j} = \lambda_s \mathbf{C}_{jj}\mathbf{a}_{js}$$

In other words, for each s the vector $\mathbf{a}_s$ defines a solution to the MCA problem, with eigenvalue λ_s, and each of these m solutions induces the same correlation matrix.

Bilinearizability has some other important consequences. A system of transformations that linearizes all regressions solves the stationary equations for any aspect of the correlation matrix. Thus, in a multivariate data matrix with bilinearizability, it does not matter which aspect we choose, because they will all give the same transformations. Another important consequence of bilinearizability is that the correlation coefficients computed by maximizing an aspect have the same standard errors as the correlation coefficients computed from known scores. This means, for example, that we can apply the asymptotically distribution-free methods of structural equation programs to optimized correlation matrices, and they will still compute the correct tests and standard errors if the data are bilinearizable (or a sample from a bilinearizable distribution).

4.3.7 Complete bilinearizability

It may be the case that there is a second set of transformations $\bar{\mathbf{y}}_j$ that satisfies Equation 4.9. Again, such a set generates m additional MCA solutions, all inducing the same correlation matrix. Moreover, $\mathbf{y}_j^\mathsf{T}\mathbf{C}_{jj}\bar{\mathbf{y}}_j = 0$ for all j, so the second set is orthogonal to the first for each variable separately. And there may even be more sets. If bilinearizability continues to apply, we can build up all MCA solutions from the solutions to Equation 4.9 and the eigenvectors of the induced correlation matrices. Another way of thinking about this is that we solve $\binom{m}{2}$ simple CA problems for each of the subtables of the Burt matrix. Equation 4.9 then says that if we have complete bilinearizability, we can patch these CA solutions together to form the MCA solution.

More precisely, suppose $\mathbf{C}$ is a Burt matrix and $\mathbf{D}$ is its diagonal. We have complete bilinearizability if there are matrices $\mathbf{K}_j$ such that $\mathbf{K}_j^\mathsf{T}\mathbf{C}_{jj}\mathbf{K}_j = \mathbf{I}$ for each j and $\mathbf{K}_j^\mathsf{T}\mathbf{C}_{j\ell}\mathbf{K}_\ell$ is diagonal for each j and ℓ. Remember that the direct sum of matrices stacks those matrices in the diagonal submatrices of a large matrix, which has all its nondiagonal submatrices equal to zero. If $\mathbf{K}$ is the direct sum of the $\mathbf{K}_j$, then $\mathbf{K}^\mathsf{T}\mathbf{D}\mathbf{K} = \mathbf{I}$ while $\mathbf{E} = \mathbf{K}^\mathsf{T}\mathbf{C}\mathbf{K}$ has the same structure as the Burt matrix, but all submatrices $\mathbf{E}_{j\ell}$ are now diagonal. This means there is a permutation matrix $\mathbf{P}$ such that $\mathbf{P}^\mathsf{T}\mathbf{K}^\mathsf{T}\mathbf{C}\mathbf{K}\mathbf{P}$ is the direct sum of correlation matrices. The first correlation matrix contains all (1,1) elements of the $\mathbf{E}_{j\ell}$, the second correlation matrix contains all (2,2) elements, and so on.

By making $\mathbf{L}$ the direct sum of the matrices of eigenvectors of these correlation matrices, we see that $\mathbf{L}^T\mathbf{P}^T\mathbf{K}^T\mathbf{CKPL}$ is diagonal, while $\mathbf{L}^T\mathbf{P}^T\mathbf{K}^T\mathbf{DKPL} = \mathbf{I}$. Thus the matrix $\mathbf{KPL}$ contains all the MCA solutions and gives a complete eigendecomposition of the Burt matrix.

This may be somewhat abstract, so let us give an important application. Suppose we perform an MCA of a standard multivariate normal, with correlation matrix $\mathbf{\Gamma}$. Because all bivariate regressions are linear, the linear transformations of the variables are a bilinearizable system, with correlation matrix $\mathbf{\Gamma}$. But the quadratic Hermite–Chebyshev polynomials are another bilinearizable system, with correlation matrix $\mathbf{\Gamma}^{(2)}$, the squares of the correlation coefficients, and so on. Thus we see that applying MCA to a multivariate normal will give m solutions consisting of polynomials of degree d, where the eigenvalues are those of $\mathbf{\Gamma}^{(d)}$, for all $d = 1,2,\ldots$.

In standard MCA we usually order the eigenvalues and keep the largest ones, often the two largest ones. The largest eigenvalue for the multivariate normal is always the largest eigenvalue of $\mathbf{\Gamma}$, but the second largest eigenvalue can be either the second largest eigenvalue of $\mathbf{\Gamma}$ or the largest eigenvalue of $\mathbf{\Gamma}^{(2)}$. If the second largest eigenvalue in the MCA is the largest eigenvalue of $\mathbf{\Gamma}^{(2)}$, then for each variable the first transformation will be linear and the second will be quadratic, which means we will find horseshoes (Van Rijckevorsel 1987) in our scatter plots. There is an example in Gifi (1990: 382–384) where two-dimensional MCA takes both its transformations from $\mathbf{\Gamma}$, which means it finds the usual NLPCA solution.

Our analysis shows clearly what the relationships are between MCA and NLPCA. In PCA we find a single set of transformations and a corresponding induced correlation matrix that is optimal in terms of an aspect. In MCA we find multiple transformations, each with its own corresponding induced correlation matrix. Only in the case of complete bilinearizability (such as is obtained in the multivariate normal) can we relate the two solutions because they are basically the same solution. MCA, however, presents the solution in a redundant and confusing manner. This gives a more precise meaning to the warning by Guttman (1941) that the additional dimensions beyond the first one in MCA should be interpreted with caution.

4.3.8 Examples of NLPCA

Our first data set is the GALO (*Groninger Afsluitingsonderzoek Lager Onderwijs*) data, taken from Peschar (1975). The objects (individuals) are 1290 school children in the sixth grade of an elementary school in the city of Groningen (the Netherlands) in 1959. The four variables and their categories are

1. Gender: (a) boys, (b) girls
2. IQ: values between 60 and 144, categorized into nine subcategories
3. Teacher's advice: (a) no further education, (b) extended primary education, (c) manual-labor education, (d) agricultural education, (e) general education, (f) secondary school for girls, (g) pre-university
4. Father's profession: (a) unskilled labor, (b) skilled labor, (c) lower white collar, (d) shopkeepers, (e) middle white collar, (f) professional

We use these data to maximize a large number of different aspects of the correlation matrix. All variables are categorical, and no monotonicity or smoothness constraints are imposed. Results are in Table 4.2. Each row of the table corresponds with a different aspect that we optimize, and thus with a different correlation matrix. The table gives the four eigenvalues of the induced correlation matrix, and in the final column the induced correlation coefficient between IQ and Advice, which are the two

Table 4.2 GALO example with eigenvalues of correlation matrices induced by maximizing different aspects.

Aspect	λ_1	λ_2	λ_3	λ_4	r_{23}
Sum of correlations	2.147	0.987	0.637	0.229	0.767
Sum of squared correlations	2.149	0.998	0.648	0.204	0.791
Sum of cubed correlations	2.139	0.934	0.730	0.198	0.796
Largest eigenvalue	2.157	0.950	0.682	0.211	0.784
Sum of two largest eigenvalues	1.926	1.340	0.535	0.198	0.795
Sum of three largest eigenvalues	1.991	1.124	0.688	0.196	0.796
Squared multiple correlation with advice	2.056	1.043	0.703	0.196	0.796
Sum of squared multiple correlations	1.961	1.302	0.538	0.199	0.795
Determinant	2.030	1.220	0.551	0.199	0.796

dominant variables in the GALO example. In the fourth row, for example, we find the eigenvalues of the correlation matrix induced by maximizing the largest eigenvalue aspect (which is also the correlation matrix induced by the first MCA dimension). And in the last row we find the eigenvalues of the correlation matrix induced by minimizing the determinant.

The largest possible eigenvalue is 2.157 (from the fourth row) and the smallest possible one is 0.196 (from the sixth row). The regression-type solutions, seeking singularities, tend to give a small value for the smallest eigenvalue. In general, the pattern of eigenvalues is very similar over the different aspects, suggesting approximate bilinearizability. We give the transformations for the aspect that maximizes the largest eigenvalue, that is, for the MCA solution, in Figure 4.3.

We can also use this example to illustrate the difference between MCA and NLPCA. Figure 4.4 has the two principal components from an MCA solution. The components come from different correlation matrices, one corresponding with linear transformations and one corresponding with quadratic ones. Thus the component scores form a horseshoe. The NLPCA solution for the same data is shown in Figure 4.5. Both components come

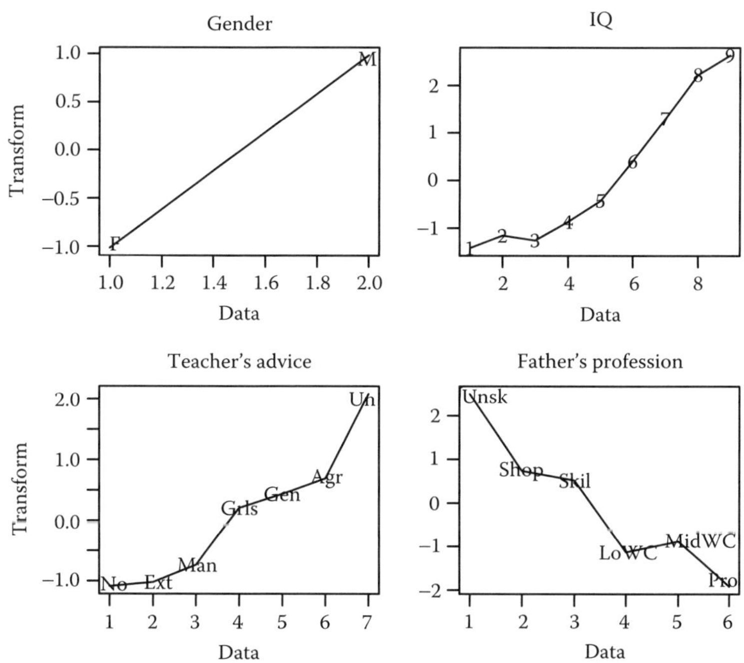

Figure 4.3. GALO data: Transformations.

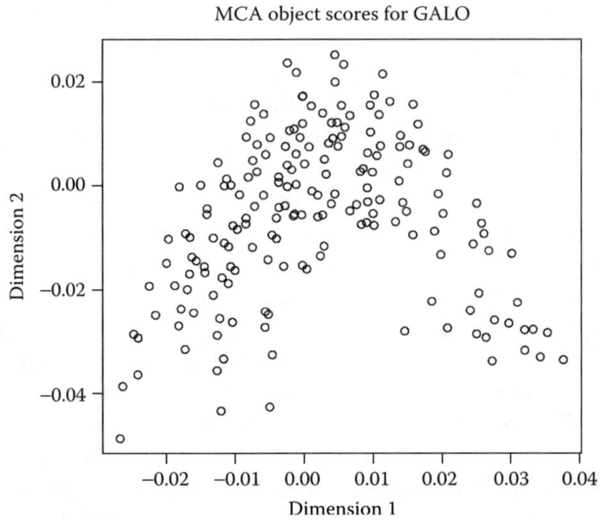

Figure 4.4 GALO data: MCA.

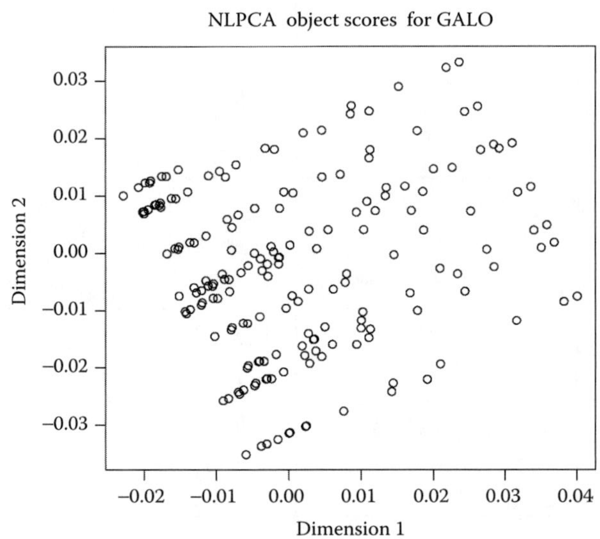

Figure 4.5 GALO data: Nonlinear PCA.

from the correlation matrix induced by the transformations in Figure 4.3. We see a completely different plot, without horseshoe, in which the discrete parallel strips of points come about because the dominant variables IQ and Advice have only a small finite number of values.

The second example of NLPCA is from Roskam (1968: 152). The Department of Psychology at the University of Nijmegen has, or had, nine different areas of research and teaching. Each of the 39 psychologists working in the department ranked all nine areas in order of relevance for their work. The areas are given in Table 4.3, and the data in Table 4.4. These are preference rank orders, and thus conditionality dictates we compute correlation coefficients among the 39 psychologists.

We first perform a linear PCA on the rank numbers, which is sometimes known as Tucker's Preference Analysis (Tucker 1960). The first two eigenvalues of (1/9) $\mathbf{R}$ are 0.374 and 0.176, which means the first two principal components capture 55% of the variation in the rank numbers. We now optimize the sum of the first two eigenvalues over all monotone transformations of the 39 rank orders. The eigenvalues increase to 0.468 and 0.297, and thus the two principal components capture 76.6% of the transformed rank numbers. For completeness, we also note that maximizing the largest eigenvalue gives 0.492 and maximizing the sum of the first three eigenvalues brings the percentage of captured variance up to 87.2%.

If we look at the plots of eigenvectors (scaled by the square roots of the eigenvalues) for the two-dimensional solution in Figure 4.6 and Figure 4.7, we see that the linear PCA produces groupings that are somewhat counterintuitive, mostly because there is so much variation left in the third and higher dimensions. The grouping in the NLPCA is clearer: psychologists in the same area are generally close together,

Table 4.3 Nine psychology areas.

Area	Plot Code
Social psychology	SOC
Educational and developmental psychology	EDU
Clinical psychology	CLI
Mathematical psychology and psychological statistics	MAT
Experimental psychology	EXP
Cultural psychology and psychology of religion	CUL
Industrial psychology	IND
Test construction and validation	TST
Physiological and animal psychology	PHY

Table 4.4 Roskam psychology subdiscipline data.

	SOC	EDU	CLI	MAT	EXP	CUL	IND	TST	PHY
1	1	5	7	3	2	4	6	9	8
2	1	3	2	7	6	4	5	8	9
3	1	6	5	3	8	2	4	7	9
4	1	5	4	7	6	2	3	8	9
5	7	1	4	3	6	8	2	9	5
6	6	1	2	5	3	7	8	4	9
7	2	1	4	5	3	8	6	7	9
8	4	1	2	8	3	5	9	6	7
9	4	1	3	5	7	6	8	2	9
10	3	1	2	4	6	8	9	7	5
11	4	1	8	3	7	6	2	5	9
12	3	2	1	5	6	8	7	4	9
13	2	9	1	6	8	3	4	5	7
14	2	7	1	4	3	9	5	6	8
15	7	2	1	3	5	8	9	4	6
16	5	7	8	1	3	9	4	2	6
17	5	9	8	1	2	7	6	3	4
18	9	6	5	1	3	7	8	2	4
19	9	6	7	2	1	8	3	4	5
20	8	3	7	2	1	9	4	5	6
21	7	2	8	5	1	9	6	4	3
22	8	7	6	3	1	9	2	5	4
23	8	6	5	2	1	9	4	7	3
24	8	7	5	2	1	9	6	4	3
25	7	3	6	2	1	9	8	4	5
26	4	7	9	5	1	8	2	3	6
27	5	6	8	2	1	9	4	7	3
28	1	8	9	2	3	7	6	4	5
29	2	5	6	4	8	1	7	3	9
30	2	5	4	3	6	1	8	7	9
31	5	3	2	9	4	1	6	7	8
32	4	5	6	2	8	7	1	3	9
33	5	7	9	3	2	8	1	4	6
34	6	3	7	2	8	5	1	4	9
35	8	5	7	4	2	9	1	3	6
36	2	6	5	4	3	7	1	8	9
37	5	8	9	2	3	7	1	4	6
38	8	7	3	4	2	9	5	6	1
39	5	6	7	2	4	9	8	3	1

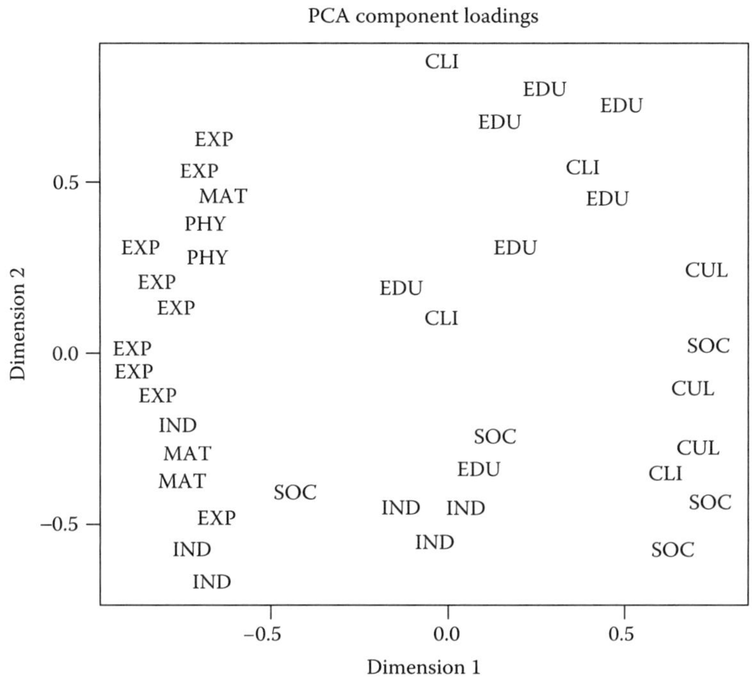

Figure 4.6 Roskam data: Linear PCA.

and there is a relatively clear distinction between qualitative and quantitative areas.

4.4 Logistic NLPCA

In the remainder of this chapter we discuss an entirely different way to define and fit NLPCA. It does not use least squares, at least not to define the loss function. The notion of correlation between variables is not used in this approach because we do not construct numerically quantified or transformed variables.

Suppose the data are categorical, as in MCA, and coded as indicator matrices. The indicator matrix $\mathbf{Z}_j$ for variable j has n rows and k_j columns. Remember that $\sum_{\ell=1}^{k_j} z_{ij\ell} = 1$ for all i and j. As in MCA, we represent both the n objects and the k_j categories of variable j as points $\mathbf{a}_i$ and $\mathbf{b}_{jl}$ in low-dimensional Euclidean space.

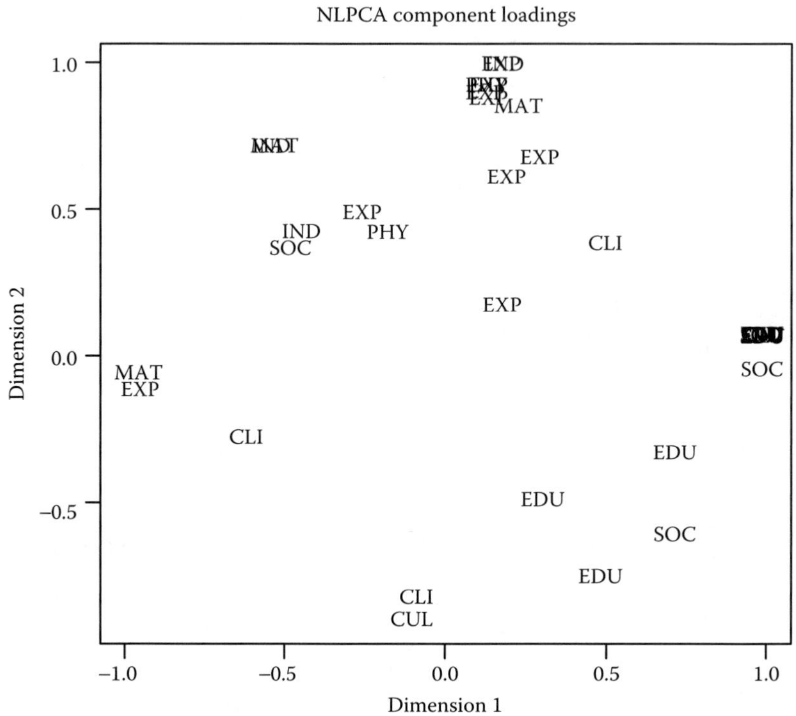

Figure 4.7 Roskam data: Nonlinear PCA.

We measure loss by using the *deviance*, or the negative log-likelihood,

$$\Delta(\mathbf{A},\mathbf{B}) = -\sum_{i=1}^{n}\sum_{j=1}^{m}\sum_{\ell=1}^{k_j} z_{ij\ell} \log \pi_{ij\ell}(\mathbf{A},\mathbf{B})$$

where

$$\pi_{ijl}(\mathbf{A},\mathbf{B}) = \frac{\exp(\eta(\mathbf{a}_i,\mathbf{b}_{j\ell}))}{\sum_{\nu=1}^{k_j} \exp(\eta(\mathbf{a}_i,\mathbf{b}_{j\nu}))}$$

For the time being, we do not specify the *combination rule* η, and we develop our results for a perfectly general combination rule. But to make matters less abstract, we can think of the inner product, $\eta(\mathbf{a}_i,\mathbf{b}_{j\ell}) = \mathbf{a}_i^{\top}\mathbf{b}_{j\ell}$, or the negative distance, $\eta(\mathbf{a}_i,\mathbf{b}_{j\ell}) = -\| \mathbf{a}_i - \mathbf{b}_{j\ell} \|$.

4.4.1 Perfect fit

In general, it will not be possible to find a perfect solution with zero deviance. We discuss under what conditions such a solution does exist. Consider the system of strict inequalities

$$\eta(\mathbf{a}_i, \mathbf{b}_{j\ell}) > \eta(\mathbf{a}_i, \mathbf{b}_{j\nu}) \tag{4.10}$$

for all (i,j,ℓ,ν) for which $z_{ij\ell} = 1$. In other words, for all i and j the largest of the $\eta(\mathbf{a}_i,\mathbf{b}_{j\nu})$ must be the one corresponding to category ℓ for which $z_{ij\ell} = 1$.

Suppose Equation 4.10 has a solution $(\hat{\mathbf{A}},\hat{\mathbf{B}})$, and suppose our combination rule η is homogeneous in the sense that $\eta(\lambda\mathbf{a}_i,\lambda\mathbf{b}_{j\ell}) = \lambda^r\eta(\mathbf{a}_i,\mathbf{b}_{j\ell})$ for some positive power r. Then by letting λ go to infinity, we see that $\pi_{ij\ell}(\lambda\hat{\mathbf{A}},\lambda\hat{\mathbf{B}})$ goes to 1 for all $z_{ij\ell} = 1$, and thus $\Delta(\lambda\hat{\mathbf{A}},\lambda\hat{\mathbf{B}})$ goes to zero. We have a perfect solution, but with all points at infinity. While generally Equation 4.10 will not be solvable, we can perhaps expect some points to move to infinity in the actual solutions we compute.

4.4.2 Geometry of combination rules

In our further analysis, we concentrate on the particular combination rule using the negative of the distance, $\eta(\mathbf{a}_i,\mathbf{b}_{j\ell}) = -\|\mathbf{a}_i - \mathbf{b}_{jl}\|$. Equation 4.10 says that we want to map objects and categories into low-dimensional space in such a way that each object is closest to the category point in which it falls.

This can be illustrated nicely by using the notion of a *Voronoi diagram* (Okabe et al. 2000). In a Voronoi diagram (for a finite number, say p, points), space is partitioned into p regions, one for each point. The cell containing the point s is the locus of all points in space that are closer to point s than to the other $p - 1$ points. Voronoi cells can be bounded and unbounded, and in the Euclidean case they are polyhedral and bounded by pieces of various perpendicular bisectors. Using the $\mathbf{b}_{j\ell}$, we can make a Voronoi diagram for each variable. Our logistic PCA, for this particular combination rule, says that each object point $\mathbf{a}_i$ should be in the correct Voronoi cell for each variable.

This type of representation is closely related to representation of categorical data in Guttman's MSA-I, discussed by Lingoes (1968). It should also be emphasized that if the data are binary, then the Voronoi diagram for a variable just consists of a single hyperplane partitioning space into two regions. Equation 4.10 now says that the "yes" responses should be on one side of the hyperplane and the "no" responses should be on the other side. This is a classical version of NLPCA, dating back to at least Coombs and Kao (1955), and used extensively in political science (Clinton et al. 2004).

To minimize the deviance, we use quadratic majorization (Böhning and Lindsay 1988; de Leeuw, 2006). We need the first and the second derivatives for a Taylor expansion of the deviance with respect to the $\eta_{ij\ell}$. We then bound the second derivatives to find the majorization function.

$$\sum_{i=1}^{n}\sum_{j=1}^{m}\sum_{\ell=1}^{k_j} [\eta_{ij\ell}(\mathbf{A},\mathbf{B}) - \tau_{ij\ell}(\tilde{A},\tilde{B})]^2 \tag{4.11a}$$

where the current target is defined by

$$\tau_{ij\ell}(\tilde{A},\tilde{B}) = \eta_{ij\ell}(\tilde{A},\tilde{B}) - 2[z_{ij\ell} - \pi_{ijl}(\tilde{A},\tilde{B})] \tag{4.11b}$$

Thus we can solve the logistic NLPCA problem by using iterative least squares. If we know how to fit $\eta_{ij\ell}(\mathbf{A},\mathbf{B})$ to a matrix by least squares, then we can also fit it logistically by maximum likelihood. In iteration k we compute the current target $\tau(A^{(k)},B^{(k)})$ by Equation 4.11b, and then we minimize (or at least improve) the least-squares loss function (Equation 4.11a) to find $A^{(k+1)}$ and $B^{(k+1)}$.

This implies immediately that for the inner product or bilinear composition rule η, we can use iterated singular-value decomposition, while for the negative distance rule we can use iterated least-squares multidimensional unfolding. In de Leeuw (2006), we give the details, and we show how the approach can easily be extended to deal with probit, instead of logit, loss functions.

As in Gifi (1990), we can construct variations on the basic technique by imposing constraints on the $\mathbf{b}_{j\ell}$. If we constrain them, for example, to be on a straight line through the origin by setting $b_{j\ell s} = z_{j\ell}\alpha_{js}$, then the bisecting hyperplanes will all be perpendicular to this line, and for each variable the space will be divided into parallel strips or bands. Objects should be in the correct strip. This is the form of NLPCA we have already discussed in the least-squares context, except that loss is measured on probabilities instead of correlations.

4.4.3 Example

The four GALO variables have a total of 24 categories, and there are 1290 individuals. Thus the metric unfolding analysis in each majorization step must fit 30,960 distances, using targets τ that can easily be negative. If we make all distances zero, which can be done by collapsing all points, then the deviance becomes 1290*(log 2 + log 9 + log 6 + log 7) = 8550. This is, in a sense, the worst possible solution, in which all probabilities are equal.

We have written some software to optimize our loss functions. It has not been tested extensively, but so far it seems to provide a convergent algorithm. It starts with the MCA solution. Remember that in MCA (Michailidis and de Leeuw 1998) we want the $\mathbf{a}_i$ to be close in the least-squares sense to the category centroids $\mathbf{b}_{j\ell}$. In the graph-drawing interpretation (de Leeuw and Michailidis 1999), where we connect each category centroid to all the objects having that category in a star pattern, we want the category stars to be small. It seems reasonable to suppose that small stars will correspond with points in the correct Voronoi cell. The MCA solution starts with a negative likelihood of 8490 and improves this to 8315.

In Figure 4.8 we draw the Voronoi cells for IQ (observe that they are all open). The category points for IQ are almost on a circle (the horseshoe closes somewhat), starting with the lowest IQ category at the bottom center, and then proceeding clockwise to the higher categories. Similar plots can be made for the other variables, but we do not present them here.

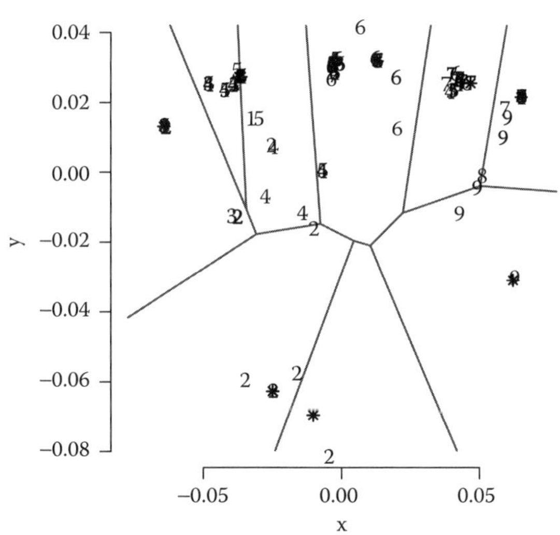

Figure 4.8 GALO data, intelligence, logistic PCA.

4.5 Discussion and conclusions

NLPCA offers many advantages over classical linear PCA because it can incorporate mixed measurement level data with ordinal, nominal, and numerical variables. It offers much flexibility in terms of admissible transformations and in terms of correlation aspects that can be maximized.

We have also seen that NLPCA has distinct advantages over MCA as an optimal scaling method. In the case of multinormal, or approximately multinormal, data, MCA will produce horseshoes and a very redundant representation of the basic information of the data. MCA can also be presented from a geometrical point of view (see Chapter 2), using the notion of chi-squared distance or minimizing the squared line length in a graph drawing of the star plots. There is no immediate generalization of the correlation aspect approach of NLPCA to these geometrical notions, although Gifi (1990) shows that NLPCA can be introduced by imposing restrictions on the location of the category quantifications in the joint MCA plots.

The solution for logistic NLPCA of the GALO data is presented somewhat tentatively because both theory and algorithm are new and will require much research and refinement. It is clear, however, that at least in principle, the basic theory and algorithms of Gifi (1990), which cover MCA, NLPCA, and various forms of nonlinear canonical analysis, can be extended to logit and probit loss functions that optimize aspects of probabilities instead of aspects of correlation coefficients.

4.6 Software Notes

There are quite a number of software options for performing the various forms of NLPCA explained in this chapter. PRINQUAL in SAS (1992) can optimize sums of the largest eigenvalues as well as the sum of correlations and the determinant aspect. Categories (Meulman and Heiser 1999) has CatPCA, which optimizes the classical eigenvalue criteria. In the R contributed packages we find the function homals from the homals package, which can perform NLPCA for categorical variables with or without ordinal constraints using the many options inherent in the Gifi system. There are also programs for NLPCA in the Guttman–Lingoes programs (Lingoes 1973).

The Gifi package for R has functions to optimize arbitrary aspects of the correlation matrix and to do the NLPCA of rank orders we applied in the Roskam example. It includes the PREHOM program discussed by Bekker and de Leeuw (1988), which finds complete bilinearizable systems of scores if they exist, and the LINEALS program discussed by de Leeuw (1988). The R code is available from the author.

Code for the logistic (and probit) versions of PCA in R is also available. The binary version has been tested quite extensively (Lewis and de Leeuw 2004) and can be compared with similar programs for IRT analysis, written mostly by educational statisticians, and for roll-call analysis, written mostly by political scientists.

SECTION II

Multiple Correspondence Analysis

CHAPTER 5

The Geometric Analysis of Structured Individuals × Variables Tables

Henry Rouanet

CONTENTS

5.1 Introduction

Geometric data analysis (GDA)—a name suggested by Patrick Suppes in 1996—is the approach to multivariate statistics initiated by J.-P. Benzécri in the 1960s, known in French-speaking literature as *Analyse des Données* (Benzécri et al. 1973; Benzécri 1982b). Beyond the "leading case" of correspondence analysis (CA), GDA includes principal component analysis (PCA), recast as a GDA method, and multiple correspondence analysis (MCA), an outgrowth of CA. The three key ideas of GDA are *geometric modeling* (constructing Euclidean clouds), *formal approach* (abstract linear algebra; GDA is properly the formal geometric approach to multivariate statistics), and *inductive philosophy* (descriptive analysis comes prior to probabilistic modeling). In applications, there are the two principles of *homogeneity* and *exhaustiveness.*

To sum up, GDA is Benzécri's tradition of multivariate statistics, with the spectral theorem as the basic mathematical tool: "All in all, doing a data analysis, in good mathematics, is simply searching eigenvectors; all the science (or the art) of it is just to find the right matrix to diagonalize." (Benzécri et al. 1973: 289). This tradition extends the geometric approach far beyond the scaling of categorical data, a fact well perceived by Greenacre (1981: 122): "The geometric approach of the French school gives a much broader view of correspondence analysis, widening its field of application to other types of data matrices apart from contingency tables."

The present chapter, in line with the book by Le Roux and Rouanet (2004a), is rooted in Benzécri's tradition. It is devoted to individuals × variables tables—a basic data set in many research studies—by either PCA (numerical variables) or MCA (categorized ones). The distinction between numerical vs. categorized matters technically, but it is not essential methodologically. As Benzécri (2003: 7) states: "One should not say: 'Continuous numerical magnitude' ≃'quantitative data' vs. 'finite number of categories' ≃ 'qualitative data.' Indeed, at the level of a statistical individual, a numerical datum is not to be taken as a rule with its full accuracy but according to its meaningfulness; and from this point of view, there is no difference in nature between age and (say) profession."

The chapter is organized as follows. I describe PCA and MCA as GDA methods in Section 5.2. Then I introduce structuring factors and structured data analysis in Section 5.3. In Sections 5.4 and 5.5, I describe two analyses of structured individuals × variables tables, embedding ANOVA (analysis of variance) techniques into the geometric framework. Finally, concluding comments are offered in Section 5.6.

5.2 PCA and MCA as geometric methods

5.2.1 PCA: from multivariate analysis to GDA

To highlight geometric data analysis, PCA is a case in point on two counts: (a) PCA preexisted as an established multivariate analysis procedure and (b) as Benzécri (1992: 57) points out: "Unlike correspondence analysis, the various methods derived from principal component analysis assign clearly *asymmetrical roles* to the individuals and the variables."

Letting n denote the number of individuals and p the number of variables, the data table analyzed by PCA is an $n \times p$ table with numerical entries. The following excerpt by Kendall and Stuart (1973: 276) nicely describes the two spaces involved: "We may set up a Euclidean space of p dimensions, one for each variable, and regard each sample set … as determining a point in it, so that our sample consists of a swarm of n points; or we may set up a space of n dimensions, one for each observation, and consider each variable in it, so that the variation is described by p vectors (lying in a p-dimensional space embedded in an n-dimensional space)." In the following discussion, these spaces will be called the "space of individuals" and the "space of variables," respectively. In conventional multivariate analysis—see, for example, Kendall and Stuart (1973) and Anderson (1958)—the space of variables is the basic one; principal variables are sought as linear combinations of initial variables having the largest variances under specified constraints. On the other hand, in PCA recast as a GDA method, the basic space is that of individuals (see, for example, Lebart and Fénelon 1971).

In PCA as a GDA method, the steps of PCA are the following:

Step 1: The distance $d(i,i')$ between individuals i and i' is defined by a quadratic form of the difference between their description profiles, possibly allowing for different weights on variables; see, for example, Rouanet and Le Roux (1993) and Le Roux and Rouanet (2004a: 131).

Step 2: The principal axes of the cloud are determined (by orthogonal least squares), and a principal subspace is retained.

Step 3: The principal cloud of individuals is studied geometrically, exhibiting approximate distances between individuals.

Step 4: The geometric representation of variables follows, exhibiting approximate correlations between variables. Drawing the *circle of correlations* has become a tradition in PCA as a GDA method.

5.2.2 MCA: a GDA method

Categorized variables are variables defined by (or encoded into) a finite set of categories; the paradigm of the individuals × categorized variables table is the $I \times Q$ table of a questionnaire in standard format, where for each question q there is a set J_q of response categories—also called *modalities*—and each individual i chooses for each question q one and only one category in the set J_q. To apply the algorithm of CA to such tables, a preliminary coding is necessary. When each question q has two categories, one of them being distinguished as "presence of property q," CA can immediately be applied after *logical coding*: "0" (absence) vs. "1" (presence); in this procedure there is no symmetry between presence and absence. The concern for symmetry—often a methodologically desirable requirement—naturally led to the coding where each categorized variable q is replaced by J_q indicator variables, that is, (0,1) variables (also known as "dummy variables"), hence (letting $J = \sum_q J_q$) producing an $I \times J$ indicator matrix to which the basic CA algorithm is applied. In this procedure, all individuals are given equal weights. In the early 1970s in France, this variant of CA gradually became a standard for analyzing questionnaires. The phrase "analyse des correspondances multiples" appears for the first time in the paper by Lebart (1975), which is devoted to MCA as a method in its own right. Special MCA software was soon developed and published (see Lebart et al. 1977).

The steps for MCA parallel the ones for PCA described above.

Step 1: Given two individuals i and i' and a question q, if both individuals choose the same response category, the part of distance due to question q is zero; if individual i chooses category j and individual i' category $j' \neq j$, the part of (squared) distance due to question q is $d_q^2(i,i') = \frac{1}{f_j} + \frac{1}{f_{j'}}$, where f_j and $f_{j'}$ are the proportions of individuals choosing j and j', respectively. The overall distance $d(i,i')$ is then defined by $d^2(i,i') = \frac{1}{Q}\sum_q d_q^2(i,i')$(see Le Roux and Rouanet 2004a). Once the distance between individuals is defined, the cloud of individuals is determined.

Steps 2 and 3: These steps are the same as in PCA (above).

Step 4: The *cloud of categories* consists of *J category points.*

Remark 1

(i) Only disagreements create distance between individuals. (ii) The smaller the frequencies of disagreement categories, the greater is the distance between individuals. Property (i) is essential; property (ii), which enhances infrequent categories, is desirable up to a certain point. Very infrequent categories of active questions need to be pooled with others; alternatively, one can attempt to put them as passive elements while managing to preserve the structural regularities of MCA; see the paper by Benali and Escofier (1987), reproduced in Escofier (2003), and the method of *specific* MCA in Le Roux (1999), Le Roux and Chiche (2004), and Le Roux and Rouanet (2004a: chap. 5).

Remark 2

There is a *fundamental property* relating the two clouds. Consider the subcloud of the individuals that have chosen category j and, hence, the mean point of this subcloud (*category mean point*); let $\bar{f}_s$ denote its sth principal coordinate (in the cloud of individuals) and g_s the sth principal coordinate of point j in the cloud of categories; then one has: $\bar{f}_s = \gamma_s\ g_s$, where γ_s denotes the sth singular value of the CA of the $I \times J$ table. This fundamental property follows from transition formulas; see Lebart et al. (1984: 94), Benzécri (1992: 410), and Le Roux and Rouanet (1998: 204). As a consequence, the derived cloud of category mean points is in a one-to-one correspondence with the cloud of category points, obtained by shrinkages by scale factors γ_s along the principal axes $s = 1, 2, \ldots, S$.

5.2.3 A strategy for analyzing individuals × variables tables

The same strategy can be applied to each table to be analyzed. The strategy is outlined in the following three phases (phrased in terms of MCA, to be adapted for PCA).

Phase 1: Construction of the individuals × variables table
- Conduct elementary statistical analyses and coding of data.
- Choose active and supplementary individuals, active and supplementary variables, and structuring factors (see Section 5.3).

Phase 2: Interpretation of axes
- Determine the eigenvalues, the principal coordinates, and the contributions of categories to axes, and then decide about how many axes to interpret.
- Interpret each of the retained axes by looking at important questions and important categories, using the contributions of categories.

- Draw diagrams in the cloud of categories, showing for each axis the most important categories, and then calculate the contributions of deviations (see Le Roux and Rouanet 1998).

Phase 3: Exploring the cloud of individuals

- Explore the cloud of individuals, in connection with the questions of interest.
- Proceed to a Euclidean classification of individuals, and then interpret this classification in the framework of the geometric space.

Each step of the strategy may be more or less elaborate, according to the questions of interest. As worked-out examples, see the "culture example" in Le Roux and Rouanet (2004a) and the data sets that are available on the Web site at http://www.math-info.univ-paris5.fr/~lerb.

5.2.4 Using GDA in survey research

In research studies, GDA (PCA or MCA) can be used to construct geometric models of individuals × variables tables. A typical instance is the analysis of questionnaires, when the set of questions is sufficiently broad and at the same time diversified enough to cover several themes of interest (among which some balance is managed), so as to lead to meaningful multidimensional representations.

In the social sciences, the work of Bourdieu and his school is exemplary of the "elective affinities" between the spatial conception of social space and geometric representations, described by Bourdieu and Saint-Martin (1978) and emphasized again by Bourdieu (2001: 70): "Those who know the principles of MCA will grasp the affinities between MCA and the thinking in terms of field."

For Bourdieu, MCA provides a representation of the two complementary faces of social space, namely the space of categories—in Bourdieu's words, the space of *properties*—and the space of individuals. Representing the two spaces has become a tradition in Bourdieu's sociology. In this connection, the point is made by Rouanet et al. (2000) that doing correspondence analyses is not enough to do "analyses à la Bourdieu," and that the following principles should be kept in mind:

1. *Representing individuals*: The interest of representing the cloud of individuals is obvious enough when the individuals are "known persons"; it is less apparent when individuals are anonymous, as in opinion surveys. When, however, there are factors structuring the individuals (education, age, etc.), the interest of depicting the individuals not as

an undifferentiated collection of points, but structured into sub-clouds, is soon realized and naturally leads to analyzing sub-clouds of individuals. As an example, see the study of the electorates in the French political space by Chiche et al. (2000).

2. *Uniting theory and methodology*: Once social spaces are constructed, the geometric model of data can lead to an *explanatory use* of GDA, bringing answers to the following two kinds of questions: How can individual positions in the social space be explained by structuring factors? How can individual positions, in turn, explain the position-takings of individuals about political or environmental issues, among others? As examples, see the studies of the French publishing space by Bourdieu (1999), of the field of French economists by Lebaron (2000, 2001), and of Norwegian society by Rosenlund (2000) and Hjellbrekke et al. (in press).

5.3 Structured data analysis

5.3.1 Structuring factors

The geometric analysis of an individuals × variables table brings out the relations between individuals and variables, but it does not take into account the structures with which the basic sets themselves may be equipped. By *structuring factors*, we mean descriptors of the two basic sets that do not serve to define the distance of the geometric space; and by *structured data*, we designate data tables whose basic sets are equipped with structuring factors. Clearly, structured data constitute the rule rather than the exception, leading to questions of interest that may be central to the study of the geometric model of data. Indeed, the set of statistical individuals, also known as units, may have to be built from basic structuring factors, a typical example being the subjects × treatments design, for which a statistical unit is defined as a pair (subject, treatment). Similarly, the set of variables may have to be built from basic structuring factors. I will exemplify such constructions for the individuals in the basketball study (see Section 5.4), and for the variables in the Education Program for Gifted Youth (EPGY) study (see Section 5.5).

In conventional statistics, there are techniques for handling structuring factors, such as analysis of variance (ANOVA)—including multivariate (MANOVA) extensions—and regression; yet, these techniques are not typically used in the framework of GDA. By *structured data analysis* we mean the integration of such techniques into GDA while preserving the GDA construction; see Le Roux and Rouanet (2004a: chap. 6).

5.3.2 From experimental to observational data

In the experimental paradigm, there is a clear distinction between *experimental factors*, or independent variables, and *dependent variables*. Statistical analysis aims at studying the *effects* of experimental factors on dependent variables. When there are several dependent variables, a GDA can be performed on them, and the resulting space takes on the status of a "geometric dependent variable."

Now turning to observational data, let us consider an educational study, for instance, where for each student, variables on various subject matters are used as active variables to "create distance" between students and thus to construct an educational space. In addition to these variables, structuring factors, such as identification characteristics (gender, age, etc.), may have been recorded; their relevance is reflected in a question such as, "How are boys and girls scattered in the educational space?" Carrying over the experimental language, one might speak of the "effect of gender," or one might prefer the more neutral language of *prediction*, hence, the question: Knowing the gender of a student ("predictor variable"), what is the position of this student in the space ("geometric variable to be predicted")? As another question of interest, suppose the results of students for some final exam are available. Taking this variable as a structuring factor on the set of individuals, one might ask: Knowing the position of a student in the space ("geometric predictor"), what is the success of this student on the exam? The geometric space is now the predictor, and the structuring factor is the variable to be predicted.

5.3.3 Supplementary variables vs. structuring factors

As a matter of fact, there is a technique in GDA that handles structured data, namely that of *supplementary variables*; see Benzécri (1992: 662), Cazes (1982), and Lebart et al. (1984). Users of GDA have long recognized that introducing supplementary variables amounts to doing regressions, and they have widely used this technique, both in a predictive and explanatory perspective. For instance, Bourdieu (1979), to build the lifestyle space of *La Distinction*, puts the age, father's profession, education level, and income as supplementary variables to demonstrate that differences in lifestyle can be explained by those status variables.

The limitations of the technique of supplementary variables become apparent, however, when it is realized that, in the space of individuals, considering a supplementary category amounts to confining attention to the mean point of a subcloud (fundamental property of MCA), ignoring the dispersion of the subcloud. Taking again *La Distinction*, this concern led Bourdieu, in his study of the upper class, to regard the fractions of this class—"the most powerful explanatory factor," as he puts it—as what we call a structuring factor in the cloud of individuals. In Diagrams 11 and 12 of *La Distinction*, the subclouds corresponding to the fractions of class are stylized as contours (for further discussion, see Rouanet et al. 2000).

To sum up, the extremely useful methodology of supplementary variables appears as a first step toward structured data analysis; a similar case can be made for the method of contributions of points and deviations developed in Le Roux and Rouanet (1998).

5.3.4 Breakdown of variances

In experimental data, the relationships between factors take the form of a *factorial design* involving relations such as nesting and crossing of factors. In observational data, even in the absence of prior design, similar relationships between structuring factors can also be defined, as discussed in Le Roux and Rouanet (1998). As in genuine experimental data, the nesting and crossing relations generate effects of factors of the following types: main effects, between-effects, within-effects, and interaction effects. In contrast, in observational data, the crossing relation is usually not orthogonal (as opposed to experimental data, where orthogonality is often ensured by design); that is, structuring factors are usually *correlated*. As a consequence, within-effects may differ from main effects. For instance, if social categories and education levels are correlated factors, the effect of the education factor within social categories may be smaller, or greater or even reversed, with respect to the main (overall) effect of the education factor ("structural effect").

Given a partition of a cloud of individuals, the mean points of the classes of the partition (category mean points) define a derived cloud, whose variance is called the between-variance of the partition. The weighted average of the variances of subclouds is called the within-variance of the partition. The overall variance of the cloud decomposes itself additively in between-variance plus within-variance. Given a set of sources of variation and of principal axes, the *double breakdown of variances* consists of calculating the parts of variance of each source on each principal axis, in the line of Le Roux and Rouanet (1984). This useful technique will be used repeatedly in the applications presented in Section 5.4 and Section 5.5.

5.3.5 Concentration ellipses

Useful geometric summaries of clouds in a principal plane are provided by *ellipses of inertia*, in the first place *concentration ellipses* (see Cramér 1946: 284). The concentration ellipse of a cloud is the ellipse of inertia such that a uniform distribution over the interior of the ellipse has the same variance as the cloud; this property leads to the ellipse with a half axis along the sth principal direction equal to $2\gamma_s$ (i.e., twice the corresponding singular value). For a normally shaped cloud, the concentration ellipse contains about 86% of the points of the cloud. Concentration ellipses are especially useful for studying families of subclouds induced by a structuring factor or a clustering procedure; see, for example, the EPGY study in Section 5.5.

Remark

In statistical inference, under appropriate statistical modeling, the family of inertia ellipses also provides *confidence ellipses* for the true mean points of clouds. For large n, the half axes of the $1 - \alpha$ confidence ellipse are $\sqrt{\chi^2_\alpha / n}\,\gamma_s$ (χ^2 with 2 d.f.); for instance, for $\alpha = .05$, they are equal to $\sqrt{5.991/n}\,\gamma_s$, and the confidence ellipse can be obtained by shrinking the concentration ellipse by a factor equal to $1.22/\sqrt{n}$. Thus, the same basic geometric construction yields descriptive summaries for subclouds and inductive summaries for mean points. For an application, see the "political space" study in Le Roux and Rouanet (2004a: 383, 388).

In the next sections, structured individuals × variables tables are presented and analyzed in two research studies: (1) *basketball* (sport); (2) *EPGY* (education). PCA was used in the first study, MCA in the second one.

Other extensive analyses of structured individuals × variables tables, following the same overall strategy, will be found in the "racism" study by Bonnet et al. (1996), in the "political space" study by Chiche et al. (2000), and in the "Norwegian field of power" study by Hjellbrekke et al. (in press). The paper by Rouanet et al. (2002) shows how regression techniques can be embedded into GDA.

5.4 The basketball study

In a study by Wolff et al. (1998), the judgments of basketball experts were recorded in the activity of selecting high-level potential players. First, an experiment was set up, where video sequences were constructed covering typical game situations; eight young players (structuring factor P) were recorded in all sequences. These sequences were submitted to nine experts (structuring factor E); each expert expressed for each player free verbal judgments about the potentialities of this player. The compound factor $I = P \times E$, made of the $8 \times 9 = 72$ (player, expert) combinations, defines the set of statistical units on which the expert judgments were recorded. Then, a content analysis of the 72 records was carried out, leading to construct 11 judgment variables about the following four aspects of performance relating to: upper body (four variables), lower body (four variables), global judgment (one variable), and play strategy (two variables: attack and defense). The basic data set can be downloaded from the site http://math-info.univ-paris5.fr/~rouanet. Hereinafter, we show how ANOVA techniques can be embedded into PCA.

Weights were allocated to the 11 standardized variables following the experts' advice, namely, 7, 3.5, 2.5, and 3 for the four aspects, respectively, yielding a total weight of 16. Then, a weighted PCA was performed on the 72×11 table. Accordingly, two axes were interpreted: axis 1 was found to be related to "dexterity," and axis 2 to "strategy" (attack and defense).

Then from the basic cloud of 72 individual points, the cloud of the eight mean points of players (indexed by factor P) and the cloud of the nine mean points of experts (factor E) were derived, and the additive cloud, i.e., the fitted cloud without interaction, was constructed. Table 5.1 shows

Table 5.1 Basketball study: Double breakdown of variances according to players (P), experts (E), interaction, and additive cloud ($P + E$).

	Variances	
	Axis 1	Axis 2
$I = P \times E$	9.298	2.104
Main P (players)	8.913	1.731
Main E (experts)	0.051	0.124
Interaction	0.335	0.250
$P + E$	8.964	1.855

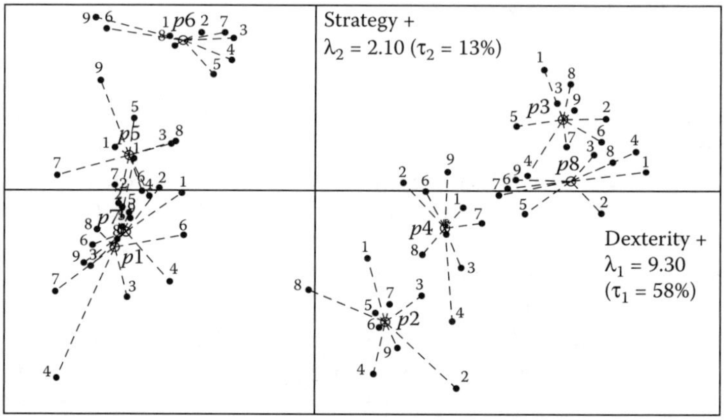

Figure 5.1 Basketball study: Basic cloud, plane 1–2; $8 \times 9 = 72$ (player, expert) points, joined to the eight player mean points.

the double breakdown of variances, for the first two axes, according to the three sources of variation: main effect of *P*, main effect of *E*, and interaction effect; it also shows the variance of the additive cloud $(P + E)$.

The table shows the large individual differences among players, the overall homogeneity of experts, and the small interaction between players and experts. Figure 5.1 shows the basic cloud, structured by the players mean points, and Figure 5.2 shows the fitted additive cloud. In Figure 5.1, the observed interaction effects between players and experts are reflected in the various locations and distances of expert points with respect to player mean points. For instance, the point of expert 6 is on the right of *p*1 but on the left of *p*6, expert 4 is on the bottom side for most players, but not for player *p*8, etc. In Figure 5.2, the pattern of experts points is the same for all players; the lines joining experts points (numbered arbitrarily from 1 to 9) exhibit the parallelism property of the additive cloud.

Remark

Structural effect and interaction effect are two different things. In the basketball study, the two factors—players and experts—are orthogonal; therefore, there is no structural effect, that is, for each axis, the sum of the two variances of the main effects *P* and *E* is exactly the variance of the additive cloud; on the other hand, the observed interaction effect between the two factors is not exactly zero.

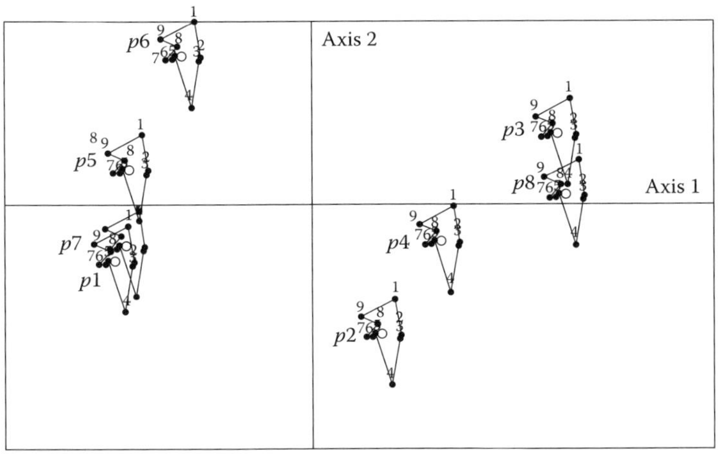

Figure 5.2 Basketball study: Additive cloud, plane 1–2; 8 × 9 = 72 (player, expert) points, showing parallelism property.

5.5 The EPGY study

The Education Program for Gifted Youth (EPGY) at Stanford University is a continuing project dedicated to developing and offering multimedia computer-based distance-learning courses in a large variety of subjects; for instance, in mathematics, EPGY offers a complete sequence from kindergarten through advanced undergraduate (see Tock and Suppes 2002).

This case study, conducted by Brigitte Le Roux and Henry Rouanet in cooperation with Patrick Suppes, deals with the detailed performances of 533 students in the third grade in the course of mathematics, with its five topics organized as *strands*, namely, Integers, Fractions, Geometry, Logic, and Measurement, and for each strand there are performance indicators of three types, namely error rates, latencies (for correct answers), and numbers of exercises (to master the concepts of the strand). The overall objective was to construct a geometric space of data exhibiting the organization of individual differences among gifted students. A specific question of interest was to investigate the trade-off between errors and latencies. In this respect, the body of existing knowledge about "ordinary students" appears to be of limited relevance, so the case study is really exploratory. The detailed study can be found in Le Roux and Rouanet (2004a: chap. 9) and on the Web site at http://epgy.stanford.edu/research/GeometricDataAnalysis.pdf.

5.5.1 Data and coding

To analyze the individuals × variables table in such a study, the set of 533 students is naturally taken as the set of individuals, whereas the set of variables is built from the two structuring factors: the set S of the five strands and the set T of the three types of performance. Crossing the two factors yields the compound factor $S \times T$, which defines the set of $5 \times 3 = 15$ variables.

In the first and indispensable phase of elementary analyses and coding of variables, we examine in detail the variables and their distributions. For error rates, the distributions differ among the strands; they are strongly asymmetric for Integers, Fractions, and Measurement, and more bell-shaped for Geometry and Logic. Latencies differ widely among strands. The number of exercises is a discrete variable. To cope with this heterogeneity, we choose MCA to construct a geometric model of data. The coding of variables into small numbers of categories (two, three, or four) aims to achieve as much homogeneity as possible, as required to define a distance between individuals.

For *error rates*, we start with a coding in three categories from 1 (low error rate) through 3 (high error rate), for each strand; hence, there are 15 categories for the five strands. Now, with this coding, category 3 has a frequency less than 1% for Integers and Fractions; therefore, we pool this category with category 2, resulting in 13 categories. For *latencies*, we take, for each strand, a four-category coding, from 1 (short) through 4 (long); hence there are $4 \times 5 = 20$ categories. For *numbers of exercises*, we code in two categories for Integers, Fractions, and Measurement, and in three categories for Geometry and Logic, from 1 (small number) through 3 (large number); hence, there are 12 categories. All in all, we get $13 + 20 + 12 = 45$ categories for 15 variables.

5.5.2 MCA and first interpretations

The basic results of MCA are the following: (1) the eigenvalues, (2) the principal coordinates and the contributions of the 45 categories to the axes (Le Roux and Rouanet 2004a: 400; Le Roux and Rouanet 2004b), (3) the principal coordinates of the 533 individuals, (4) the geometric representations of the two clouds (categories and individuals).

Eigenvalues and modified rates

There are $J - Q = 45 - 15 = 30$ eigenvalues, and the sum of eigenvalues $(J - Q)/Q$ is equal to 2. How many axes to interpret? Letting $\lambda_m = 1/Q$

Table 5.2 EPGY study: Eigenvalues, raw rates, and modified rates for the first two axes

	Axis 1	Axis 2
Eigenvalues (λ)	.3061	.2184
Raw rates of inertia	15.3%	10.9%
Benzécri's modified rates	63.1%	25.4%
Greenacre's modified rates	55.5%	21.8%

(mean eigenvalue, here .067) and $\lambda' = (\lambda - \lambda_m)^2$, we have calculated modified rates by Benzécri's formula $\lambda'/\Sigma\lambda'$, the sum being taken over the eigenvalues greater than λ_m (Benzécri 1992: 412). The modified rates indicate how the cloud deviates from a spherical cloud (with all eigenvalues equal to the mean eigenvalue). As an alternative to Benzécri's formula—which Greenacre (1993a) claims to be too optimistic — we have also calculated modified rates using Greenacre's formula (Table 5.2) (see Section 2.3.4). At any rate, one single axis is not sufficient. In the following discussion, we will concentrate on the interpretation of the first two axes.

From the basic table of the contributions of the 45 categories to axes (not reproduced here), the contributions of the 15 variables can be obtained by adding up their separate contributions (Table 5.3). The more compact contributions of the three types of performance can be similarly derived, as well as the contributions of the five strands; see Table 5.3 (contributions greater than average are in boldface).

Making use of the $S \times T$ structure, the cloud of 45 categories can be subdivided into subclouds. For instance, for each type of performance, we can construct and examine the corresponding subcloud. As an example, Figure 5.3 depicts the subcloud of the 12 numbers-of-exercises categories in plane 1–2; this figure shows for this type of performance the coherence between strands, except for Geometry.

5.5.3 *Interpretation of axes*

Interpretation of axis 1 ($\lambda_1 = .3061$)

There are 20 categories whose contributions to axis 1 are greater than average ($1/Q = 1/45 = .022 = 2.2\%$), to which we will add the low-error-rate category for Logic; these 21 categories account for 81% of the variance of the axis, on which we base the interpretation of axis 1. The opposition between high error rates (right of axis) and low error

Table 5.3 EPGY study: Contributions to axes of the 15 variables and of the three types of performance.

Contributions			
Performance	Variables	Axis 1	Axis 2
Error rate	Integers	**.083**	.043
	Fraction	.051	.034
	Geometry	**.104**	.035
	Logic	**.096**	.053
	Measurement	**.098**	.018
	(Total)	**.433**	.184
Latency	Integers	.048	**.159**
	Fraction	.043	**.157**
	Geometry	.059	**.123**
	Logic	.066	**.131**
	Measurement	.054	**.106**
	(Total)	.271	**.677**
Exercises	Integers	.065	.022
	Fraction	.028	.028
	Geometry	.023	.023
	Logic	**.097**	.044
	Measurement	**.084**	.022
	(Total)	.296	.139
Total		1	1

Note: For the 15 variables, contributions greater than 1/15 = .067 are in boldface. For the three types of performance, total contributions greater than 1/3 = 0.333 are in bold.

rates (left) accounts for 35% of the variance of axis 1 (out of the 43% accounted for by all error-rate categories). The contributions of short latency categories to the five strands are greater than the average contribution. These five categories are located on the right of origin (see Figure 5.4), exhibiting the link between high error rates and short latencies. The opposition between low error rates and short latencies accounts for 28% of the variance of axis 1, and the one between small and large numbers of exercises for accounts for 24%. The opposition

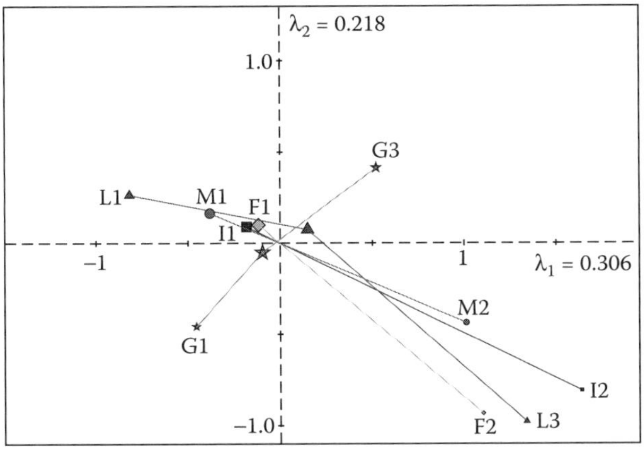

Figure 5.3 EPGY study: Space of categories, plane 1–2, subcloud of the 12 numbers-of-exercises categories. Sizes of markers reflect frequencies.

between the 7 categories on the left and the 14 ones on the right accounts for 67% of the variance of axis 1.

Note that the first axis is the *axis of error rates and numbers of exercises*. It opposes on one side low error rates and small numbers of exercises and on the other side high error rates and large numbers of exercises, the latter being associated with short latencies.

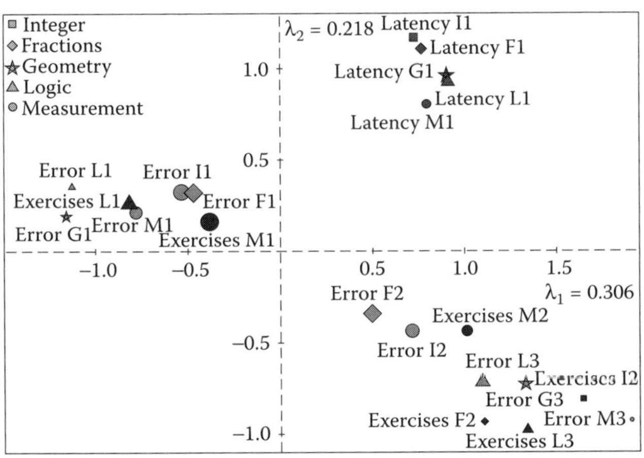

Figure 5.4 EPGY study: Cloud of categories, plane 1–2. Interpretation of axis 1: 21 categories contributing most to axis. Sizes of markers reflect frequencies.

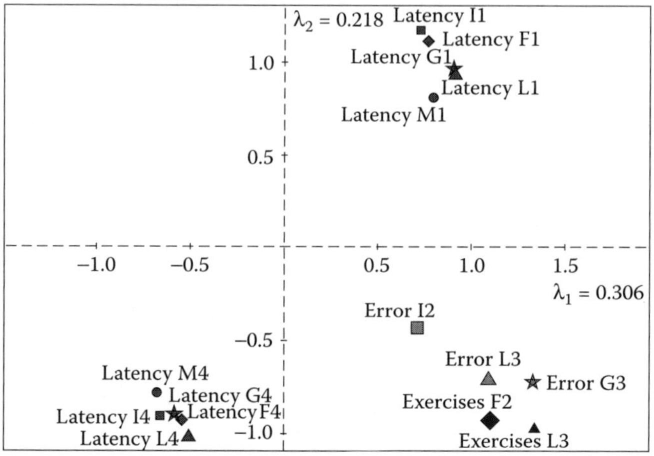

Figure 5.5 EPGY study: Cloud of categories, plane 1–2. Interpretation of axis 2: 15 categories contributing most to axis. Sizes of markers reflect frequencies.

Interpretation of axis 2 ($\lambda_2 = .2184$)

Conducting the analysis in the same way, we look for the categories contributing most to the axis; we find 15 categories that can be depicted again in plane 1-2 (see Figure 5.5). The analysis leads to the conclusion:

Note that the second axis is the *axis of latencies.* It opposes short latencies and long latencies, the latter being associated with high error rates and large numbers of exercises.

5.5.4 Cloud of individuals

The cloud of individuals (533 students) is represented in Figure 5.6; it consists of 520 observed response patterns, to which we add the following four extreme response patterns:

Pattern 11111 11111 11111 (point A, representing low error rates, short latencies, and small number of exercises)
Pattern 11111 44444 11111 (point B, representing low error rates, long latencies, and small number of exercises)
Pattern 22332 11111 22332 (point D, representing high error rates, short latencies, and large number of exercises)
Pattern 22332 44444 22332 (point C, representing high error rates, long latencies, and large number of exercises).

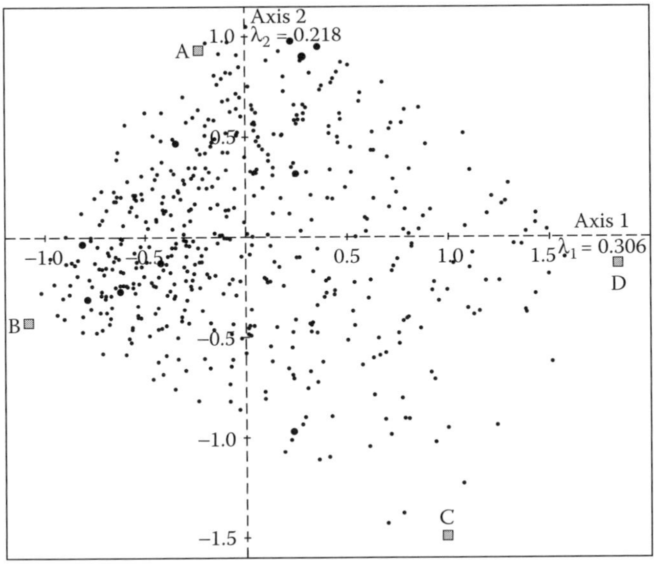

Figure 5.6 EPGY study: Cloud of individuals with extreme (unobserved) patterns A, B, C, D.

None of the 533 individuals matches any one of these extreme patterns, which will be used as landmarks for the cloud of individuals.

The individuals are roughly scattered inside the quadrilateral ABCD, with a high density of points along the side AB and a low density along the opposed side. This shows that there are many students who make few errors, whatever their latencies. On the other hand, students with high error rates are less numerous and very dispersed.

In structured individuals × variables tables, the cloud of individuals enables one to go further than the cloud of categories to investigate compound factors. As an example, let us study, for the Measurement strand, the crossing of error rates and latencies. There are 3 × 4 = 12 composite categories, and for each of them there is associated a subcloud of individuals, each one with its mean point. Figure 5.7 shows the 3 × 4 derived cloud of mean points. The profiles are approximately parallel. There are few individuals with high error rates (3) (dotted lines) whose mean points are close to side CD. As one goes down along the AB direction, latencies increase, while error rates remain about steady; as one goes down along the AD direction, error rates increase, while latencies remain about steady.

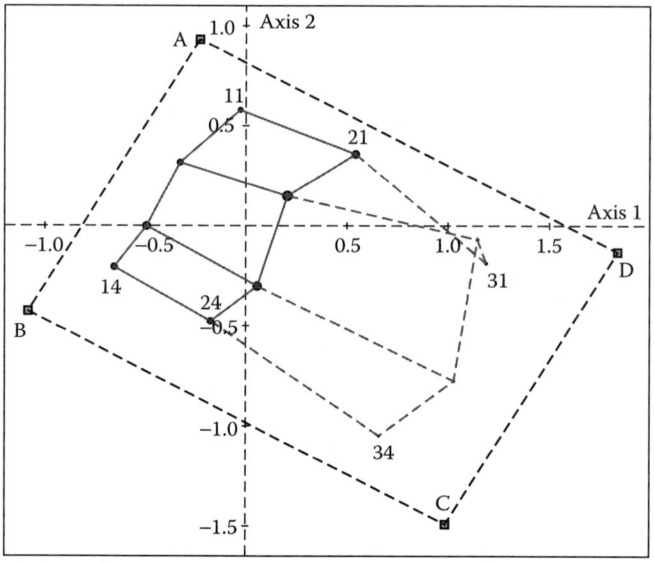

Figure 5.7 EPGY study: Space of individuals, plane 1–2. For the Measurement strand: Mean points of the 12 composite categories error × latencies. For example, point 14 is the mean point of individuals with low error rates (1) and long latencies (4) for Measurement.

To illustrate the study of external factors in the cloud of individuals, we sketch the joint analysis of age and gender, allowing for missing data (57 and 46, respectively). Crossing age (in four classes) and gender (283 boys and 204 girls) generates 4 × 2 = 8 classes. A large dispersion is found within the eight classes. In plane 1–2, the within-variance is equal to .4738, and the between-variance is only .0405 (see Table 5.4). The

Table 5.4 EPGY study: Double breakdown of variances for the crossing Age × Gender.

	Axis 1	Axis 2	Plane 1–2
Between (Age × Gender)	.0306	.0099	.0405
Age	.0301	.0067	.0368
Gender	.0000	.0012	.0012
Interaction	.0006	.0016	.0022
Within (Age × Gender)	.2683	.2055	.4738
Total variance (n = 468)	.2989	.2154	.5143

Table 5.5 EPGY study: Between- and within-variances for the five-class partition.

	Axis 1	Axis 2
Between-variance	.1964	.1277
Within-variance	.1097	.0907

variances of the two main effects and of the interaction between age and gender are also shown on this table. The crossing of age and gender is nearly orthogonal; therefore, there is virtually no structural effect; that is, for each axis, the sum of the two main-effect variances is close to the difference "between-variance of crossing minus interaction variance."

5.5.5 Euclidean classification

Distinguishing classes of gifted students was one major objective of the EPGY study. We have made a Euclidean classification, that is, an ascending hierarchical clustering with the inertia (Ward) criterion. The procedure first led to a six-class partition, from which a partition into five classes (c1, c2, c3, c4, c5) was constructed and retained as a final partition. The between- and within-variances on the first two axes of this final partition are given in Table 5.5. A synopsis of this partition is presented in Table 5.6.

There are two compact classes of highly performing students, as seen in Figure 5.8. One is class c1, close to point A, with short latencies and medium error rates; the other is class c4, close to point B, with rather low error rates (especially in Geometry and Logic) and medium-to-long latencies.

Table 5.6 EPGY study: Synopsis of final five-class partition.

	Frequencies	Error Rates	Latencies	Exercises
c1	111	—	short	small except in geometry
c2	25	high	—	rather large
c3	142	high	—	—
c4	178	low	—	rather small
c5	77	—	long	rather small

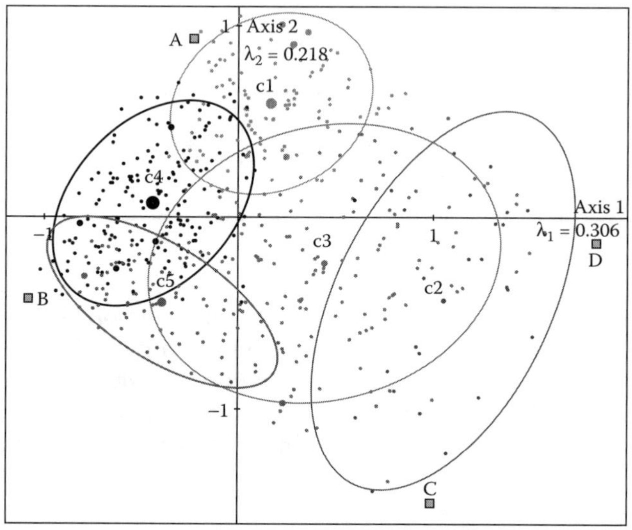

Figure 5.8 EPGY study: Final Five-class partition (C).

5.5.6 Conclusion of EPGY study

The geometric study shows the homogeneity of matters (strands), except for Geometry. It also shows how the differences among gifted students are articulated around two scales: that of error rates and numbers of exercises, and that of latencies. Within the geometric frame, differentiated clusters have been identified. The highly performing class c4, with low error rates and (comparatively) long latencies, is of special interest insofar as its profile is hardly reinforced by the current standards of educational testing!

5.6 Concluding comments

In geometric data analysis, individuals × variables tables, beyond technical differences (i.e., choice of PCA vs. MCA), can be analyzed with a common strategy, with the joint study of the cloud of variables (or of categories) and the cloud of individuals.

Structured data arise whenever individuals or variables are described by structuring factors. A cloud of individuals with structuring factors is no longer an undifferentiated set of points; it becomes a complex, meaningful geometric object. The families of subclouds induced by structuring factors can be studied not only for their mean points, but also for their dispersions. Concentration ellipses can be constructed, and various effects (between, within, interaction) can be investigated both geometrically and numerically.

Software note

To carry out the analyses, we used SPSS for the elementary descriptive statistics and the data codings. Programs in ADDAD format (written by B. Le Roux, P. Bonnet, and J. Chiche) were then used to perform PCA and MCA. Finally, starting from principal coordinates calculated with ADDAD, the exploration of clouds and the double breakdowns of variance were made with EyeLID. The EyeLID software, for the graphical investigation of multivariate data, was developed by Bernard et al. (1988); it combines two original features: a language for interrogating data (LID), which designates relevant data sets in terms of structuring factors and constitutes a command language for derivations, and the visualization ("Eye") of the clouds designated by EyeLID requests. For illustrations of the command language, see Bernard et al. (1989) and Bonnet et al. (1996). Concentration ellipses were determined by the Ellipse program by B. Le Roux and J. Chiche, which prepares a request file for the drawings done by the freeware WGNUPLOT.

An extensive (though limited in data size) version of ADDAD (Association pour le Développement et la Diffusion de l'Analyse des Données) and a DOS version of EyeLID are available on the following ftp site: ftp.math-info.univ-paris5.fr/pub/MathPsy/AGD.

The ADDAD, EyeLID, and Ellipse programs can be downloaded from the Brigitte Le Roux home page at http://www.math-info.univ-paris5.fr/~lerb (under the "Logiciels" heading).

Acknowledgments

This chapter is part of an ongoing joint project on "the comparison of the French and Norwegian social spaces," headed by B. Le Roux and O. Korsnes and funded by the Aurora-program (CNRS France and NC Norway). I thank those who have encouraged and helped me to write this chapter, especially my dearly departed friend Werner Ackermann.

CHAPTER 6

Correlational Structure of Multiple-Choice Data as Viewed from Dual Scaling

Shizuhiko Nishisato

CONTENTS

6.1 Introduction

In dual scaling (Nishisato 1980), special attention has been paid to the correlational structure of multiple-choice data. The aim of this chapter is to summarize scattered pieces of information on this topic, and then to identify and discuss future directions. We begin with an introduction to different approaches to scaling of multiple-choice data, primarily to identify characteristics peculiar to dual scaling (DS). One of them is its ability to capture nonlinear relations, as well as linear relations, between variables. Unlike principal component analysis (PCA), DS provides a correlation matrix for each component (see also Chapter 3). Thus, the relation between two variables consists of component-to-component relations, which may be linear or nonlinear. To capture such multidimensional relations, we need a special framework to discuss whether it is possible to derive a single summary statistic of correlation for a set of relations. This topic will be addressed, and a proposal will be presented in the chapter. Taking advantage of the forced-classification procedure of DS, we will transform the intervariable relation in multidimensional space into reduced space, with no loss of information. We will show that the proposed coefficient, derived from reduced space, is identical to Cramér's coefficient of association V (Cramér 1946). In the derivation of the coefficient, the statistic, called the "squared item-component correlation" in DS or the "discrimination measure" in homogeneity analysis (Gifi 1990), plays an important role in interpreting multiple-choice data. We will look at a number of interesting roles that this statistic plays in quantification theory, such as its relation to the generalized internal consistency reliability (Cronbach 1951), singular values, and eigenvalues.

6.2 Permutations of categories and scaling

In the social sciences, Likert scales (Likert 1932) are often used when response categories are ordered. For instance, values of 1, 2, 3, 4, and 5 are assigned to such categories as "strongly disagree," "moderately disagree," "neither disagree nor agree," "moderately agree," and "strongly agree," respectively. When the chosen category scores are added over a number of questions for each respondent and used as the respondent's score, the procedure is often referred to as the method of summated ratings (MSR). For this type of data, one can also use those chosen category scores as item scores and subject the data to PCA. Another alternative is DS. To provide a comparison of these three procedures, let us use a simple example. Suppose we ask 120 respondents the following eight questions:

1. RT: What do you think of the current *retirement* policy? (disagree, agree): 1, 2
2. BP: How would you rate your *blood pressure*? (low, medium, high): 1, 2, 3
3. MG: Do you get *migraines*? (rarely, sometimes, often): 1, 2, 3
4. AG: What is your *age* group? (20–34, 35–49, 50–65): 1, 2, 3
5. AX: How would you rate your daily level of *anxiety*? (low, medium, high): 1, 2, 3

6. WT: How would you rate your *weight*? (light, medium, heavy): 1, 2, 3
7. HT: What about your *height*? (short, medium, tall): 1, 2, 3
8. WK: When can you *work* best? (before breakfast, between breakfast and lunch, between lunch and supper, after supper): 1, 2, 3, 4

Item scores and total scores for the first three respondents are given in Table 6.1. Notice that in PCA unknown weights A, B, C,...,H are given to items 1 to 8, respectively, and in DS unknown option weights a, b for item 1; c, d, e for for item 2; f, g, and h for item 3; and so on until u, v, w, x for item 8.

Under MSR, no scaling is involved, and we simply sum the prescribed scale values. Under PCA, eight weights A to H are determined to maximize the between-respondent sum of squares. Under DS, 24 weights for categories are determined, also to maximize the between-respondent sum of squares. Of these three procedures, MSR is not of interest from the quantification point of view, so we will concentrate only on PCA and DS.

6.3 Principal component analysis and dual scaling

Table 6.2 lists only the first five respondents' responses, out of the sample of 120, as well as the response patterns in indicator, or dummy variable, format. PCA (left-hand side of Table 6.2) and DS (right-hand side) yield results that are fundamentally different from each other. PCA seeks only clusters of linearly related variables, such as a cluster of BP, AG, and AX (i.e., the older one gets, the higher the blood pressure and level of anxiety), and fails to capture clusters of nonlinearly related variables, such as a relation between BP and MG as follows: when blood pressure is either high or low, migraines are experienced often. DS captures nonlinear relations as well as linear ones. DS attains this extended search through the use of more dimensions (16 dimensions) than PCA (8 dimensions).

The most ingenious aspect of DS, therefore, seems to lie in the fact that it deals with nonlinearity by expanding dimensions of the space, since it scales each of the categories. In contrast, PCA handles the quantification problem by providing a single weight to all categories of each item, thus using the given category order and intervals as fixed (e.g., 1, 2, 3). In the traditional PCA framework, nonlinearity could be investigated by increasing the dimensionality of the data, for example,

Table 6.1 Item and total scores under three procedures for first three respondents.

	Resp. No.	RT	BP	MG	AG	AX	WT	HT	WK	Total
MSR	1	1	1	3	3	3	1	1	4	17
	2	2	1	3	1	3	2	3	1	16
	3	1	3	3	3	3	1	3	2	19
PCA	1	A	B	3C	3D	3E	F	G	4H	A+B+3C+3D+3E+F+G+4H
	2	2A	B	3C	D	3E	2F	3G	H	2A+B+3C+D+3E+2F+3G+H
	3	A	3B	3C	3D	3E	F	3G	2H	A+3B+3C+3D+3E+F+3G+2H
DS	1	a	c	h	k	n	o	r	x	a+c+h+k+n+o+r+x
	2	b	c	h	i	n	p	t	u	b+c+h+i+n+p+t+u
	3	a	e	h	k	n	o	t	v	a+e+h+k+n+o+t+v

Table 6.2 Likert scores and response patterns for first five respondents.

Resp.	Likert Score								Response Pattern							
No.	RT	BP	MG	AG	AX	WT	HT	WK	RT	BP	MG	AG	AX	WT	HT	WK
1	1	1	3	3	3	1	1	4	10	100	001	001	001	100	100	0001
2	2	1	3	1	3	2	3	1	01	100	001	100	001	010	001	1000
3	1	3	3	3	3	1	3	2	10	001	001	001	001	100	001	0100
4	1	3	3	3	3	1	1	3	10	001	001	001	001	100	100	0010
5	2	2	1	2	2	3	2	4	01	010	100	010	010	001	010	0001

Table 6.3 Correlation matrix under fixed-interval scoring.

RT	1.00	—	—	—	—	—	—	—
BP	−.43	1.00	—	—	—	—	—	—
MG	−.04	−.07	1.00	—	—	—	—	—
AG	−.52	.66	.23	1.00	—	—	—	—
AX	−.20	.18	.21	.22	1.00	—	—	—
WT	−.13	.17	−.58	−.02	.26	1.00	—	—
HT	.27	−.21	.10	−.30	−.23	−.30	1.00	—
WK	.07	.01	−.13	−.07	.03	.13	.02	1.00
	RT	BP	MG	AG	AX	WT	HT	WK

by augmenting the data table with columns consisting of cross-products of pairs of items, squares of items, triple products of items, and so on. However, in practice, these added variables often contribute very little and appear on minor principal axes. This provides a stark contrast to DS, where categories are scaled in relation to the data at hand, thus wasting no variables in describing the data.

Before leaving this section, let us look at the matrices of correlation for PCA (Table 6.3) and DS (Table 6.4 to Table 6.6). In DS, we obtain 16 correlation matrices, corresponding to 16 components, one for each component, of which the first three matrices are presented. For example, in Table 6.3, the correlation between BP and MG for PCA is −0.07, showing very little (linear) relation between them. The same correlation under DS in Table 6.4 is 0.99. This large difference reflects the fact that BP is nonlinearly related to MG, as captured by DS. Because PCA detects only linear relations, PCA of Table 6.3 can never

Table 6.4 Correlation matrix for the first component of DS.

RT	1.00	—	—	—	—	—	—	—
BP	.10	1.00	—	—	—	—	—	—
MG	.03	.99	1.00	—	—	—	—	—
AG	.24	.62	.57	1.00	—	—	—	—
AX	.16	.47	.50	.69	1.00	—	—	—
WT	−.04	.43	.41	.05	−.34	1.00	—	—
HT	.02	.53	.56	.14	.20	.22	1.00	—
WK	.07	.11	.10	.09	.07	.01	.09	1.00
	RT	BP	MG	AG	AX	WT	HT	WK

Table 6.5 Correlation matrix for the second component of DS.

RT	1.00	—	—	—	—	—	—	—
BP	.41	1.00	—	—	—	—	—	—
MG	–.04	.25	1.00	—	—	—	—	—
AG	.51	.62	–.14	1.00	—	—	—	—
AX	.27	.37	–.06	.29	1.00	—	—	—
WT	.23	.30	.29	.18	–.16	1.00	—	—
HT	.26	.35	.32	.29	.03	.33	1.00	—
WK	.17	.20	–.07	.16	.16	–.01	.14	1.00
	RT	BP	MG	AG	AX	WT	HT	WK

capture such a nonlinear relation, even if all possible components are considered.

Here we encounter an interesting question: How should we define correlation between two categorical variables? This is one question that this chapter is intended to answer. The current example shows that the correlation between two variables is scattered over 16 dimensions.

On the topic of linear and nonlinear correlation, there is a well-known measure of association, called Cramér's *V*, a measure that is related to the total inertia in correspondence analysis (see Chapter 1). If we calculate Cramér's *V* for the current data, we obtain a single matrix of coefficients (Table 6.7). Then, a natural question is whether each of these coefficients can be related to the corresponding 16 measurements of correlation from DS. This is another question that this chapter intends to answer.

Table 6.6 Correlation matrix for the third component of DS.

RT	1.00	—	—	—	—	—	—	—
BP	.38	1.00	—	—	—	—	—	—
MG	.03	.05	1.00	—	—	—	—	—
AG	.47	.48	.25	1.00	—	—	—	—
AX	.02	–.14	.59	.15	1.00	—	—	—
WT	–.12	.11	.56	–.02	.27	1.00	—	—
HT	–.28	–.15	.26	–.26	.19	.33	1.00	—
WK	.15	.14	.19	.08	.21	.15	.07	1.00
	RT	BP	MG	AG	AX	WT	HT	WK

Table 6.7 Correlation matrix of Cramér's *V*.

RT	1.00	—	—	—	—	—	—	—
BP	.44	1.00	—	—	—	—	—	—
MG	.00	.71	1.00	—	—	—	—	—
AG	.52	.61	.45	1.00	—	—	—	—
AX	.27	.44	.57	.54	1.00	—	—	—
WT	.24	.37	.50	.40	.32	1.00	—	—
HT	.27	.46	.45	.25	.20	.40	1.00	—
WK	.17	.18	.16	.13	.21	.15	.16	1.00
	RT	BP	MG	AG	AX	WT	HT	WK

6.4 Statistics for correlational structure of data

It is well known that two categorical variables with numbers of categories J_q and J_s can be fully accounted for by $(p - 1)$ components (solutions), where p is the smaller number of J_q and J_s and that the full information is contained in the singular values, $\rho_1, \rho_2, \ldots, \rho_{p-1}$ of the matrix of standardized residuals for the relative frequencies (see Chapter 1). Thus, we can consider singular values to describe the measure of correlation between two categorical variables.

Related to the singular value is the square of the item-total (or item-component) correlation, also called "discrimination measure" in homogeneity analysis (Gifi 1990). For a set of items, this statistic is related to the square of the singular value, or correlation ratio, η^2:

$$\eta^2 = \frac{\sum_{q=1}^{Q} r_{qt}^2}{Q} \tag{6.1}$$

where r_{qt}^2 is the squared correlation between the optimally quantified variable q and the total score t (see Chapter 2, where η^2 is denoted by λ, the principal inertia of the indicator matrix in multiple correspondence analysis). Table 6.8 contains the squared item-component (total) correlation of eight items over the first five components (out of 16). Each row sum divided by Q (= 8) is the correlation ratio η^2.

Notice that the total contribution of an item to the total space is equal to the number of categories minus 1. If the sum of the squared-item total correlation over all possible dimensions is equal to 1 (i.e., two categories), one can capture only linear relations between variables; if

Table 6.8 Squared item-total correlation for 16 components.

Dimension	RT	BP	MG	AG	AX	WT	HT	WK	Sum
1	.03	.91	.89	.57	.42	.10	.35	.03	3.30
2	.50	.71	.05	.60	.19	.21	.31	.06	2.62
3	.00	.02	.77	.08	.52	.51	.17	.17	2.24
4	.00	.02	.03	.39	.37	.81	.21	.02	1.85
5	.00	.03	.01	.02	.05	.00	.20	.81	1.12
.	.	.	.	.	.	.	.	.	.
.	.	.	.	.	.	.	.	.	.
16	.	.	.	.	.	.	.	.	.
Sum	1.00	2.00	2.00	2.00	2.00	2.00	2.00	3.00	16.00

the sum is 2 (i.e., three categories), one can describe linear and quadratic relations between variables. Thus, the sum in total space indicates the degree of the polynomial function up to which nonlinear functional relations can be captured. Therefore, if we categorize continuous variables and subject the data to DS, we can capture only a limited number of nonlinear relations between variables.

The number of dimensions required to accommodate a set of items increases as a function of Q and the number of categories for each item. Thus, if we are interested in assessing the relation between two variables using r^2_{qt}, Table 6.8 for 16 dimensions is generally not convenient. There are two ways to overcome this "size" problem. One is the procedure of forced classification of DS, which alters the distribution of the statistic r^2_{qt}. The other is to look at the contingency table of two categorical variables. Let us now briefly look at forced classification of DS.

6.5 Forced classification

Multiple correspondence analysis, or DS of multiple-choice data, determines the weights of item options to maximize the correlation ratio, defined over the entire data set. Thus, the option weights of one item are likely to change if we discard another item from the data set or add new items to the data set. This means that the sum of squares of one item for a given dimension changes, depending on what items are included in the data set. The following questions arise: Can we carry out quantification such that the characteristics of one item can uniquely be determined independently of the other items that might be included in the data set? Can we maximize the contribution of one particular item to the first component such that the correlation of items with a particularly chosen item is maximal? Forced classification offers an answer to these questions.

Forced classification of DS (Nishisato 1984; Nishisato and Gaul 1990; Nishisato and Baba 1999) is a simple procedure for discriminant analysis of categorical data. Consider the following modified indicator matrices associated with the ordinary indicator matrix $\mathbf{Z}$:

$$[\mathbf{Z}_1, \mathbf{Z}_2, \mathbf{Z}_3, \ldots, \mathbf{Z}_q, \mathbf{Z}_q, \ldots, \mathbf{Z}_q, \mathbf{Z}_{q+1}, \ldots, \mathbf{Z}_Q] \quad (6.2)$$

where $\mathbf{Z}_q$ is repeated w times for some constant w.

$$[\mathbf{Z}_1, \mathbf{Z}_2, \mathbf{Z}_3, \ldots, w\mathbf{Z}_q, \ldots, \mathbf{Z}_Q] \quad (6.3)$$

where the elements of $\mathbf{Z}_q$ are multiplied by a constant w.

These two matrices yield an identical set of singular values, except for possibly additional singular values of 0 from the first matrix. Furthermore, as the constant w increases to plus infinity, DS of the above matrices shows the following results:

1. For the first $q - 1$ components, $r^2_{qt} = 1$. These components are called *proper solutions*.
2. For the remaining components, $r^2_{qt} = 0$. These components are called *conditional solutions*.

In other words, we are forcing the scaling such that the first $q - 1$ components are analyzed in the subspace of variable q. In fact, when we consider only proper solutions, the forced classification of DS is equivalent to DS of the following two matrices:

1. $\mathbf{P}_q\mathbf{Z} = \mathbf{Z}_q(\mathbf{Z}_q^{\mathsf{T}}\mathbf{Z}_q)^{-1}\mathbf{Z}_q^{\mathsf{T}}\mathbf{Z}$
2. $\mathbf{Z}_q^{\mathsf{T}}\mathbf{Z}^*$

where $\mathbf{Z}^*$ is the matrix $\mathbf{Z}$ obtained by deleting $\mathbf{Z}_q$, that is, the contingency table of the categories of variable q and categories of the remaining variables. As for conditional solutions, one must deal with $\mathbf{I} - \mathbf{P}_q\mathbf{Z}$ to obtain them by procedure 1, and one cannot generate conditional solutions with procedure 2.

In forced classification, variable q is called the criterion, and as we noted previously, we obtain $q - 1$ proper solutions. As an example, consider forced classification with the criterion being AG, which changes the

Table 6.9 Squared correlation r_{qt}^2 under forced classification with the criterion being age.

Dimension	RT	BP	MG	AG	AX	WT	HT	WK	Sum
1	.11	.49	.31	1.00	.53	.07	.08	.02	2.61
2	.16	.30	.10	1.00	.06	.25	.04	.02	1.93
3	.02	.66	.60	.00	.12	.52	.59	.03	2.54
4	.	.	.	.00	.	.	.	.	.
5	.	.	.	.00	.	.	.	.	.
.	.	.	.	.	.	.	.	.	.
.	.	.	.	.	.	.	.	.	.
16	.	.	.	.00	.	.	.	.	.
Sum	1.00	2.00	2.00	2.00	2.00	2.00	2.00	3.00	16.00

squares of item-total correlation from the earlier ones in Table 6.8 to the ones in Table 6.9. Under forced classification, the variable AG (age) is completely explained by the first 2 dimensions rather than the 16 dimensions. The correlational information between AG and BP, for example, must be contained in these proper dimensions of AG, for the remaining dimensions do not contain any information about the criterion variable AG. Thus, we are now ready to consider correlation between two categorical variables in reduced space.

6.6 Correlation among categorical variables

Let us first study the distribution of information under forced classification. Paying attention only to those proper dimensions, we obtain the following summary of the relevant results that were obtained by specifying each item in turn as the criterion, that is, eight forced-classification outcomes. Remember that (a) the number of proper solutions is equal to the number of categories of the criterion variable minus 1 and (b) the numbers of categories of the eight items are 2, 3, 3, 3, 3, 3, 3, and 4, respectively.

Consider two forced-classification analyses, one using variable q as the criterion and the other q' as the criterion. Then, there exist the following relations among $J_q - 1$ and $J_{q'} - 1$ proper solutions:

$$\sum_{k=1}^{J_q-1} r_{qt(k)}^2 = \sum_{k=1}^{J_{q'}-1} r_{q't(k)}^2 = T(q,q'), \text{ say,} \tag{6.4}$$

Table 6.10 Squared correlation r^2_{qt} of eight forced classification analyses.

CR	RT	BP	MG	AG	AX	WT	HT	WK
RT	1.00	.19	.00	.27	.08	.06	.08	.03
BP	.00	1.00	1.00	.35	.23	.20	.33	.01
	.19	1.00	.00	.45	.15	.08	.09	.06
sum	.19	—	1.00	.80	.38	.28	.42	.07
MG	.00	1.00	1.00	.34	.27	.18	.33	.01
	.00	.00	1.00	.07	.37	.32	.07	.04
sum	.00	1.00	—	.41	.64	.50	.40	.05
AG	.11	.49	.31	1.00	.53	.07	.08	.02
	.16	.30	.10	1.00	.06	.25	.04	.02
sum	.27	.79	.41	—	.59	.32	.12	.04
AX	.02	.23	.51	.46	1.00	.13	.08	.03
	.06	.16	.13	.13	1.00	.07	.00	.06
sum	.08	.39	.64	.59	—	.20	.08	.09
WT	.03	.16	.42	.08	.07	1.00	.15	.03
	.03	.11	.07	.24	.13	1.00	.17	.01
sum	.06	.27	.49	.32	.20	—	.32	.04
HT	.00	.37	.33	.02	.02	.14	1.00	.02
	.07	.05	.07	.11	.06	.17	1.00	.03
sum	.08	.42	.40	.13	.08	.31	—	.05
WK	.03	.04	.03	.02	.05	.01	.02	1.00
	.00	.01	.01	.01	.02	.03	.01	1.00
	.00	.03	.01	.00	.01	.01	.02	1.00
sum	.03	.08	.05	.03	.08	.05	.05	—

where $r^2_{qt(k)}$and $r^2_{q't(k)}$ are the squared item-component correlations of the noncriterion variables associated with the proper solutions. For example, look at items RT and BP (the sum of 0.19 in row 1, column 2 as well as in row 4, column 1 of Table 6.10), or items HT and WK (the sum of 0.05), or items RT and WK (the sum of 0.03). When the numbers of categories are different, it is interesting to see what happens when one variable is projected to the other variable: the identical sum is distributed over the respective dimensions (e.g., 0.02 and 0.03 of WK over the space of HT; 0.02, 0.01, and 0.02 of WK over the space of HT). One might be curious about the relation between BP and MG, showing 1.00 and 0.00 of MG over the space of BP and 1.00 and 0.00 of BP over the space of MG. If we look at the 3 × 3 contingency table of categories of BP and MG, we can understand the distribution of correlations as noted previously. Although the original contingency table is 3 × 3, we can collapse it to 2 × 2 with frequencies only in the main diagonal positions, leading to the value of 1.00 for the first dimension, and 0 for the second dimension. This is a rare case in which a 3 × 3 table can be collapsed into a 2 × 2 table.

There is another interesting relation: the sum of the squared item-component correlation of a noncriterion item over the corresponding proper solutions of the criterion item cannot exceed the smaller number of categories of the two variables minus 1, that is,

$$T(q,q') \le \min(J_q, J_{q'}) - 1 = p - 1 \tag{6.5}$$

Thus, the following measure of correlation between two categorical variables is proposed:

$$\nu_{qq'} = \sqrt{\frac{T(q,q')}{p-1}} \tag{6.6}$$

where $T(q,q')$ is given by Equation 6.6. For instance,

$$\nu_{\text{BP,MG}} = \sqrt{\frac{1.00}{3-1}} = 0.71, \quad \text{and} \quad \nu_{\text{MG,AG}} = \sqrt{\frac{0.41}{3-1}} = 0.45 \tag{6.7}$$

The measure of correlation proposed here is bounded between 0 and 1, unlike the familiar product-moment correlation. This can be interpreted as a way in which the scaling is conducted, that is, to maximize the correlation (if the value is negative, reverse the sign of weights).

We now show that the proposed measure is equal to Cramér's *V*. Suppose that we analyze the contingency table, crossing variables q and q', which yields the correlation ratios (or squared singular values)

$$1 \ge \eta_1^2 \ge \eta_2^2 \ge \ldots \ge \eta_{p-1}^2 \tag{6.8}$$

where the first correlation ratio of 1 is the trivial solution, and p is the smaller value of J_q and $J_{q'}$. Then, there exists the following relation:

$$\sum_{k=1}^{p-1} \eta_k^2 = \sum_{q=1}^{J_q} r^2_{qt(k)} = \frac{\chi^2}{n} \tag{6.9}$$

where n is the total frequency of the contingency table. Recall that Cramér's V (Cramér 1946) is given by

$$V = \sqrt{\frac{\chi^2}{n(p-1)}} \tag{6.10}$$

From these relations, we conclude that the proposed coefficient is equal to Cramér's V, as illustrated by comparing the results in Equation 6.7 with Table 6.8.

6.7 Properties of squared item-total correlation

When h standardized continuous variables are subjected to PCA, we have the following relevant relations:

$$\sum_{j=1}^{h} r^2_{jt(k)} = \lambda_k = \text{eigenvalue of component } k \tag{6.11}$$

$$\sum_{k=1}^{h} r^2_{jt(k)} = 1 \quad \text{for any item } j \tag{6.12}$$

When multiple-choice items with J_q categories are subjected to DS, we have

$$\sum_{q=1}^{J_q} r^2_{qt(k)} = Q\eta_k^2 \quad \text{for dimension } k \tag{6.13}$$

$$\sum_{k=1}^{T} r^2_{qt(k)} = J_q - 1 \quad \text{for any item } q \tag{6.14}$$

$$\sum_{q=1}^{Q}\sum_{k=1}^{T} r^2_{qt(k)} = Q\sum_{k=1}^{T} \eta_k^2 = J - Q \tag{6.15}$$

where $T = J - Q$, J being the total number of categories of Q items, and η_k^2 is the maximized correlation ratio of component k. The internal-consistency reliability, often called "Cronbach's alpha" (Cronbach 1951), indicated by α, is given (see Nishisato 1980) by

$$\alpha = 1 - \frac{1-\eta_k^2}{(n-1)\eta_k^2} \tag{6.16}$$

Because η^2 is given by

$$\eta_k^2 = \frac{\sum_{q=1}^{Q} r_{qt}^2}{Q} \tag{6.17}$$

we obtain that

$$\alpha = \frac{Q-1}{Q} \frac{\sum_{q=1}^{Q} r_{qt(k)}^2}{\left(\sum_{q=1}^{Q} r_{qt(k)}^2 - 1\right)} \tag{6.18}$$

Thus, α attains its maximum of 1 when all Q items are perfectly correlated with the total score, and it becomes negative when the sum of squared item-component correlations becomes less than 1. This expression also implies the statement by Nishisato (1980) that α becomes negative when η_k^2 is less than $1/Q$. We can also derive that the average correlation ratio of all possible correlation ratios associated with multiple-choice data is equal to $1/Q$ (see Nishisato 1994, 1996).

The singular values and the item-total correlations are closely related. In this context, we can mention another interesting aspect of quantification and its relation to singular values, found in the process of the method of reciprocal averages (Horst 1935). Given the contingency table of two variables q and q', that is, $\mathbf{Z}_q^{\mathsf{T}}\mathbf{Z}_{q'} = \mathbf{C}_{qq'}$, we form a sequence of products,

$$\frac{\mathbf{C}_{qq'}\mathbf{a}_0}{|k_1|} = \mathbf{b}_1 \quad \frac{\mathbf{b}_1^{\mathsf{T}}\mathbf{C}_{qq'}}{|k_1'|} = \mathbf{a}_1^{\mathsf{T}} \cdots \frac{\mathbf{C}_{qq'}\mathbf{a}_{s-1}}{|k_s|} = \mathbf{b}_s \quad \frac{\mathbf{b}_s^{\mathsf{T}}\mathbf{C}_{qq'}}{|k_s'|} = \mathbf{a}_s^{\mathsf{T}} \tag{6.19}$$

where $\mathbf{a}_0$ is an arbitrary nonnull vector, and $|k_s|$ and $|k_s'|$ are the largest absolute values of the elements of $\mathbf{C}_{qq'}\mathbf{a}_s$ and $\mathbf{b}_s^{\mathsf{T}}\mathbf{C}_{qq'}$, respectively. As s increases, the process generally converges to two distinct constants $|k_s| = k$ and $|k_s'| = k'$, say, and $kk' = \eta_1^2$.

6.8 Structure of nonlinear correlation

Using the numerical example, we derived 16 DS correlation matrices and 16 sets of squared-item total correlation. Although we opted for forced classification to arrive at a measure of correlation between categorical variables, the question arises whether the same information is contained in those 16 correlation matrices and 16 sets of r_{qt}^2. The use of forced classification is based on the idea of projecting one variable onto the space of the other variables, which made it possible to work in a reduced space. The task remains to find the relationship between the matrix of Cramér's V coefficients and the correlation matrices in the 16 dimensions.

When data are decomposed into components, the correlation between two variables in each component depends on other variables, since the weights of categories of each variable are determined in terms of the entire data set. We have 16 sets of r_{qt}^2. The correlation between two specific variables from these 16 matrices must be equal to that obtained from 12 matrices obtained by discarding, for example, RT and WK, or that of four matrices obtained by discarding four other remaining items as well. Thus, finding the correlation from these standard analyses looks trickier than from forced classification. Unfortunately, we must leave this matter for future research.

6.9 Concluding remarks

When variables are normally distributed, we can dismiss the search for nonlinear relations; when they are not distributed normally, principal component analysis, for example, becomes an analysis of data linearity and not of the data *per se*. In this chapter, we have looked at data analysis without any distributional assumption.

In data quantification, we commonly impose order constraints on a set of categories: for example, "low," "medium," "high" for BP (blood pressure) and "rarely," "sometimes," "often" for MG (migraine headaches). As we have already seen, however, quantification of these categories under the order constraints loses significance if we are constrained to describe the nonlinear relation between BP and MG. If we impose a strong- or a weak-order constraint on each set of the categories, then we cannot arrive at the conclusion that migraines often occur when blood pressure is either low or high. This is an important point to note, for the general practice seems to be that when the adjectives describing the various categories are ordered, we must impose an order constraint on them in quantification as well. Such a practice has the effect of filtering out information of nonlinear relations among variables from analysis. In a broader context, the assumption of a normal distribution underlying the ordinal variables also filters out nonlinear relations. Whether we should impose an order constraint on ordinal categories is a matter of opinion, but this chapter argues that if we are to capture as much information as possible, particularly nonlinear relations, then an order constraint is a hindrance to information retrieval.

Adoption of the correlation matrix associated with the first component, which is always positive definite or semidefinite, was suggested by Nishisato and Hemsworth (2004). However, the analysis in this chapter shows that such a matrix reflects only a portion of the relations. Furthermore, this approach has a logical drawback in that the correlation between two items is affected by the other items that are included in the data set. In other words, the category weights of a variable are determined based on their ability to optimize the objective function involving all of the items in the data set; hence, the correlation is also affected by other items in the data set. This chapter shows that the correlation between two categorical variables must be defined in terms of all proper solutions of forced classification.

Given this conclusion, let us continue our discussion to see what effects we should expect when the assessment of correlation is carried out only in terms of the first eigenvalue or in the first few dimensions, rather than in total space. It is well known in PCA that the correlation between continuous variables is defined as the cosine of the angle between them. Suppose now that the variables span three-dimensional space and that we project the data onto two-dimensional space. Then, the angle between the two variables becomes smaller in two-dimensional space than in three-dimensional space. The consequence of this projection is that the correlation defined as the cosine of the angle between axes increases, leading to an overestimated coefficient of correlation. Yet the general practice in data analysis is to interpret data in reduced space, that is, in terms of data structure associated with overestimated correlation. The higher the correlation, the clearer is the isolation of data clouds. So, does the general practice of interpreting data in reduced dimension present an overly simplified description of data structure? This matter is worthy of further investigation.

DS is now at a new phase of development in two ways: the first extends DS for use in categorizing continuous variables, and the second suggests a new look at the definition of information in data analysis. The motivation for the first aspect is a venture to capture more information than what most linear models for continuous variables can typically tap into. However, further research is needed to understand how continuous variables can be categorized in a rational and justifiable way. Categorization of a continuous variable, no matter how it is done, involves some loss of information. Thus, the main concern should be to compromise between two opposing objectives, one to enhance the ability to capture nonlinear relations with ease by DS of categorized variables and the other to minimize loss of information contained in the original variable through categorization.

New statistics must be developed to measure both linear and nonlinear information in data beyond a traditional and popular definition of information by the sum of eigenvalues. This involves not only inclusion of nonlinear relations, but also a shift from the concept of the sum of the variances to the sum of covariances. This last point is based on the belief that a set of homogeneous variables contains less information than a set of heterogeneous variables. An attempt to define information in this new context is underway (Nishisato 2002, 2003a, 2003b).

Acknowledgments

This work is supported by a research grant from the Natural Sciences and Engineering Council of Canada.

CHAPTER 7

Validation Techniques in Multiple Correspondence Analysis

Ludovic Lebart

CONTENTS

7.1 Introduction

Multiple correspondence analysis (MCA) provides useful data visualizations that highlight associations and patterns between several categorical variables, for example, in socioeconomic surveys and marketing. However, the outputs (parameter estimates and graphical displays) are often difficult to assess. In an era of computer-intensive techniques, we cannot content ourselves with the criterion of "interpretability" of the results, an approach that was widely used during the first phases of the upsurge of data-analytic methods 30 years ago.

Let us begin by briefly recalling that in both cases of simple and multiple correspondence analysis, most of the theoretical results provided by mathematical statistics are not applicable. The hypothesis of independence between rows and columns of a matrix is often too strict to be realistic. Under this hypothesis, in two-way correspondence analysis applied to an $I \times J$ contingency table, the eigenvalues are asymptotically those obtained from a Wishart matrix with $I - 1$ and $J - 1$ degrees of freedom (Lebart 1976). As a consequence, always under the hypothesis of independence, the relative values of the eigenvalues (percentages of variance) are statistically independent of their sum, which follows the usual chi-square distribution with $(I - 1)(J - 1)$ degrees of freedom. In the case of MCA, or more generally in the case of binary data, the distribution of eigenvalues is more complex; their sum does not have the same meaning as in CA; and the percentages of variance are misleading measures of information (Lebart et al. 1984; Chapter 2 of this volume). The delta method, one of the classical methods of asymptotic statistics (Gifi 1990), allows us to observe the consequences of perturbations of the data on eigenvalues and eigenvectors under much more realistic hypotheses.

In this chapter, we will focus on the two following issues:

1. External validation, involving external data or metadata (usually considered as supplementary elements, also called passive or illustrative) and allowing for some classical statistical tests, including cross-validation procedures in the scope of supervised learning, and
2. Internal validation, based on resampling techniques such as the bootstrap and other Monte Carlo methods.

7.2 External validation

External validation is the standard procedure in the case of supervised learning models for classification. Once the parameters of the model have been estimated (learning phase), external validation serves to assess the model (generalization phase), usually with cross-validation methods (see, e.g., Hastie et al. 2001). However, external validation can be used as well in the unsupervised context of MCA in the two following practical circumstances:

1. When the data set can be split into two or more parts, one part being used to estimate the model, the other part(s) serving to check the adequacy of that model, and
2. When some metadata or external information is available to complement the description of the elements to be analyzed.

7.2.1 *External validation in the context of MCA*

We will assume that external information has the form of supplementary elements (extra rows or columns of the data table that are utilized afterward; see Benzécri et al. 1973; Cazes 1982; Gower 1968). In data-analysis practice, the supplementary elements are projected afterward onto the principal visualization planes. For each projection and each principal axis, a "test value" (explained below) is computed that converts the coordinate on the axis into a standardized normal variable (under the hypothesis of independence between the supplementary variable and the axis).

The principle of this assessment procedure is as follows. Suppose that a supplementary category j contains n_j individuals (respondents or rows). The null hypothesis is that the n_j individuals are chosen at random among the n individuals in the analysis, i.e., among the n rows of the $n \times J$ indicator matrix $\mathbf{Z}$. Under these circumstances, on a specific principal axis s, the coordinate g_{js} of the supplementary category j is a random variable, with mean 0, and variance $v(j)$ given by

$$v(j) = \frac{1}{n_j}\left(\frac{n-1}{n-n_j}\right)$$

The test value is the standardized coordinate $t_s(j) = g_{js}/\sqrt{v(j)}$ with mean 0 and variance 1 and, moreover, is asymptotically normally distributed. Thus, using the usual approximation, an absolute test value greater than 2 indicates a significant position of the corresponding category j on axis s (at a two-sided significance level of 0.05). In an exploratory approach, particularly in the case of survey data processing, numerous supplementary elements could be projected, leading to as many test values (Lebart 1975).

As a part of the software output, all of the significant supplementary variables can be sorted afterward according to the strengths of their links with each principal axis. Such strengths are aptly described by the test values: the larger the test value, the more significant the link between the corresponding variable and the axis. These series of test statistics entail the unavoidable problem of multiple comparisons, which will be dealt with briefly in the next section.

7.2.2 Multiple comparisons

The simultaneous computation of numerous test values runs into the obstacle of multiple comparisons, a permanent problem in data-mining and text-mining applications. Suppose that the corpus under study is perfectly homogeneous and, thus, that the hypothesis of independence holds. Under these conditions, given a significance level of 5%, out of 100 calculated test values, there are, on average, five that are significant. In fact, the 5% threshold only makes sense for a single test and not for a series of tests. In other words, when testing repeatedly, the unsuspecting user will always find "something significant" at the 5% level.

A practical way to solve this difficulty is to choose a stricter threshold for the p-value, that means a larger threshold for the test value t (one must keep in mind that $p = 0.05$ corresponds to $t = 1.96$; $p = 0.001$ to $t = 3.09$). In the context of analysis of variance, several procedures have been devised to overcome this difficulty. As a pioneering example, the Bonferroni method, also known as α-adjusting, recommends dividing the probability threshold by the number of tests (number of comparisons in the case of the design of experiments). This reduction of the probability threshold is generally considered too strict (Hochberg 1988). When the tests are not independent, which is often the case in the context of MCA, the Bonferroni solution is obviously too conservative. An intermediate value of t (such as the average of the usual value and the Bonferroni value) provides the user with a convenient order of magnitude. Classical overviews and discussions about multiple comparisons are found in Hsu (1996) and Saville (1990).

7.3 Internal validation (resampling techniques)

In the context of principal axes techniques such as principal component analysis (PCA), simple (CA), or multiple (MCA) correspondence analysis, bootstrap resampling techniques (Efron 1979) are used to produce confidence regions on two-dimensional displays. The bootstrap replication scheme allows one to draw confidence ellipses or convex hulls for both active and supplementary categories, as well as supplementary continuous variables.

In computing the precision of estimates, the bootstrap method is useful because (i) the classical approach is both unrealistic and analytically complex; (ii) the bootstrap makes almost no assumption about the underlying distributions; and (iii) the bootstrap offers the possibility of mastering every statistical computation for each sample replication and therefore of dealing with parameters computed through the most complex algorithms.

7.3.1 Basic principles of the bootstrap

The first phase consists of drawing a sample of size n, with replacement, of the n statistical units (the rows of the data matrix $\mathbf{Z}$) and of computing the parameter estimates of interest such as means, variances, and eigenvectors, on the new "sample" obtained. A simple, uniform pseudorandom generator provides n independent drawings of one integer between 1 and n, from which the n "bootstrap weights" are easily derived. This phase is repeated K times. The value of K can vary from 10 to several thousand, according to the type of application (see Efron and Tibshirani 1993). We have, at this stage, K samples (the replicates) drawn from a new "theoretical population" defined by the empirical distribution of the original data set and, as a consequence, K estimates of the parameters of interest. Under rather general assumptions, it has been proved that we can estimate the variance of these parameters (or other summary statistics) directly from the set of their K replicates.

7.3.2 Context of principal component analysis

In the PCA case, there are variants of bootstrap for active variables and supplementary variables, both continuous and nominal. In the case of numerous homogeneous variables, a bootstrap on variables is also proposed, with examples of application to the case of semimetric data (Lebart et al. 2003). Numerous papers have contributed to select the relevant number of axes, and these have proposed confidence intervals for points in the subspace spanned by the principal axes. These parameters are computed after the realization of each replicated sample and involve constraints that depend on these samples. The sth eigenvector of a replicated correlation matrix is not necessarily the homologue of the sth eigenvector of the original matrix; mismatches are possible because of rotations, permutations, or changes of sign. In addition, the expectations of the eigenvalues of the replicated matrices are distinct from the original eigenvalues. This is exemplified by a classical result: let us suppose that the theoretical covariance matrix is the unit matrix **I** (that is, all theoretical eigenvalues take the value 1, and all theoretical covariances take the value 0). However, the largest eigenvalue from the PCA of any finite sample drawn from that population will be greater than 1. Similarly, in the context of bootstrap, the expectation of the first eigenvalue of the PCA of the replicated matrices is markedly greater than its "theoretical" counterpart, calculated on the observed covariance matrix.

Several procedures have been proposed to overcome these difficulties (Chateau and Lebart 1996): partial replications using supplementary elements (partial bootstrap), use of a three-way analysis to process simultaneously the whole set of replications, filtering techniques involving reordering of axes, and Procrustean rotations (Markus 1994a; Milan and Whittaker 1995).

The partial bootstrap, which makes use of projections of replicated elements on the original reference subspace provided by the eigendecomposition of the observed covariance matrix, has several advantages. From a descriptive standpoint, this initial subspace is better than any subspace undergoing a perturbation by a random noise. In fact, unlike the eigenvalues, this subspace is the expectation of all the replicated subspaces having undergone perturbations. The plane spanned by the first two axes, for instance, provides an optimal point of view on the data set. In this context, to apply the usual bootstrap to PCA, one can project the K replicates of variable points in the common reference subspace and compute confidence regions (ellipses or convex hulls) for the locations of these replicates (Greenacre 1984).

7.3.3 Context of CA and MCA

Gifi (1990) and Greenacre (1984) did pioneering work in addressing the problem in the context of simple CA and MCA. As mentioned previously in the case of PCA, it is easier to assess eigenvectors than eigenvalues that are biased replicates of the theoretical ones (Alvarez et al. 2004). In fact, as far as bootstrapping is concerned, the context of MCA is identical to that of PCA. All that we have said above about total and partial bootstrap applies to MCA. A specific replication can be generated by a drawing with replacement of the n individuals (rows of the indicator matrix **Z**). Each replication k leads to a Burt contingency table $\mathbf{C}_k$, whose rows (or columns) can be projected as supplementary elements onto the initial principal axes (partial bootstrap). K replicates $(j_1, j_2, \ldots, j_k, \ldots, j_K)$ are obtained for each category point j. Then, J PCAs are performed in the two-dimensional space spanned by a chosen pair of axes to draw the J confidence ellipses. The lengths of the two principal radii of these ellipses are normatively fixed to two standard deviations (i.e., twice the square roots of the eigenvalues, for each PCA). Empirical evidence suggests that $K = 30$ is acceptable for the number of replicates and, in such a case, that the corresponding ellipses contain approximately 86% of the replicates. Alternatively, the confidence ellipses can be replaced by convex hulls, as displayed in the forthcoming examples. Note that the two ways of visualizing the uncertainty around each category point (ellipses or convex hulls) are complementary: the former taking into account the density of the replicated points, the latter pinpointing the peripheral points.

7.4 Example of MCA validation

7.4.1 Data

The data set is the British section of a multinational survey conducted in seven countries in the late 1980s (Hayashi et al. 1992), and it can be downloaded from http://www.lebart.org (example TDA2 accompanying the software DTM described in the Software Notes at the end of the chapter). It deals with the responses of $n = 1043$ individuals and comprises objective characteristics of the respondents (age, status, gender, facilities). Other questions relate to attitudes or opinions.

7.4.2 Active questions, global analysis

In this example we focus on a set of $Q = 6$ attitudinal questions, with a total of $J = 24$ response categories, that constitutes the set of active variables. The external information (supplementary categories) is provided by a single categorical variable constructed by crossing age with education (nine categories).

Table 7.1 gives the wording of the 6 questions and the 24 corresponding categories, together with the basic results of the MCA of the indicator matrix crossing the 1043 respondents (rows) with the 24

Table 7.1 Frequencies, coordinates, contributions, and squared cosines (relative contributions) of active categories on axes 1 and 2.

		Coordinates		Test Values	
Active Categories	Frequencies	Axis 1	Axis 2	Axis 1	Axis 2
1. Change in the Global Standard of Living Last Year					
Std.liv/much better	223	−0.96	0.51	9.6	3.6
Std.liv/lit better	417	−0.17	−0.30	0.6	2.3
Std.liv/the same	164	0.34	−0.42	0.9	1.8
Std.liv/lit.worse	156	0.78	−0.02	4.5	0.0
Std.liv/v.much worse	83	1.30	1.00	6.5	5.1
2. Change in Your Personal Standard of Living Last Year					
SL.pers/much better	250	−0.94	0.47	10.4	3.4
SL.pers/lit better	317	−0.07	−0.42	0.1	3.5
SL.pers/the same	283	0.28	−0.23	1.0	0.9
SL.pers/lit.worse	123	0.82	0.14	3.9	0.1
SL.pers/v.much worse	70	1.11	0.93	4.0	3.7
3. Change in Your Personal Standard of Living Next 5 Years					
SL.next/much better	123	−0.57	0.56	2.0	2.5
SL.next/lit.better	294	−0.42	0.35	2.5	2.3
SL.next/the same	460	0.07	−0.49	0.1	6.8
SL.next/lit.worse	134	1.16	0.34	9.1	1.0
SL.next/v.much worse	32	1.21	1.00	4.1	3.0
4. Will People Be Happier in Years to Come?					
People/happier	188	−1.02	0.76	9.2	6.7
People/less happy	535	0.61	0.20	9.4	1.3
People/the same	320	−0.42	−0.78	2.7	11.8
5. Will People's Peace of Mind Increase?					
P.of.mind/increases	180	−0.76	0.72	4.9	5.8
P.of.mind/decreases	618	0.40	0.18	4.7	1.2
P.of.mind/no change	245	−0.46	−0.97	2.4	14.3
6. Will People Have More or Less Freedom?					
More freedom	443	−0.40	0.47	3.3	5.9
Less freedom	336	0.70	0.15	7.8	0.5
Freedom/the same	264	−0.23	−0.97	0.6	15.4

categories (columns): frequency of the responses, coordinates on the two first principal axes, and contributions to these axes (also called "absolute contributions"). The first two eigenvalues of 0.342 and 0.260 (data not shown) account for 12.1% and 9.2% of the total inertia, respectively. It is widely recognized that these values give a pessimistic idea of the quality of the description provided by the MCA, and several corrected or adjusted formulas for these percentages have been suggested by Benzécri (1979) and Greenacre (1994). Figure 7.1 shows the categories in the principal plane obtained from that MCA.

The most positive and optimistic responses are grouped in the upper left side of Figure 7.1 (answers "much better" to the three questions about "standard of living" and answers "people happier," "peace of mind increases," and "more freedom" to the next three questions). The most negative and pessimistic responses occupy the upper right side of the display, in which the three "very much worse" items relating to the three first questions form a cluster markedly separated from the remaining categories. All neutral responses ("the same," "no change") are located in the lower part of the display, together with some moderate responses such as "little better."

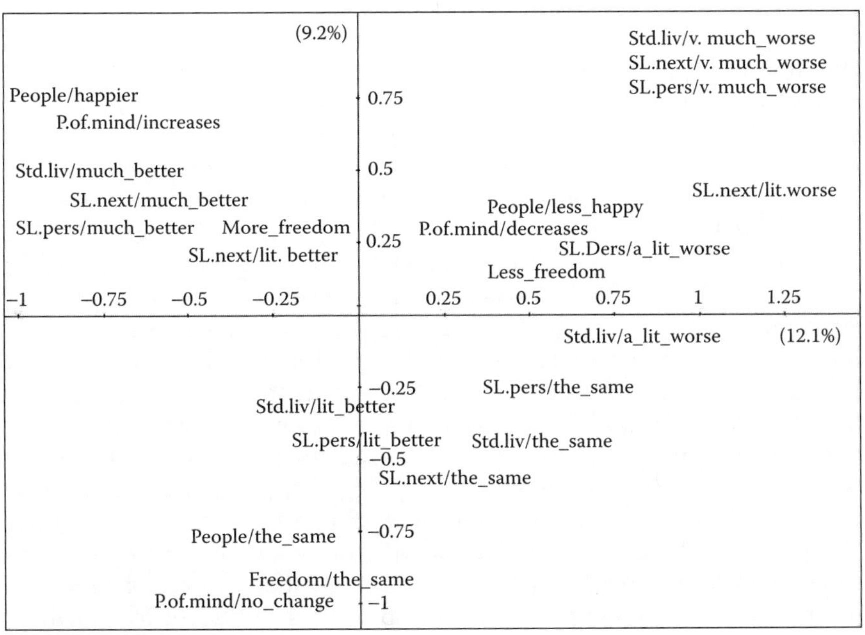

Figure 7.1 Plane of axes 1 and 2 from the MCA of the 1043 × 24 indicator matrix.

While the first axis can be considered as defining (from the right to the left) a particular scale of "optimism, satisfaction, happiness," the second axis appears to oppose moderate responses (lower side) to both extremely positive and extremely negative responses. Such a pattern of responses suggests what is known in the literature as the "Guttman effect" or "horseshoe effect" (Van Rijckevorsel 1987). However, we are not dealing here with a pure artifact of the MCA. In the following discussion, we will see that some particular categories of respondents are significantly devoted either to moderate responses or to systematic extreme responses.

Another possible pattern visible in Figure 7.1 is the "battery effect," often observed in survey analysis. When several questions have the same response categories, which is the case for the three first questions, the respondent often chooses identical answers without sufficiently considering the content of the questions (see Chapter 20). That could account for the tight grouping of the responses "very much worse" in the upper right part of the display.

7.4.3 Supplementary categories and test values

The unique supplementary variable comprising nine categories is built via cross-tabulating age (three categories: <30, 30–55, +55) and educational level (three categories: low, medium, high). Table 7.2 gives the identifiers of these nine categories together with the frequencies of the responses, the corresponding coordinates, and test values defined in Section 7.2.1.

Considering the test values allows us to make conclusions in the framework of classical statistical inference. The young educated respondents (<30 years, high level of education) are significantly optimistic in their answers to these attitudinal questions: their location is at −2.8 standard deviations ($t = -2.8$) from the mean point (origin) on axis 1. Likewise, the younger respondents with a medium level of education are on the optimistic side of the axis, but closer to the origin. The category that best characterizes this optimistic side of the first axis is the category (30–55, medium level of education) ($t = -4.8$). On the opposite side, among pessimistic categories, persons over 55 having a low level of education occupy a highly significant location ($t = 6.7$). On axis 2, persons between 30 and 55 with a low level of education are located on the side of extreme responses, opposed to all the categories of respondents over 55.

Table 7.2 Frequencies, coordinates, and test values of supplementary categories on axes 1 and 2.

Supplementary		Coordinates		Test Values	
Categories	Frequencies	Axis 1	Axis 2	Axis 1	Axis 2
Age and Educational Level					
<30/low	18	–0.08	–0.01	–0.3	–0.1
30–55/low	226	0.08	0.21	1.4	3.6
+55/low	237	0.38	–0.16	6.7	–2.7
<30/medium	187	–0.16	0.09	–2.4	1.3
30–55/medium	159	–0.35	0.08	–4.8	1.0
+55/medium	72	0.19	–0.24	1.6	–2.1
<30/high	61	–0.35	0.12	–2.8	1.0
30–55/high	61	–0.18	–0.24	–1.4	–2.0
+55/high	22	–0.18	–0.60	–0.8	–2.8

Owing to the limited incidence of multiple comparisons (only nine tests), we have chosen a moderately conservative threshold here ($t > 2.3$), which corresponds to the intermediate value suggested in Section 7.2.2. The effect of multiple comparisons is even more limited in the case of bootstrap validation, which provides simultaneous confidence intervals for all the categories, taking into account the structure of correlations between the categories.

7.4.4 Partial bootstrap in MCA for active variables

Figure 7.2 is an example of partial bootstrap applied to the previous MCA. To obtain a legible display, we had to select a subset of all possible confidence areas. Over this representation are drawn confidence ellipses corresponding to the categories of the active variables: "Change in your personal standard of living last years" (categories: much better, little better, the same, little worse, much worse) and "Will people be happier in years to come" (categories: happier, less happy, the same). The smaller the confidence area, the more precise is the location of the corresponding category. The sizes of the ellipses (as measured by the radius of a circle having the same area) are approximately proportional to the inverse square root of the frequencies.

Only $K = 30$ replications of the original sample have been used. In practice, 10 replications would suffice to produce ellipses having

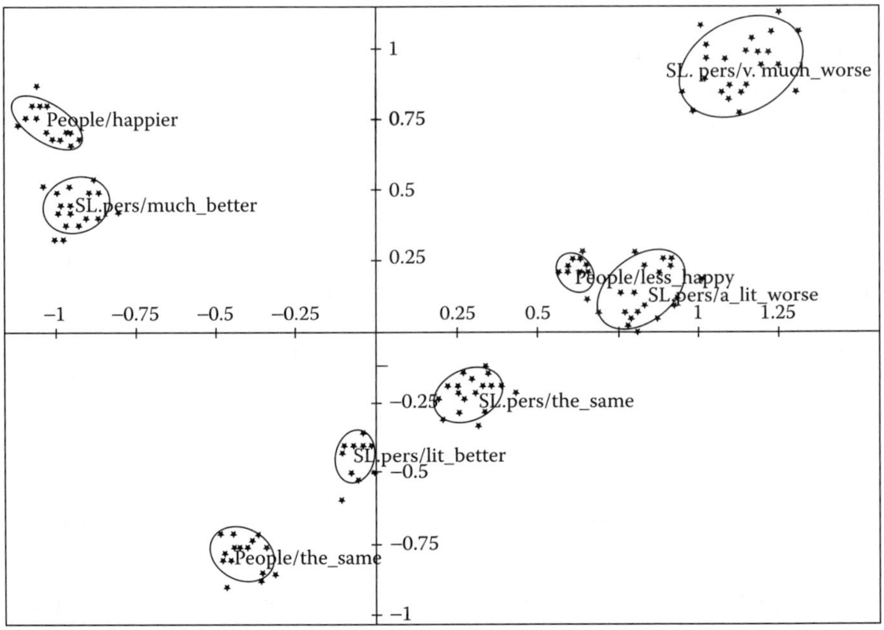

Figure 7.2 Partial bootstrap confidence ellipses for eight active categories points in the MCA principal plane (same plane as Figure 7.1). Five categories concern question 2 ("Change in your personal standard of living last year": identifiers beginning with "SL.pers"), and three categories concern question 4 ("Will people be happier in years to come?": identifiers beginning with "People").

similar shapes and sizes. Each replication involves n = 1043, drawing with replacements of the n respondents, and leads to a new positioning of the categories. Most of these replications are visible as dots within (and sometimes around) the confidence zones of each category on the display.

The response categories belonging to a specific question appear to be significantly distinct. Although they are not all displayed here, the confidence areas of the homologous categories of the three first questions have similar sizes, with most sets of identical items relating to distinct questions overlap. For example, the three categories "very much worse" relating to the three first questions are characterized by largely overlapping ellipses; their locations are not significantly distinct.

7.4.5 Total bootstrap in MCA for active variables

The total bootstrap is a conservative validation procedure. In this context, each replication leads to a separate MCA, but the absence of a common subspace of reference may induce a pessimistic view of the variances of the coordinates of the replicates on the principal axes. To remedy the arbitrariness of the signs of the axes, the orientations of the replicated axes could be *a posteriori* changed (if necessary) to maintain a positive correlation with the axes of the original MCA having the same rank.

No permutations relating to the first two axes have been observed for the example. The occurrence of rotation has been considered as a sign of instability of the axes, and no correction has been applied. Figure 7.3 shows the first principal plane that superimposes, after the possible changes of sign mentioned above, the principal planes of the 30 MCAs performed on the replicated data. Evidently, the ellipses are markedly larger than that of the partial bootstrap, and the scale of the display had to be modified in consequence. Note that the stability of the underlying structure is clearly established by the overly strict trial of the total bootstrap.

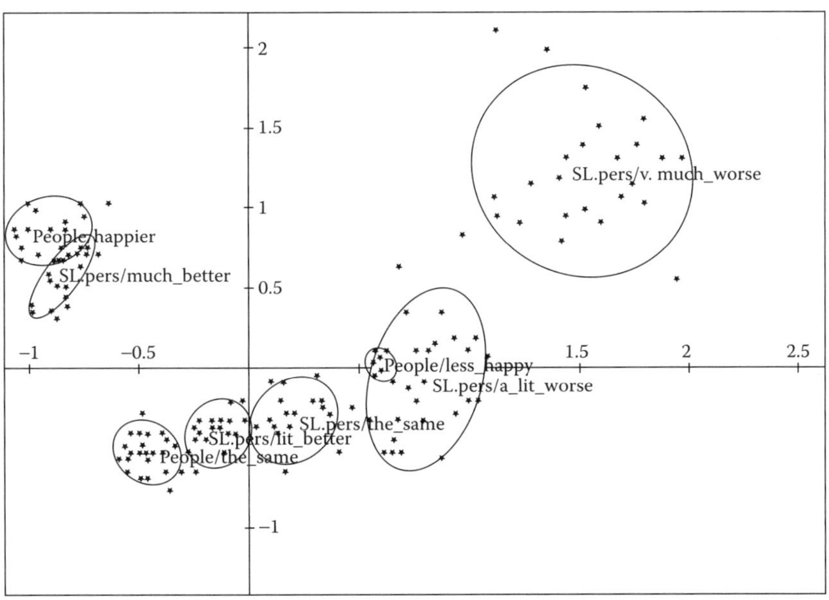

Figure 7.3 Total bootstrap confidence ellipses for eight active categories points in the MCA principal plane. (The categories are the same as in Figure 7.1. Note that the scale has changed because of the larger dispersion of the replicates.)

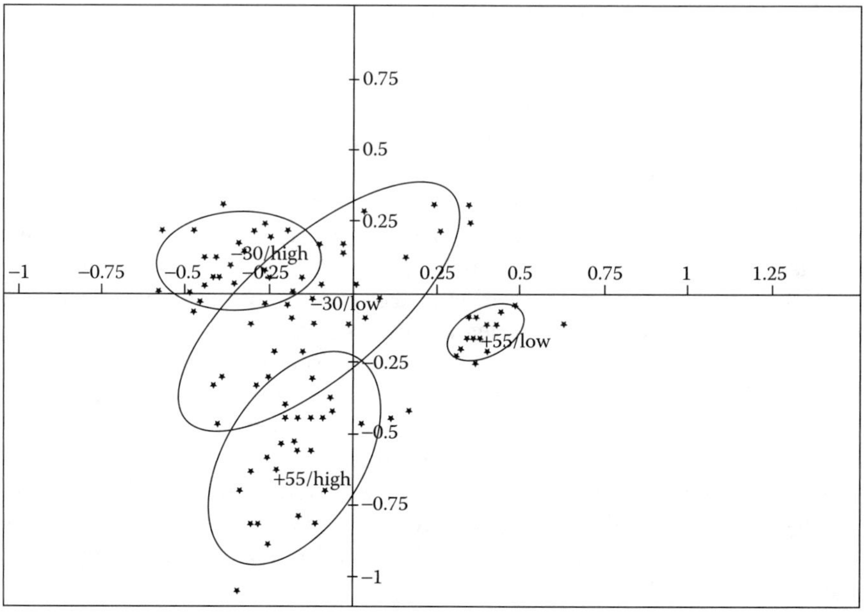

Figure 7.4 Bootstrap confidence ellipses for four supplementary category points in the MCA principal plane (same plane as Figure 7.1). For the sake of clarity, only the four categories involving two extreme categories of age (–30 and +55) and two extreme categories of education (low and high) are presented.

7.4.6 Partial bootstrap for external information in MCA

In the case of external information, the partial bootstrap is fully justified, since the sociodemographic categories are not used as active variables in the MCA. Figure 7.4 represents the first principal plane of the initial MCA in which a first selection of four (out of nine) "external" categories cross-tabulating the respondent's age and educational level are plotted afterward as supplementary elements. These four categories involve only extreme age categories and educational level.

Figure 7.4 highlights again the opposition between *young educated* people (<30/high, optimistic) and *old uneducated* respondents (+55/low, rather pessimistic). The position of the *young uneducated* (<30/low) cannot be considered as significant (the ellipse, rather large, contains the origin of the axes). *Old educated* people (+55/high) occupy a position significantly linked with the vertical axis, on the moderate side. Finally, the location of *old uneducated* respondents (+55/low) appears to be the most reliable, on the side of pessimistic responses.

Figure 7.5 shows the locations of the intermediate categories, involving either the intermediate level of education (medium) or the intermediate class of age (30–55). These locations were expected to be less significant than those of extreme categories presented previously in Figure 7.4. This is by no means the case: the confidence areas of intermediate categories also deserve to be interpreted. In particular, the category "30–55/low" is the most significant one on the upper part of axis 2. The respondents belonging to that category use extreme items of response because of being either pessimistic or optimistic. If we superimpose Figure 7.4 and Figure 7.5, we note that the three categories over 55 are all located in the "moderate" lower part of the display. The three categories with a high level of education are in the "optimistic" left-hand side of the display.

Figure 7.6 aims merely at illustrating the second technique for visualizing the shape of the set of replicates: the convex hull. The background

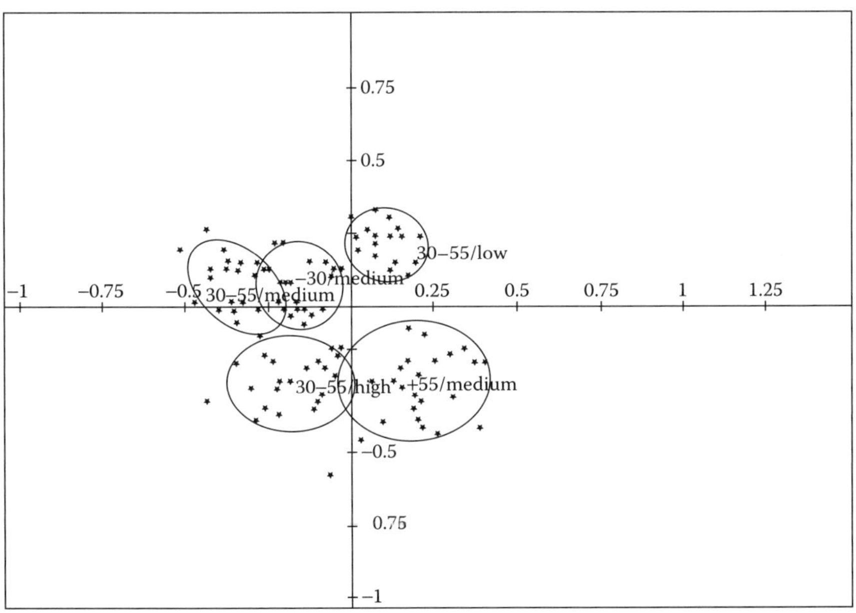

Figure 7.5. Bootstrap confidence ellipses for five remaining supplementary category points in the MCA principal plane (same plane as Figure 7.1).

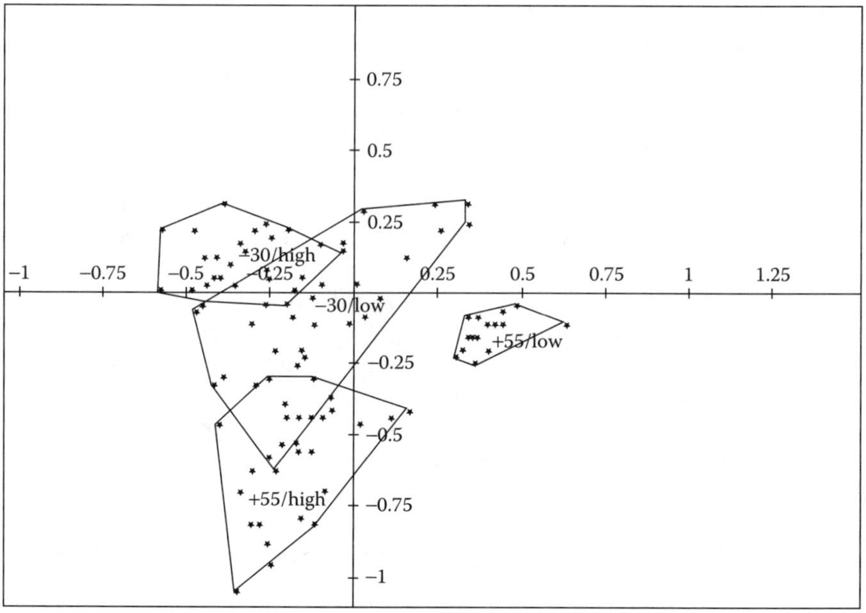

Figure 7.6 Bootstrap convex hulls for the four supplementary category points (presented in Figure 7.4) in the MCA principal plane (same plane as Figure 7.1).

of Figure 7.6 is identical to that of Figure 7.4. In this particular case, the conclusions about the significance of the location of the category points are unchanged. Note that the convex hull adapts better to the shape of the subcloud and can reveal the presence of outliers among the replicates.

7.5 Conclusion

The previous example pinpoints the convergence of both bootstrap and test values in the interpretation of the axes. Note that the test values characterize each axis separately and refer to the hypothesis of independence between the external category and the coordinates of the individual points on a particular axis. The bootstrap stipulates that the observed sample can serve as an approximation of the population: it takes into account the multivariate nature of the observations and involves simultaneously all the axes. We can then compare and analyze the proximities between pairs of categories, without reference to a specific axis.

In the framework of MCA, bootstrapping can also be used to process weighted data (circumstances occurring in most sample surveys) and to draw confidence intervals around the location of supplementary continuous or numerical variables in MCA. In the case of multilevel samples (for example, a sample of colleges and samples of students within the colleges), the replications can involve the different levels separately, allowing one to study the different components of the observed variance.

From a statistical perspective, in the context of the processing of sample surveys, categorical data tables often have the following characteristics in common: they are large, high-dimensional, with a high level of noise. MCA and its variants can help to describe the associations between the categories of a homogeneous set of active questions. They can also consider the principal space of the first dimensions as a predictive map that purports to visualize all of the remaining information contained in the survey data file (set of supplementary questions). In fact, the second approach, closely related to multiple regression and analysis of variance, is widely used by practitioners. In both cases, validation procedures are difficult to carry out in a classical statistical framework. External validation and resampling techniques (mainly bootstrap in the case of unsupervised approaches) possess all of the properties needed to provide the user with the ability to supplement appealing visualizations with inferential tools.

Software Notes

The software used to process all phases of the application is DTM (data and text mining). It comprises clustering algorithms and principal axes techniques (PCA, CA, MCA) designed to process large survey data (up to 12,000 individuals and 1,000 variables), together with some textual data-analysis methods dedicated to the processing of responses to open-ended questions. The software can be freely downloaded from the Web site: http://www.lebart.org.

CHAPTER 8

Multiple Correspondence Analysis of Subsets of Response Categories

Michael Greenacre and Rafael Pardo

CONTENTS

8.1 Introduction

In the social sciences, the principal application of multiple correspondence analysis (MCA) is to visualize the interrelationships between response categories of a set of questions in a questionnaire survey, for example a set of statements to which the respondents answer on the following scale: "strongly agree," "somewhat agree," "neither agree nor disagree," "somewhat disagree," "strongly disagree." There are often many nonresponses as well, and this absence of a response is a potential category that also needs to be considered. Once the relationships between the questions, or items, are visualized in a spatial map and interpreted, the method additionally allows the display of explanatory demographic variables such as age, education, and gender in order to enrich the interpretation.

It may be interesting from a substantive point of view to focus on a particular subset of response categories, for example the categories of agreement only. Or we might want to focus on the categories of agreement and disagreement alone, excluding both the nonresponses and the fence-sitting "neither agree nor disagree" responses. A further analysis of interest would be just of the nonresponses by themselves, to understand how these are correlated between items as well as how item nonresponse is correlated with demographic variables. The response category "neither agree nor disagree" provides another interesting subset that could be analyzed alone, to see if there are specific questions to which respondents are giving this unsure response and how the pattern of these responses is related to the demographic characteristics.

MCA is generally defined in two practically equivalent ways: either as (a) the simple correspondence analysis (CA) of the individual response data in the format of an indicator matrix, where all response categories form the columns of the indicator matrix, or (b) the CA of all cross-tabulations concatenated in the so-called Burt matrix, a symmetric matrix that has the response categories as rows and columns (see Chapter 2). The maps obtained by MCA are frequently overpopulated with points, making their printing difficult and interpretation complicated. There are strategies for rectifying this situation, such as not plotting points that contribute weakly to the principal axes of the map, but this would be undesirable when we are truly interested in each category point across all the questions. Furthermore, it is commonly found that the principal dimensions of MCA tell an obvious and unsurprising story about the data at hand while the more interesting patterns are hidden in higher dimensions. Exploring further dimensions is not a simple task because all the category points appear on and contribute to every dimension, to a greater or lesser extent. The basic problem is that the MCA map is trying to show many different types of relationships simultaneously, and these relationships are not isolated to particular dimensions. While the technique does its best to visualize all the response categories, the maps may not be easily conducive to visualizing those relationships of particular interest to the researcher.

The methodology we expose in this chapter allows subsets of categories to be analyzed and visualized, thus focusing the map on relationships within a chosen subset or between a subset and another subset. Thus, this approach would allow, for example, a direct analysis and interpretation of the nonresponses, how they interrelate, and how they relate to other response categories and to demographic variables.

We shall illustrate the methodology on a number of questions from the survey on Family and Changing Gender Roles II in the International Social Survey Program (ISSP 1994). We shall use the German data from this study as an example, for both (former) West and East Germany, involving a total sample of 3291 respondents (a few respondents had to be omitted owing to missing data for the demographic variables of interest). We consider 11 questions (Table 8.1) related to the issue of single or dual earners in the family, mostly concerning the question of women working or not, which we shall call the substantive variables (see Chapter 20 for an analysis of the same questions). To simplify our presentation, we have combined the two response categories of agreement, "strongly agree" and "agree somewhat," into one, and similarly have combined the two corresponding categories of disagreement. In Table 8.1 we also list five demographic variables, referred to as the exogenous variables, that will be used to interpret the patterns of response found among the 11 substantive variables. The raw response data of interest are thus of the form given in Table 8.2a, showing the first four substantive and first two exogenous variables as examples, while Table 8.2b shows the same data coded as zero–one dummy variables in the columns of an indicator matrix. The Burt matrix corresponding to these data would be equal to the transpose of the indicator matrix multiplied by itself; part of the Burt matrix is shown in Table 8.2c.

There are two possible analytical strategies in this situation: first, using MCA, that is, CA of the indicator matrix of dummy variables corresponding to the substantive variables (or the corresponding cross-tabulations in the Burt matrix), with the categories of the exogenous categories displayed as supplementary points (see Chapter 2, Section 2.4); or second, CA of the cross-tabulations of the variables with the exogenous variables, that is, an analysis of several concatenated tables (see Chapter 1 and, for example, Greenacre 1994). Here we treat the former case of MCA, which is more concerned with the interrelationships between the substantive variables, with the exogenous variables visualized *a posteriori*.

In order to motivate our approach, first consider the usual MCA map of these data in Figure 8.1, showing all the categories of the substantive variables and of the exogenous variables, the latter displayed as supplementary points. This result is typical of analyses of survey data such as these, where nonresponse categories have been introduced into the analysis: the nonresponse categories are highly associated across questions and have separated out from the actual response categories. The latter response categories form a diagonal strip of points, with the "polar" categories of agreement (1) and disagreement (3) generally at the extremes and the unsure categories (?) in the middle, while the supplementary points (indicated by a diamond symbol, without labels) form another band of points just to the right of the response categories. In many similar examples, the nonresponse categories are aligned with the first principal axis, and we can thus eliminate most of their effect by mapping the points with respect to the plane of the second and third

Table 8.1 List of variables used in this study.

	Survey Questions
A	A working mother can establish just as warm and secure a relationship with her children as a mother who does not work
B	A pre-school child is likely to suffer if his or her mother works
C	All in all, family life suffers when the woman has a full-time job
D	A job is all right, but what most women really want is a home and children
E	Being a housewife is just as fulfilling as working for pay
F	Having a job is the best way for a woman to be an independent person
G	Most women have to work these days to support their families
H	Both the man and woman should contribute to the household income
I	A man's job is to earn money; a woman's job is to look after the home and family
J	It is not good if the man stays at home and cares for the family and the woman goes out to work
K	Family life often suffers because men concentrate too much on their work
	Exogenous Variables
Region	DW (West Germany), DE (East Germany)
Sex	M, F
Age	A1 (up to 25), A2 (26–35), A3 (36–45) A4 (46–55), A5 (56–65), A6 (66 and over)
Marital status	MA (married), WI (widowed), DI (divorced), SE (separated), SI (single)
Education	E0 (none), E1 (incomplete primary), E2 (primary), E3 (incomplete secondary), E4 (secondary), E5 (incomplete tertiary), E6 (tertiary)

Note: Response scales for each question: strongly agree, agree somewhat, neither agree nor disagree, disagree somewhat, strongly disagree. In our application, we have merged the two categories of agreement into one and the two categories of disagreement into one.

Source: Survey on Family and Changing Gender Roles II, as part of the International Social Survey Program (ISSP 1994).

Table 8.2 Raw data in two different but equivalent forms: (a) original response data for the first four questions (A, B, C, and D) and the first two exogenous variables (region and sex), and (b) the corresponding indicator (dummy variable) form of coding; (c) a part of the Burt matrix showing some cross-tabulations with question A only.

(a)

	A	B	C	D	...	Region	Sex	...
↑	1	1	1	1	...	1	1	...
	3	1	1	3	...	1	1	...
	1	1	1	1	...	1	1	...
	1	3	3	3	...	1	2	...
3291	1	2	2	3	...	1	2	...
cases	1	1	3	2	...	1	1	...
	.	.	.	.	...	.	.	...
	.	.	.	.	...	.	.	...
	.	.	.	.	...	.	.	...
↓	.	.	.	.	...	.	.	...

(b)

	A				B				C				D				...	Region		Sex		...
	+	?	−	x	+	?	−	x	+	?	−	x	+	?	−	x		DW	DE	M	F	
↑	1	0	0	0	1	0	0	0	1	0	0	0	1	0	0	0	...	1	0	1	0	...
	0	0	1	0	1	0	0	0	1	0	0	0	0	0	1	0	...	1	0	1	0	...
	1	0	0	0	1	0	0	0	1	0	0	0	1	0	0	0	...	1	0	1	0	...
	1	0	0	0	0	0	1	0	0	0	1	0	0	0	1	0	...	1	0	0	1	...
3291	1	0	0	0	0	1	0	0	0	1	0	0	0	0	1	0	...	1	0	0	1	...
cases	1	0	0	0	1	0	0	0	0	0	1	0	0	1	0	0	...	1	0	1	0	...
	.	.	.	.	.	.	.	.	.	.	.	.	.	.	.	.	...	.	.	.	.	...
	.	.	.	.	.	.	.	.	.	.	.	.	.	.	.	.	...	.	.	.	.	...
	.	.	.	.	.	.	.	.	.	.	.	.	.	.	.	.	...	.	.	.	.	...
↓	.	.	.	.	.	.	.	.	.	.	.	.	.	.	.	.	...	.	.	.	.	...

Note: Response categories for questions A–K are: 1, strongly agree or agree (combined responses [+]); 2, neither agree nor disagree [?]; 3, disagree or strongly disagree (combined responses [−]); 4, nonresponse [x]. Data are shown only for first six respondents (out of $n = 3291$). Response categories for two exogenous variables (region and sex) are as follows. Region: 1, former West Germany (DW); 2, former East Germany (DE). Sex: 1, male (M); 2, female (F).

Table 8.2 (Continued.)

(c)

		A				B				...	Region		...
		+	?	–	x	+	?	–	x		DW	DE	
A	+	2675	0	0	0	1328	374	901	72	...	1685	990	...
	?	0	111	0	0	85	17	8	1	...	92	19	...
	–	0	0	525	0	472	13	31	9	...	461	64	...
	x	0	0	0	110	61	3	4	42	...	86	24	...
	.	.	.	.	.	.	.	.	.	...	.	.	...
	.	.	.	.	.	.	.	.	.	...	.	.	...
	.	.	.	.	.	.	.	.	.	...	.	.	...
	.	.	.	.	.	.	.	.	.	...	.	.	...

dimensions. In this particular example, however, there is an added complication that the nonresponses separate out diagonally in the plane, which would necessitate a rotation to "reify" the solution, an operation that is perfectly feasible in correspondence analysis but almost never done or incorporated into CA software.

In order to investigate the spread of points in the cloud of points in the upper left-hand side of Figure 8.1, we have several possible courses of action. One possible strategy would be to remove all cases that have nonresponses, called "listwise deletion," but this would entail a reduction in sample size from 3291 to 2479, that is, a loss of 812 cases, or 25% of the sample. A second strategy would be to try to isolate the effect of the nonresponses on the first principal axis by rotating the solution appropriately, as mentioned above, and then considering dimensions from two onward. But the nonresponse points will still contribute, albeit much less, to these subsequent dimensions, thereby complicating the interpretation. A third way to improve the visualization of the data would be to omit the nonresponse categories from the indicator matrix and then apply CA. But then the totals of the rows of the indicator matrix are no longer equal, and the profiles have different nonzero values (and different masses), depending on the number of nonmissing responses.

A better solution, as we intend to show, will be provided by applying a variant of CA, called subset correspondence analysis, to the indicator

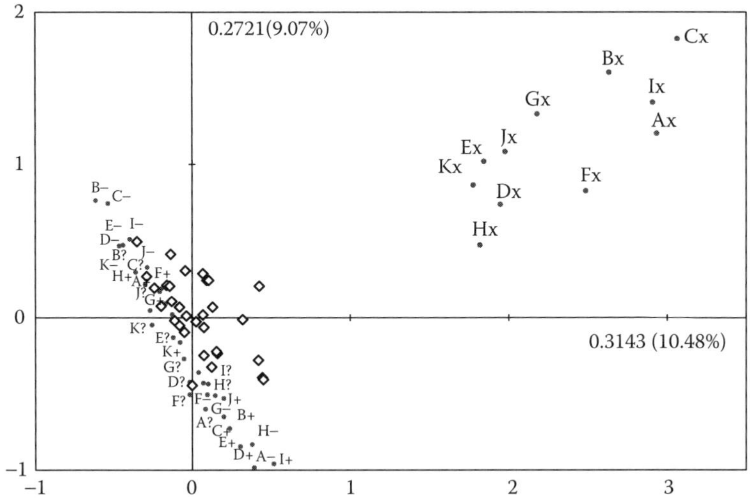

Figure 8.1 MCA map of Table 8.2, showing the four response categories for each of the 11 questions A to K (see Table 8.1). The unlabeled points with diamond symbols are the supplementary points for the exogenous variables.

matrix or to the Burt matrix. Greenacre and Pardo (in press) showed how subset correspondence analysis can improve the interpretability of CA maps by focusing on subsets of points. Our proposal is thus to apply this subset methodology to the present case of MCA. This approach can be applied to *any* subset of the response categories, but usually a subset will consist of the same response categories across the questions: for example, one might simply want to analyze all the categories excluding nonresponses, or the "agreement" categories alone, or even just the nonresponse categories by themselves. The main idea is to analyze the subset of the original profile matrix, in this case the row profile matrix, and not reexpress the profiles relative to their new totals within the subset. Furthermore, the row and column masses used in any subset are the same as those masses in the original data matrix of which a subset is being analyzed. A further benefit of this approach is that, if we partition the categories completely into mutually exclusive subsets, then we obtain a decomposition of the total inertia of the indicator matrix into parts accounted for by each subset. This decomposition of total inertia into parts is even more interesting when we think of MCA as an analysis of the Burt matrix rather than of the indicator matrix.

8.2 Correspondence analysis of a subset of an indicator matrix

CA, and thus MCA too, is a particular case of weighted principal component analysis (see, for example, Greenacre 1984: chapter 3). In this general scheme, a set of multidimensional points exists in a high-dimensional space in which distance is measured by a weighted Euclidean metric and the points themselves have differential weights, these latter weights being called masses to distinguish them from the dimension weights. A two-dimensional solution (in general, low-dimensional solution) is obtained by determining the closest plane to the points in terms of weighted least-squared distance, and then projecting the points onto the plane for visualization and interpretation. The original dimensions of the points can also be represented in the plane by projecting unit vectors onto the plane; these are usually depicted as arrows rather than points, since they can be considered as directions in the biplot style of joint interpretation of row and column points (Gower and Hand 1996; Greenacre 1993b, 2004). In the context of MCA, however, when the rows represent many, often thousands, of respondents, we are generally interested in the column points only along with groups of respondents, for example, age groups or social class groups represented as supplementary column points or, equivalently, by the centroids of the respondent points that fall into these groups.

The most general problem and solution are as follows. Suppose that we have a data matrix $\mathbf{Y}$ $(n \times m)$, usually centered with respect to rows or columns or both. We assume that the rows represent respondents and that the columns represent variables, which in our context are categories of response. Let $\mathbf{D}_r$ $(n \times n)$ and $\mathbf{D}_w$ $(m \times m)$ be diagonal matrices of row masses and column weights, respectively, where the masses give differentiated importance to the rows and the column weights and serve to normalize the contributions of the variables in the weighted Euclidean distance function between rows. With no loss of generality, the row masses are presumed to have a sum of 1. The rows of $\mathbf{Y}$ are thus presumed to be points with varying masses, given by the diagonal of $\mathbf{D}_r$, in an m-dimensional Euclidean space, structured by the inner product and metric defined by the weight matrix $\mathbf{D}_w$. The solution, a low-dimensional subspace that fits the points as closely as possible using weighted least squares, minimizes the following function:

$$\text{In}(\mathbf{Y} - \hat{\mathbf{Y}}) = \sum_{i=1}^{n} r_i \, (\mathbf{y}_i - \hat{\mathbf{y}}_i)^{\mathsf{T}} \mathbf{D}_w \, (\mathbf{y}_i - \hat{\mathbf{y}}_i) \tag{8.1}$$

where $\hat{\mathbf{y}}_i$, the ith row of $\hat{\mathbf{Y}}$, is the closest low-dimensional approximation of $\mathbf{y}_i$ (equivalently, $\hat{\mathbf{Y}}$ is the best optimal low-rank matrix approximation of $\mathbf{Y}$). The function In(·) stands for the *inertia*, in this case the inertia of the difference between the original and approximated matrices. The *total inertia*, a measure of dispersion of the points in the full m-dimensional space, is equal to In(**Y**).

The solution can be obtained compactly and neatly using the generalized singular-value decomposition (GSVD) of the matrix **Y** (see Greenacre 1984: appendix A). Computationally, using the ordinary singular-value decomposition (SVD) algorithm commonly available in software packages such as R (Venables and Smith 2003), the steps in finding the solution are to first transform the matrix **Y** by pre- and postmultiplying by the square roots of the weighting matrices, then calculate the SVD, and then postprocess the solution using the inverse transformation to obtain principal and standard coordinates. The steps are summarized as follows:

Step 1: $$\mathbf{S} = \mathbf{D}_r^{1/2}\mathbf{Y}\mathbf{D}_w^{1/2} \tag{8.2}$$

Step 2: $$\mathbf{S} = \mathbf{U}\boldsymbol{\Delta}\mathbf{V}^{\mathsf{T}} \tag{8.3}$$

Step 3: $$\text{Principal coordinates of rows: } \mathbf{F} = \mathbf{D}_r^{-1/2}\mathbf{U}\boldsymbol{\Delta} \tag{8.4}$$

Step 4: $$\text{Principal coordinates of columns: } \mathbf{G} = \mathbf{D}_w^{1/2}\boldsymbol{\Delta}\mathbf{V} \tag{8.5}$$

Step 2 is the SVD, with the (positive) singular values in descending order in the diagonal matrix $\boldsymbol{\Delta}$, and left and right singular vectors in the matrices **U** and **V**, respectively. A two-dimensional solution, say, would use the first two columns of **F** and **G**, where the principal coordinates in **F** and **G** are the projections of the rows (respondents) and columns (variables) onto principal axes of the solution space. An alternative scaling for the columns is to plot standard coordinates, that is, Equation 8.5 without the postmultiplication by $\boldsymbol{\Delta}$; the points are then projections onto the principal axes of the unit vectors representing the column variables and are usually depicted by vectors from the origin of the map to the points. The total inertia is the sum of squares of the singular values $\delta_1^2 + \delta_2^2 + \ldots$, the inertia accounted for in a two-dimensional solution is the sum of the first two terms $\delta_1^2 + \delta_2^2$, while the inertia not accounted for (minimized in Equation 8.1) is the remainder of the sum: $\delta_3^2 + \delta_4^2 + \ldots$.

Regular MCA (see Chapter 2) is the above procedure applied to an indicator matrix $\mathbf{Z}$ of the form illustrated by the left-hand matrix in Table 8.1b, that is, of the $Q = 11$ questions (variables). The matrix $\mathbf{Y}$ is this indicator matrix divided by Q (i.e., the row profile matrix), centered with respect to the averages of its columns. The averages of the columns are the column totals of $\mathbf{Z}$ divided by $\mathbf{Z}$'s grand total nQ, where n is the number of rows (respondents) and, hence, are exactly the proportions of respondents giving the corresponding categories of response, divided by Q. Thus, if a particular category of response is given by 2675 of the total of 3291 respondents (as in the case of A+, see Table 8.2c), and since there are 11 questions, then the corresponding column average is equal to the proportion of response 2675/3291 = 0.8128 divided by 11, i.e., 0.0739. In matrix notation:

$$\mathbf{Y} = \frac{1}{Q}\mathbf{Z} - \frac{1}{nQ}\mathbf{1}\mathbf{1}^{\mathsf{T}}\mathbf{Z} = \left(\mathbf{I} - \frac{1}{n}\mathbf{1}\mathbf{1}^{\mathsf{T}}\right)\left(\frac{1}{Q}\mathbf{Z}\right)$$

The row masses in this case are all equal, being the row totals of $\mathbf{Z}$, all equal to Q, divided by its grand total nQ; hence, the masses are all $1/n$. The column weights used (inversely) in the chi-square distance function are in the vector of column averages $(1/nQ)\mathbf{1}^{\mathsf{T}}\mathbf{Z}$.

We now wish to analyze and visualize a chosen subset of the indicator matrix. The subset version of simple CA of Greenacre and Pardo (in press), applied to the indicator matrix, implies that we maintain the same row and column weighting as in classical MCA described above, but the matrix to be analyzed is the chosen subset of the profile matrix $\mathbf{Y}$, not of the original indicator matrix. That is, suppose that $\mathbf{H}$ is a selected subset of the columns of $\mathbf{Y}$, already centered, and that the corresponding subset of column weights (masses) is denoted by $\mathbf{h}$. Then subset MCA is defined as the principal component analysis of $\mathbf{H}$ with row masses $\mathbf{r}$ in $\mathbf{D}_r$, as before, and metric defined by $\mathbf{D}_h^{-1}$, where $\mathbf{D}_h$ is the diagonal matrix of $\mathbf{h}$. Hence, the subset MCA solution is obtained using steps 1 through 4, with $\mathbf{Y}$ equal to $\mathbf{H} - \mathbf{1}\mathbf{h}^{\mathsf{T}} = (\mathbf{I} - \mathbf{1}\mathbf{r}^{\mathsf{T}})\mathbf{H}$, $\mathbf{D}_r$ equal to the present $\mathbf{D}_r$, and $\mathbf{D}_w$ equal to $\mathbf{D}_h^{-1}$. The matrix (Equation 8.2) that is decomposed is thus:

$$\mathbf{S} = \mathbf{D}_r^{1/2}(\mathbf{I} - \mathbf{1}\mathbf{r}^{\mathsf{T}})\mathbf{H}\mathbf{D}_h^{-1/2} \tag{8.6}$$

and the row and column principal coordinates from Equation 8.4 and Equation 8.5 are thus:

$$\mathbf{F} = \mathbf{D}_r^{-1/2}\mathbf{U}\Delta \quad \mathbf{G} = \mathbf{D}_h^{-1/2}\mathbf{V}\Delta \tag{8.7}$$

All the usual numerical diagnostics (or contributions) of ordinary CA apply as previously, since the total inertia, equal to the sum of squares of Equation 8.6, can be broken down into parts corresponding to points and to principal axes, thanks to the SVD decomposition (see Greenacre 2004).

8.3 Application to women's participation in labor force

Figure 8.2 shows the subset MCA map of the response categories agree (+), neither agree nor disagree (?), and disagree (–) for the questions A to K, omitting the nonresponses (x). All the unsure (?) points line up on the vertical second axis of the solution, while most of the agreements

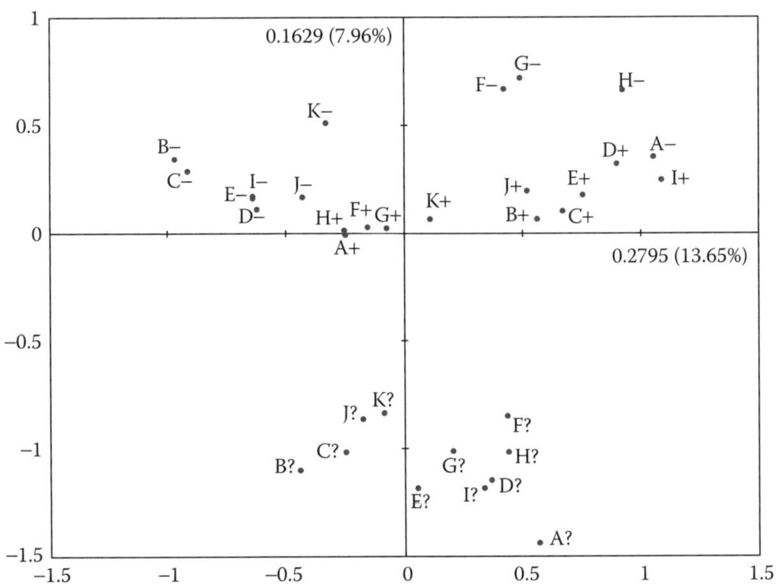

Figure 8.2 Subset MCA map of the response categories omitting the nonresponse categories.

and disagreements are spread horizontally along the first axis. Among the 11 statements, there are four that are worded in an opposite sense compared with the others: in statements A, F, G, and H, agreement represents a favorable attitude to women working, whereas in the other statements it is the opposite. Notice that in Figure 8.2 the disagreements for these four statements are on the side of the agreements with the others, which is what we would expect.

In order to interpret the agreement and disagreement categories without having to cope with the dimension of "neither … nor" responses, Figure 8.3 shows the subset MCA map of what we shall loosely call the "polar" responses, that is, the agreements (+) and disagreements (–), without the "nonpolar" responses, that is, without nonresponses (x) and "neither … nor"s (?). The spectrum of agreements and disagreements is now easier to interpret, and we see that there is a closely correlated group of items B, C, E, I, and to a lesser extent J, stretching horizontally across the map and that these are being responded to differently

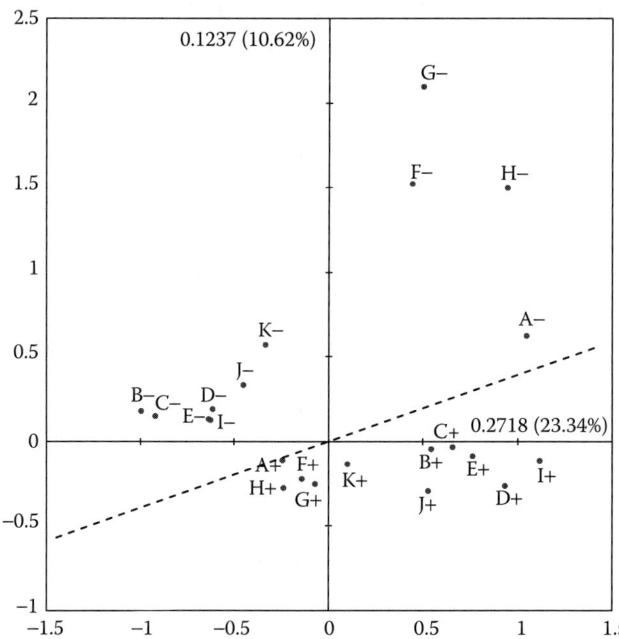

Figure 8.3 Subset MCA map of the agreement and disagreement categories only (A+, A– to K+, K–), without NSRs ("neither agree nor disagree" and nonresponses). The dashed line shows a direction lining up attitudes from lower left (favorable to women working) to upper right (unfavorable to women working).

Table 8.3 Percentages of co-occurrences of agreements (+), unsures (?), and disagreements (–) between the two sets of statements with opposite wording with respect to women working: A, F, G, and H are favorable to women working, while B, C, D, E, and I are unfavorable.

		B, C, D, E, I		
		+	?	–
	+	30.1%	10.6%	38.2%
A, F, G, H	?	3.9%	1.7%	2.9%
	–	7.3%	1.2%	4.0%

Note: Nonresponses were omitted (using pairwise deletion) when compiling these percentages.

than F, G, H, and to a lesser extent A, which are running more or less vertically in the map. (The solution could be slightly rotated anticlockwise to line up these two sets of points with the two axes of the map.) These two groups of questions are worded in an opposite sense: the former group favorable to women staying at home, and the latter group favorable to women working. (In the context of a woman in a partnership or family, these items are labeled as "single earner" and "dual earner," respectively.) From the subset MCA map we can see that agreement to a question of one set is not necessarily associated with disagreement to a question in the other set. Otherwise A+, F+, G+, and H+ should lie among the bunch of "–" categories of the other statements, and similarly A–, F–, G–, and H– should lie among the "+" categories of the others.

To check this observation in the data, we constructed a table where we counted how many times respondents had agreed or disagreed to statements in the group A, F, G, and H compared with their agreement or disagreement to the statements in the group formed by B, C, D, E, and I (Table 8.3). If agreements to the first set of questions generally co-occurred with disagreements to the second set, then we would expect much larger percentages in the top right and bottom left of the table, compared with their corresponding margins. However, this table shows that whereas 38.2% of co-occurring responses are in the upper right corner of the table (agreements to A, F, G, H and disagreements to B, C, D, E, I), there are as many as 30.1% co-occurring agreements to statements from both sets. There is a similar effect in the last row: although the overall level of disagreement to A, F, G, H is lower, there are relatively many co-occurrences of disagreements to statements from both sets (4.0%) compared with the response patterns where

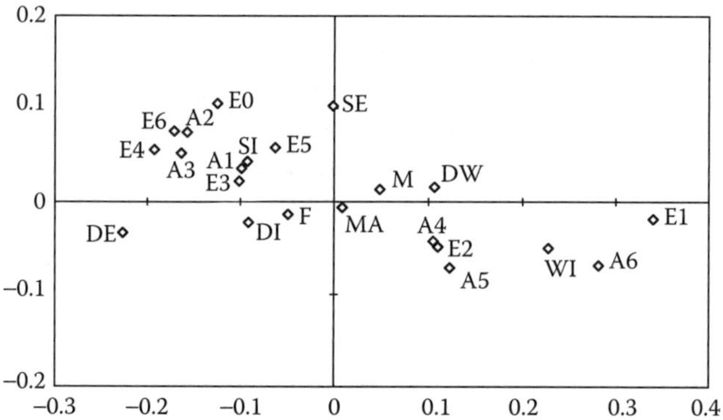

Figure 8.4 Positions of the supplementary points in the map of Figure 8.3.

disagreement to statements in the first set coincides with agreement to statements in the second set (7.3%).

From the map in Figure 8.3, a direction of spread that scales attitudes regarding women working would lie diagonally across the map, drawn informally as a dashed line. The projection of the categories onto this "axis" lines up attitudes favorable to women working, with most favorable toward lower left and least favorable toward upper right. Figure 8.4 shows the categories of the exogenous demographic variables, which are represented as supplementary points with respect to the dimensions of Figure 8.3, but displayed separately. The points East Germany (DE) and West Germany (DW) as well as males (M) and females (F) follow the line of the dashed-line "axis" drawn in Figure 8.4, with East Germans more in favor of women working than West Germans, and females more in favor than males. Taking an example from each set of points in the original data, we find that 11.0% of East Germans think that "a man's job is to earn money; a women's job is to look after the home and family" (statement I), compared with 36.3% for West Germans; and 90.2% of East Germans think that "a working mother can establish just as warm and secure a relationship with her children as a mother that works" (statement A), compared with 72.5% for West Germans. In the case of females versus males, the difference is less pronounced, as seen in Figure 8.4, but in the same direction: for example, using the same statements as above, 25.7% of females agree with statement I, compared with 30.6% for males, and 82.0% of females agree with statement A, compared with 74.4% for males. When it comes to the education and age groups in Figure 8.4, the spread is more from upper left to lower right, that is, in the direction of the former group of questions (worded in favor of women not working) rather than the latter. To verify this, consider the youngest and oldest age groups (A1 and A6, respectively) and the same statements as above. In the case of statement I, only 18.1% of the youngest group agrees, compared with 53.5% of the oldest group, which is a huge difference. But in the case of statement A, the differences are modest: 75.4% of the youngest group agree, compared with 72.6% of the oldest.

All of the interesting detail in the above interpretation has been made possible thanks to the use of the subset analysis, since none of these features was apparent in the previous maps that included more categories. The interpretation using supplementary points could be further enriched by indicating the positions of interactive subgroups of respondents, for example, male East Germans, female East Germans, male West Germans, and female West Germans, or East Germans in different age groups, and so on.

Finally, we performed a subset MCA of the nonresponse (x) categories alone, which we had originally omitted, and the resulting map is shown in Figure 8.5. All the category points are on the positive side

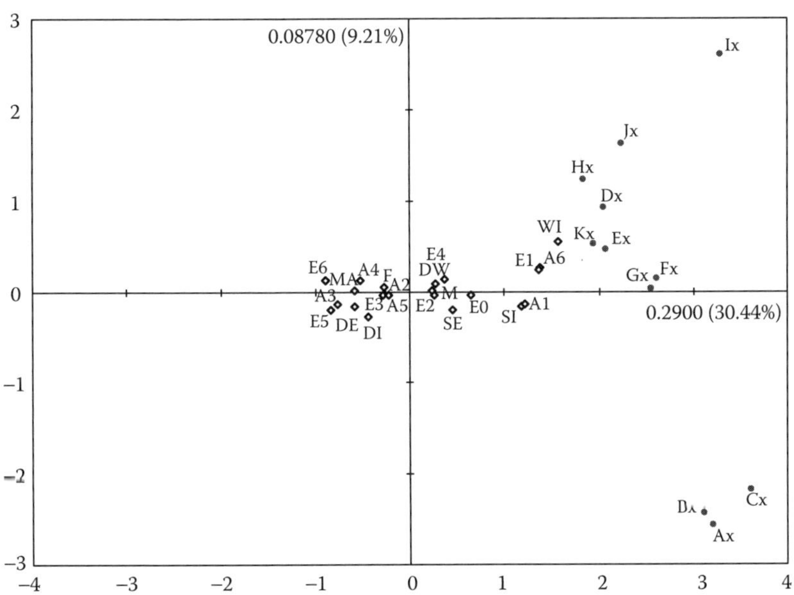

Figure 8.5 Subset MCA map of the nonresponse categories only (Ax to Kx), showing supplementary demographic categories.

of the first axis, so that the first axis is a dimension of overall nonresponses and would be highly correlated with a count of nonresponses for each respondent. However, we see an interesting dichotomy in the items, with nonresponses to A, B, and C clearly separated from the rest; these are exactly the items that include the word "mother" and relate to the working woman's relationship with her children and family, for which there must exist a special pattern of interrelated nonresponses in some respondents. The origin of the map in Figure 8.5 represents the average nonresponse point for all 11 questions. Demographic categories are also shown in Figure 8.5, and those to the left of center will thus have fewer than average nonresponses and categories to the right more than average. Hence, higher educational groups have fewer nonresponses, as do East Germans compared with West Germans. Both the youngest and oldest age groups have higher than average nonresponses, and so on. Similar to Figure 8.3 and Figure 8.4, the contrast in attitudes reflected by the second dimension is not correlating so strongly with the demographic categories. The group widowed (WI) is the only one that distinguishes itself slightly in the specific direction of nonresponses to questions D to K, lying on the upper side of the vertical dimension, while there are no demographic groups that are separating out specifically in the direction of nonresponses to A through C on the lower side.

Thanks to the fact that the margins in each subset analysis are always determined from the full data table, there is an interesting decomposition of total inertia across the subset analyses. A subset MCA of the complete indicator matrix, which is just an ordinary MCA, has a total inertia equal to $(J - Q)/Q = (44 - 11)/11 = 3$, where Q = number of questions and J = total number of categories. In the analysis of the subset without the nonresponse categories (Figure 8.2), the total inertia is 2.046, while in the analysis of the nonresponse categories alone (Figure 8.5), the total inertia is 0.954. Hence, in subset MCA, the total inertia of all categories is decomposed into parts for each of the two mutually exclusive but exhaustive subsets (i.e., "complete disjunctive" subsets). This breakdown is summarized in Table 8.4, including the percentages of inertia on the first two dimensions of each analysis reported previously. When the "neither...nor"s (?) are removed from the analysis of Figure 8.2, the total inertia for the "polar" responses (Figure 8.3) is equal to 1.163. From these results one can deduce that the inertia of the "neither...nor" categories is equal to $2.046 - 1.163 = 0.833$. Thus, the inertia from "nonpolar responses" (? and x) is equal to $0.954 + 0.833 = 1.837$, more than half of the inertia in the original MCA.

Table 8.4 Total inertias of different subsets of categories and the percentages of inertia along the first two dimensions of the analyses reported in Figures 8.1, 8.2, 8.3, and 8.5.

		Percentages	
Subset Analyzed	Total Inertia	Axis 1	Axis 2
All response categories (+, ?, –, x) (Figure 8.1)	3.000	10.4 (50.5)	9.1 (34.3)
Without nonresponses (+, ?, –) (Figure 8.2)	2.046	13.7	8.0
Polar responses (+, –) (Figure 8.3)	1.163	23.4	10.7
Nonresponses (x) (Figure 8.5)	0.954	30.1	9.3

Note: For the first analysis of all the categories (MCA), the adjusted percentages are given in parentheses.

The percentages of inertia indicated in each map are all underestimates of the true variance explained, which is the same issue that affects the percentages of inertia in MCA. In Figure 8.1, where all categories are analyzed and which is thus a regular MCA, the principal inertias (eigenvalues) can be adjusted (see Chapter 2). Table 8.4 shows the adjusted percentages in the first row of the table. When analyzing the subset, we could still calculate the percentages of inertia explained conditional on the coordinates obtained in the subset analysis, but it is not clear whether adjusted estimates can be obtained easily from the subset MCA solution, as in MCA.

8.4 Subset MCA applied to the Burt matrix

The relationship between subset MCA of the indicator matrix and a subset analysis of the corresponding Burt matrix gives another perspective on the problem. As an example, suppose that we divide the question response categories into "polar responses" (PRs) and "nonpolar responses" (NPRs): as in our application, the categories "+"and "–" on the one hand, and the categories "?" and "x" on the other. Then the Burt matrix **B** can be partitioned into four parts:

	PR	NPR
PR	$\mathbf{B}_{11}$	$\mathbf{B}_{12}$
NPR	$\mathbf{B}_{21}$	$\mathbf{B}_{22}$

Table 8.5 Total inertias of different subsets of categories in the analysis of the submatrices of the Burt matrix and the percentages of inertia along the first two dimensions.

		Percentages	
Subset Analyzed	Total Inertia[a]	Axis 1	Axis 2
All response categories (+, ?, −, x)	0.3797	26.0 (50.5)	19.5 (34.3)
Polar responses (+, −)	0.1366	54.1	11.2
Nonpolar responses (?, x)	0.2077	41.9	9.0
PRs by NPRs	0.0177	35.1	15.7

Note: For the first analysis of all the categories (MCA), the adjusted percentages are given in parentheses.

[a] It is readily checked that 0.3797 = 0.1366 + 0.2077 + 2(0.0177), as described in the text, corresponds to the inertias of the submatrices of the Burt matrix.

If $\lambda_1, \lambda_2, \ldots, \lambda_n$ are the principal inertias of the indicator matrix **Z**, with sum $(J - Q)/Q)$, then $\lambda_1^2, \lambda_2^2, \ldots, \lambda_n^2$ are the principal inertias of the Burt matrix **B** (see Chapter 2). Table 8.5 gives the total inertias and first two percentages of inertia for several subset MCAs of **B**, including the complete MCA in the first line. For example, in the complete MCA in Figure 8.1, the first principal inertia is 0.3143, which if squared is equal to 0.09878. This is the first principal inertia of the Burt matrix and represents 26.0% of the total inertia 0.3797 of **B**, which agrees with Table 8.5. This property carries over to the subset analyses as well. For example, in Table 8.4, the first principal inertia in the subset MCA of the PRs is 0.2718, which if squared is equal to 0.07388. Expressed as a percentage of the inertia 0.1366 of the submatrix $\mathbf{B}_{11}$ of the Burt matrix, which analyzes the PRs, a percentage of 54.1% is obtained, again agreeing with the percentage reported in Table 8.5. The connection—between the principal inertias in the subset MCA of the indicator matrix and the subset MCA of the corresponding part of the Burt matrix—holds for exactly the same reason as in the complete MCA: the matrix $\mathbf{B}_{11}$ analyzed in the latter case is exactly $\mathbf{Z}_1^\mathsf{T}\mathbf{Z}_1$, where $\mathbf{Z}_1$ is the submatrix of the indicator matrix analyzed in the former case.

From the above partitioning of **B** into four submatrices, the total inertia of **B** is equal to the sum of inertias in the subset analyses of $\mathbf{B}_{11}$, $\mathbf{B}_{22}$, $\mathbf{B}_{12}$, and $\mathbf{B}_{21}$. (Notice that the last two are transposes of each other, so we could just count the inertia of one of them twice; the footnote for Table 8.5 gives the calculation to verify this assertion.) As we have just noted, the subset analyses of $\mathbf{B}_{11}$ and $\mathbf{B}_{22}$ give results whose principal inertias are exactly the squares of those in the respective subset analyses of the corresponding indicator matrices of PR and NPR responses. But there is an "extra" subset analysis, namely that of $\mathbf{B}_{12}$, that is manifest in the Burt matrix but not in the indicator matrix. In our illustration, the submatrix $\mathbf{B}_{12}$ captures the associations between PRs and NPRs. In Table 8.5, which gives the corresponding decomposition of inertia for these subset analyses, we can see that the level of association between the PRs and NPRs is much less than within PRs and within NPRs. It should be remembered, however, that all these components of inertia are inflated by fragments of the "diagonal blocks" from the Burt matrix, as in the complete MCA. In the case of $\mathbf{B}_{11}$ and $\mathbf{B}_{22}$, these are inflated by subsets of the diagonal matrices on the diagonal of $\mathbf{B}$. In the case of $\mathbf{B}_{12}$ or $\mathbf{B}_{21}$, the elements of the matrix corresponding to the same question account for the inflation and consist of blocks of zeros because there is zero contingency between the PRs and NPRs of the same question. It is a question of continuing research whether there are simple ways for adjusting the eigenvalues and their percentages, as is possible in the MCA of the complete Burt matrix.

8.5 Discussion and conclusions

One of Benzécri et al.'s (1973) basic principles of *Analyse des Données* is that one should analyze all the available information, a principle that implies that every possible category of response, including missing responses, should be analyzed together. In the case of MCA, this means analyzing the so-called *tableau disjonctif complet* (the indicator matrix, in our terminology), which has as many ones in each row as there are variables indicating the categories of response. When analyzing several variables, however, it is almost always the case that the interpretation is hampered by the large number of category points in the map, all of which load to a greater or lesser extent on every dimension, so that interpretation and conclusions are limited to broad generalities. We have shown that there is great value in restricting the analysis to subsets of categories, which can be visualized separately and, thus, with better quality than would have been the case in a complete MCA. The method allows exploration of several issues of prime importance to social scientists:

- Analyzing substantive responses only, ignoring nonresponses
- Studying the pattern of nonresponses by themselves and how they relate to demographic variables

- Focusing on the role played by neutral responses, how they are related to one another, how they are related to the nonresponses, and whether any patterns correlate with the demographic variables

In the first visualization of the data (Figure 8.1), that is, the complete MCA, the points representing the missing data were so strongly associated that they forced all the other categories into a group on the opposite side of the first axis. Even though the frequency of nonresponses was fairly low, they dominate this map, leaving little remaining "space" to understand the relationships between the other responses categories. The subset analysis allowed this effect to be removed, showing the separation of the "nonmissing" response categories more clearly in Figure 8.2. The neutral categories could also be removed (Figure 8.3) to further clarify the associations between the agreement and disagreement poles of questions that were worded in a favorable and unfavorable direction toward working women. The subset could also consist of just one response category across the questions, as illustrated by the map in Figure 8.5, which showed the missing data categories only. In all cases, supplementary points could be added to show the relationship between the analyzed response categories and the demographic variables (Figure 8.4 and Figure 8.5).

The subset variant of simple CA has been extended here to MCA and maintains the geometry of the masses and chi-square distances of the complete MCA, the only difference being that we do not reexpress the elements of the subset with respect to their own totals, but maintain their profile values with respect to the totals of the complete data set. This approach ensures the attractive property that the total inertia is decomposed into parts for each of the subsets of categories. The same idea has already been used in an MCA context by Gifi (1990) to exclude nonresponses, where the approach is called "missing data passive" (see also Michailidis and de Leeuw 1998), and similarly by Le Roux and Rouanet (2004a). These uses are limited to the exclusion of missing data, whereas in our application we consider a much wider number of possibilities, including the analysis of missing data alone, where all the substantive responses are excluded.

Acknowledgment

The support of the BBVA Foundation (Fundación BBVA) in Madrid in sponsoring this research is gratefully acknowledged.

Software notes

Programs for CA, MCA, subset CA, and subset MCA are available as functions in the R language (www.r-project.org) as well as in XLSTAT (www.xlstat.com). See the computational appendix at the end of this book.

CHAPTER 9

Scaling Unidimensional Models with Multiple Correspondence Analysis

Matthijs J. Warrens and Willem J. Heiser

CONTENTS

9.1 Introduction

This chapter discusses the application of multiple correspondence analysis (MCA) as a method of scaling. For this type of application of MCA, several well-known unidimensional models from the psychometric literature will be considered. All of these models are characterized by item and person parameters. The objective of this chapter is to determine what information on these parameters can be obtained from applying MCA when the unidimensional models are used as gauges. "Gauge" is a term introduced by Gifi (1990) to denote benchmark data sets, or mechanisms to generate such data sets, for studying the behavior of general-purpose data analysis methods, such as MCA and simple correspondence analysis (CA). Some of the basics on this topic are reviewed, but many new insights and results are also presented.

The models discussed in this chapter can be classified by three different aspects, i.e., each model is either

Deterministic or probabilistic
Dichotomous or polytomous
Monotonic or unimodal

It follows that there are $2 \times 2 \times 2 = 8$ possible categories to classify a model, but only six of them will actually be discussed. For the categories corresponding to the probabilistic and unimodal options, no successful applications of MCA are currently available. Only for the two deterministic unimodal models are there some ideas for scaling with MCA (see Section 9.6). For each of the four deterministic categories, only one candidate model seems available, while for the probabilistic categories there are numerous possible models that can be selected. However, for three of the deterministic models, an additional distinction is presented between different possible forms of the models. This distinction provides several new insights into the question of when and especially how to apply MCA.

With MCA, CA, or related methods, the structure of the multivariate data is often visualized in a two-dimensional representation. A common conviction is that when one applies a technique such as MCA to data satisfying a unidimensional model, the resulting two-dimensional representation will reflect some sort of horseshoe (see van Rijckevorsel 1987; Gifi 1990; Hill and Gauch 1980). More precisely, the phenomenon will be called an arch if the graph reflects a quadratic function. If, in addition, the ends of this arch bend inward, the phenomenon will be called a horseshoe. It will be shown that unidimensionality is not a sufficient condition for a horseshoe, that is, the data may be unidimensional in terms of a model, but a method such as MCA will not produce a horseshoe in two or any other dimensions. Furthermore, it will be shown that, with an appropriate analysis, most relevant information in terms of item and person parameters can be found in the first MCA dimension.

We end this section with some words on notation. The quantification of category j $(1,\ldots,J_q)$ of item q $(1,\ldots,Q)$ on dimension s $(1,\ldots,S)$ is denoted by y_{qs}^{j}. In our context, J_q is the same for all items, and we use the notation J for this constant number of categories per item. (This is different from standard MCA notation, where J is the total number of categories for all items.) The vector $\mathbf{y}_s^j$ then denotes the quantifications of the jth category for all items on dimension s, and $\mathbf{y}_s$ denotes all quantifications on dimension s. Furthermore, let $\mathbf{x}_s$ denote the

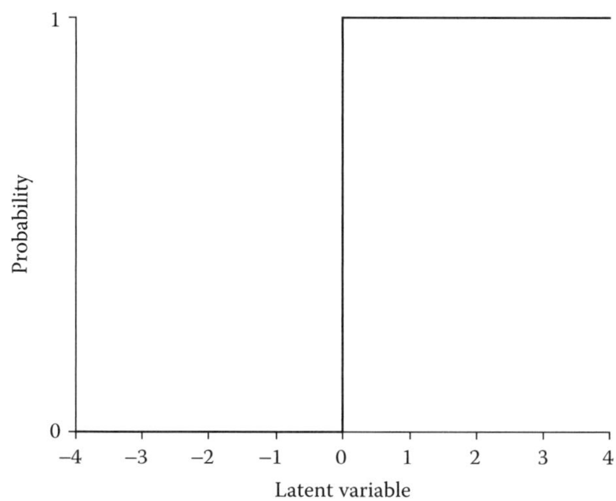

Figure 9.1 Item function of the DGS.

person scores on dimension s. Let u denote a latent variable, and δ_q a location parameter of item q. Explicit formulas for the item functions that relate u to δ_q will not be given; only their shapes will be shown in the figures.

9.2 The dichotomous Guttman scale

The simplest and oldest model considered in this chapter is the dichotomous Guttman scale (DGS), named after the person who popularized the model with the method of scalogram analysis. With the DGS, each item is characterized by a step function, as is shown in Figure 9.1.

Guttman (1950b, 1954) advocated the practical properties of the DGS, but earlier, other authors (for example, Walker 1931) also noted the special structure of the data matrix and the possibilities of ordering both persons and items. Parameters of the DGS are only unique in terms of their order, that is, they form an ordinal scale. Often-used estimates are the sum score as an index for persons and the proportion correct for items.

Both the DGS and the application of MCA to the DGS were thoroughly studied by Guttman (1950b, 1954). Guttman (1950b) derived that all relevant information for the ordinal properties of the DGS is contained in $\mathbf{y}_1$ for item categories and $\mathbf{x}_1$ for persons. Guttman

(1950b, 1954) also studied the quantifications and scores for dimensions $s > 1$. With Q items there are $Q + 1$ possible score patterns in a DGS. The DGS can be referred to as *complete* if all possible score pattern are present and *uniform* if all present score patterns occur equally often. The matrix **M** below contains an example of a complete and uniform DGS of three items.

$$\mathbf{M} = \begin{bmatrix} 1 & 1 & 1 \\ 2 & 1 & 1 \\ 2 & 2 & 1 \\ 2 & 2 & 2 \end{bmatrix} \Rightarrow \mathbf{Z} = \begin{bmatrix} 1 & 0 & 1 & 0 & 1 & 0 \\ 0 & 1 & 1 & 0 & 1 & 0 \\ 0 & 1 & 0 & 1 & 1 & 0 \\ 0 & 1 & 0 & 1 & 0 & 1 \end{bmatrix} \rightarrow \begin{bmatrix} 1 & 1 & 1 & 0 & 0 & 0 \\ 0 & 1 & 1 & 1 & 0 & 0 \\ 0 & 0 & 1 & 1 & 1 & 0 \\ 0 & 0 & 0 & 1 & 1 & 1 \end{bmatrix}$$

The matrix **Z** is the indicator matrix resulting from **M**, and the right matrix shows **Z** with the columns permuted such that the elements of $\mathbf{y}_1^1$ and $\mathbf{y}_1^2$ come together and are ordered. The data matrix **M** reflects a scalogram structure, whereas the (permuted) indicator matrix **Z** reflects a parallelogram structure.

For the DGS, Guttman (1950b) showed that

The ordering of the proportion correct is reflected in both $\mathbf{y}_1^1$ and $\mathbf{y}_1^2$
The ordering of the sum score is reflected in the person scores $\mathbf{x}_1$
$\mathbf{x}_2$ has a quadratic relation to $\mathbf{x}_1$

Guttman (1950b, 1954) considered even higher polynomial relations between $\mathbf{x}_1$ and $\mathbf{x}_s$ for $s > 2$, but these are outside the scope of this chapter. Note that $\mathbf{y}_1$ and $\mathbf{y}_2$ do not have a precise quadratic relation, although the relation will have resemblance to a horseshoe or an arch (see Figure 9.2). In fact there are two superimposed arches, one for each category, a result proven by Schriever (1985).

The above result shows that applying MCA to the DGS provides category quantifications that reflect the ordering of items and scores that reflect the ordering of persons. The same result would be obtained by using the proportion item correct and the sum score for indices, respectively. However, all these indices give an ordering for items and persons separately, that is, the indices for items and persons do not imply a simultaneous ordering. However, for a special case of the DGS, a stronger result can be obtained.

For the complete and uniform DGS, Guttman showed that the person scores are proportional to the sum scores. If there are Q δ_q's

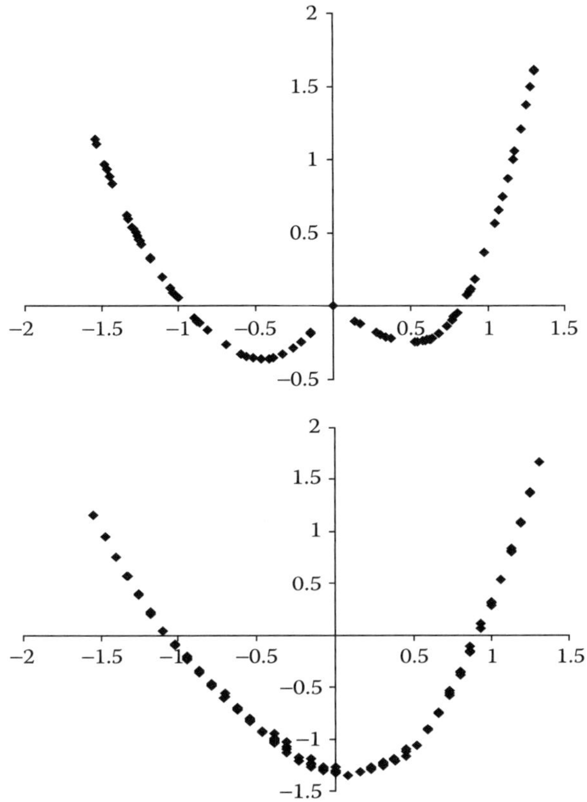

Figure 9.2 First two axes of MCA of DGS for $Q = 40$, $J = 2$; category quantifications (above) and person scores (below).

representing location parameters, there are $Q + 1$ u's indicating the possible person scores. For each pair u_q and u_{q+1}, a δ_q is required that satisfies $u_q \le \delta_q \le u_{q+1}$. A desirable value for this location parameter δ_q would be

$$\hat{\delta}_q = \frac{u_q + u_{q+1}}{2} \tag{9.1}$$

The following proposition shows how this estimate for δ_q can be obtained from y_q^1 and y_q^2, the MCA quantifications of the categories of item q on the first dimension.

Proposition. Under the condition of a complete uniform DGS, the estimate in Equation 9.1 can be obtained by taking $\delta_q = y_q^1 + y_q^2$.

Proof. With a uniform DGS, each score pattern occurs T times. Without loss of generality we can take $T = 1$. Because with a complete uniform scale the u's are proportional to the sum scores (Guttman 1950b), they are, for convenience, not put in standard scores, but are expressed as real integers, that is, $u_q = q$, for $q = 1,\ldots,Q + 1$, minus the grand mean. For a set of $Q + 1$ score patterns, the grand mean value is then given by $\frac{1}{2}(Q + 2)$. The category quantifications of the item that discriminates between u_q and u_{q+1} can be expressed as

$$y_q^1 = \tfrac{1}{2}(q+1);$$

$$y_q^2 = \tfrac{1}{2}(q+1) + \tfrac{1}{2}(Q+1)$$

before the grand mean is subtracted. Centering around the grand mean gives

$$u_q = q - \tfrac{1}{2}(Q+2);$$

$$y_q^1 = \tfrac{1}{2}(q+1) - \tfrac{1}{2}(Q+2)$$

$$= \tfrac{1}{2}q - \tfrac{1}{2}Q - \tfrac{1}{2}$$

$$y_q^2 = \tfrac{1}{2}(q+1) + \tfrac{1}{2}(Q+1) - \tfrac{1}{2}(Q+2)$$

$$= \tfrac{1}{2}q$$

Hence,

$$y_q^2 + y_q^1 = q - \tfrac{1}{2}Q - \tfrac{1}{2}$$

$$= \delta_q$$

9.3 The Rasch model

A probabilistic generalization of the DGS is the model proposed by Rasch (1960). In the Rasch model, an item is characterized by a logistic function instead of a step function. Similar to the DGS, the Rasch model is a location family, or a holomorphic model, meaning that the item functions have the same shape but are translations of each other (see Figure 9.3).

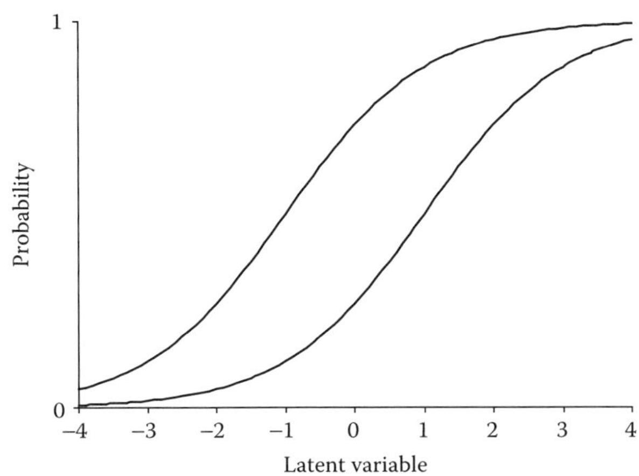

Figure 9.3 Two item functions of the Rasch model.

With a probabilistic dichotomous model, there are 2^Q possible score patterns for Q items. However, depending on how the Rasch model is specified, some score patterns are more likely to occur than others. For example, if one specifies the Rasch model such that it has relatively steep slopes, then the score patterns will have close resemblance to the score patterns of the DGS. On the other hand, if the Rasch model is specified such that the slopes are not steep, the score patterns will look close to random. The existence of multiple forms of the Rasch model has important consequences for the application of MCA and related methods. A Rasch model with steep slopes will provide a two-dimensional solution that is similar to the arch for the DGS. For a Rasch model with shallow slopes, the two-dimensional solution will look like random data without structure (see Figure 9.4).

So, MCA has a limited usefulness in detecting probabilistic unidimensional models. However, instead of looking at both $\mathbf{y}_1$ and $\mathbf{y}_2$ or $\mathbf{x}_1$ and $\mathbf{x}_2$, several results indicate that most relevant information for the Rasch models is obtained in $\mathbf{y}_1$ and $\mathbf{x}_1$.

Schriever (1985) showed that, if the item functions of a dichotomous model are monotonic and have monotone likelihood ratio (Lehmann 1966), then the ordering is reflected in both $\mathbf{y}_1^1$ and $\mathbf{y}_1^2$. The item functions of the Rasch model satisfy this more general property. Furthermore, because the category quantifications reflect the ordering of the items, the reciprocal averages, that is, the person scores, will reflect a reasonable ordering of the persons. These ordering properties are

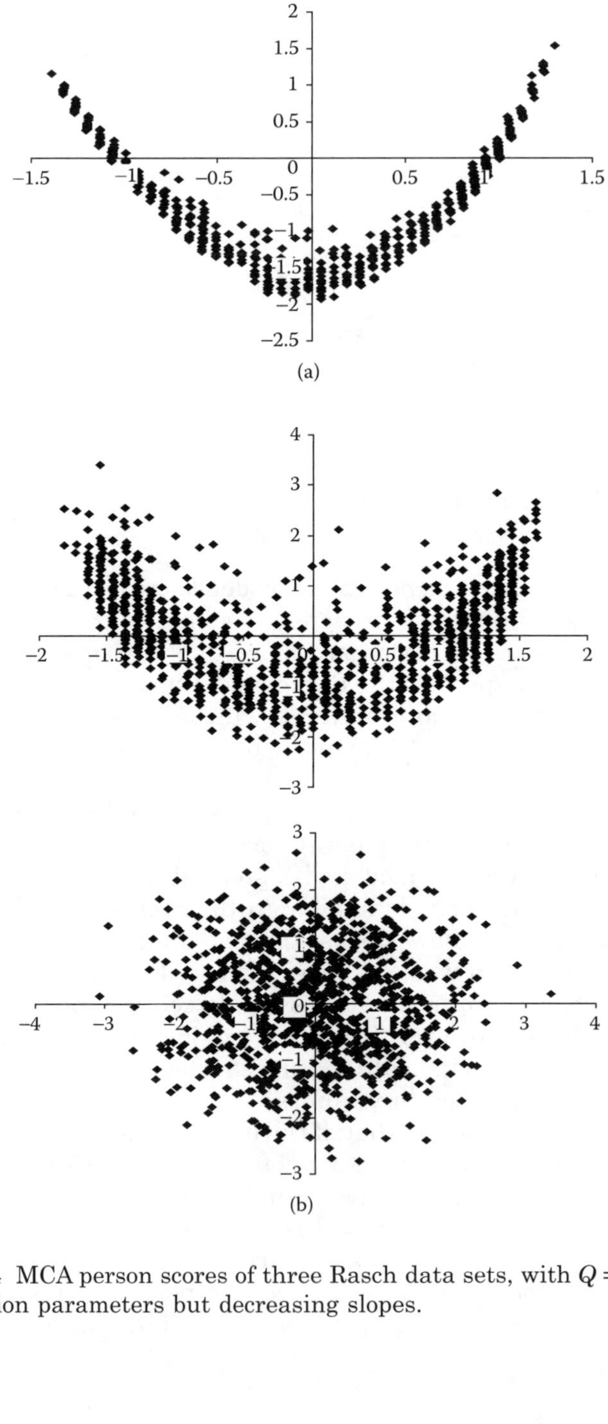

Figure 9.4 MCA person scores of three Rasch data sets, with $Q = 40$ and the same location parameters but decreasing slopes.

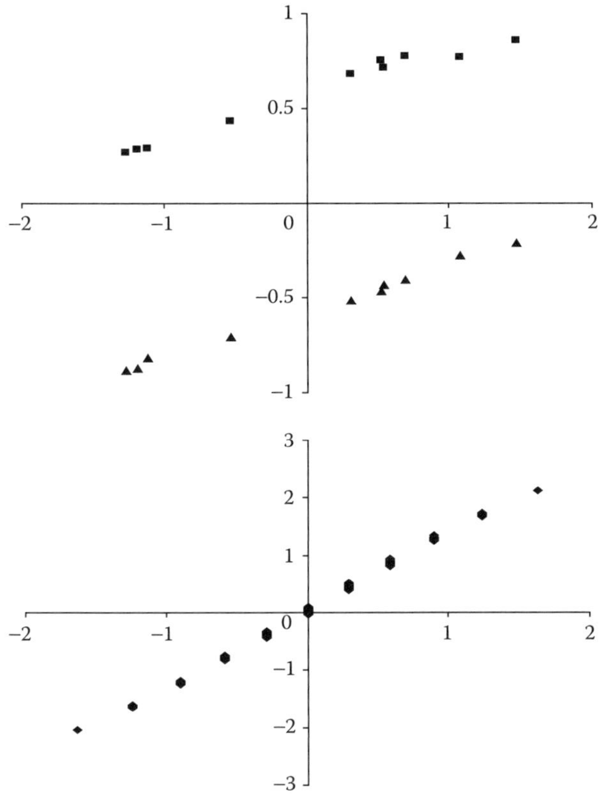

Figure 9.5 The first plot shows the MCA quantifications (vertical) of the two sets of categories plotted vs. item-response-theory estimates for the location parameters (horizontal) of a Rasch data set with $Q = 10$. Both sets of quantifications reflect the ordering of the location parameters. The second plot shows the MCA person scores (vertical) vs. the item-response-theory person estimates (horizontal).

visualized in Figure 9.5, where the MCA parameters are plotted against Rasch parameter estimates. (The latter were obtained using the Multilog software by Thissen et al. 2003.)

The ordering of the items is reflected in both sets of category quantifications. Furthermore, the person scores give a similar ordering compared with the sum score or item-response-theory estimates. Because MCA assigns different scores to different score patterns with the same sum score, the person scores and sum scores are close but not the same.

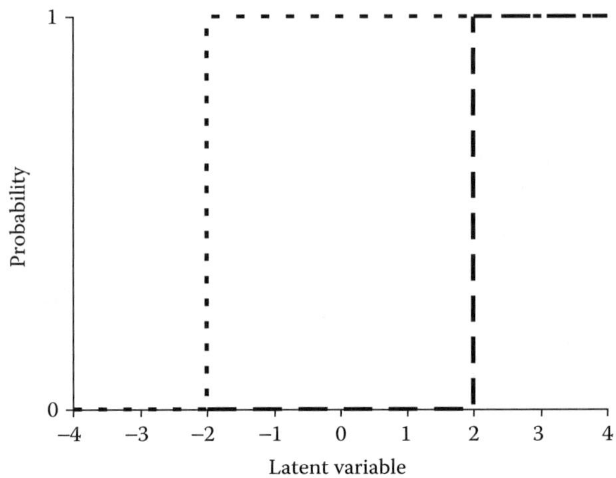

Figure 9.6 Two item-step functions of a PGS.

9.4 The polytomous Guttman scale

The extension of the DGS to more than two categories is the polytomous Guttman scale (PGS). An item is now characterized by two or more item-step functions, as shown in Figure 9.6.

The number of possible score patterns for a PGS is $Q(J-1)+1$. With $J > 2$, various combinations of score patterns can form a PGS, so the PGS is not unique. Three interesting PGSs can be identified in the case $Q = 2$ and $J = 3$:

$$\mathbf{M}_1 = \begin{bmatrix} 1 & 1 \\ 2 & 1 \\ 2 & 2 \\ 3 & 2 \\ 3 & 3 \end{bmatrix} \qquad \mathbf{M}_2 = \begin{bmatrix} 1 & 1 \\ 2 & 1 \\ 3 & 1 \\ 3 & 2 \\ 3 & 3 \end{bmatrix} \qquad \mathbf{M}_3 = \begin{bmatrix} 1 & 1 \\ 2 & 1 \\ 2 & 2 \\ 2 & 3 \\ 3 & 3 \end{bmatrix}$$

The score patterns of $\mathbf{M}_1$ are such that the easier item steps of both items are taken first. Hence, there is an ordering of items within item steps (IWIS). A property of an IWIS PGS is that it has a maximum number of entries of the middle categories and a minimum number of entries of the two outer categories.

The score patterns of $\mathbf{M}_2$ are such that both item steps of the first item are taken first, and then the item steps of the second item are taken. Thus, there is a complete ordering by items: all item steps of the first item are easier to take than the item steps of the second item. Hence, there is an ordering of item steps within items (ISWI). A property of an ISWI PGS is that it has a maximum number of entries of the two outer categories and a minimum number of entries of the middle categories.

The score patterns of $\mathbf{M}_3$ are such that all item steps of one item lie between the item steps of the other item, that is, item within item (IWI).

Similar to the different forms of the Rasch model, the various forms of the PGS also have consequences for the application of MCA. Only for the IWIS PGS can the rows and columns of the indicator coding be reshuffled such that there are consecutive ones for both rows and columns. The consecutive-ones property for rows indicates that, in a binary pattern, there is only one row of ones and either one or two rows of zeros. A similar property can be formulated for the columns.

$$\mathbf{M}_1 = \begin{bmatrix} 1 & 1 \\ 2 & 1 \\ 2 & 2 \\ 3 & 2 \\ 3 & 3 \end{bmatrix} \Rightarrow \mathbf{Z}_1 = \begin{bmatrix} 1 & 0 & 0 & 1 & 0 & 0 \\ 0 & 1 & 0 & 1 & 0 & 0 \\ 0 & 1 & 0 & 0 & 1 & 0 \\ 0 & 0 & 1 & 0 & 1 & 0 \\ 0 & 0 & 1 & 0 & 0 & 1 \end{bmatrix} \rightarrow \begin{bmatrix} 1 & 1 & 0 & 0 & 0 & 0 \\ 0 & 1 & 1 & 0 & 0 & 0 \\ 0 & 0 & 1 & 1 & 0 & 0 \\ 0 & 0 & 0 & 1 & 1 & 0 \\ 0 & 0 & 0 & 0 & 1 & 1 \end{bmatrix}$$

Hence, MCA can order both categories and persons of an IWIS PGS (see Hill 1974; Heiser 1981), which results in a parallelogram pattern, which in turn gives an arch in the two-dimensional solution. In Figure 9.7, the two-dimensional MCA person scores of an IWIS, ISWI, and IWI PGS, all complete and uniform, are plotted. Each PGS consists of scores of 13 persons on three items with five categories.

The ordering of the persons of an IWIS PGS is clearly illustrated with the arch in Figure 9.7. However, for the ISWI PGS, MCA cannot distinguish between several score patterns: in Figure 9.7 some person scores in the two-dimensional solution for the ISWI PGS are the same, although the original score patterns are different. Although it is not shown here, it is interesting to note that several item categories of the ISWI PGS also could not be distinguished. Combinations of these item categories and the corresponding persons obtain the same scores on different dimensions. In Figure 9.7, the original 13 ISWI score patterns are clustered in five groups.

It is easily shown that there exist no permutations for rows and columns such that the indicator matrix of the IWI PGS will have a

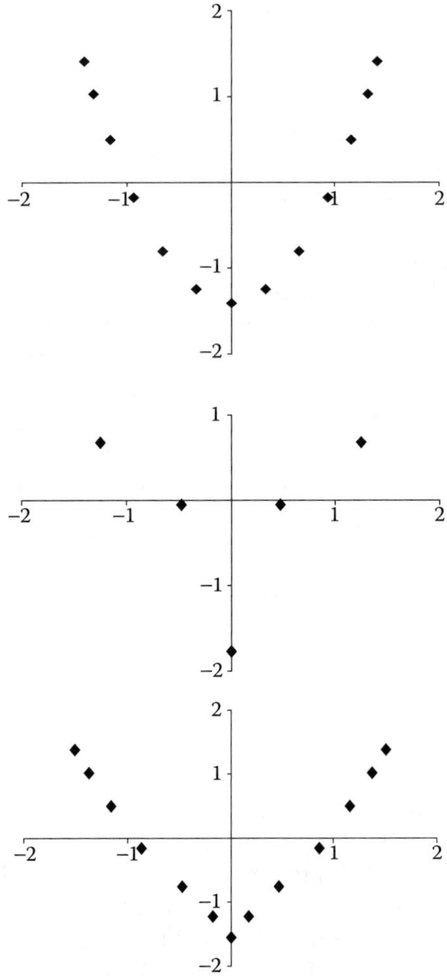

Figure 9.7 Plots of 13 MCA person scores in two dimensions for IWIS, ISWI, and IWI PGSs; $Q = 3$, $J = 5$.

parallelogram pattern, although it is possible to obtain a pattern that is very close to a parallelogram. This may account for the two-dimensional visualization presented in Figure 9.7 of the IWI PGS. The IWI arch in Figure 9.7 is pointed, but it does reflect the correct ordering of the persons.

To obtain the same strong results for the PGS that hold for the DGS discussed in Section 9.2, one should analyze the item steps and not the original data matrix. It is easily shown that any PGS can be made into a DGS by recoding the data matrix of a PGS into item steps. (The authors are not aware of any work in the literature where this property is explicitly stated.)

$$\mathbf{M}_2 = \begin{bmatrix} 1 & 1 \\ 2 & 1 \\ 3 & 1 \\ 3 & 2 \\ 3 & 3 \end{bmatrix} \Rightarrow \mathbf{M}_2^{\mathrm{IS}} = \begin{bmatrix} 1 & 1 & 1 & 1 \\ 2 & 1 & 1 & 1 \\ 2 & 2 & 1 & 1 \\ 2 & 2 & 2 & 1 \\ 2 & 2 & 2 & 2 \end{bmatrix} \Rightarrow \mathbf{Z}_2^{\mathrm{IS}} = \begin{bmatrix} 1 & 1 & 1 & 1 & 0 & 0 & 0 & 0 \\ 0 & 1 & 1 & 1 & 1 & 0 & 0 & 0 \\ 0 & 0 & 1 & 1 & 1 & 1 & 0 & 0 \\ 0 & 0 & 0 & 1 & 1 & 1 & 1 & 0 \\ 0 & 0 & 0 & 0 & 1 & 1 & 1 & 1 \end{bmatrix}$$

When one analyzes the item-step data matrix, the individual item steps are analyzed as if they were dichotomous items, and all results of the dichotomous case apply.

9.5 The graded response model

A possible generalization of both the Rasch model to more than two categories, as well as the PGS to the probabilistic case, is the graded response model (GRM) proposed by Samejima (1969). An item with three response categories is characterized by two item-step functions (see Figure 9.8).

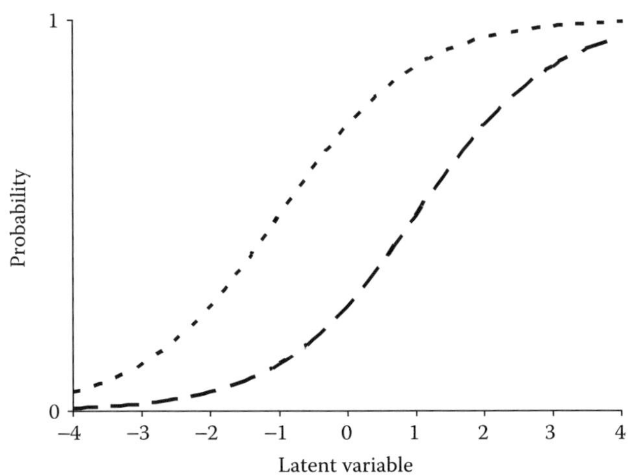

Figure 9.8 Two item-step functions of the GRM.

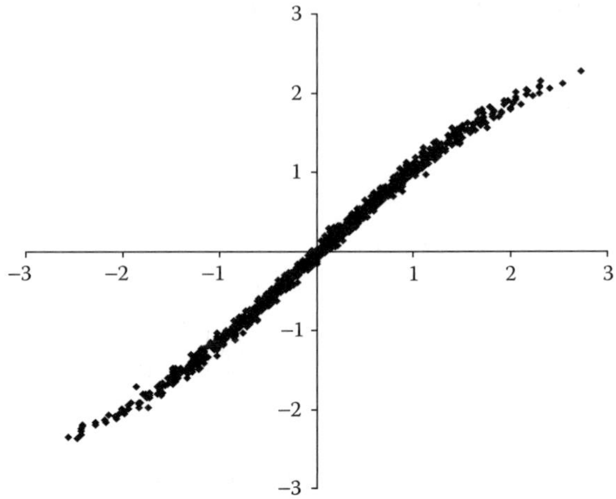

Figure 9.9 Plot of MCA person scores (vertical) vs. item-response-theory estimates (horizontal) for GRM.

Some results from Section 9.3 on the Rasch model also apply to the GRM. Depending on the steepness of the slopes, results very similar to those depicted in Figure 9.4 can be obtained for the GRM. Also, the category quantifications corresponding to the two outermost categories can be shown to be ordered under the same conditions as derived by Schriever (1985) for the Rasch model. An interesting graph shows the item-response-theory person estimates plotted against the MCA person scores of the first dimension (see Figure 9.9). This figure illustrates that the MCA person score is a reasonable approximation of the latent variable (see McDonald 1983; Cheung and Mooi 1994).

9.6 Unimodal models

It was Coombs (1964) who popularized the unidimensional unfolding technique and his method of parallelogram analysis. For the dichotomous Coombs scale (DCS), it holds that each item function is of unimodal shape instead of a monotonic one (see Figure 9.10).

With Q items, the complete DCS has $2Q + 1$ score patterns. Its extension to more than two categories, the polytomous Coombs scale (PCS), will have $2Q(J - 1) + 1$ possible score patterns in the complete case, i.e., the highest category has one category function, whereas the lower categories, except for the lowest, have category functions on both sides of the highest.

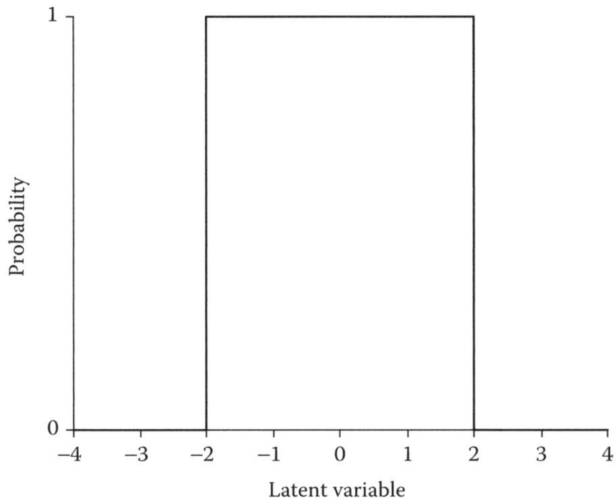

Figure 9.10 Item function of the DCS.

The list of literature on scaling the DCS or PCS with MCA is very short, or perhaps nonexistent. Heiser (1981: 120) demonstrated for a PCS that MCA does not find the intended regularity: neither the order of persons, nor the order of the items, nor the order of the categories within items are recovered in any simple way.

Where the DGS has one item step, which is also the item function, the item function of the DCS consists of two item steps, one left and one right. Some results have been obtained for the restricted DCS, i.e., the DCS where the left and right item steps have the same ordering across items. The restricted DCS can be analyzed by not assigning weights to both categories, but just to the category of interest, that is, simple CA of the (0,1)-table. Then it holds that the CA solution of the restricted DCS provides an ordering for both items and persons (see Hill 1974; Heiser 1981).

$$\mathbf{M} = \begin{bmatrix} 1 & 1 & 1 \\ 2 & 1 & 1 \\ 2 & 2 & 1 \\ 1 & 2 & 1 \\ 1 & 2 & 2 \\ 1 & 1 & 2 \\ 1 & 1 & 1 \end{bmatrix} \Rightarrow \begin{bmatrix} 0 & 0 & 0 \\ 1 & 0 & 0 \\ 1 & 1 & 0 \\ 0 & 1 & 0 \\ 0 & 1 & 1 \\ 0 & 0 & 1 \\ 0 & 0 & 0 \end{bmatrix} \rightarrow \mathbf{M}^0 = \begin{bmatrix} 1 & 0 & 0 \\ 1 & 1 & 0 \\ 0 & 1 & 0 \\ 0 & 1 & 1 \\ 0 & 0 & 1 \end{bmatrix}$$

A possible generalization of the above coding to the PCS, called "conjoint coding" (as opposed to "disjoint" or "indicator coding" used in ordinary MCA), has been proposed by Heiser (1981). Alternatively, to successfully apply MCA to the DCS or PCS, one should make the distinction between the lower categories that lie on the left and on the right of the highest category. This is possible for both the DCS and PCS, because both models have deterministic data structures. The following is an example for a PCS.

$$\mathbf{M} = \begin{bmatrix} 1 & 1 & 1 \\ 2 & 1 & 1 \\ 2 & 2 & 1 \\ 2 & 2 & 2 \\ 3 & 2 & 2 \\ 3 & 3 & 2 \\ 3 & 3 & 3 \\ 2 & 3 & 3 \\ 2 & 2 & 3 \\ 2 & 2 & 2 \\ 1 & 2 & 2 \\ 1 & 1 & 2 \\ 1 & 1 & 1 \end{bmatrix} \Rightarrow \mathbf{M}^{\mathrm{D}} = \begin{bmatrix} 1 & 1 & 1 \\ 2 & 1 & 1 \\ 2 & 2 & 1 \\ 2 & 2 & 2 \\ 3 & 2 & 2 \\ 3 & 3 & 2 \\ 3 & 3 & 3 \\ 4 & 3 & 3 \\ 4 & 4 & 3 \\ 4 & 4 & 4 \\ 5 & 4 & 4 \\ 5 & 5 & 4 \\ 5 & 5 & 5 \end{bmatrix}$$

This idea is discussed for applications in the field of item-response theory by several authors (Verhelst and Verstralen 1993; van Schuur 1993c; Andrich 1996; Roberts and Laughlin 1996).

9.7 Conclusion

As shown in the previous sections of this chapter, the application of MCA to monotonic unidimensional models has a vast potential for producing interesting results. The application of MCA to unimodal models is somewhat less fruitful although some results could be obtained for the deterministic Coombs scales. The idea that one needs to consider what coding is appropriate for the multivariate categorical data at hand, before applying MCA, is probably the most important point of this chapter. A choice must be made between coding items or item steps, or between disjoint or conjoint coding of the data. But given this choice, MCA optimizes a general-purpose criterion, not a model-specific one.

In Section 9.6, a coding scheme was discussed that made a distinction between the lower categories that lie on the left and on the right of the highest category of a PCS. What the reader may not have noted is that, by making this distinction, the PCS with J categories becomes a PGS with $2(J - 1) + 1$ categories. Furthermore, all results from Section 9.4 on the PGS now also apply to the PCS. Even the DCS can be considered a PGS after recoding the data. Note that the item steps coding from Section 9.4 can be applied as well, in which case all properties from Section 9.2 also apply to the DCS and PCS. Thus, after the appropriate codings, the PGS, DCS, and PCS can be analyzed as if they were DGSs. A similar result holds for the probabilistic GRM. If one applies the item-steps coding to the GRM data, applying MCA becomes the same as analyzing dichotomous item steps with the Rasch model.

From the figures in this chapter, it is clear that the common conviction—that applying MCA to data corresponding to a unidimensional model always results in a horseshoe or an arch—is not true. This finding even holds for relatively less complex unidimensional models such as the PGS or the Rasch model. Even though an arch is not necessarily obtained with probabilistic models, Figure 9.5 and Figure 9.9 demonstrate that MCA contains relevant information on the person and sometimes the location parameters of a monotonic model in its first solution. For each of the deterministic models, a vast number of possible probabilistic generalizations exist. This chapter has been limited to only a few of them. What has not been shown here, but what is interesting to note, is that models that are more complex than the basic Rasch model, for example, the two-dimensional plot (or higher-dimensional plots), are very unclear and hard to interpret. However, even if the two-dimensional plot looks like random data, the first MCA dimension contains relevant information on at least the person parameters of monotonic unidimensional models (see Figure 9.5 and Figure 9.9). The person score seems a reasonable approximation of the latent variable for a wide variety of models. This latter property should be interpreted by the MCA community as a critical note against the sometimes blind use of two-dimensional plots.

CHAPTER 10

The Unfolding Fallacy Unveiled: Visualizing Structures of Dichotomous Unidimensional Item–Response–Theory Data by Multiple Correspondence Analysis

Wijbrandt van Schuur and Jörg Blasius

CONTENTS

10.1 Introduction

The aim of this chapter is to apply multiple correspondence analysis (MCA), particularly its graphical representation, to sets of dichotomous data conforming to one of two different types of unidimensional item-response theory (IRT) models: dominance or cumulative models and proximity or unfolding models. We shall show that any two sets of cumulative IRT data can be interpreted as a single set of unfolding data (but not the other way around). In the case in which the two sets of cumulative data each measure the *opposite* latent trait, this is not a great problem because the interpretation remains the same. But in the case in which the two cumulative sets denote two *independent* cumulative scales, the interpretation as a single unfolding scale is fallacious.

In MCA, values of variables—response categories—are represented as points in low-dimensional space, often a plane. In unidimensional IRT models, in contrast, response categories are represented as closed areas along the unidimensional continuum. In IRT models it is the *thresholds between different values* that are represented as points in the space. The MCA representations of data generated by the two types of IRT models are so different (Gifi 1990; Greenacre 1984; Heiser 1981; Chapter 9, this volume) that they can be used to distinguish between the two IRT models in case of doubt. Before we apply MCA to the empirical data to be discussed below, we will show the patterns we expect by simulating perfect IRT data and showing their structures in a plane using MCA.

10.2 Item response models for dominance data

IRT models for dominance data are well known by such names as the (deterministic) Guttman scale, the (probabilistic) nonparametric Mokken scale, or the parametric Rasch scale and its successors. For reasons of simplicity, we will restrict ourselves to discussing data that consist of dichotomous items only. Such dichotomous items are used to measure a latent ability (e.g., knowledge) or a latent attitude (e.g., religiosity). A positive response to an item (the response "1") indicates the presence of the ability or attitude, and the negative response (the response "0") indicates the absence of the ability or attitude.

As an example of a good (nonparametric) cumulative scale, the following seven items, taken from the World Values Study 1990, form a scale that measures the strength of religious belief, where the items are ordered according to their "difficulty" (see Van Schuur 2003). In the perfect cumulative case, the first item would be the "easiest" (say, "I believe in God"), i.e., this item has the highest number of positive responses. The second item is the second-"easiest" item (say, "I believe that people have a soul"). This item has the second-highest number of positive responses. If the two items belong to a perfect cumulative scale, a real subset of positive respondents from item 1 answer positive to item 2 (and none of those who give a negative response to item 1 give a positive response to item 2). The same holds for the successive items; each following item is more "difficult" than the one before. A possible order of items measuring religious beliefs (taken from the World Value Survey) is

1. Do you believe in God?
2. Do you believe people have a soul?
3. Do you believe in sin?
4. Do you believe in a life after death?
5. Do you believe in heaven?
6. Do you believe the Devil exists?
7. Do you believe in hell?

All questions have the response possibilities "yes" and "no." The purpose of this scale is to measure strength of religious belief by the number of items to which the positive response was given. Before we discuss this example in more detail, we show the structure of perfect cumulative data. Responses of eight respondents to seven hypothetical items that conform to the deterministic scale in the cumulative order from top to bottom are shown in Table 10.1.

Table 10.1 shows the structure of a perfectly cumulative scale where a "1" shows a positive response and a "0" a negative one. In the given example, person 1 (P1) gave negative responses to all items, person 2 (P2) answered only item A positive and all others negative, while person 8 (P8) gave positive responses to all items. Further, item A is the "easiest" one; only person 1 gave a negative response, The most "difficult" item is G, and only person 8 gave a positive response

Table 10.1 Perfect cumulative Guttman scale with eight respondents and seven items.

	A	B	C	D	E	F	G
P1	0	0	0	0	0	0	0
P2	1	0	0	0	0	0	0
P3	1	1	0	0	0	0	0
P4	1	1	1	0	0	0	0
P5	1	1	1	1	0	0	0
P6	1	1	1	1	1	0	0
P7	1	1	1	1	1	1	0
P8	1	1	1	1	1	1	1

Note: Value 1 = positive response; value 0 = negative response.

to it. Following the order of zeros and ones in a person's perfectly ordered response pattern, i.e., a deterministic cumulative Guttman scale, there is no 0 between two 1s, and there is no 1 between two 0s. In the given case, all combinations of possible responses are considered.

The probabilistic cumulative model can be described as follows. Each item i and each subject s have a location, a scale value, along the dimension, indicated by δ_i and θ_s, respectively. The larger θ_s, the higher the value of subject s on the latent dimension (e.g., the higher the subject's ability or attitude). The larger δ_i, the greater the ability or attitude it takes a subject to give the positive response to the item. Note that a dichotomous item has only one item parameter, δ_i. This parameter marks the location where the negative response and the positive response are in equilibrium. If person s (with scale value θ_s) and item i (with scale value δ_i) have the same value, then the probability of a positive response is 0.50. If $\theta_s > \delta_i$, then subject s dominates item i, and if $\theta_s < \delta_i$, then subject s is dominated by item i. This latter statement is deterministically true in the Guttman model and probabilistically in the other, more recent, IRT models for dominance data. In the Rasch model, the probability of the positive response is given as

$$P(x=1 \mid \theta_s, \delta_i) = \frac{e^{(\theta_s - \delta_i)}}{1 + e^{(\theta_s - \delta_i)}}$$

To compare the Rasch model with the cumulative Guttman scale, we simulated (probabilistic) Rasch data with seven variables (with δ_i equidistant with scale values −2.7, −1.9, −1.1, −0.3, 0.5, 1.3, and 2.1; N = 1000; uniformly distributed between −3 and +3) and cumulative Guttman data with seven variables (see Table 10.1).

10.3 Visualizing dominance data

What happens when unidimensional dominance data, e.g., the religious belief data as discussed above, are submitted to an MCA analysis, in which both the positive and the negative response to all items as well as all subjects are represented as points in the space? Figure 10.1 shows the result of simulated data that conform to the deterministic cumulative Guttman scale, and Figure 10.2 shows the MCA map for the Rasch data. Figure 10.3 shows the MCA map for the N = 1017 Dutch respondents of the 1990 World Values Study (the seven variables measuring religious beliefs, given in Section 10.2). The three patterns look virtually identical. Note that the two categories (0 = no, 1 = yes) of all items are both shown in these MCA maps.

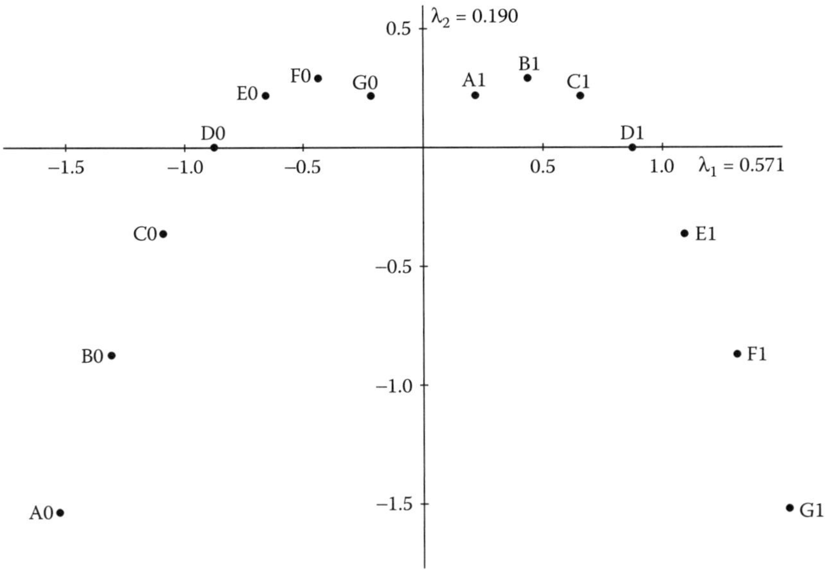

Figure 10.1 MCA map of Guttman data of Table 10.1.

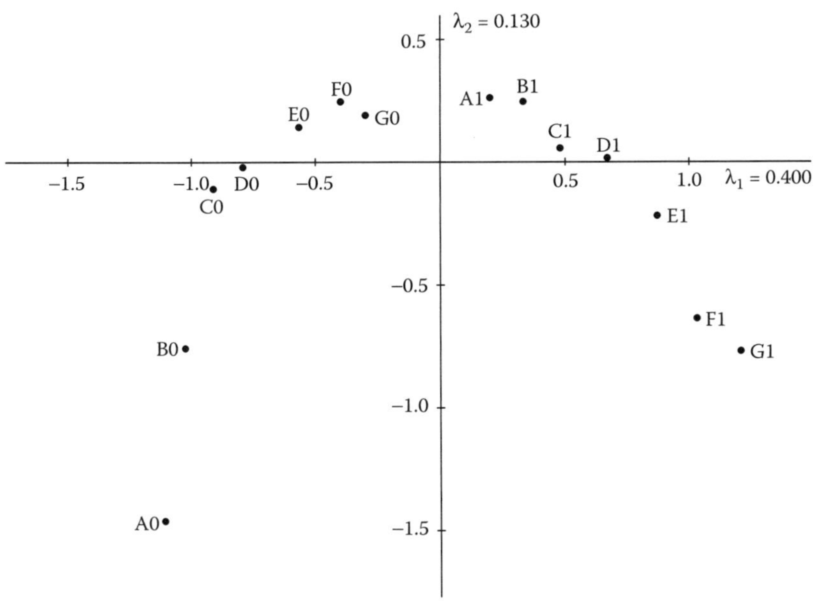

Figure 10.2 MCA map of Rasch items.

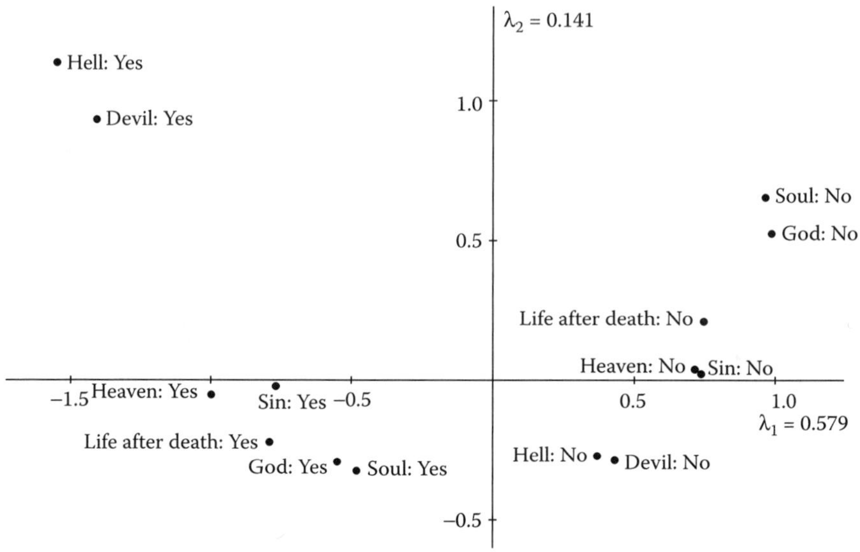

Figure 10.3 MCA map of religious beliefs.

Figure 10.1 shows the representation of seven simulated dichotomous Guttman items in order from "easy" to "difficult." Among the positive responses (A1 to G1), the positive response to item A (A1) is the most popular, and among the negative responses (A0 to G0), the response to item G (G0) is the most popular. These responses are represented closest to the origin of the representation (see Chapter 1). Less frequent responses, such as the negative response to the easiest item (A0) or the positive response to the most difficult item (G1), are represented bottom left, and bottom right, respectively. In fact, in the MCA map each pair of categories on each dimension is located one opposite the other, and a line connecting them goes through the origin of the representation in all dimensions. The distance of each pair of categories to the origin is a function of their marginals: the higher the weight of a category (i.e., the more cases belonging to it), the closer the distance to the origin.

Both the positive and the negative response categories of the items form two parabolic curves. This representation is known as the "Guttman gauge" in the literature (Gifi 1990). This shape has also been discussed in Heiser (1981) and Greenacre (1984); it is explained in more detail and extended to the polytomous case by Warrens and Heiser (see Chapter 9, this volume).

Figure 10.2 shows a very similar shape for the simulated Rasch data with seven items; further examples including the respondents' scores are given in Chapter 9. Figure 10.3 shows the variables categories for the "religious belief" items from the World Values Study discussed above. According to their shape, the items can be ordered along the first dimension as follows: "belief in soul," "belief in God," "belief in sin," "belief in life after death," "belief in heaven," "belief in the devil," and "belief in hell." According to the beliefs of the respondents, it is easiest to believe in the soul and in God, and most difficult to believe in the devil and in hell. Whereas the first five and the last two items are clearly separated, the three items in the middle of the scale are somewhat overlapping; the beliefs in life after death, sin, and heaven seem to be in a tied order. In terms of MCA interpretation, the first axis mirrors the extent of religious beliefs, with strong beliefs on the left (respondents who also believe in the devil and in hell) and strong nonbeliefs on the right part (respondents who believe neither in God nor in a soul). The second dimension can be interpreted to reflect the contrast between strong believers and strong nonbelievers on the positive part versus the somewhat believers on the negative part.

As a reminder, the unidimensional visualization of the deterministic data in IRT form looks much like that in Figure 10.4. The thresholds between the "yes" and "no" values for each item are ordered such that the most dominant subject (the strongest believer) is located in the rightmost area, and the least dominant subject (the strongest nonbeliever) in the leftmost area of this unidimensional representation. Person 1, for instance, with scale value θ_1, will respond positively to the "easier" items 1 to 3 and negatively to the more "difficult" items 4 to 6, and will therefore have the response pattern 111000.

We also exposed two simulated independent Guttman scales and two simulated independent Rasch scales to the same procedure (see Figure 10.5 and Figure 10.6) to show the similarity between their MCA representations. The Guttman data of the two independent scales with

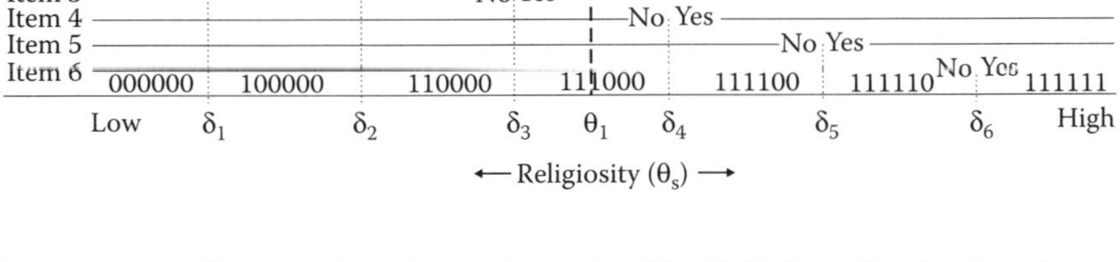

Figure 10.4 Six questions from the 1990 World Values Study that form a Guttman scale.

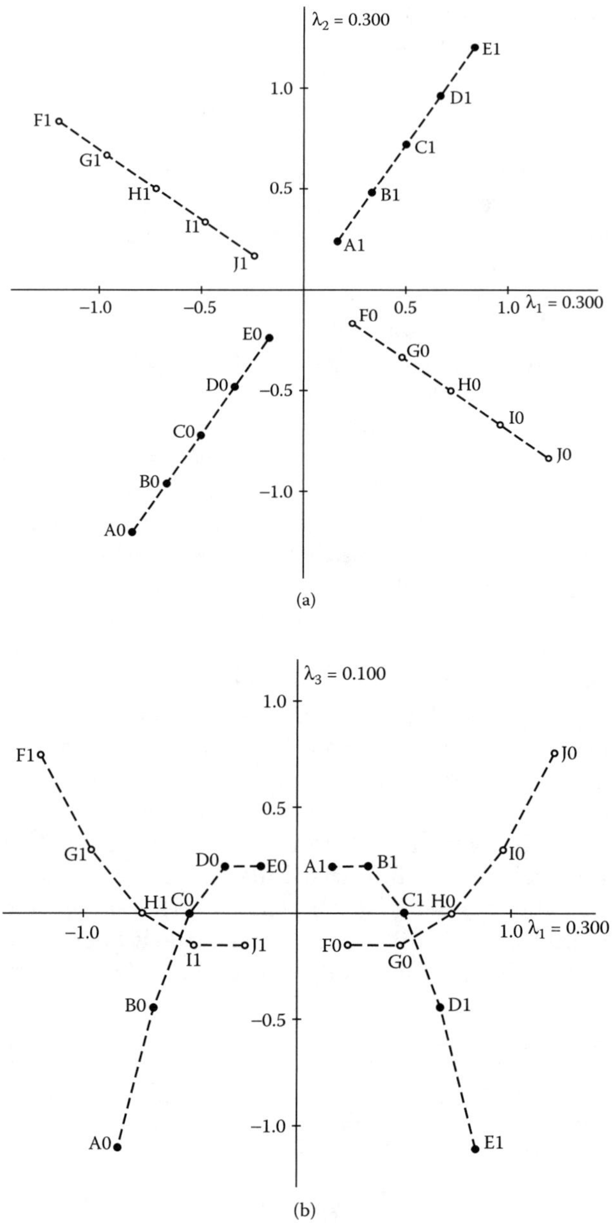

Figure 10.5 MCA map of two independent cumulative scales: (a) dimension 1 versus 2; (b) dimension 1 versus 3.

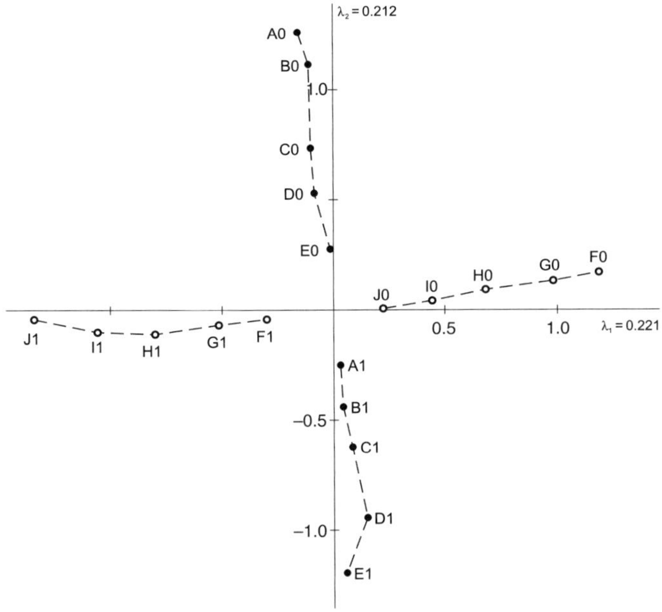

(a)

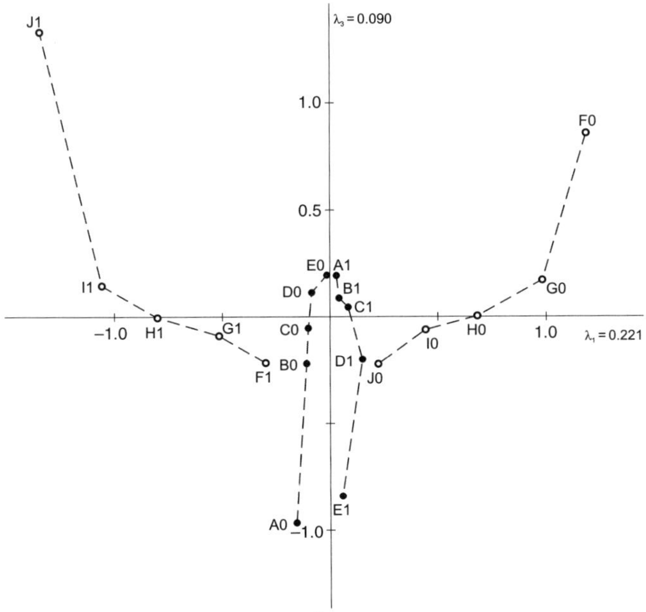

(b)

Figure 10.6 MCA map of Rasch items, two independent sets: (a) dimension 1 versus dimension 2; (b) dimension 1 versus dimension 3.

ten items (A to E, and F to J) and 36 respondents are given in Table 10.2. The table consists of two sets of five variables and six respondents; each case from the one set is combined with each case from the other set, resulting in 6 × 6 = 36 respondents. The two independent Rasch scales are constructed with $N = 1000$ respondents, uniformly distributed between −3 and +3. In both scales the items are equidistant, the first with five items with scale values −2.7, −1.5, −0.3, 0.9, 2.1, and the second with five items with scale values −2.1, −0.9, 0.3, 1.5, 2.7. The responses to the two sets of items were simulated in two separate and independent simulations for the same respondents.

For the representation of these structures, we show the first three dimensions. With respect to the first two of these dimensions, the positive and negative categories of the items line up in a straight line, each covering the opposite quadrant, with the most popular categories closest to the origin and the order of the other categories in their expected cumulative order. The representation of two such Guttman scales form a perfect orthogonal cross in the first two dimensions. Dimensions one and three form the horseshoes for the two independent scales when the representations of the positive and negative response categories are combined.

10.4 Item response models for proximity data

Recently, more attention has been given to other types of dichotomous data in which the positive response is given by respondents whose scale value is *close* to that of the item, but where the negative response is given by respondents who are represented far away from that item in opposite directions. For instance, people like to take a shower in water that is not too hot but also not too cold. They like some milk in their coffee: not too little, but also not too much. They like to vote for a party that is not too extremely leftist but also not too extremely rightist. Such items are called proximity items. One possible model formulation for such items, derived from the Rasch model, is the following formulation by Andrich (1988):

$$P(x=1 \mid \theta_s, \delta_i) = \frac{1}{1+e^{(\theta_s-\delta_i)^2}}$$

A subject s (with scale value θ_s) who has the same location as item i (with scale value δ_i) has the highest probability of giving the positive

Table 10.2 Two perfect, independent, cumulative Guttman scales.

	A	B	C	D	E	F	G	H	I	J
P1	0	0	0	0	0	1	1	1	1	1
P2	0	0	0	0	0	1	1	1	1	0
P3	0	0	0	0	0	1	1	1	0	0
P4	0	0	0	0	0	1	1	0	0	0
P5	0	0	0	0	0	1	0	0	0	0
P6	0	0	0	0	0	0	0	0	0	0
P7	0	0	0	0	1	1	1	1	1	1
P8	0	0	0	0	1	1	1	1	1	0
P9	0	0	0	0	1	1	1	1	0	0
P10	0	0	0	0	1	1	1	0	0	0
P11	0	0	0	0	1	1	0	0	0	0
P12	0	0	0	0	1	0	0	0	0	0
P13	0	0	0	1	1	1	1	1	1	1
P14	0	0	0	1	1	1	1	1	1	0
P15	0	0	0	1	1	1	1	1	0	0
P16	0	0	0	1	1	1	1	0	0	0
P17	0	0	0	1	1	1	0	0	0	0
P18	0	0	0	1	1	0	0	0	0	0
P19	0	0	1	1	1	1	1	1	1	1
P20	0	0	1	1	1	1	1	1	1	0
P21	0	0	1	1	1	1	1	1	0	0
P22	0	0	1	1	1	1	1	0	0	0
P23	0	0	1	1	1	1	0	0	0	0
P24	0	0	1	1	1	0	0	0	0	0
P25	0	1	1	1	1	1	1	1	1	1
P26	0	1	1	1	1	1	1	1	1	0
P27	0	1	1	1	1	1	1	1	0	0
P28	0	1	1	1	1	1	1	0	0	0
P29	0	1	1	1	1	1	0	0	0	0
P30	0	1	1	1	1	0	0	0	0	0
P31	1	1	1	1	1	1	1	1	1	1
P32	1	1	1	1	1	1	1	1	1	0
P33	1	1	1	1	1	1	1	1	0	0
P34	1	1	1	1	1	1	1	0	0	0
P35	1	1	1	1	1	1	0	0	0	0
P36	1	1	1	1	1	0	0	0	0	0

Note: Value 1 = positive response; value 0 = negative response.

response to that item (50%, according to this model from Andrich 1988). However, this percentage can differ with different models (see Andrich and Luo 1993; Hoijtink 1990; Roberts and Laughlin 1996). To the extent that subjects are located farther away from the item, in either direction, the probability of giving the positive response decreases.

A data matrix in which every respondent gives positive responses to exactly three out of nine items ("pick 3/9 data") that conforms to the deterministic proximity model is given in Table 10.3. The unfolding scale might be called the "Coombs scale," after the originator of this model (Coombs 1950, 1964). This data matrix bears some resemblance to the data matrix of Table 10.1, in the sense that the positive response (the value 1) is given to items that are adjacent to each other. However, there are now two sets of negative responses (the value 0): they are given not only to items to the right of the items to which the positive response is given, but also to items to the left of them.

The visual representation of one deterministic proximity item (e.g., "Can you imagine voting for this political party?" with response categories "yes" and "no") according to the unfolding model is given in Figure 10.7 (upper part). The positive response is given by respondents who are represented in an area along the scale between two item steps: the left-sided item step, which indicates the change between the negative and the positive response, and the right-sided item step, which indicates the change between the positive and the negative response. The area between these two item steps is called the "latitude of acceptance." The left-sided negative response is given by respondents who will not vote for the party because it is too right-wing for them.

Table 10.3 Perfect unfolding data: "pick 3/9" data.

	A	B	C	D	E	F	G	H	I
P1	1	1	1	0	0	0	0	0	0
P2	0	1	1	1	0	0	0	0	0
P3	0	0	1	1	1	0	0	0	0
P4	0	0	0	1	1	1	0	0	0
P5	0	0	0	0	1	1	1	0	0
P6	0	0	0	0	0	1	1	1	0
P7	0	0	0	0	0	0	1	1	1

Note: Value 1 = positive response; value 0 = negative response.

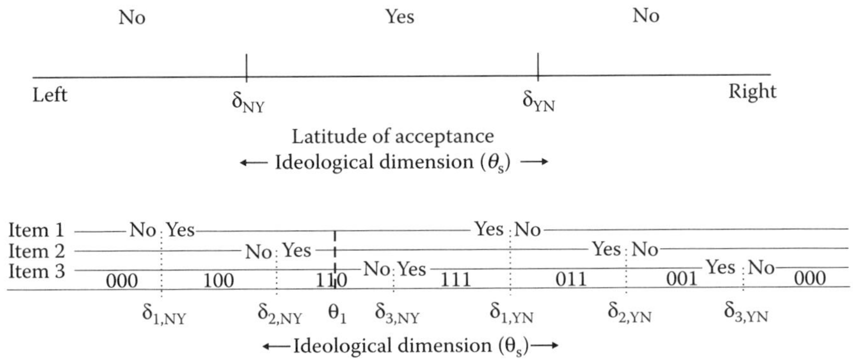

Figure 10.7 Representation of the positive and negative responses to a proximity item.

The right-sided negative response is given by respondents who find the party too left-wing.

A (deterministic) representation of three political parties, X, Y, and Z, on a left–right scale (e.g., a leftist Green party, a centrist Social-Democratic party, and a rightist Christian-Democratic party) is given in Figure 10.7 (lower part). The parties are represented with overlapping latitudes of acceptance, which structures their representation. The respective latitudes of acceptance are given as horizontal lines between the relevant item steps. Person 1, represented by θ_1, will give the positive response to items 1 and 2, but the negative response to item 3.

10.5 Visualizing unfolding data

The unfolding data from Table 10.3 are visualized with MCA in Figure 10.8. The positive response categories (A1 to I1), taken separately, form a horseshoe around the origin from bottom left via top middle to bottom right. The negative response categories, also taken separately, form a reverse horseshoe closer to the origin, from A0 (close to the origin) via bottom middle to I0 (close to the origin). This latter result stems from the fact that the negative response is given by two sets of subjects: for the first set of subjects the item was too far to the right, for the second set of subjects the item was to far to the left.

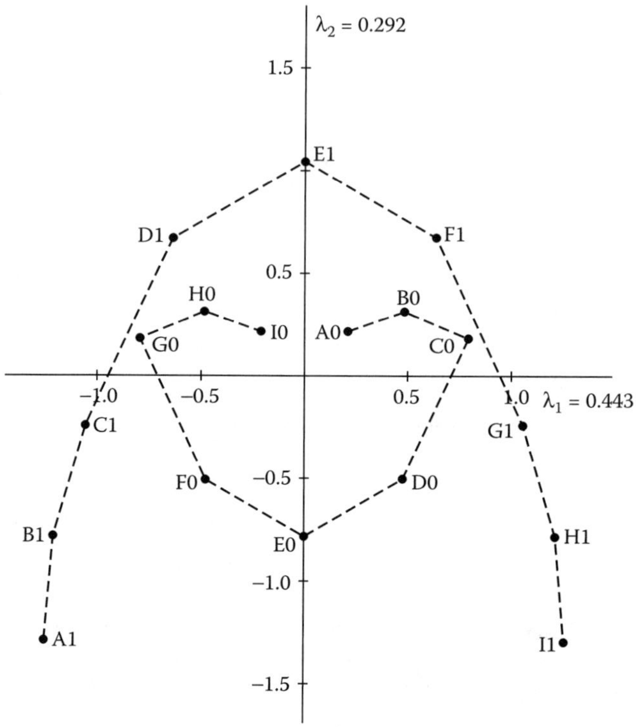

Figure 10.8 MCA map of unfolding "pick 3/9" data of Table 10.3.

Hence, there is this compromise solution of the negative response close to the origin.

In terms of common negative responses, response E0 is given along with D0 and/or F0 by four of the seven subjects (P1, P2, P6, and P7, see Table 10.3). Since responses B0, C0, G0, and H0 are only mentioned twice together with response E1, from this point of the scale, response E1 should be closest to the responses A0 and I0. Because we analyze dichotomous data—i.e., a perfect negative association between positive and negative responses holds for all items—the negative responses, taken by themselves, form a heart shape that is inverse to the horseshoe of positive answers. This representation is also known from the literature—for the statistical details, see Heiser (1981); for the extension to the polytomous case, see Chapter 9, this volume.

Table 10.4 "Pick any/*m*" data, constructed by stacking "pick *k*/*m*" for $k = 2$ to 6 and $m = 9$.

	A	B	C	D	E	F	G	H	I
P1	1	1	0	0	0	0	0	0	0
P2	0	1	1	0	0	0	0	0	0
P3	0	0	1	1	0	0	0	0	0
P4	0	0	0	1	1	0	0	0	0
P5	0	0	0	0	1	1	0	0	0
P6	0	0	0	0	0	1	1	0	0
P7	0	0	0	0	0	0	1	1	0
P8	0	0	0	0	0	0	0	1	1
P9	1	1	1	0	0	0	0	0	0
P10	0	1	1	1	0	0	0	0	0
P11	0	0	1	1	1	0	0	0	0
P12	0	0	0	1	1	1	0	0	0
P13	0	0	0	0	1	1	1	0	0
P14	0	0	0	0	0	1	1	1	0
P15	0	0	0	0	0	0	1	1	1
P16	1	1	1	1	0	0	0	0	0
P17	0	1	1	1	1	0	0	0	0
P18	0	0	1	1	1	1	0	0	0
P19	0	0	0	1	1	1	1	0	0
P20	0	0	0	0	1	1	1	1	0
P21	0	0	0	0	0	1	1	1	1
P22	1	1	1	1	1	0	0	0	0
P23	0	1	1	1	1	1	0	0	0
P24	0	0	1	1	1	1	1	0	0
P25	0	0	0	1	1	1	1	1	0
P26	0	0	0	0	1	1	1	1	1
P27	1	1	1	1	1	1	0	0	0
P28	0	1	1	1	1	1	1	0	0
P29	0	0	1	1	1	1	1	1	0
P30	0	0	0	1	1	1	1	1	1

Note: Value 1 = positive response; value 0 = negative response.

Picking "pick any/*m*" data rather than "pick *k*/*m*" data does not change the MCA visualization at all. Table 10.4 gives a data set where a "pick any/*m*" data set is built up from a stacked set of "pick *k*/*m*" data sets, where $m = 9$ and k increases from 2 to 6. Figure 10.9 shows that the resulting visualization is essentially the same as that of Figure 10.8.

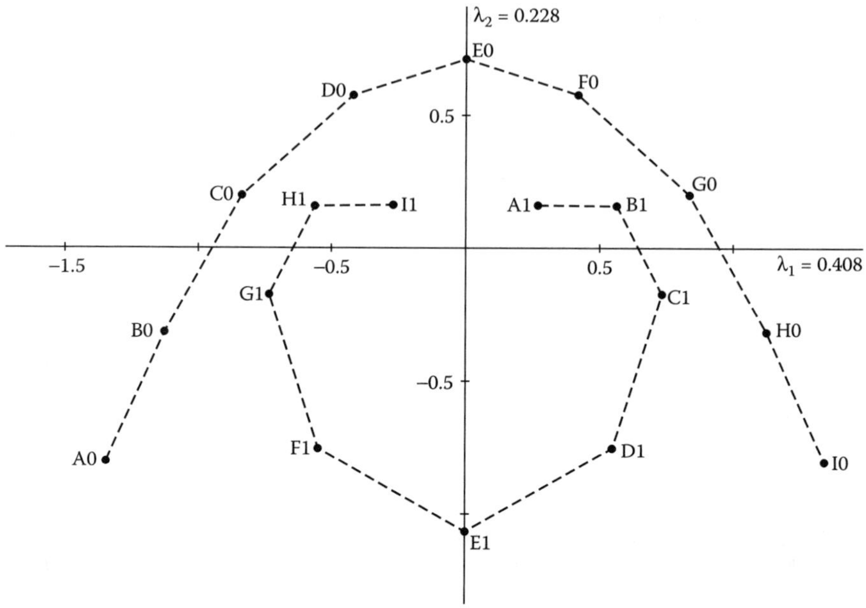

Figure 10.9 MCA map of unfolding "pick any *k*/*m*" data of Table 10.4.

10.6 Every two cumulative scales can be represented as a single unfolding scale

An unfolding analysis can be done with "pick *k*/*m*" data as well as with "pick any/*m*" data. But a cumulative scale analysis, however, can only be done with "pick any/*m*" data. If a subject could pick (respond positively to) only *k* items, the cumulative model predicts that every subject would pick the same *k* easiest items. But how does one distinguish a representation of unfolding "pick any/*m*" data from cumulative "pick any/*m*" data?

If we go back to Table 10.2, with the two independent cumulative scales, we can see how the response pattern of each of the 36 subjects can be regarded as an unfolding response pattern. Every response pattern in the first set of cumulative items (A to E) can be combined with every pattern in the second set of cumulative items (F to J) to produce these unfolding response patterns. This data set is therefore indistinguishable from a data set for a deterministic unfolding scale (Wolters 1982; Van Schuur 1998).

The MCA visualization of this data set is identical to the visualization discussed earlier, in Figure 10.5 and Figure 10.6. We observe, without any further proof, that this relationship holds not only for two deterministic cumulative scales forming a single deterministic unfolding scale, but also for two parametric or nonparametric cumulative scales that together can be represented as a single parametric or nonparametric unfolding scale.

To summarize, cumulative data (deterministic and probabilistic) provide us with a horseshoe when we take the positive and negative response categories together, whereby positive and negative categories are divided by the first dimension, i.e., the positive responses are on one part of the first axis and the negative responses are on the other part. The order of the items on both parts of the dimension is the same. In contrast, unfolding data provides a separate horseshoe for the positive responses and a second horseshoe, in an inverted position, for the negative responses. The items are ordered along the first axis according to the position of the respondents to the underlying dimension; in the hypothetical example, the items are ordered from left-wing to right-wing politics. The fundamental distinction between both representations should be clear.

10.7 Consequences for unfolding analysis

There are applications of unfolding analysis to data for which it is at least questionable whether the items should be regarded as proximity items. An important reason for questioning whether items can be regarded as proximity items exists if no two convincing opposite reasons can be given for the negative response. We will give two examples of data sets that have been recognized as unfolding data in reviewed scientific publications. The first example is the "traffic data" set, collected in 1991 ($N = 600$; see Hoijtink 1993a for further information) and used for a number of different unfolding models. Respondents were asked whether they agreed or disagreed with the following statements (among others):

A: Car use cannot be abandoned. Some pressure on the environment has to be accepted.

B: A cleaner environment demands sacrifices like decreasing car use.

C: Putting a somewhat higher tax burden on car driving is a step in the direction of a healthier environment.

D: The environmental problem justifies a tax burden on car driving so high that people quit using a car.
E: It is better to deal with other forms of environmental pollution than car driving.
F: Instead of environmental protection measures with respect to car use, the road system should be extended.
G: Technically adapted cars do not constitute an environmental threat.
H: Considering the environmental problems, people should decide for themselves how often they should use their cars.
I: People who keep driving a car are not concerned with the future of our environment.
J: Car users should have to pay taxes per mile driven.

Items B, C, D, I, and J are formulated in favor of environmental protection and against car use, whereas items A, E, F, G, and H are formulated in favor of car use and against environmental protection. According to Croon (1993), Forman (1993), Hoijtink (1993b), Post and Snijders (1993), Van Blokland-Vogelesang (1993), Van Schuur (1993b), and Verhelst and Verstralen (1993), this set of data formed an acceptable unfolding scale. However, if we apply the litmus test of determining whether the negative response to each of these items can have two opposite meanings, the answer is no. For instance, it is difficult to interpret a person's attitude as protective toward the environment and in favor of car use if that person disagrees that car users should have to pay taxes per mile driven. However, such interpretation ought to be possible for an unfolding scale.

Applying MCA to these data, the first dimension reflects the contrast between environmental protection and supporting car use (Figure 10.10). From the subdivision of positive and negative categories, we can conclude that the data do not conform to the unidimensional unfolding model. We do not see separate nested horseshoes for the positive or negative response categories. The representation does, however, conform to the requirements of a single cumulative scale, in which the items B, C, D, I, and J have the opposite interpretation of the items A, E, F, G, and H. A nonparametric cumulative scale analysis corroborates this interpretation. The two sets of items are indeed each other's opposites and form a homogeneous unidimensional cumulative scale together, once the responses of one set of items is recoded in the reverse direction.

A second example comes from data that were collected in 1992 ($N =$ 166; see Van Schuur 1993a for further information) to test the Groningen

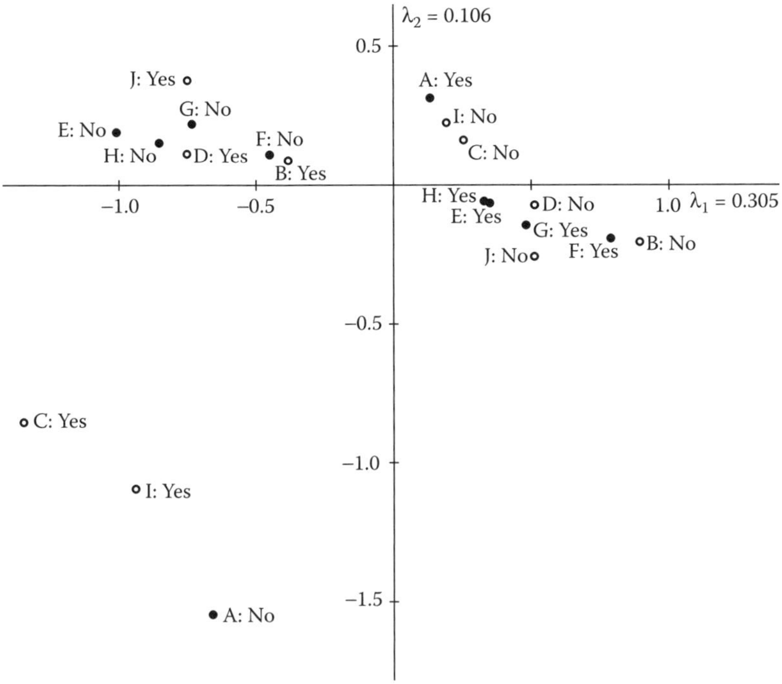

Figure 10.10 MCA map of traffic data.

androgyny scale, in which respondents were asked whether the following personality descriptive terms were applicable to themselves: individualistic, sharp, independent, sober, rational, self-confident, vulnerable, emotional, romantic, oversensitive, female, sentimental, and motherly. Here again, a negative response to any of these items can have only one interpretation. The first six items were dubbed "masculine" (from "individualistic" to "self-confident"), and the last seven items were dubbed "feminine" (from "vulnerable" to "motherly"). Both the masculine items and the feminine items form a good cumulative scale, as had in fact already been reported by Sanders and Hoijtink (1992). In contrast to the traffic data, however, recoding one of the two subsets of items does not result in a single cumulative scale: masculinity and femininity are two independent characteristics; they are not a single bipolar concept, as the unfolding results would have it. The MCA visualization of this data set confirms this interpretation; there is a clear subdivision of the two sets of items (see Figure 10.11). The first dimension contrasts masculine and feminine characteristics,

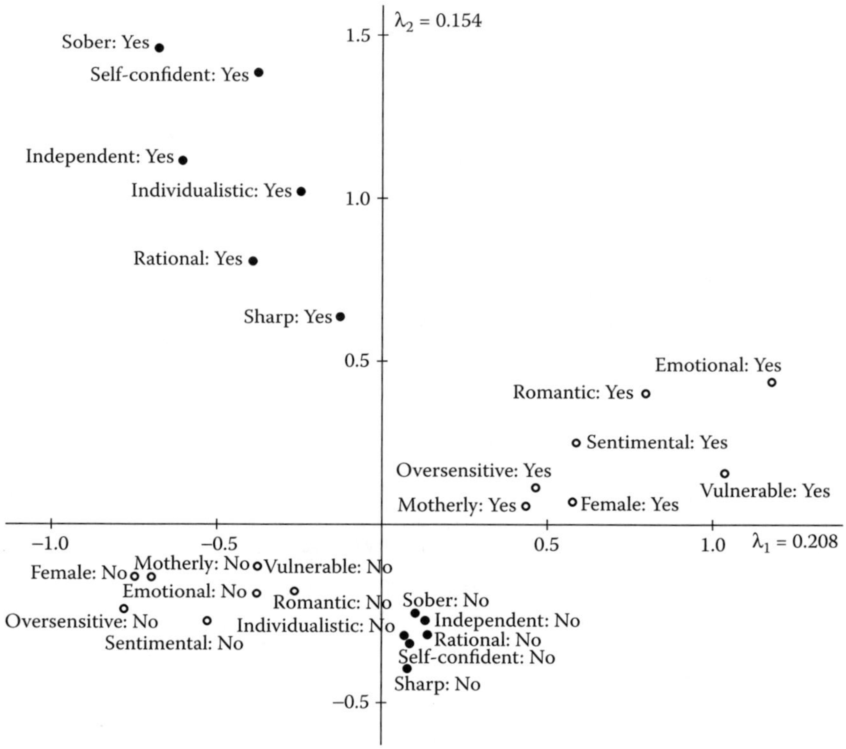

Figure 10.11 MCA map of Andrich data.

while the second dimension contrasts positive and negative responses. When two new orthogonal dimensions are superimposed, one reflects the masculine characteristics ("yes" versus "no"), and the other reflects the feminine characteristics.

10.8 Discussion

This chapter shows how MCA can be helpful in determining the nature of a data set. MCA visualizes response categories of variables as points. However, for the data discussed in this chapter, which conform to IRT models for dominance or proximity data, response categories are better represented as closed areas and the thresholds between them as points. In the case of the dichotomous proximity model, the negative response to an item is even represented as *two* separate closed areas; in an MCA visualization, such negative responses are often (misleadingly) represented close to the origin of the space. One way to overcome this problem is by ignoring the negative responses altogether and focusing on only the positive responses (see Heiser 1981).

As shown in this chapter, MCA can shed light on a problem in the analysis of proximity data that—observing the literature—seems not to be well understood. The problem is to determine whether any two cumulative scales can form a single proximity scale for "pick any/*m*" data. When the two cumulative scales measure the identical but opposite trait, MCA shows the single "Guttman gauge." However, when the two cumulative scales are independent, MCA shows this in the first two dimensions.

One possible solution to the problem of creating adequate unfolding data is to construct double-barreled items in which the negative response is explicitly expected to have the required double meaning. For this purpose, Roberts et al. (2000) used data to measure the bipolar negative–positive feelings toward abortion. Examples of antiabortion items are

1. Abortion should be illegal, except in cases involving incest or rape.
2. Abortion is basically immoral, except when the woman's physical health is in danger.

Examples of proabortion items are

3. Abortion should be a woman's choice, but should never be used simply due to its convenience.
4. Abortion should generally be legal, but should never be used as a conventional method of birth control.

It turns out, however, that both the first and the last two items belong to one of two sets of good cumulative items, which form a single cumulative scale once one of the two sets of responses has been reversed.

In fact, none of the unfolding data that we analyzed by MCA simultaneously showed the expected horseshoe for the positive responses and the heart for the negative responses. This is not for lack of individual characteristics, where subjects prefer optima over maxima. Biological characteristics, such as temperature, humidity, and the amount of food, drink, or sleep, come first to mind. Coombs

(1964) characterized proximity data as based on two interacting processes: "good things satiate" and "bad things escalate." These two processes together form a single-peaked, but not necessarily symmetrical, item-response function. Nonpreference for an item can therefore have two independent reasons: not enough good things versus too many bad things. The difficulty in identifying the exact reason for a preference may well have to do with the way in which we have presented the item.

For unfolding analysis, Coombs (1964) advocated the use of rank-order data, or "pick any/*m*" data, in which the researcher first gets an overview of the full set of items before he or she reacts. The subsequent reaction is then the comparison of the items to the "ideal point." This is similar to procedures advocated by Thurstone (1927) and Thurstone and Chave (1929): first determine the scale values of items by a separate judgment approach (paired comparisons or summated ratings), and then let the subject select the items with which he or she agrees the most. But in all of our "unfolding" data, such a procedure has never been used. Rating data, in which the subject rates the items, have been interpreted as "pick any/*m* data." But this is a false premise because the subject, in rating the data, evaluates each item in sequence and is never exposed to the whole set before this evaluation. Therefore, it can be seriously questioned whether the response process to these data has been a "proximity response process," as is needed in an unfolding analysis.

This explanation clarifies why it is so difficult to find unfolding items in survey data. Item writers insist that answers to questions should be unambiguous. However, the negative response to unfolding items is ambiguous by its very nature, which explains why item writers do not like such items and do not include them in their survey questions.

Another fundamental difference between a genuine unidimensional unfolding scale and the unfolding representation of two cumulative scales lies in the distribution of the subjects. In the case of two cumulative scales this distribution must be single peaked, with most subjects located close to the "easiest" items. In the case of a single unfolding scale there is no reason for any specification of the subject distribution. This can—and may well be—multimodal.

CHAPTER 11

Regularized Multiple Correspondence Analysis

Yoshio Takane and Heungsun Hwang

CONTENTS

11.1 Introduction

Multiple correspondence analysis (MCA) is a useful technique for the structural analysis of multivariate categorical data (Greenacre 1984; Lebart et al. 1984; Nishisato 1980). MCA assigns scores to rows (representing the subjects) and columns (representing the response categories) of a data matrix, yielding a graphical display of the rows and the columns of the data matrix. The graphical display facilitates our intuitive understanding of the relationships among the categories of the variables.

MCA, however, has remained largely descriptive, although there have been some attempts to make it more inferential (Gifi 1990;

Greenacre 1984, 1993a; Markus 1994b). These attempts mostly focused on the assessment of stability of solutions using a bootstrap resampling technique (Efron 1979). However, inferential data analysis is not limited to assessing stability, but is also intended for use in estimating population characteristics using sample data. The quality of solutions should be assessed in terms of how close the estimates are to the corresponding population quantities.

An interesting question arises from this perspective. Is the conventional method of MCA the best method for obtaining estimates of parameters? In MCA, the conventional method is well established, is computationally simple, and has been in use almost exclusively in data analysis involving MCA. However, does it provide estimates that are on average closest to the population parameters? The answer is "not always." In this chapter, we propose an alternative estimation procedure for MCA, called regularized MCA (RMCA), and demonstrate that in some cases it provides estimates that are on average closer to population parameters than the conventional estimation method. This method is easy to apply and is also computationally simple, almost as simple as the conventional method.

The basic idea of RMCA comes from ridge regression, which has proved to be useful in mitigating the multicollinearity problems often encountered in multiple regression analysis (Hoerl and Kennard 1970). Let $\mathbf{X}$ and $\mathbf{y}$ denote a matrix of predictor variables and an observed criterion vector, respectively, and let $\mathbf{b}$ denote a vector of regression coefficients. Then, the ordinary least-squares (LS) estimates of regression coefficients are obtained by

$$\hat{\mathbf{b}} = (\mathbf{X}^{\mathsf{T}}\mathbf{X})^{-1}\mathbf{X}^{\mathsf{T}}\mathbf{y} \tag{11.1}$$

In ridge regression, on the other hand, regression coefficients are estimated by

$$\mathbf{b}^{*} = (\mathbf{X}^{\mathsf{T}}\mathbf{X} + \lambda\mathbf{I})^{-1}\mathbf{X}^{\mathsf{T}}\mathbf{y} \tag{11.2}$$

where the additional quantity, λ, is a regularization (or ridge) parameter, which typically takes a small positive value. The LS estimation provides the best (minimum variance) estimates among all linear unbiased estimates (BLUE) under mild distributional assumptions on errors. However, it can provide poor estimates of regression coefficients (associated with large variances) when the matrix $\mathbf{X}^{\mathsf{T}}\mathbf{X}$ is ill conditioned (nearly singular) due to multicollinearity (high correlations among the predictor variables). The ridge estimator, on the other hand, is biased but is more robust against multicollinearity. A small positive number added to the diagonals of $\mathbf{X}^T\mathbf{X}$ works almost magically to provide estimates that are more stable than the ordinary LS estimates.

The quality of parameter estimates is measured by the squared Euclidean distance between the estimates and parameters. If we take the expected value of the squared distance over replicated samples of data, we obtain mean squared error (MSE). MSE can be decomposed into two distinct parts. One is the squared bias (the squared distance between the population parameters and the mean of the estimates over replications), and the other is the variance (the average distance between individual estimates and the mean of the estimates). The LS squares estimates have zero bias, but they may have large variances (as is typically the case in the presence of multicollinearity). The ridge estimates, on the other hand, although often biased, are also typically associated with a smaller variance, and most importantly, if the variance is small enough, ridge estimates may well have a smaller MSE than their LS counterparts. That is, in spite of their bias, they are on average closer to the population values. Indeed, for a certain range of values of λ, it is known that ridge estimators always have a smaller MSE than the ordinary LS estimates (Hoerl and Kennard 1970), regardless of the existence of multicollinearity problems. We exploit this fact to obtain better estimates in MCA.

To see the effect of regularization in MCA, let us look at Figure 11.1, which shows a two-dimensional configuration of response categories

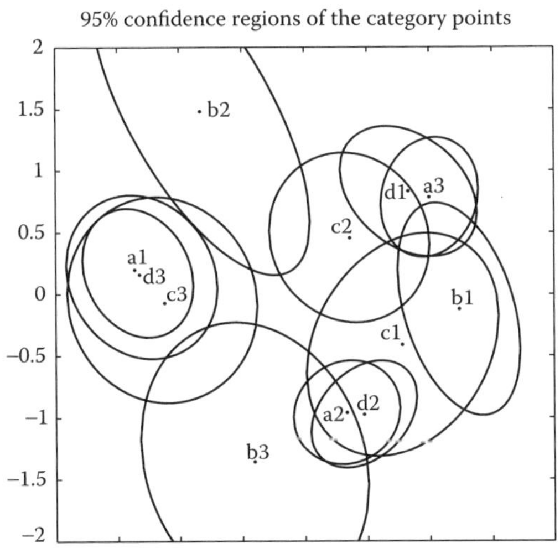

Figure 11.1 Nonregularized MCA of Nishisato's data: category points and 95% confidence regions.

obtained by the usual (nonregularized) MCA of Nishisato's (1994) small survey data (see Table 11.1). In this data set, 23 subjects responded to four multiple-choice items, each having three response categories. The questionnaire items, the associated response categories, and the data are given in Table 11.1. In Figure 11.1, each response category is labeled using an alphabet/number pairing, where the letter indicates an item and the number indicates a category number within the item.

Table 11.1 Data from small survey.

	Item			
Respondent	a	b	c	d
1	3	1	2	1
2	2	1	3	2
3	2	1	2	2
4	1	2	2	3
5	3	1	2	2
6	1	3	1	2
7	2	1	2	2
8	2	1	1	2
9	1	2	3	1
10	3	1	2	1
11	1	2	2	3
12	2	1	1	1
13	2	1	3	3
14	3	1	2	1
15	1	1	2	3
16	3	1	2	1
17	3	1	1	1
18	2	3	2	2
19	3	1	2	1
20	2	1	2	2
21	1	3	3	3
22	2	1	2	2
23	1	3	3	3

a: Your age? (1 = 20–29, 2 = 30–39, 3 = over 40). b: Children these days are not as well disciplined as children when you were a child (1 = Agree, 2 = Don't agree, 3 = Can't say which). c: Children today are not as happy as children when you were a child (1 = Agree, 2 = Don't agree, 3 = Can't say which). d: Religion should be taught at school (1 = Agree, 2 = Disagree, 3 = Don't care).

Source: Nishisato, (1994). With permission

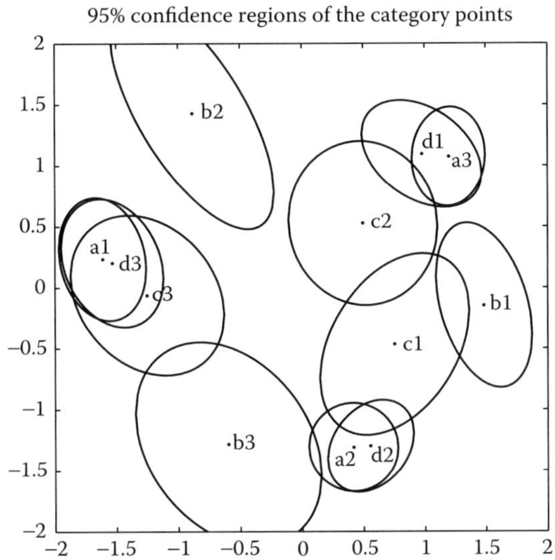

Figure 11.2 Regularized MCA of Nishisato's data: category points and 95% confidence regions.

Ellipses surrounding the points are 95% confidence regions obtained by the bootstrap procedure (Efron 1979). The smaller the ellipse, the more reliably is the point estimated. The ellipses are fairly large in all cases, indicating that the category points are not reliably estimated. This is understandable, given that the sample size in this data set is rather small. Now let us look at Figure 11.2 for comparison. This figure is essentially the same as Figure 11.1, except that it was derived from the regularized MCA (RMCA). As can be seen, ellipses are almost uniformly smaller compared with those in Figure 11.1, indicating that the point locations are more reliably estimated by RMCA. This exemplifies the type of benefit we might expect as a result of regularization. (This example is discussed further in Section 11.3.1.)

11.2 The method

11.2.1 Multiple correspondence analysis

In this section we briefly discuss ordinary MCA as an introduction to RMCA, which we develop in the next section.

Let $\mathbf{Z}_k$ ($k = 1,\ldots, K$) denote an n (cases) by p_k (categories) matrix of raw indicator variables for the kth categorical variable. We use $\mathbf{Z}$ to denote a block matrix formed by arranging $\mathbf{Z}_k$ side by side. That is, $\mathbf{Z} = [\mathbf{Z}_1,\ldots, \mathbf{Z}_K]$. Define a block diagonal matrix $\tilde{\mathbf{D}}$ consisting of $\tilde{\mathbf{D}}_k = \mathbf{Z}_k^{\mathsf{T}}\mathbf{Z}_k$ as the kth diagonal block. Let $\mathbf{X}$ denote a columnwise centered data matrix obtained from the raw data matrix by $\mathbf{X} = \mathbf{Q}_n\mathbf{Z} = \mathbf{Z}\mathbf{Q}_{1_p/\tilde{D}}$, where $\mathbf{Q}_n$ is the centering matrix of order n (i.e., $\mathbf{Q}_n = \mathbf{I}_n - \mathbf{1}_n\mathbf{1}_n^{\mathsf{T}}/n$, where $\mathbf{1}_n$ is the n-element vector of ones), and $\mathbf{Q}_{1_p/\tilde{D}}$ is the block diagonal matrix with $\mathbf{Q}_{1_{p_k}/\tilde{D}_k} = \mathbf{I}_{p_k} - \mathbf{1}_{p_k}\mathbf{1}_{p_k}^{\mathsf{T}}\ \tilde{\mathbf{D}}_k/n$ as the kth diagonal block where $\mathbf{1}_{p_k}$ is the p_k-element vector of ones. We assume that $\mathbf{X}$ is partitioned in the same way as $\mathbf{Z}$ (i.e., $\mathbf{X} = [\mathbf{X}_1,\ldots, \mathbf{X}_k]$). Define $\mathbf{D} = \tilde{\mathbf{D}}\mathbf{Q}_{1_p/\tilde{D}}$, which is the block diagonal matrix with $\mathbf{D}_k = \mathbf{X}_k^{\mathsf{T}}\mathbf{X}_k$ as the kth diagonal block.

In MCA, we find the matrix of column scores (weights) $\mathbf{W} = [\mathbf{W}_1,\ldots, \mathbf{W}_K]$ partitioned in the same way as $\mathbf{X}$ that maximizes

$$\phi(\mathbf{W}) = \mathrm{tr}(\mathbf{W}^{\mathsf{T}}\mathbf{X}^{\mathsf{T}}\mathbf{X}\mathbf{W}) \tag{11.3}$$

subject to the orthonormalization restriction that $\mathbf{W}^{\mathsf{T}}\mathbf{D}\mathbf{W} = \mathbf{I}_A$, where A is the dimensionality of the representation space. This leads to the following generalized eigenequation

$$\mathbf{X}^{\mathsf{T}}\mathbf{X}\mathbf{W} = \mathbf{D}\mathbf{W}\boldsymbol{\Delta}^2 \tag{11.4}$$

where $\boldsymbol{\Delta}^2$ is the diagonal matrix of generalized eigenvalues in descending order of magnitude, and $\mathbf{W}$ is the matrix of generalized eigenvectors of $\mathbf{X}^{\mathsf{T}}\mathbf{X}$ with respect to $\mathbf{D}$. Once $\mathbf{W}$ and $\boldsymbol{\Delta}^2$ are obtained by solving the above eigenequation, the matrix of row scores, $\mathbf{F}$, can be obtained by $\mathbf{F} = \mathbf{X}\mathbf{W}\boldsymbol{\Delta}^{-1}$.

Essentially the same results can also be obtained by the generalized singular-value decomposition (GSVD) of $\mathbf{X}\mathbf{D}^-$ with row metric $\mathbf{D}$. (This is based on the well-known relationship between the generalized eigenvalue decomposition and the GSVD; see, for example, Takane 2002.) This is written as $\mathrm{GSVD}(\mathbf{X}\mathbf{D}^-)_{I_n,D}$, where $\mathbf{D}^-$ indicates a generalized inverse (g-inverse) of $\mathbf{D}$. (For definition and computation of GSVD, see Greenacre 1984 or Takane and Hunter 2001.) Let $\mathrm{GSVD}(\mathbf{X}\mathbf{D}^-)_{I_n,D}$ be denoted by $\mathbf{X}\mathbf{D}^- = \mathbf{F}^*\boldsymbol{\Delta}^*\mathbf{W}^{*\mathsf{T}}$, where $\mathbf{F}^*$ is the matrix of left singular vectors such that $\mathbf{F}^{*\mathsf{T}}\mathbf{F}^* = \mathbf{I}_r$, $\mathbf{W}^*$ is the matrix of right-generalized singular vectors such that $\mathbf{W}^{*\mathsf{T}}\mathbf{D}\mathbf{W}^* = \mathbf{I}_r$, and $\boldsymbol{\Delta}^*$ is the positive-definite diagonal matrix of generalized singular values arranged in descending order of magnitude. Here, r is the rank of the columnwise-centered data matrix, **X**. Matrix **W** in Equation 11.4 is obtained from **W*** by retaining only the first A columns of **W*** corresponding to the A largest generalized singular values, and matrix $\boldsymbol{\Delta}$ is obtained from $\boldsymbol{\Delta}^*$ by retaining only the first A rows and columns. The matrix of row scores **F** can be similarly obtained by retaining only the first A columns of **F***.

Whether we use Equation 11.4 or GSVD to obtain MCA solutions, we can replace **D** in the formulae by $\tilde{\mathbf{D}}$ (and $\mathbf{D}^-$ by $\tilde{\mathbf{D}}^{-1}$) without affecting the computational results. This simplifies the computation considerably because $\tilde{\mathbf{D}}$ is diagonal and can be directly calculated from the raw data, whereas the computation of **D** requires an additional step. Also, $\tilde{\mathbf{D}}^{-1}$ is much easier to compute than $\mathbf{D}^-$, which is a g-inverse of **D** of a specific kind (**D** is necessarily of rank deficient because each diagonal block of **D** is of rank deficient) to obtain $\mathbf{W}_k$ that satisfies $\mathbf{1}_{p_k}^\mathsf{T}\mathbf{D}_k\mathbf{W}_k = \mathbf{0}^\mathsf{T}$ for every k, which is a standard requirement in MCA solutions. The use of $\tilde{\mathbf{D}}$ and $\tilde{\mathbf{D}}^{-1}$ automatically ensures this requirement.

11.2.2 Regularized MCA

We now introduce a regularization procedure. Define

$$\tilde{\mathbf{D}}(\lambda) = \tilde{\mathbf{D}} + \lambda\mathbf{J}_p \tag{11.5}$$

where λ is a regularization parameter (whose value is determined by some cross-validation method, as will be discussed in the next section), and $\mathbf{J}_p$ is a block diagonal matrix with $\mathbf{J}_{p_k} = \mathbf{X}_k^\mathsf{T}(\mathbf{X}_k\mathbf{X}_k^\mathsf{T})^-\mathbf{X}_k$ as the kth diagonal block. (Matrix $\mathbf{J}_{p_k}$ is the orthogonal projector onto the row space of $\mathbf{X}_k$.) Also, define

$$\mathbf{D}(\lambda) = \tilde{\mathbf{D}}(\lambda)\mathbf{Q}_{1_p/\tilde{D}} = \tilde{\mathbf{D}}\mathbf{Q}_{1_p/\tilde{D}} + \lambda\mathbf{J}_p\mathbf{Q}_{1_p/\tilde{D}} = \mathbf{D} + \lambda\mathbf{J}_p \tag{11.6}$$

Note that $\mathbf{J}_p\mathbf{Q}_{1_p/\tilde{D}} = \mathbf{J}_p$. In RMCA,we maximize

$$\phi_\lambda(\mathbf{W}) = \mathrm{tr}(\mathbf{W}^\mathsf{T}(\mathbf{X}^\mathsf{T}\mathbf{X} + \lambda\mathbf{J}_p)\mathbf{W}) \tag{11.7}$$

with respect to **W**, subject to the orthonormalization restriction that $\mathbf{W}^\mathsf{T}\mathbf{D}(\lambda)\mathbf{W} = \mathbf{I}_A$. This criterion is an extension of Equation 11.3. The above criterion leads to the following generalized eigenequation analogous to Equation 11.4:

$$(\mathbf{X}^\mathsf{T}\mathbf{X} + \lambda\mathbf{J}_p)\mathbf{W} = \mathbf{D}(\lambda)\mathbf{W}\boldsymbol{\Delta}^2 \tag{11.8}$$

To derive the GSVD equivalent to the above generalized eigenequation, we need to define a special metric matrix, $\mathbf{M}(\lambda)$, as follows:

$$\mathbf{M}(\lambda) = \mathbf{I}_n + \lambda(\mathbf{X}\mathbf{X}^{\mathsf{T}})^{+} \tag{11.9}$$

where $(\mathbf{X}\mathbf{X}^{\mathsf{T}})^{+}$ indicates the Moore-Penrose inverse of $\mathbf{X}\mathbf{X}^{\mathsf{T}}$. Note that using $\mathbf{M}(\lambda)$, $\mathbf{X}^{\mathsf{T}}\mathbf{X} + \lambda\mathbf{J}_p$ can be rewritten as

$$\mathbf{X}^{\mathsf{T}}\mathbf{X} + \lambda\mathbf{J}_p = \mathbf{X}^{\mathsf{T}}\mathbf{M}(\lambda)\mathbf{X} \tag{11.10}$$

assuming that $\mathbf{X}_k$'s are disjoint (i.e., rank($\mathbf{X}$) = $\sum_{k=1}^{K}$ rank($\mathbf{X}_k$)). This condition is usually met in practical data analysis situations. See Takane and Hwang (2004) for more details of the derivation. The equivalent GSVD problem can now be stated as GSVD$(\mathbf{X}\mathbf{D}(\lambda)^{-})_{\mathbf{M}(\lambda),\mathbf{D}(\lambda)}$.

As in the nonregularized case, $\mathbf{D}(\lambda)$ and $\mathbf{D}(\lambda)^{-}$ in Equation 11.8 or in the above GSVD problem can be replaced by $\tilde{\mathbf{D}}(\lambda)$and $\tilde{\mathbf{D}}(\lambda)^{-1}$, respectively. Again, this simplifies the computation. The exact rationale that allows this replacement can again be found in Takane and Hwang (2004).

The criterion, Equation 11.7, can be further generalized into

$$\phi_{\lambda}^{(L)}(\mathbf{W}) = \mathrm{tr}(\mathbf{W}^{\mathsf{T}}(\mathbf{X}^{\mathsf{T}}\mathbf{X} + \lambda\mathbf{L})\mathbf{W}) \tag{11.11}$$

which is maximized with respect to $\mathbf{W}$, subject to the restriction that $\mathbf{W}^{\mathsf{T}}(\mathbf{D} + \lambda\mathbf{L})\mathbf{W} = \mathbf{I}_A$, where $\mathbf{L}$ is a block diagonal matrix with $\mathbf{L}_k$ as the kth diagonal block. Matrix $\mathbf{L}_k$ could be any symmetric nonnegative definite matrix such that Sp($\mathbf{L}_k$) = Sp($\mathbf{X}_k$), where Sp($\mathbf{Y}$) indicates the space spanned by the column vectors of $\mathbf{Y}$. This criterion leads to a solution of the following generalized eigenequation:

$$(\mathbf{X}^{\mathsf{T}}\mathbf{X} + \lambda\mathbf{L})\mathbf{W} = \mathbf{D}(\mathbf{L})\mathbf{W}\boldsymbol{\Delta}^2 \tag{11.12}$$

where $\mathbf{D}(\mathbf{L}) = \mathbf{D} + \lambda\mathbf{L}$. This generalization is often useful when we need a regularization term more complicated than $\lambda\mathbf{J}_p$. Such cases arise, for example, when we want to incorporate, by way of regularization, certain degrees of smoothness in the function to be approximated (see, e.g., Ramsay and Silverman 1997). Adachi (2002) used this form of regularization to modulate the degree of smoothness of the trajectory describing changes in responses to a categorical variable over a period of time. In this case, $\mathbf{M}(\lambda)$ defined in Equation 11.9 should also be generalized into

$$\mathbf{M}^{(L)}(\lambda) = \mathbf{I}_n + \lambda(\mathbf{X}\mathbf{L}^{-}\mathbf{X}^{\mathsf{T}})^{+} \tag{11.13}$$

Properties similar to those for $\mathbf{M}(\lambda)$ hold for $\mathbf{M}^{(L)}(\lambda)$ as well.

11.2.3 The choice of regularization parameter λ

In this section, we first discuss a cross-validation procedure for selecting an optimal value of the regularization (ridge) parameter, λ. We then briefly discuss other nonparametric procedures that help make MCA more inferential.

We should note at the outset that a wide range of values of λ exists for which the regularization method works reasonably well, so that we do not have to be overly concerned about its choice. As will be shown in section 11.3.2, any value between 2 and 20 works substantially better than $\lambda = 0$, and they all give similar results. In this sense, the proposed regularization method is robust. Having said that, we may still want to have an objective procedure that can determine a near-optimal value of λ. We can use some kind of cross-validation method such as the bootstrap or the G-fold cross-validation method. We use the latter here without any good reason to favor one over the other. In this method, the data set at hand is randomly divided into G subsamples. One of the subsamples is set aside, and estimates of parameters are obtained from the remaining data. These estimates are then used to predict the cases in the sample that was set aside to assess the amount of prediction error. We repeat this process G times, each time setting aside one of the G subsamples in turn.

Let $\mathbf{X}^{(-g)}$ denote the data matrix with data in sample g eliminated from $\mathbf{X}$. We denote the data in sample g (that are eliminated) by $\mathbf{X}^{(g)}$. We apply RMCA to $\mathbf{X}^{(-g)}$ to obtain $\mathbf{W}^{(-g)}$, from which we calculate $\mathbf{X}^{(g)}\mathbf{W}^{(-g)}\mathbf{W}^{(-g)\mathsf{T}}$. This gives the cross-validation prediction of $\mathbf{X}^{(g)}\mathbf{D}(\lambda)^{-}$. We repeat this for all G subsamples and collect all cross-validated predictions, $\mathbf{X}^{(g)}\mathbf{W}^{(-g)}\mathbf{W}^{(-g)\mathsf{T}}$, in matrix $\widehat{\mathbf{X}\mathbf{D}}(\lambda)^{-}$. We then calculate

$$\varepsilon(\lambda) = \mathrm{SS}(\mathbf{X}\mathbf{D}(\lambda)^{-} - \widehat{\mathbf{X}\mathbf{D}}(\lambda)^{-})_{M(\lambda),D(\lambda)} \tag{11.14}$$

as an index of prediction error, where $\mathrm{SS}(\mathbf{Y})_{M(\lambda),D(\lambda)} = \mathrm{tr}(\mathbf{Y}^{\mathsf{T}}\mathbf{M}(\lambda)\mathbf{Y}\mathbf{D}(\lambda))$. We compare the value of $\varepsilon(\lambda)$ for different values of λ (e.g., $\lambda = 0, 1, 2, 5, 10, 20, 30$) and then choose the value of λ associated with the smallest value of $\varepsilon(\lambda)$.

When G is taken equal to the size of the original data set, this procedure is called the leaving-one-out (LOO) or jackknife method. The LOO method can be avoided for large data sets because it requires $G = N$ solutions of RMCA and could be quite time consuming, although it can still be used for smaller data sets. In most cases, however, the G-fold cross-validation method with $G < N$ gives results similar to the LOO method.

The above procedure for determining an optimal value of λ presupposes that we already know the number of dimensions in the RMCA solution. For dimensionality selection, we can use permutation tests similar to the one used by Takane and Hwang (2002) in generalized constrained canonical correlation analysis. Although the permutation tests can also be affected by the value of λ, our experience indicates that the best dimensionality is rarely, if ever, affected by the value of the regularization parameter. Thus, the permutation tests can be applied initially with $\lambda = 0$, by which a tentative dimensionality is selected, and the G-fold cross-validation method is applied to select an optimal value of λ. We can then reapply the permutation tests using the selected optimal value of λ to ensure that the best dimensionality remains the same. General descriptions of the permutation tests for dimensionality selection can be found in Legendre and Legendre (1998) and ter Braak (1990).

We can also use a bootstrap method (Efron 1979) to assess the reliability of parameter estimates derived by RMCA. In this procedure, random samples (called bootstrap samples) of the same size as the original sample are repeatedly sampled from the original sample with replacement. RMCA is applied to each bootstrap sample to obtain successive estimates of parameters. We then calculate the mean and the variance-covariance of the estimates across the bootstrap samples, from which we calculate estimates of standard errors of the parameter estimates, or draw confidence regions to indicate how reliably parameters are estimated. The latter is done under the assumption of asymptotic multivariate normality of the parameter estimates. Figure 11.1 and Figure 11.2, presented in Section 11.1 for Nishisato's (1994) data, were obtained in this way.

When the assumption of asymptotic normality is suspect, we can use a nonparametric method to construct confidence regions (e.g., Markus 1994b). Alternatively, we can simply plot as many point estimates (as obtained by the bootstrap procedure) in the configuration of category points (Greenacre 1984, 1993b). This is usually good enough to give a rough indication of how tightly or loosely category points have been estimated. (See discussion in Section 11.3.2 as well as Figures 11.8 and 11.9.)

Significance tests of the elements of **W** can also be performed as by-products of the bootstrap method described above. We simply count the number of times bootstrap estimates "cross" the value of zero (i.e., if the original estimate obtained from the original sample is positive, we count the number of times the corresponding bootstrap estimates turn out to be negative, and vice versa). If the relative frequency (the p-value) of the crossover is smaller than a prescribed α level, we conclude that the corresponding parameter is significantly different from zero.

11.3 Examples

We report two sets of numerical results in this section. One involves Nishisato's (1994) small survey data, as discussed in Section 11.1. The other pertains to Greenacre's (1993a) car-purchase data. The latter is a huge data set (over half a million cases) representing the total population of car purchases made in some country during a certain period of time. Because it is a population data set, we can sample data of varying sizes from the population and directly estimate MSE as well as the bias and variance of the estimates that result from a certain estimation procedure. We can directly compare these quantities across different values of λ by systematically varying its value.

11.3.1 Analysis of Nishisato's data continued

As described previously, this data set consists of 23 subjects responding to four items, each having three response categories. Information regarding this data set is given in Table 11.1.

Permutation tests indicated that the first dimension (corresponding to the singular value of 2.59) was highly significant ($p = 0$), while the second dimension (corresponding to the singular value of 1.79) was only marginally so ($p = 0.07$). The p-values reported for only $\lambda = 2$ were found to be optimal for this data set. However, a similar pattern of significance (p-values) was observed for other values of λ. All subsequent analyses on this data set assumed $A = 2$.

The LOO method was then applied to find an optimal value of the regularization parameter. (The LOO method was feasible because this was a small data set.) The estimate of prediction error (ε) was found to be .391 for $\lambda = 0$, .383 for $\lambda = 1$, .382 for $\lambda = 2$, and .391 for $\lambda = 5$. Thus, an optimal value of λ was found to be 2.

A bootstrap method was used to assess the reliability of the parameter estimates. This was done for both $\lambda = 0$ (nonregularized MCA) and the optimal value of $\lambda = 2$ (RMCA) for comparison. One thousand bootstrap samples were generated, parameter estimates were obtained for each sample, and the mean and the variance–covariance estimates of the estimated parameters were calculated, from which the 95% confidence regions depicted in Figure 11.1 and Figure 11.2 were drawn. As we have seen already, confidence regions are almost uniformly smaller for the parameter estimates obtained by RMCA (with the optimal value of $\lambda = 2$) than for those obtained by nonregularized MCA (under $\lambda = 0$), indicating that the parameters are more reliably estimated in the former. The regularization tends to shrink the estimates. To avoid getting smaller confidence regions just because of the shrinking effects, the configuration obtained by RMCA was sized up to match the size of the configuration obtained by the ordinary MCA, and the variance–covariance estimates of the former were adjusted accordingly.

Confidence regions were also drawn for subject points (row scores). This was done as follows. The column scores (**W**) derived from each bootstrap sample were applied to the original data to derive estimates of the coordinates of subject points. The mean and the variance-covariance estimates of the subject points were calculated over the bootstrap samples, and the confidence regions were drawn in the same way as before. Figure 11.3 and Figure 11.4 display the configurations of the

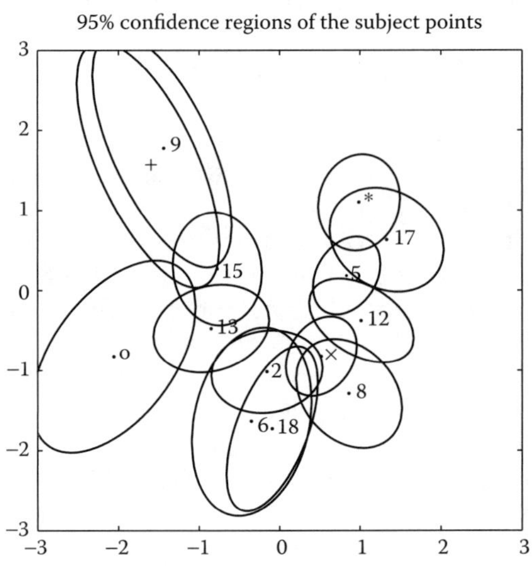

Figure 11.3 Nonregularized MCA of Nishisato's data: subject points and 95% confidence regions. Subjects 1, 10, 14, 16, and 19—having an identical response pattern—are indicated by *; similarly, subjects 3, 7, 20, and 22 are indicated by ×; subjects 4 and 11 by +; and subjects 21 and 23 by o.

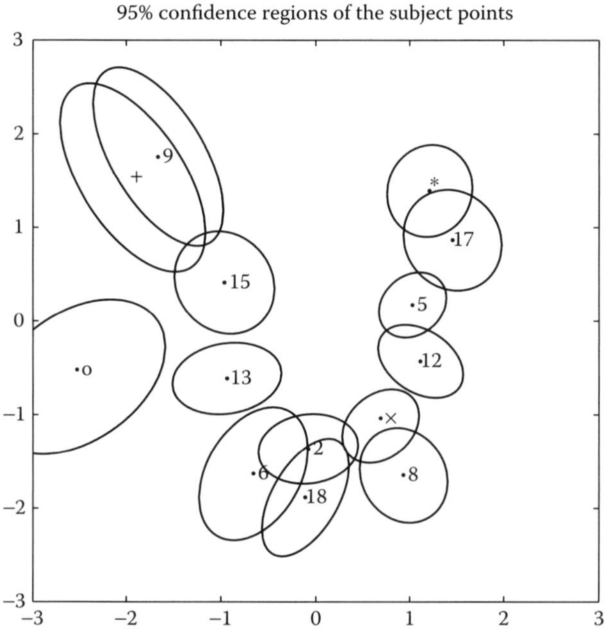

Figure 11.4 Regularized MCA of Nishisato's data: subject points and 95% confidence regions. Subjects 1, 10, 14, 16, and 19—having an identical response pattern—are indicated by *; similarly, subjects 3, 7, 20, and 22 are indicated by ×; subjects 4 and 11 by +; and subjects 21 and 23 by o.

subject points along with the 95% confidence regions obtained by MCA and RMCA, respectively. Again, we find that the subject points were more reliably estimated by RMCA. Confidence regions are almost uniformly smaller in Figure 11.4 than in Figure 11.3. This corroborates our finding in Figure 11.1 and Figure 11.2.

11.3.2 Analysis of Greenacre's car-purchase data

The second data set we analyze comes from Greenacre (1993a: 124–125). The total number of 581,515 cars purchased in the U.S. during the last quarter of 1988 were cross-classified in terms of 14 size classes of car and purchaser profiles, and these are reported in the form of a two-way contingency table. The purchaser profiles were defined by the age of the oldest person in the household of the purchaser and by the income of the household. The age variable was classified into seven groups, and the income variable into nine levels,

Table 11.2 Variables (size class, age, income) and categories in car-purchase data.

Categories					
Symbol	Size Class	Symbol	Age[a]	Symbol	Income
A	Full-size standard	a1	18–24 years	i1	\$75,000 or more
B	Full-size luxury	a2	25–34	i2	\$50,000–\$74,999
C	Personal luxury	a3	35–44	i3	\$35,000–\$49,999
D	Intermediate regular	a4	45–54	i4	\$25,000–\$34,999
		a5	55–64	i5	\$20,000–\$24,999
E	Intermediate specialty	a6	65–74	i6	\$15,000–\$19,999
		a7	75 or older	i7	\$10,000–\$14,999
F	Compact regular			i8	\$8,000–\$9,999
G	Compact specialty			i9	Less than \$8,000
H	Subcompact regular				
I	Subcompact specialty				
J	Passenger utility				
K	Import economy				
L	Import standard				
M	Import sport				
N	Import luxury				

[a] Age is that of the oldest person in the household.

Source: Greenacre (1993a). With permission.

which are factorially combined to create 63 categories. Detailed descriptions of the categories in the three variables (size classes, age, and income) can be found in Table 11.2.

Because the data are population data, there is no inferential problem with which to contend. On the other hand, this provides a golden opportunity to perform various sampling experiments on the data to examine the quality of estimates against population parameters. In particular, we can directly estimate MSE as well as its breakdown into squared bias and variance by applying RMCA to data sampled from this population. We can compare MSE across different values of the regularization parameter, including the nonregularized case ($\lambda = 0$), to assess the effect of regularization. We can also systematically vary the sample size.

In order to apply MCA, the data were first rearranged into a multiple-choice data format. We took the three variables (size classes, age, and income), all constituting column categories defining profiles

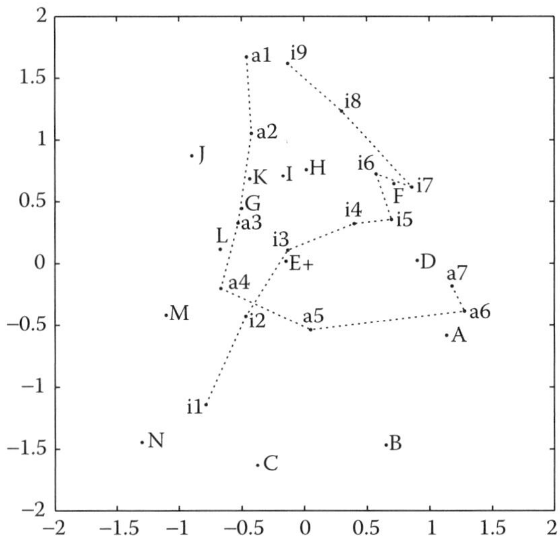

Figure 11.5 Two-dimensional population configuration for the car data.

of purchases. Rows of this data represent 882 (= 14 × 63) distinct profiles of purchases indicated by patterns of size classes of cars purchased, age groups, and income levels of the household of purchasers. Figure 11.5 presents the two-dimensional population configuration of the 30 (= 14 + 7 + 9) category points obtained by applying the ordinary MCA to this data set. Size classes of car are indicated by uppercase alphabetic characters (A through N), age groups by a lowercase "a" followed by the number indicating the age group (a1 through a7), and income levels by a lowercase "i" followed by the number indicating the income level (i1 through i9). Age groups are connected by dotted line segments, and so are income levels.

The direction that goes from the bottom right corner to the top left corner roughly corresponds with the size dimension, with the bottom right representing larger cars and the top left smaller cars. The direction perpendicular to this dimension (going from the bottom left corner to the top right corner) roughly corresponds with the price dimension, with the left bottom corner representing more expensive cars, as opposed to the top right corner representing more economical cars. The age variable is more or less linearly related to the size dimension, with older people tending to prefer larger cars. Its relation to the price dimension, however, is quadratic, with middle-aged people preferring more expensive cars, while younger and older people prefer more economical cars. The higher end of the income variable is linearly related to the price dimension, while the lower end is more in line with the smaller side of the size dimension. These results are similar to those of Greenacre's (1993a), which were obtained by an analysis of data in contingency table form, except that the current configuration is rotated about 45°.

We then obtained 100 samples each of varying sizes ($N = 200$, 500, 1000, 2000, and 5000) from this data set, applied RMCA to those sampled data with the value of regularization parameter systematically varied (λ = 0, 5, 10, 20, 30, 40, and 50), and calculated MSE, squared bias, and variance. Because this is a multiparameter situation, these quantities have to be aggregated across parameters. The simple sum of squared discrepancies was taken as an aggregate measure of discrepancies. These aggregated measures of discrepancies are then averaged across the samples. Let θ denote the vector of parameters, $\hat{\theta}_i$ their estimate from sample i, and $\bar{\theta}$ the mean of $\hat{\theta}_i$ across samples. Then,

$$\text{Squared Bias} = (\bar{\theta} - \theta)^{\mathsf{T}}(\bar{\theta} - \theta),$$

$$\text{Variance} = (1/I)\sum_{i=1}^{I}(\hat{\theta}_i - \bar{\theta})^{\mathsf{T}}(\hat{\theta}_i - \bar{\theta}),$$

$$\text{MSE} = (1/I)\sum_{i=1}^{I}(\hat{\theta}_i - \theta)^{\mathsf{T}}(\hat{\theta}_i - \theta),$$

where I is the number of sampled data sets within each condition. In the present case, I = 100 in all conditions.

Figure 11.6 displays average MSEs plotted as a function of the sample size and the value of regularization parameter. Average MSE goes down dramatically from the nonregularized case ($\lambda = 0$) to regularized cases in all sample sizes. MSE takes the smallest value for λ between 10 and 20 in all cases. A few remarks are in order. First of all, if we do not regularize, we need a much larger sample size to achieve the same degree of MSE than when we regularize with a near-optimal value of the regularization parameter. For example, to achieve without regularization the level of MSE (with $\lambda = 20$) achieved for a sample size of $N = 500$, we need roughly four times as many observations ($N = 2000$). This ratio diminishes as the sample size increases. However, it can still be substantially larger than 1 for the sample size as large as $N = 5000$. Second, MSE does not go up drastically even if we overshoot its value, that is, if we happen to choose too large a value of λ by mistake. This

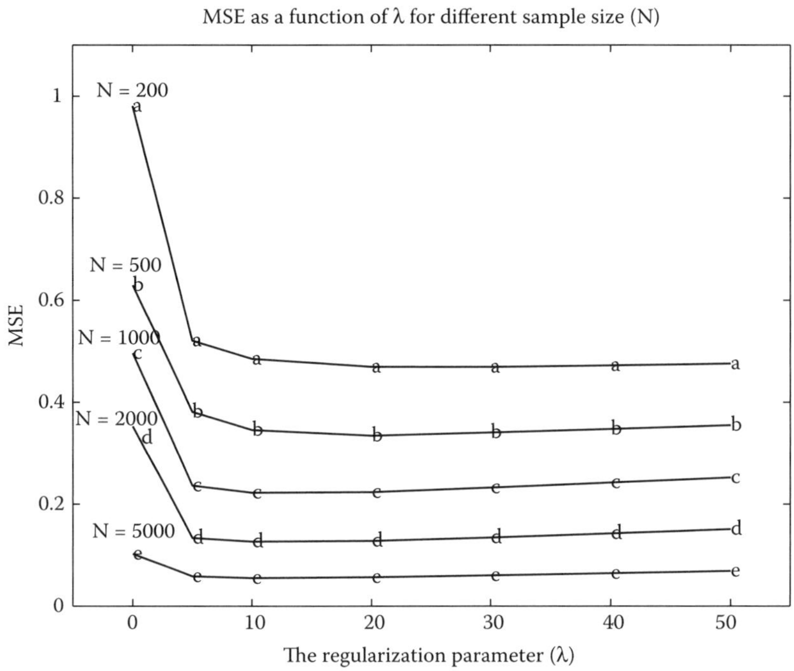

Figure 11.6 MSE as a function of the regularization parameter (λ) and sample size (N): a, $N = 200$; b, $N = 500$; c, $N = 1000$; d, $N = 2000$; e, $N = 5000$.

indicates that we do not have to be overly concerned about the choice of the value of the regularization parameter. This tendency holds across different sample sizes. The computation of MSE in Figure 11.6 presupposed a two-dimensional configuration of category points. However, essentially the same results hold for other dimensionalities, including the single dimensional case.

Figure 11.7 breaks down the MSE presented in Figure 11.6 into squared bias and variance components for $N = 500$. The squared bias tends to go up, while the variance goes down, as the value of λ increases. The sum of these quantities, MSE, takes the smallest value somewhere in the mid range. This is the characteristic of the MSE function that Hoerl and Kennard (1970) theoretically derived in the context of ridge regression. It is reassuring to find a similar tendency in RMSA, although the curves depicted in Figure 11.6 were derived empirically and not theoretically. Although only the results for $N = 500$ are presented, essentially the same tendency was observed for other

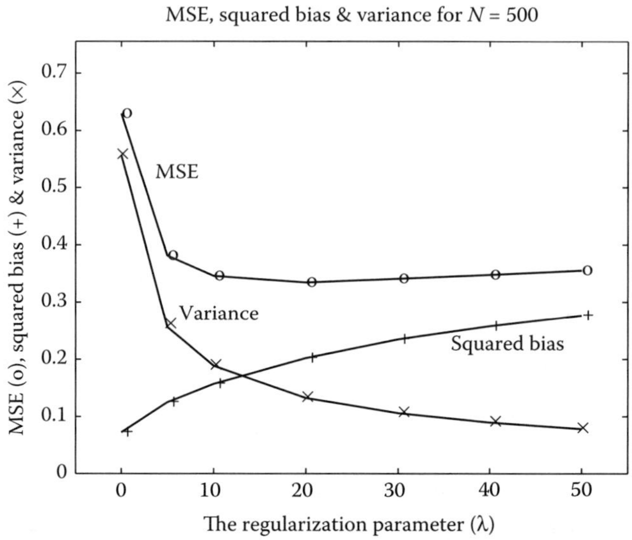

Figure 11.7 MSE, squared bias and variance as functions of λ for $N = 500$: o, MSE; +, squared bias; ×, variance.

sample sizes. Also, these curves were derived in all cases from two-dimensional solutions. As before, a similar tendency holds for other dimensionalities.

To compare the quality of estimates obtained from RMCA with those obtained from the nonregularized MCA, two categories of the size-class variable were picked out and subjected to further examination. Those two categories are Class C and Class A cars. The former has a relatively small marginal frequency (9,626 purchases out of 581,515), and the latter has a relatively large observed frequency (68,977). These translate into less than 2% and nearly 12% of the total purchases, respectively. It is anticipated that the sampling error is much larger in the small-frequency category.

Figure 11.8a plots estimates of the point representing Class C cars obtained by the nonregularized MCA from 100 samples of size $N = 500$ from the population. A small circle indicates the population location, while the × indicates the mean of the 100 estimates. (The plus symbol indicates the origin of the representation space.) It can be seen that the estimates are quite widely scattered, indicating that their reliability is relatively low. The difference between o and × indicates the bias in the estimation. Figure 11.8b, on the other hand, displays the RMCA

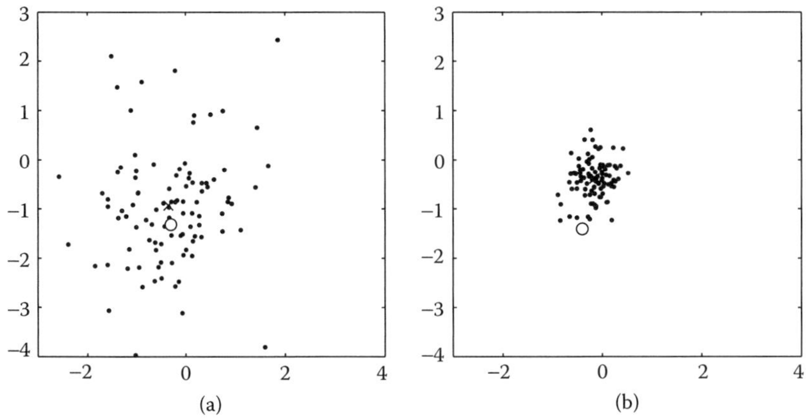

Figure 11.8 Estimates of point location by (a) nonregularized and (b) regularized MCA for car class C over 100 replicated samples of size 500.

solutions with a near-optimal value of the regularization parameter ($\lambda = 20$). Estimates of the point are much less scattered than in Figure 11.8a, although the bias is slightly larger. In total, we get much smaller MSE for the estimates.

Figure 11.9 presents essentially the same information as the previous figure for Class A cars, which received a much larger observed frequency. A similar tendency as in the previous figure can also be observed in this figure, but on a much smaller scale. Estimates obtained from the nonregularized MCA are not as scattered as those for Class C cars (cf. Figure 11.8a and Figure 11.9a). However, RMCA is still an improvement over the nonregularized case. It is associated with a slightly larger bias, but also with a smaller variance and a smaller MSE than the nonregularized case. This indicates that regularization is most effective when we have categories with small observed frequencies. Note that it is not harmful to regularize categories with relatively large frequencies; indeed, regularization in such cases may still improve the conventional estimation method.

11.4 Concluding remarks

Regularization techniques similar to the one developed in this chapter have been investigated in many other statistical methods, including regression analysis (Hoerl and Kennard 1970; Groß 2003), canonical correlation analysis (Ramsay and Silverman 1997; Vinod 1976),

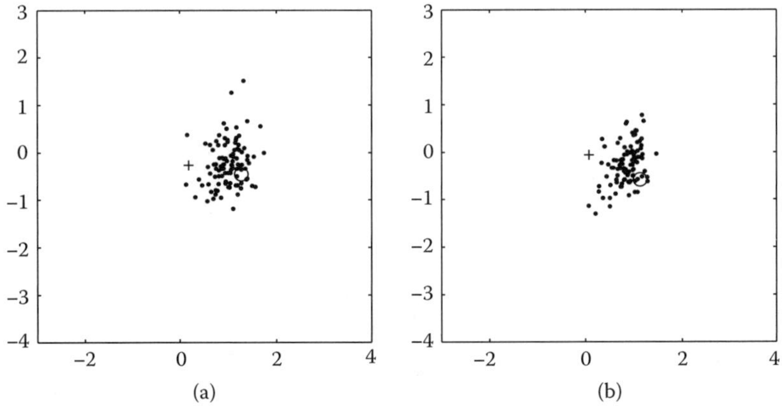

Figure 11.9 Estimates of point location by (a) nonregularized and (b) regularized MCA for car class A over 100 replicated samples of size 500.

discriminant analysis (DiPillo 1976; Friedman 1989; Hastie et al. 2001), and PCA (principal component analysis). This chapter is the first demonstration of its usefulness in MCA. Its usefulness was demonstrated through numerical experiments involving actual data sets. A similar regularization technique can be incorporated into many other multivariate data analysis techniques, for example, generalized canonical correlation analysis (Takane and Hwang 2004), redundancy analysis (Takane and Yanai 2003), hierarchical linear models (HLM), logistic regression and discrimination, generalized linear models, log-linear models, and structural equation models (SEM).

Incorporating prior knowledge is essential in many data analysis situations. Information obtained from the data is never sufficient and must be supplemented by prior information. In regression analysis, for example, the regression curves or surfaces (the conditional expectation of the criterion variable, **y**, as a function of predictor variables, **X**) are estimated for the entire range of **X** based on a finite number of observations. In linear regression analysis, this is enabled by the prior knowledge (assumption) that the regression curves and surfaces are linear within a certain range of **X**. Regularization can be viewed as a way of incorporating prior knowledge in data analysis. In the ridge type of regularization, the prior knowledge takes the form that any parameters in the model (category points in MCA) should not be too far away from 0 (the origin). That is, the effect of λ is to shrink the estimates toward zero in the light of this prior knowledge.

In a broader perspective, the results presented in this chapter cast serious doubts about the adequacy of the conventional estimation methods (such as the maximum-likelihood estimation method) that rely on asymptotic rationale. For problems with small to moderate sample sizes, there are better estimation procedures in the sense of achieving smaller MSE. It is not easy to prove theoretically the superiority of the regularization method, and little theoretical work has been done outside regression analysis. However, the kinds of numerical experiments used in the present chapter are easy to implement in a variety of contexts other than MCA, and it is expected that similar results will be obtained.

Software Notes

MATLAB® programs used to obtain the results reported in this chapter can be obtained by sending a request to takane@takane2.psych.mcgill.ca.

Acknowledgments

The work reported in this chapter is supported by Grant A6394 for the first author by the Natural Sciences and Engineering Research Council of Canada. Thanks are due to Michael Hunter of the University of Victoria and Ringo Ho of McGill University for their valuable comments on an earlier draft of this chapter.

SECTION III

Analysis of Sets of Tables

CHAPTER 12

The Evaluation of "Don't Know" Responses by Generalized Canonical Analysis

Herbert Matschinger and Matthias C. Angermeyer

CONTENTS

12.1 Introduction

Attitudes are often measured by means of items where the respondent is asked to evaluate them with respect to ordinal categories. It is implicitly assumed that the items are related to a latent dimension and that the respondents are familiar with the subject addressed in the questionnaire. If it is expected that this is not the case for a large minority of the respondents, often an extra "don't know" category is explicitly employed in order to prevent a large amount of missing values or an uncontrollable bias of the sample. Although this category is distinguished from real missing values, these responses are very often treated "per fiat" as neutral categories or simply as missing, neither of which can be considered a satisfactory solution (Blasius and Thiessen 2001a).

Much effort has been expended in deciding empirically whether a "don't know" response should be taken for an answer, particularly in the framework of the "attitude–nonattitude" discussion (Converse 1964; Duncan and Stenbeck 1988; Schuman and Converse 1971). It was pointed out that respondents answering "don't know" may nevertheless have an underlying attitude (Gilljam and Granberg 1993). These respondents are usually called "false negatives." The opposite might also be true, as respondents who actually lack an attitude may express an opinion (Converse 1970; Smith 1984); therefore they are often called "false positives." Furthermore, we should be aware that the specific wording of the items may result in a "don't know" response, which is then related to the latent attitude measured by the set of items. These sources result in at least three specific dependencies of the probability of a "don't know" category:

1. The probability of responding with "don't know" may depend on the "true value" of a respondent with respect to the latent dimension to be measured. Listwise deletion of respondents with respect to these "don't know" responses may result in a considerable bias of the sample (deletion of the false negative) because "don't know" may not be an indicator for "no opinion."
2. The probability of giving a "don't know" response may depend on the wording of a particular indicator since, for instance, answering "definitely not" or "not at all" and "don't know" might turn out to be very similar responses if the "don't know" response represents a critical attitude with respect to a positively worded item. It is questionable whether the "don't know" response results from "item ambiguity or response uncertainty" (Coombs and Coombs 1976). Again, "don't know" in this case may not be an indicator for "no opinion."
3. The probability of a "don't know" response depends on the more general willingness to respond, which is affected by personal characteristics such as education, political interest, and gender (Thiessen and Blasius 1998). If there are a substantial number of "don't knows," listwise deletion would lead to a biased sample.

The main goal of this chapter is to investigate the meaning of a "don't know" category for a particular item measuring one latent dimension in order to find out whether a "don't know" response should actually be treated as a nonresponse. If the existence of a latent dimension can be justified in a psychometric sense, a sound interpretation of both the ordinal (Likert type) and the nominal ("don't know") categories becomes possible. We will investigate whether the assumption of "ordinal" items holds with respect to both the particular wording of an item and the latent dimension to be measured (Blasius and Thiessen 2001b). The willingness to respond will be measured by the number of "don't know" answers produced by each respondent when answering a set of items. The impact of this figure on the location of respondents on the dimension of interest as well as on the quantification of the other categories and the eigenvalues will be investigated. The item-specific "don't know" responses are not independent of the sum of "don't know" answers, and vice versa, so the quantification of the item categories should be made independent of the individual sum of "don't know" responses, and the components of interest should be orthogonal to this amount. The relationship of "don't know" to individual characteristics such as gender, education, or specific attitudes (Thiessen and Blasius 1998) is beyond the scope of this chapter. Finally, we will investigate whether the mapping of the dimensions can be improved by employing the sum of "don't know" answers as a categorical variable.

12.2 Method

12.2.1 Sample and instrument

In 1990 a representative survey was conducted in both parts of Germany, involving German citizens aged 18 years and older who were living in private households. In 2001 another representative survey was conducted in Germany. We will use data from both surveys for the analysis. In total, 5809 individuals were interviewed face to face, of whom 3.37% did not completely fill out the questionnaire, so we end up with a total of 5613 observations.

In both surveys the respondents were asked to express their views on the effect of psychotropic drugs with respect to ten five-point items, five of them covering positive effects and five covering negative effects (see Table 12.1). The response categories range from 1 (strongly agree) to 5 (do not agree at all). Only the categories 1 and 5 where labeled explicitly. The items allege the effects of psychotropic drugs in a very apodictic manner, which might have hampered a more differentiated judgment necessary to adopt the five response categories for an

Table 12.1 Items to measure attitudes toward psychotropic drugs.

Item No.	Item
1	*Drug treatment is the best way of treating mental illness.*
2	The cause of mental illness cannot be dealt with by drug treatment.
3	Psychotropic drugs carry a high risk of dependency.
4	Drug treatment can only calm patients down.
5	*In severe mental illness, drug treatment is the only proper treatment.*
6	*Drug treatment is the most reliable way of preventing relapse.*
7	Taking drugs helps one to see everything through rose-tinted spectacles, leaving basic problems unchanged.
8	*Drug treatment is the treatment most likely to bring about rapid improvement.*
9	*The benefit brought about by drug treatment far outweighs the risk associated with it.*
10	In the end, psychotropic drugs make one more ill than one was before.

Note: Positively worded items are in italics.

evaluation. An extra "don't know" category was provided and labeled explicitly. This category was chosen quite frequently by the respondents, and a listwise deletion with respect to "don't know" answers would have led to a reduction of the sample size from 5613 observations to 2921. A total of 302 respondents answered exclusively with "don't know," and these are also included in the analysis.

12.2.2 Generalized canonical analysis

The method described below is a generalized form of homogeneity analysis or multiple correspondence analysis (MCA). Similar to canonical correlation analysis, it focuses on the relation between sets of variables in a low-dimensional space. In the more common form of canonical correlation analysis, the correlation between only two sets of variables is estimated. The approach used here, briefly described below, allows for more than two sets. The contribution of a particular variable to the solution is independent of all the other variables in the same set. Additionally, restrictions with respect to the quantifications of the categories can be imposed to map nominal, ordinal, or even metric variables simultaneously. If there is only one variable in each set and no restrictions are imposed on the category quantifications, we are back at homogeneity analysis or MCA. The category quantifications are then called "multiple nominal," because a different quantification is estimated for each dimension. The generalized canonical approach is extensively described elsewhere (Bekker and de Leeuw 1988; Gifi 1990; van der Burg 1988), and to be consistent with this literature, this approach is referred to as OVERALS in the following discussion.

We adopt the feature of independent contributions of the variables in each set to partial out the sum of "don't know" responses. This is done by generating as many copies of this variable as there are variables to measure the latent dimension (van der Burg et al. 1988). The data subjected to canonical analysis then contain ten sets with two variables each, one being the variable of interest (see Table 12.1) and the other being one of the copies of the sum of "don't know" answers. This "sum variable" has 11 categories, varying from 0 "don't know" responses up to 10, where all the items are answered "don't know." The aim of this approach is to generate a (first) dimension that only and perfectly maps the sum variable with an eigenvalue of 1. All the other ten variables measuring the attitude toward psychotropic drugs should have loadings and quantifications of zero on the first dimension, just as the sum variable should have loadings and quantifications of zero on all except the first dimension. If no restriction on the quantification of the categories were imposed, this analysis would require ten extra dimensions with eigenvalues of exactly 1 for each of them (Verdegaal 1986).

To keep the projection as simple as possible with respect to the number of extra dimensions, the so-called single nominal quantification is adopted, where category quantifications are still considered to be nominal but are projected on a line through the origin. By imposing this restriction, one single quantification holds for all dimensions. This restriction is adopted only for the ten copies of the sum variable. The other ten variables (measuring the attitudes toward psychotropic drugs) are treated just as in ordinary MCA. By restricting the quantifications for the sum variable to be equal for all dimensions except for a multiplicative constant, only one extra dimension is necessary. As said previously, the first axis discriminates only between the number of "don't know" responses. The second and the third axes then represent the dimensions of interest. The quantifications of the "don't know" categories are independent of the individual number of "don't know" responses.

12.2.3 Stability of the analysis

The stability of the results is investigated by means of the naïve bootstrap (Efron and Tibshirani 1993; Markus 1994b), drawing 1000 bootstrap samples with replacement (see also Chapter 7 in this volume), each with the same number of 5613 observations. To regard each respondent of the original sample equally, we adopt the algorithm for a "balanced bootstrap" (Gleason 1988), where each observation appears exactly 1000 times in the total of all the samples. The 1000 quantifications for the six categories of a particular variable are represented by means of six overlaid scatterplot matrices, as will be seen later in this chapter (Chambers et al. 1983; Cleveland 1994).

Investigating the stability of the solution for each item separately permits a closer look at the effect of item wording (positive versus negative) on the location of "don't know" responses with respect to the latent attitude toward psychotropic drugs.

12.3 Results

The distribution of the number of "don't know" responses for each item is shown as part of Table 12.2, varying from 713 (12.7%) to 1528 (27.2%). Furthermore, only 52% of the respondents never use "don't know," while more than 5% respond to all ten items with "don't know." As outlined previously, the whole sample is subjected to OVERALS, including even those 302 observations with only "don't know" answers.

12.3.1 Original sample

To facilitate the interpretation of the category quantifications for the first three dimensions, both the MCA and the OVERALS solutions are shown. Figure 12.1 shows the quantification without taking into account the sum of "don't know" responses, that is, regular MCA. The eigenvalues for the three axes are 0.552, 0.417, and 0.308. Because the second axis is the most important one, it is plotted horizontally in both scatterplots. The first number indicates the item (1–10), and the second number indicates the categories for each item (1–6), with category 6 representing the "don't know" response. As expected, the first dimension is dominated by the "don't know" categories. This dimension discriminates between respondents who answered "don't know" and those actually using one of the other response categories. Unfortunately, the quantifications for the first dimension of the categories 1

Table 12.2 Category quantifications on the second dimension.

Item	Categories					
	Strongly Agree 1	2	3	4	Do Not Agree at All 5	Don't Know 6
P1	−0.879	−0.838	−0.346	0.263	1.297	0.047
	436	936	1286	828	1036	1091
N2	0.817	−0.120	−0.341	−0.744	−0.642	−0.101
	1644	971	819	672	546	961
N3	0.666	−0.345	−0.800	−1.020	−0.526	−0.182
	2335	1437	750	264	114	713
N4	0.947	−0.100	−0.904	−0.652	−0.093	−0.470
	1459	1437	1134	550	197	836
P5	−0.591	−0.634	−0.183	0.238	1.402	0.102
	629	1100	1227	735	711	1211
P6	−0.831	−0.807	−0.317	0.203	1.369	0.116
	376	872	1251	802	784	1528
N7	0.933	−0.150	−0.569	−1.011	−0.533	−0.116
	1527	1507	995	409	193	982
P8	−0.651	−0.346	−0.735	0.061	1.291	0.036
	413	990	1203	881	1022	1104
P9	−0.658	−0.810	−0.290	0.250	1.502	0.097
	400	935	1360	820	717	1381
N10	1.395	0.255	−0.297	−0.727	−0.757	−0.071
	828	1018	1267	788	500	1212

P = positively worded items; N = negatively worded items. Frequencies are given below each quantification.

to 5 do not lie perfectly on only one line orthogonal to the first axis. The quantifications still depend on the "don't know" responses, as some of the categories 1 to 5 are closer to the respective "don't know" category 6 than others. The right plot shows the well-known horseshoe, where the categories indicating those observations that are strongly in favor of psychotropic drugs and the cluster of categories that denote an extremely critical attitude toward these drugs are marked "for" and "against," respectively. With respect to the second and third axis, the "don't know" categories are located in the center of the graph, so we would assume at a first glance that these responses do not contribute to the second and third dimensions.

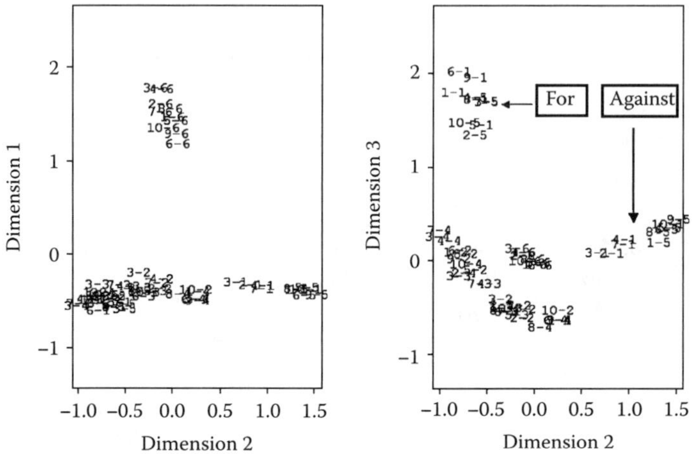

Figure 12.1 Category quantifications by MCA for the first three axes without copies of the sum of "don't know" responses.

Figure 12.2 represents the OVERALS solution, where each of the ten sets of variables now contains a copy of the sum of "don't know" responses. Controlling for the individual sum of "don't know" response, we again obtain a three-dimensional solution, but this one is no longer dominated by the item-specific "don't know" responses. The quantifications

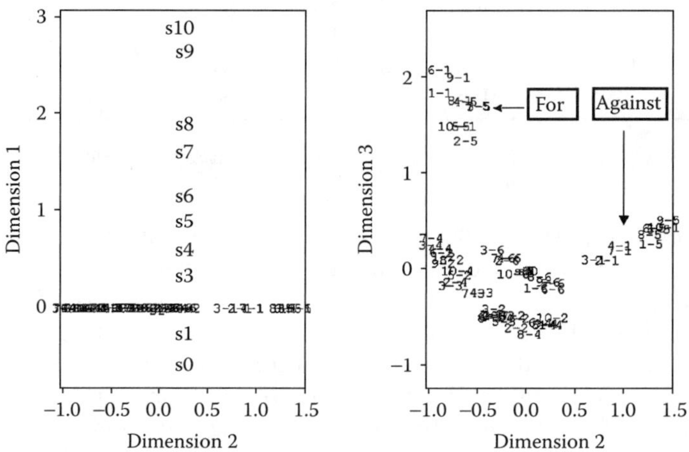

Figure 12.2 Category quantifications by OVERALS for the first three axes with ten copies of the sum of "don't know" responses.

of all these ten sum variables are identical, so only one of them is shown in Figure 12.2. The eigenvalues for the axes are 1, 0.419, and 0.309, respectively. The "don't know" categories of the ten variables now only load on the second and third dimension. The left graph also shows that the quantifications of the second axis are completely independent of the first: all the quantifications for the first dimensions are exactly zero. Since we employed a "single nominal" quantification for the 11 categories of the sum of "don't know" responses, the category quantifications for the copies of the sum variable all lie on one line through the origin, orthogonal to the second and third dimensions. On the other hand, the loadings of the ten variables on the first axis are zero without exception. Now the quantifications of the "don't know" categories are distributed between –0.5 and 0.5 along the second dimension instead of clustering around the origin.

To evaluate the item-specific meaning of "don't know" for each item separately, it becomes necessary to inspect more closely the category centroids (i.e., the means of the objects scores for those respondents who chose the same response category). Because the second dimension is of central importance, only the centroids for this dimension are reported in Table 12.2. Here we have to take into account that the object score runs from negative values, indicating the more positive attitude in favor of psychotropic drugs, to positive values, which portray the more critical attitude. We present results only for the OVERALS solution. The item-specific location of the "don't know" responses within the range of the ordinal categories shows that both the latent dimension and the wording of the items exert an effect (Table 12.2). The relative locations of the "don't know" categories show that this reaction indicates a more critical appraisal of the different effects of psychotropic drugs. This effect is more prominent for the negatively worded items, which address the undesirable effects of psychotropic drugs (items 2, 3, 4, 7, and 10). Furthermore, we have to take into account that the order of categories 4 and 5 (indicating disagreement) is reversed for four of these items.

For the negatively worded items, category 6 ("don't know") is located between categories 1 and 2, indicating agreement rather than a neutral attitude. For the positively worded items (1, 5, 6, 8, and 9), the quantifications are always located between categories 3 and 4, also indicating a more critical view, though not as prominent as for the other set of items.

The order of the categories along the second axis is not always as expected. This holds particularly for the negatively worded items if the respondents want to express an attitude in favor of psychotropic drugs. In that case categories 4 and 5 are always reversed. We might speculate that rejecting a negatively worded item is a complex task prone to misunderstanding. Negatively worded items are much less reliable indicators for the attitudes toward psychotropic drugs than items formulated in favor of antipsychotic medication. However, for the positively worded items 5 and 9, the same phenomenon can be observed for categories 1 and 2. These are the two positively worded items that do not begin with the words "drug treatment" (see Table 12.1). Furthermore, for item 8, the middle category is seriously misplaced. Additionally, we deduce from Table 12.2 that the "don't know" quantifications for the negatively worded items are negative and positive for those questions that are worded positively. So the effect of the latent dimension on the probability of answering "don't know" is moderated by the wording of a particular question. To respond with a "don't know" on a positively worded item indicates a critical appraisal of these drugs; the same response to a negatively worded question only separates those with a critical attitude (sometimes extremely critical attitude) from all the other members of the sample. Therefore, it seems questionable whether this also should be interpreted as an indicator of a critical attitude toward psychotropic drugs, but rather be addressed as an indicator for nonattitude (Blasius and Thiessen 2001b).

12.3.2 Stability of the OVERALS solution

The similarity of the quantifications will be inspected by a bootstrap analysis (see also Chapter 7). This analysis will provide information about whether the categories really can be treated as ordinal, or whether a different generating mechanism should be presumed. The white numbers in the center of the plots mark the quantifications of the original data set. For these scatterplot matrices, the aspect ratio has not been preserved, since only the demarcation of the clouds in the two-dimensional spaces or their projections on the axes are of interest, so distances should not be interpreted. The matrix plot of the quantifications for the 1000 replications will be shown only for four items, two positively worded (items 1 and 6) and two negatively (items 3 and 7). We present the plot for only the second and third axes, since the first axis is of no interest at all.

Figure 12.3 contains the bootstrap analysis for items 1 and 6. We see that all the quantifications are well separated and that, for item 1, the cloud for category 6 ("don't know") is located between categories 3 and 4 (neutral and disagreement), while for item 6 it is nearly between categories 4 and 5. Furthermore, by projecting the clouds for categories 1 and 2 (agreement) on the second axis, it is shown that these categories do not discriminate between the observations with respect to

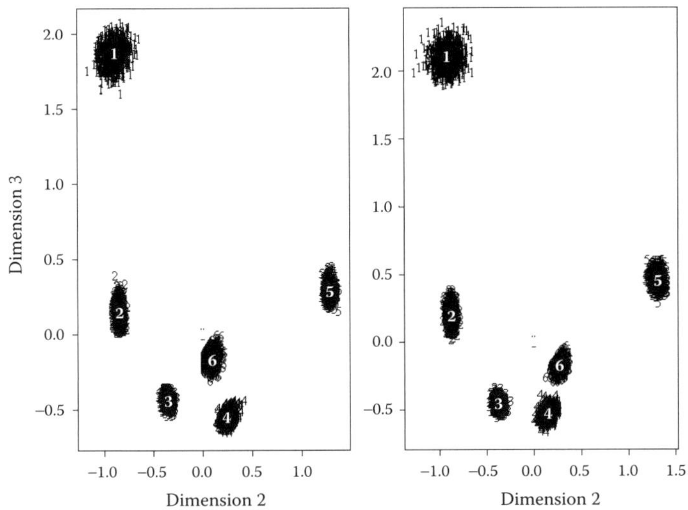

Figure 12.3 Bootstrap replications of categories for (left) item 1 ("Best way of treating mental illness") and (right) item 6 ("Most reliable way of preventing relapse").

this dimension. The most critical appraisal of the positive aspects of the application of psychotropic drugs (category 5) is well separated. This structure holds, in general, for the five positively worded items. As expected, the 11 quantifications for the sum of "don't know" responses (0 to 10) all lie exactly in the center of the graph. This variable does not contribute at all to the solution with respect to the second and third dimension.

Positively worded items that claim the positive impact of psychotropic drugs turn out to evoke relatively stable reactions compared with those caused by negatively worded items, which explicitly address the undesirable side effects. Figure 12.4 shows the bootstrap plots for the negatively worded items 3 and 7, respectively. These items evoke much more "dichotomous" reactions, since only the "strongly agree" (1) is an unmistakable stimulus, whereas the opposite category ("don't agree at all") is sometimes confusing due to double negation.

However, we must not neglect that the stability of a category quantification, evaluated by a naïve bootstrap, also depends on the frequency of this category. The frequencies of all the categories 5 for the negatively worded items are quite low (see Table 12.2). The bootstrap analysis for items 2 and 10 would show a much smaller cloud for the categories 5, but the "don't know" category is located between

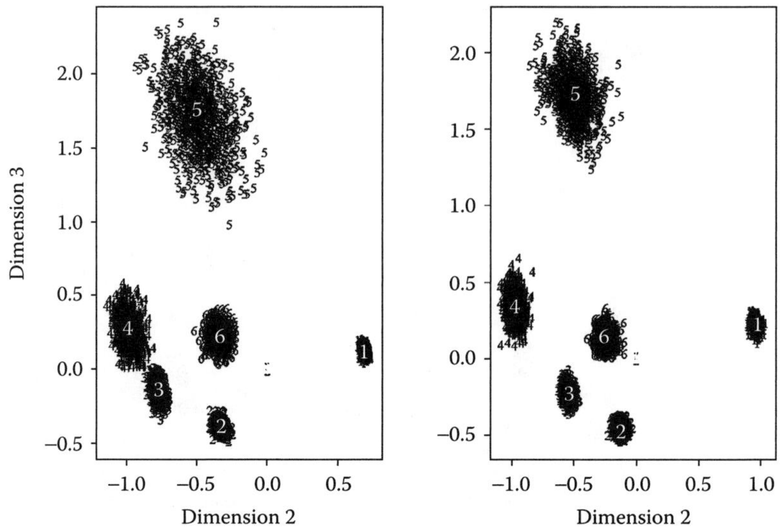

Figure 12.4 Bootstrap replications of categories for (left) item 3 ("High risk of dependency") and (right) item 7 ("Helps one to see everything through rose-tinted spectacles").

the categories 2 and 3, just as in Figure 12.4. The precision of the different responses portrayed by the bootstrap analysis is in line with the latent dimension to be measured: the more critical attitude toward psychotropic drugs results in responses that are much more distinct.

12.3.3 Relation between first and second axes

To show the improvement of the solution by controlling for the sum of "don't know" responses, the relation between the object score for the first and second axes is presented as a final step. Figure 12.5 shows this relation without controlling for the sum of "don't know" responses. The 11 groups for the different number of "don't know" responses show some small variation with respect to the first dimension, which is due to the impact of the sum of "don't know" responses.

In Figure 12.6, it is confirmed that the first dimension only portrays the amount of "don't know" responses if the effect of this sum is controlled for in the manner described previously. Within each group, characterized by a particular amount of "don't know" responses, the object scores have no variance at all, thus lying perfectly on a line parallel to

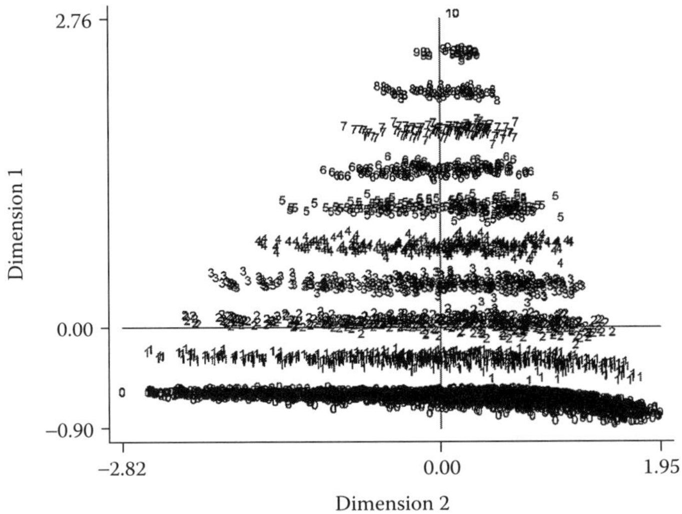

Figure 12.5 Respondent scores by MCA for first and second axes labeled by the number of "don't know" responses.

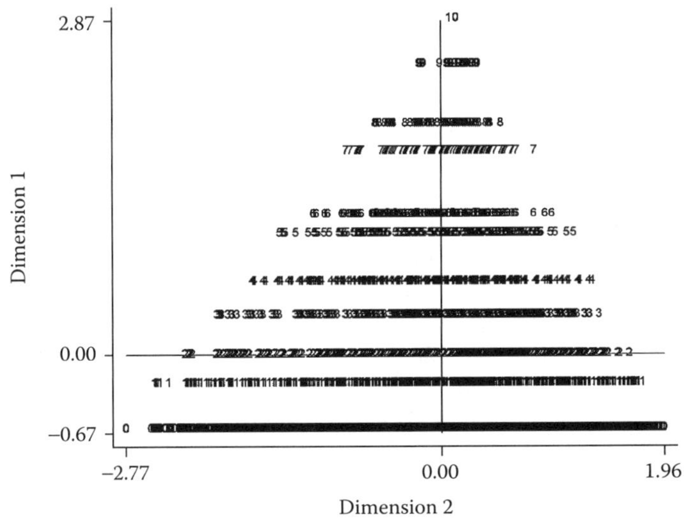

Figure 12.6 Respondent scores by OVERALS for first and second axes labeled by the number of "don't know" responses

the horizontal axis. The variance of the scores for the second dimension depends on the number of different patterns in each slice. Because there is only one pattern if ten missing values are observed, all of the 302 subjects responding to all ten questions with "don't know" have the same object score. These respondents are located close to the center of the scale, indicating, that this group does not contribute much to the solution. Given the condition that no "don't know" responses are observed ($n = 2921$), the variance of the object score for the second axis is highest because the number of possible patterns is the maximum.

12.4 Discussion

In this chapter we have focused on the problem of evaluating the meaning of "don't know" responses with respect to a single latent dimension and its relation to the other item categories. By comparing the category quantifications or the category centroids of this response for each variable with the centroids of the other response categories of the same variable, it became possible to evaluate whether a "don't know" response should be treated as nonsubstantive. Because uniquely defined meanings were assigned to the categories in advance, the order of all categories ("don't know" included) along the latent dimension allows for a sound interpretation of the "don't know" response considering the wording of a particular item. It was shown that the wording of the items exerts a considerable effect both on the stability and reliability of the items and on the location of the "don't know" responses. Items worded in favor of psychotropic drugs generate a much more stable solution, resulting in a more consistent and reliable mapping of the underlying dimension. Items that explicitly deny any effect or explicitly address undesirable side effects generate less stable quantifications. In the MCA of the ten variables, the first axis is dominated by the "don't know" categories. Using the sum of "don't know" response as an additional variable in each set and applying the generalized canonical analysis (OVERALS) improves the solution with respect to the relation of the first and the two following dimensions. Now the first dimension totally isolates this sum variable.

It was shown that using copies of "don't know" responses not only forces the eigenvalue of the first axis to be 1, but also makes the dimensions of interest (and the object scores) independent of this indicator of the willingness to respond. The relative location of a "don't know" response additionally serves as an indicator for the distortion of the sample resulting from listwise deletion. In the case of the analysis presented here, respondents with a more critical attitude toward psychotropic drugs would have been discarded.

Finally, we point out that the strategy adopted here is similar in spirit to the idea of "focusing" (Greenacre 1984: 224–226, 275–280) and "forced classification" (Nishisato 1984; Chapter 6 of this volume). In both these approaches, a variable is made to coincide with the first principal axis by increasing the weight assigned to that variable. The objective can be to ensure that all subsequent dimensions are orthogonal to the chosen variable, as is our objective here.

CHAPTER 13

Multiple Factor Analysis for Contingency Tables

Jérôme Pagès and Mónica Bécue-Bertaut

CONTENTS

13.1 Introduction

In the social sciences, data often come from different sources requiring particular methodologies. For example, when a survey is performed in different countries, specific problems arise, ranging from obtaining equivalent questionnaires in the different languages to managing the comparison of social features, even though the socioeconomic groups have different sizes.

In the example used in this chapter, people from three cities (Tokyo, Paris, New York) were asked an open-ended question, "Which dishes do you like and eat often?" These data have been analyzed from another point of view by Lebart (1995) and Lebart et al. (1998). In each city, respondents were gathered into six groups by crossing gender (male, female) and age (in three age intervals: under 30, between 30 and 50, over 50). Then, for each city, frequency tables are constructed by crossing the six groups and the most frequently used words: in each table t ($1 = 1,\ldots,3$), the general term (row i, column j) is the quotation frequency of word j by respondents belonging to group i in city t.

We present here a methodology to study a set of contingency tables having the same rows, but possibly different columns (or vice versa), situated in the framework of correspondence analysis (CA), which is a reference for analyzing contingency tables (Benzécri et al. 1973; Escofier 2003; Greenacre 1984; Lebart et al. 1977).

CA is usually extended to multiple contingency tables by using simple CA on concatenated tables (see Figure 13.1) (Benzécri 1982a; Cazes 1980; Van der Heijden 1987; Van der Heijden et al. 1989). In the "stacked tables" approach of Chapter 1 (see Section 1.5), this strategy works well because the tables have the same margins and thus the same centroids and zero "between tables" inertia. But in our case this solution presents some drawbacks, as both "between tables" and "within tables" inertias have an influence on the results.

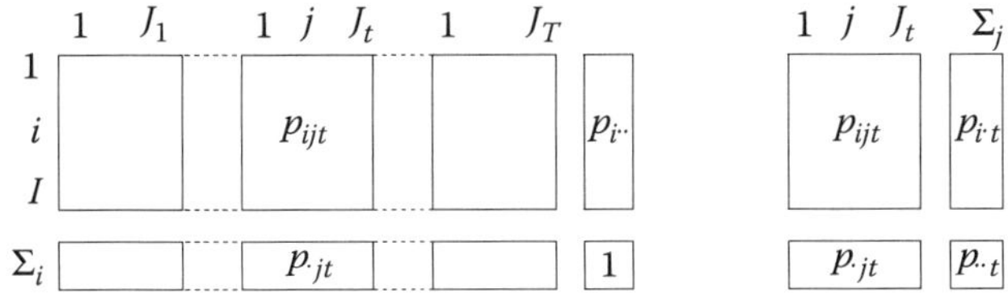

Figure 13.1 Notation for multiple contingency table: T contingency tables, having the same rows, are concatenated row-wise. On the left, the whole table $\mathbf{P}$ and its margins. On the right, the table $\mathbf{P}_t$ and its margins. In the example, $T = 3$; $I = 6$; $J_1 = 176$, $J_2 = 100$, $J_3 = 102$; p_{ijt} is the quotation proportion of word j by respondents belonging to group i in city t.

When the sums along the rows (row margins) are not proportional through all the tables, the column subclouds do not have the same centroids and the inertia between subclouds is not zero. This between-tables inertia, frequently due to different quotas imposed on the samples, as in our example, must not intervene in the study of the association between rows and columns. Furthermore, some tables can play a dominating role, which contradicts the aim of a simultaneous analysis to study the tables jointly in an equitable manner. Lastly, the representation of rows corresponds to the whole set of concatenated tables, without any reference to the row structure induced by each table.

To solve these difficulties, we have proposed (Bécue-Bertaut and Pagès 1999, 2001) a methodology for a global analysis of several contingency tables, called multiple factor analysis for contingency tables (MFACT). MFACT combines internal correspondence analysis (ICA), as a solution to the difference between margins, and multiple factor analysis (MFA) in order to balance the influence of the different tables in a global analysis as well as to provide tools to compare the row structures induced by each table in a Procrustean way. A detailed theoretical presentation of this methodology can be found in Bécue-Bertaut and Pagès (2004).

13.2 Tabular conventions

13.2.1 Notation

The T contingency tables (in the following example $T = 3$) have the same I rows ($I = 6$): each row is a group of respondents. Columns are not homologous through the tables (they correspond to words belonging to different languages). These T tables are concatenated into a two-way table as in Figure 13.1. Henceforth, we call table **P** the "complete" table transformed into proportions (relative to the total of all T tables) with subtables $\mathbf{P}_t$. We denote by p_{ijt} the proportion, in table $\mathbf{P}_t$, with which row i ($i = 1,\ldots, I$) is associated with column j ($j = 1,\ldots, J_t$; $\sum_t J_t = J$) where, in our example, i is a respondent group, j is a word, and t is a city.

$$\sum_{ijt} p_{ijt} = 1$$

A point replaces an index on which a summation is carried out. Thus we denote

$p_{i\cdot\cdot} = \sum_{jt} p_{ijt}$, row margin of table **P**

$p_{\cdot jt} = \sum_{j} p_{ijt}$, column margin of table **P**

$p_{i\cdot t} = \sum_{j} p_{ijt}$, row margin of table $\mathbf{P}_t$ as a subtable of table **P**

$p_{\cdot\cdot t} = \sum_{ij} p_{ijt}$, sum of the terms of table $\mathbf{P}_t$ as a subtable of table **P**

13.2.2 Row and column profiles

In CA, data are considered through conditional frequencies, which are called *profiles*. In our table, the ith row profile is $\{p_{ijt}/p_{i\cdot\cdot}; j = 1,\ldots, J_t; t = 1,\ldots,T\}$ and the jth column profile (j belonging to table t) is $\{p_{ijt}/p_{\cdot jt}; i = 1,\ldots, I\}$.

13.3 Internal correspondence analysis

To compare contingency tables having different margins, Benzécri (1983) as well as Escofier and Drouet (1983) introduced intratables CA. Cazes and Moreau (1991, 2000) proposed a generalization of this methodology, which they called internal CA (ICA), to handle tables with partition or graph structures on the rows and columns.

13.3.1 Equivalence between CA and a particular principal component analysis

Correspondence analysis of a single table having general term p_{ij} can be viewed as a weighted principal component analysis (PCA) performed on a table having the general term:

$$\frac{p_{ij} - p_{i\cdot}p_{\cdot j}}{p_{i\cdot}p_{\cdot j}}$$

using $\{p_{i\cdot\cdot}; i = 1,\ldots, I\}$ as row weights (and as a metric in the column space) and $\{p_{\cdot j}; j = 1,\ldots,J\}$ as column weights (and as a metric in the row space). So, the CA of the table **P** is equivalent to the weighted PCA of the table having the general term:

$$\frac{p_{ijt} - p_{i\cdot\cdot}\, p_{\cdot jt}}{p_{i\cdot\cdot}\, p_{\cdot jt}} \tag{13.1}$$

using $\{p_{i\cdot\cdot}; i = 1,\ldots, I\}$ as row weights (and as a metric in the column space) and $\{p_{\cdot jt}; j = 1,\ldots, J_t; t = 1,\ldots, T\}$ as column weights (and as a metric in the row space).

The inertia decomposed in this analysis is the usual mean-square contingency between the observed distribution $\{p_{ijt}\}$ and the distribution corresponding to the independence model $\{p_{i\cdot\cdot}\, p_{\cdot jt}\}$ corresponding to the whole table, the latter being the reference distribution. This inertia is given by

$$\phi^2 = \sum_t \sum_j \sum_i \frac{(p_{ijt} - p_{i\cdot\cdot}\, p_{\cdot jt})^2}{p_{i\cdot\cdot}\, p_{\cdot jt}} \tag{13.2}$$

13.3.2 Principle of internal correspondence analysis

The previous paragraph highlights the role of the independence model in CA. Escofier (1984, 2003) proposed a generalization of CA to any model having the same margin as the data table. ICA can be viewed as such a generalization in which the general independence model is replaced by the independence between rows and columns within each table t (intratables independence model).

The subcloud of column profiles corresponding to the contingency table $\mathbf{P}_t$ is denoted by $N(J_t)$. Geometrically, in ICA, each $N(J_t)$ is relocated so that its centroid coincides with the global centroid.

Using the equivalence between CA and the PCA of a conveniently transformed table (with suitable weights for rows and columns), the results of ICA can also be obtained by performing a weighted PCA on the table having the following general term:

$$\frac{p_{ijt} - \dfrac{p_{i\cdot t}}{p_{\cdot\cdot t}}\, p_{\cdot jt}}{p_{i\cdot\cdot}\, p_{\cdot jt}} = \frac{1}{p_{i\cdot\cdot}} \left[\frac{p_{ijt}}{p_{\cdot jt}} - \frac{p_{i\cdot t}}{p_{\cdot\cdot t}} \right] \tag{13.3}$$

Table 13.1 Coefficients relative to table t in two analyses: separate CA of table t and internal CA (ICA) applied to table P.

	General Term	Row Wt. = Metric in Col. Space	Col. Wt. = Metric in Row Space
CA applied only to $N(J_t)$	$\dfrac{p_{ijt} - \dfrac{p_{i\cdot t}}{p_{\cdot\cdot t}} p_{\cdot jt}}{\dfrac{p_{i\cdot t}}{p_{\cdot\cdot t}} p_{\cdot jt}}$	$\dfrac{p_{i\cdot t}}{p_{\cdot\cdot t}}$	$\dfrac{p_{\cdot jt}}{p_{\cdot\cdot t}}$
Internal CA applied to $N(J)$	$\dfrac{p_{ijt} - \dfrac{p_{i\cdot t}}{p_{\cdot\cdot t}} p_{\cdot jt}}{p_{i\cdot\cdot}\, p_{\cdot jt}}$	$p_{i\cdot\cdot}$	$p_{\cdot jt}$

using $\{p_{i\cdot\cdot};\ i = 1,\ldots,I\}$ as row weights (and as a metric in the column space) and $\{p_{\cdot jt};\ j = 1,\ldots,J_t;\ t = 1,\ldots,T\}$ as column weights (and as a metric in the row space).

Remark 1

There is a difference between the $N(J_t)$ in its separate CA and in ICA: the metric of the column space differs and thus the shape of the cloud $N(J_t)$ is modified. Table 13.1 recapitulates the differences between the same data in the two analyses from the PCA point of view.

Remark 2

When the row margins are proportional between all the tables, ICA is equivalent to CA applied to the concatenated tables.

13.3.3 Small illustrative example of internal correspondence analysis

To show the interest of centering the subclouds, ICA is applied to the two data matrices of Table 13.2, which differ only in the interchanging of the two first rows and, consequently, have the same eigenvalues and column margins but different row margins. Separately or simultaneously, they are perfectly represented in a two-dimensional space.

Table 13.2 Comparison of internal CA (ICA) and of CA of concatenated table applied to two small data matrices.

Table 13.2a Data matrix 1.

	A	B	C	Total
C1	120	60	20	200
C2	20	60	20	100
C3	10	30	60	100
Total	150	150	100	400

Table 13.2b Data matrix 2.

	a	b	c	Total
C1	20	60	20	100
C2	120	60	20	200
C3	10	30	60	100
Total	150	150	100	400

Table 13.2c Frequencies as given by independence model.

	A	B	C	Total
C1	75	75	50	200
C2	37.5	37.5	25	100
C3	37.5	37.5	25	100
Total	150	150	100	400

Table 13.2d Frequencies as given by independence model.

	a	b	c	Total
C1	37.5	37.5	25	100
C2	75	75	50	200
C3	37.5	37.5	25	100
Total	150	150	100	400

Table 13.2e Matrix analyzed in CA of concatenated tables (cf. Equation 13.1).

	A	B	C	a	b	c
C1	1.13	0.07	–0.47	–0.64	0.07	–0.47
C2	–0.64	0.07	–0.47	1.13	0.07	–0.47
C3	–0.73	–0.20	1.40	–0.73	–0.20	1.40

Table 13.2f Matrix analyzed in internal CA (cf. Equation 13.3).

	A	B	C	a	b	c
C1	0.80	–0.27	–0.80	–0.31	0.40	–0.13
C2	–0.31	0.40	–0.13	0.80	–0.27	–0.80
C3	–0.73	–0.20	1.40	–0.73	–0.20	1.40

The row margin of table $\mathbf{P}_t$ corresponds to the center of gravity of the subcloud of columns $N(J_t)$. In the CA of the concatenated tables, the difference between these margins induces a deviation between the two subclouds highlighted here only along axis 2 (more precisely, 37.5% of the inertia of this axis is due to between-subclouds inertia, i.e., inertia of $\{g, G\}$). When applying ICA, this between-subclouds inertia is removed by centering the subclouds $N(J_t)$ (cf. Figure 13.2, where g and G are confounded with the origin).

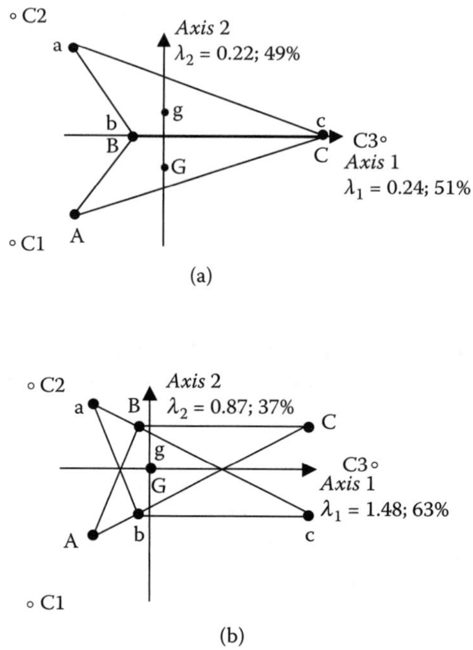

Figure 13.2 Principal planes obtained by applying, respectively, CA or internal CA (ICA).

Centering each subcloud on its own margins helps to compare the internal structures of both tables, that is, the deviations from the independence model within each table. For example, in the CA of the concatenated tables, *b* and *B* are conflated due to their identical profiles. In ICA, these points are separated because these same profiles must be compared with different references: the intra-table independence. Thus, *B* is associated with C2 more than in the model of independence given by Table 13.2c (observed frequency: 60; theoretical frequency: 37.5): C1 less than in the model of independence (observed frequency: 60; theoretical frequency: 75). For *b*, the reference is the model of independence intra-table Table 13.2d. Thus, in the data, *b* is particularly associated with C1 (observed frequency: 60; theoretical frequency: 37.5). This example illustrates how the usual CA applied to concatenated tables depends on both the between and within inertias of the columns' subclouds: these two inertias "cancel" one another until confounding *B* and *b* and, on the contrary, are "added" to strongly separate *a* and *A*. These effects could be difficult to disentangle by the user. By recentering every subcloud, ICA solves this difficulty.

13.4 Balancing the influence of the different tables

To discuss the need of balancing the tables, we shall use a set of three small data matrices (see Table 13.3). Data matrix 3 has a stronger structure ($\phi^2 = 0.3611$) than data matrix 4 ($\phi^2 = 0.0787$).

13.4.1 Predominant role of a table

ICA presents the drawback that a single table can play a predominant role and thereby determine by itself the first principal axis. The predominant role of a table $\mathbf{P}_t$ can have two origins:

1. A stronger structure of table $\mathbf{P}_t$, i.e., a stronger relationship between rows and columns. For example, when analyzing concatenated data matrices 3 and 4 (see Table 13.3) using ICA, data matrix 3 dominates the analysis, as we obtain a first factor very close to the first factor of a separate CA of data matrix 3 (correlation = 0.99; see Table 13.3d).
2. A high global frequency ($p_{..t}$), which can be due, for example, to a larger sample size. When applying ICA to the concatenated data matrices 3 and 5 (data matrix 5 is data matrix 4 multiplied by 10 (see Table 13.3)), the first factor is now very close to the first factor of a separate CA of data matrix 5 (correlation = 0.98; see Table 13.3d).

These two important effects must be removed in a global analysis. Thus, it is necessary to balance the influence of the table $\mathbf{P}_t$ in the construction of the first axes. One solution is to adopt the multiple factor analysis (Escofier and Pagès 1994, 1998) point of view and reweight the columns according to the table to which they belong.

13.4.2 Multiple factor analysis

Multiple factor analysis (MFA) analyzes data in which a single set of individuals is described by several sets of variables, quantitative or qualitative (Escofier and Pagès 1994, 1998). The core of MFA is a PCA applied to the union of all the sets of variables (global analysis), but

Table 13.3 Data matrices 3, 4, and 5. Correlation coefficients between the first-row factors of separate CA, CA of the concatenated tables, and MFA.

Table 13.3a
Data matrix 3.

	A	B	C
C1	60	30	10
C2	20	60	20
C3	10	30	60

Table 13.3b
Data matrix 4.

	a	b	c
C1	40	20	40
C2	35	35	30
C3	20	20	60

Table 13.3c
Data matrix 5
(= matrix 4 × 10).

	aa	bb	cc
C1	400	200	400
C2	350	350	300
C3	200	200	600

Table 13.3d Correlation coefficients between the first-row factors.

	$F_1(3)$ CA of matrix 3	$F_1(4)$ CA of matrix 4 (or matrix 5)	$F_1(3\&4)$ Concatenated CA of matrices 3 and 4	$F_1(3\&4 \times 10)$ Concatenated CA of matrices 3 and 4 × 10
$F_1(4)$	.66	—	—	—
$F_1(3\&4)$	.99	.75	—	—
$F_1(3\&4\times 10)$	.80	.98	.87	—
(MFA)	.90	.92	.95	.98

Note: The symbol "&" indicates that the two matrices are concatenated. In all these analyses, the rows have the same weight (1/3): CA of concatenated tables and ICA are equivalent.

done in such a way that the influences of the various sets of variables are balanced to prevent a single set from dominating the construction of the first axis.

There are two main ways of balancing the inertias of subtables in a global analysis: total inertia and maximum axial inertia. The first approach seems more natural but presents the following drawback: in the case of a subtable having a high dimensionality, the inertia in any direction is weak; this subtable will have little influence on the first axis in comparison with a subtable having a low dimensionality, which has its inertia concentrated in the first dimensions.

The second approach is used in MFA: the highest axial inertia of each set is normalized to 1 by dividing the weight of the columns belonging to set t by λ_1^t (denoting λ_1^t as the first eigenvalue of PCA applied to the set t). This reweighting of variables by $1/\lambda_1^t$ has several properties. For example:

- Due to the reweighting, every separate analysis has the first eigenvalue equal to 1.
- The intra-tables structures are not modified.
- A high-dimensional table influences more axes of the global analysis than a low-dimensional one.
- Except for very peculiar cases, the first axis of the global analysis cannot be generated by a unique table.
- The contribution of table $\mathbf{P}_t$ to axis s is between 0 and 1; it is equal to 1 if axis s corresponds to the first axis of the separate analysis of table $\mathbf{P}_t$.
- The first eigenvalue of the global analysis is between 1 and T: it is equal to 1 if every pair of variables belonging to different sets is uncorrelated; it is equal to T if all the sets have the same first factor.

MFA provides classical results of PCA: coordinates, contributions, and squared cosines of individuals; correlation coefficients between factors; and continuous variables. But beyond this global analysis, MFA offers various tools to compare the sets of variables, in particular:

- MFA can be viewed as a particular multicanonical analysis: it highlights dispersion axes common to all the sets and dispersion axes specific to some of them.
- It provides a superimposed representation of the T clouds of individuals, called partial clouds, corresponding to separate analysis of table $\mathbf{P}_t$. This aspect of MFA is close to the aim of generalized Procrustes analysis (Gower 1975).

These tools, as well as the interpretation aids that they provide, turn MFA into a convenient method of comparing several sets of variables, analyzing their relationships, and visualizing the different descriptions obtained through the global analysis as well as through all the separate analyses.

13.5 Multiple factor analysis for contingency tables

13.5.1 Combining ICA and MFA

Abdessemed and Escofier (1996) have already proposed an extension of MFA to compare several contingency tables, but the method is restricted to the case in which row margins of the different tables are identical. We generalize this extension to the case of tables having different row margins. The basic idea is to combine ICA (which deals with the problem of the differences between rows margins) and MFA (which balances the influence of the subtables in the global analysis). This method is therefore called MFA for contingency tables (MFACT) and can be viewed as a PCA:

- Analyzing the same table as ICA, whose general term is given in Equation 13.3
- Preserving the row weights used in ICA ($p_{i..}$)
- Modifying the column weights to balance the influence of the tables t

A reweighting carries out this modification of the column weights:

- In an identical way for all the columns of a table $\mathbf{P}_t$ (so as not to modify the structure of table $\mathbf{P}_t$)
- In order to balance the maximum axial inertia of each subcloud $N(J_t)$

This reweighting consists of multiplying the weight ($p_{.jt}$) of the column (j,t) in ICA by $1/\lambda_1^t$, where λ_1^t is the first eigenvalue of the separate PCA of the subtable $\mathbf{P}_t$ whose general term is given in Equation 13.3, with the weights used in ICA ($p_{i..}$ for row i, $p_{.jt}$ for the column (j,t); cf. Table 13.1). We call this restricted analysis "pseudoseparate CA," the "separate CA" being the usual simple CA of table $\mathbf{P}_t$.

The choice of $p_{i..}$ as weight for row i seems natural: as in CA, the group i has a high weight when representing many observations. In the case of tables with notable higher frequencies than others, the former can strongly dominate those weights. Then a question arises: should the balancing of the tables also intervene in defining the row weights? There is no unique answer, and different users can have different opinions, depending on the data set (see Chapter 14, where the row weight for each table t is taken into account). Alternatively, to balance the tables' influence on the row weights, the data can first be transformed into

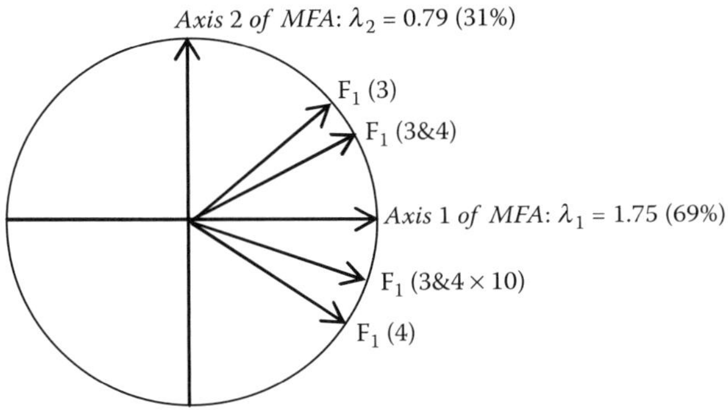

Figure 13.3 Representation of the first factors issued from the different analyses on the MFACT first plan (see also Table 13.3d): F1(3) [respectively F1(4)], first factor issued from the analysis of only data matrix 3 [respectively, data matrix 4 F1(3&4)] [(respectively, F1(3&4 × 10)], first factor issued from the analysis of data matrix 3 and data matrix 4 [respectively, data matrix 3 and data matrix 5 = data matrix 4 × 10] juxtaposed row-wise.

proportions ($n_{ijt}/n_{\cdot\cdot t}$ instead of n_{ijt}) before the concatenation. The formulation of MFACT does not depend on this transformation.

13.5.2 Illustration of balancing

MFACT has been applied to the concatenated data matrices 3 and 4 of Table 13.3. The graph of Figure 13.3 represents the correlations between the first two factors of MFACT and the first factor of

- Separate CA of matrices 3 and 4
- CA of concatenated matrices 3 and 4: CA(3&4)
- CA of concatenated matrices 3 and 5: CA(3&5) = CA(3&4 × 10)

(Because the row margins of these three matrices are proportional, ICA and simple CA are the same).

CA(3&4) is mainly influenced by matrix 3; CA(3&4 × 10) is mainly influenced by matrix 4; the first factor obtained with MFACT is clearly a compromise between the first axes of both separate CAs. Therefore, the reweighting of the tables allows us to reach the objective of giving the same importance to each subtable in a global analysis.

13.5.3 The two steps of the MFACT methodology

First step: pseudoseparate analyses

"Correspondence analysis," but using weights ($p_{i\cdot\cdot}$) and ($p_{\cdot jt}$), is applied to each table $\mathbf{P}_t$ in order to calculate the eigenvalues λ_1^t. These rows and columns weights are distinct from those used in separate CA, which are, respectively, $p_{i\cdot t}/p_{\cdot\cdot t}$ and $p_{\cdot jt}/p_{\cdot\cdot t}$ (see Table 13.1).

Second step: global analysis

This step provides global results that are

- Analogous to those of CA applied to concatenated tables (mainly, representations of rows and columns).
- Specific to multiple tables (mainly superimposed representations of row–group structures induced by each table $\mathbf{P}_t$, i.e., representations of factors derived from pseudoseparate analyses).

The main properties are close to those of CA. They are presented and illustrated using the example in the following section.

13.6 MFACT properties

13.6.1 Objectives of the example

The general objective is to relate groups and vocabulary, as summarized by the following questions:

- Which words characterize a given group?
- Which groups characterize a given word?
- Which groups are similar from the point of view of vocabulary?
- Which words are similar from their user group point of view?

The distinction between the three cities presented in this example also suggests additional questions specific to multiple tables:

- Which groups are similar, whatever the cities?
- Which groups are similar in one city and different in the others?

The main results of the separate CA of each table $\mathbf{P}_t$ are gathered in Figure 13.4. In the three separate analyses, the first axis ranks age

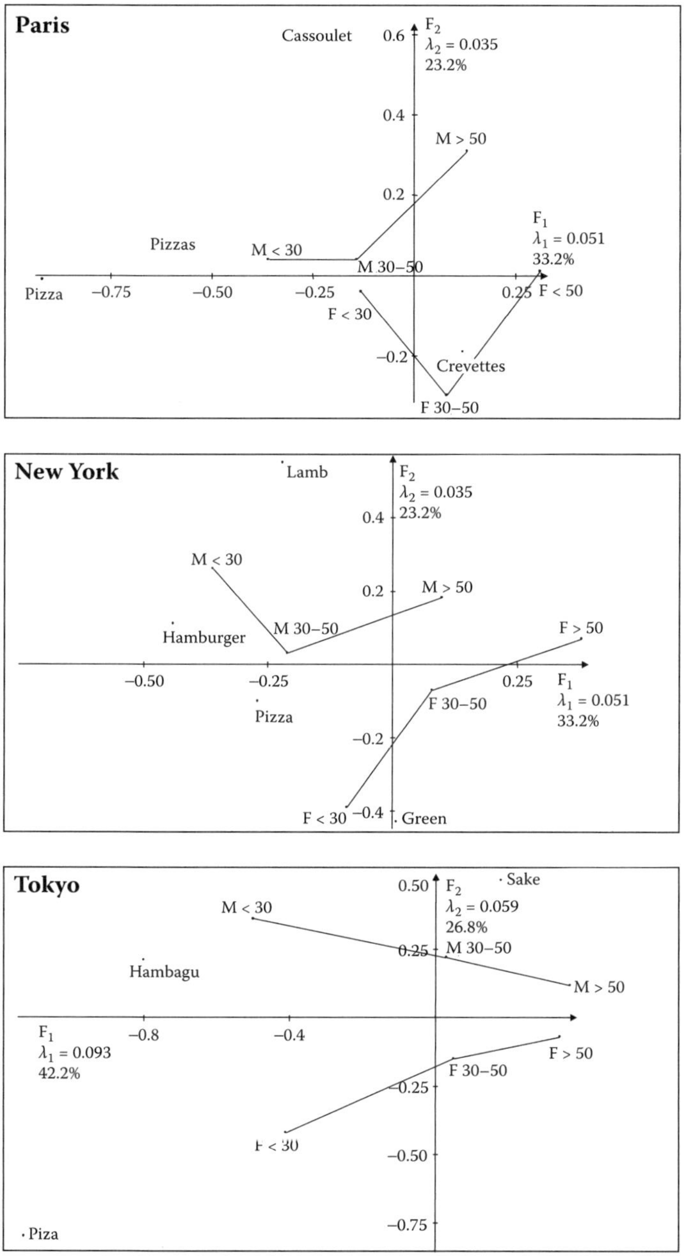

Figure 13.4 Separate CA for the three cities: gender–age structure and a few words as shown in planes of first and second axes.

intervals in their natural order, while the second axis contrasts genders. However, while in the CA of Tokyo data, the first plan accounts for almost 70% of the inertia, three axes are required in the two other cities to reach the same global quality of representation. In fact, we can note that in Paris, males aged 30 to 50 (M 30–50) and females younger than 30 years (F < 30), as well as females aged 30 to 50 (F 30–50) in New York, are poorly represented.

13.6.2 Distances between rows (groups) and between columns (words)

As with CA (or ICA) of table **P**, MFACT first provides a representation of rows and columns (see Figure 13.5 and Figure 13.6). We now give details about how proximities are interpreted in MFACT.

Distance between respondent groups (rows)

The distance between groups can be interpreted as a resemblance in the groups' use of words through all the tables. More precisely, the squared distance between groups i and i', calculated from coordinates given in Equation 13.3 for the variable j of set t having the weight $p_{\cdot jt}/\lambda_{1t}$, is

$$d^2(i,l) = \sum_t \sum_{j \in J_t} \frac{1}{\lambda_1^t p_{\cdot jt}} \left[\left(\frac{p_{ijt}}{p_{i\cdot\cdot}} - \frac{p_{i'jt}}{p_{i'\cdot\cdot}} \right) - \frac{p_{\cdot jt}}{p_{\cdot\cdot t}} \left(\frac{p_{i\cdot t}}{p_{i\cdot\cdot}} - \frac{p_{i'\cdot t}}{p_{i'\cdot\cdot}} \right) \right]^2 \qquad (13.4)$$

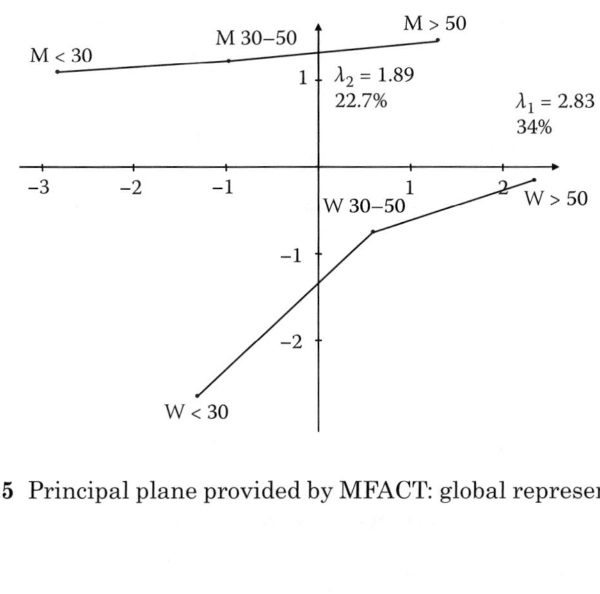

Figure 13.5 Principal plane provided by MFACT: global representation of the groups.

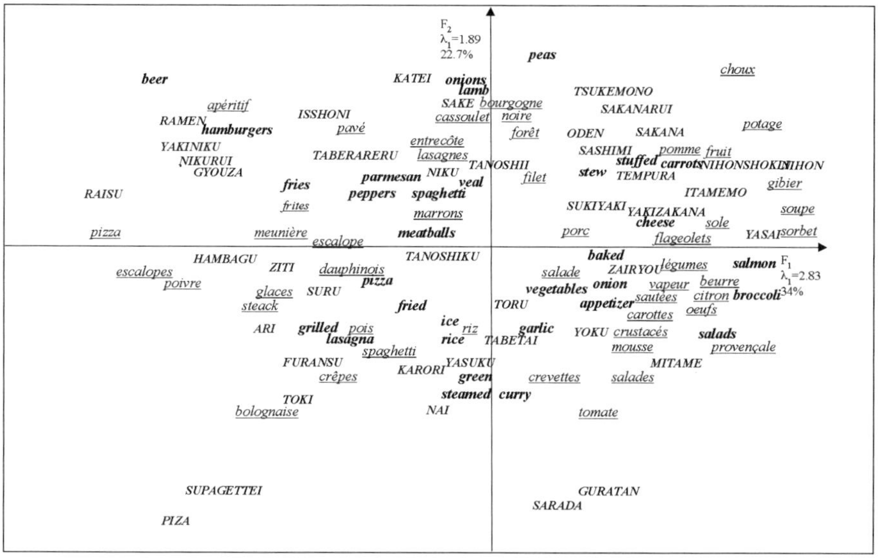

Figure 13.6 Principal plane provided by MFACT: excerpt of the representation of the words—pizza (word used in Paris), ***pizza*** (word used in New York City), and *PIZA* (word used in Tokyo).

due to

$$\sum_t \sum_{j \in J_t} \frac{1}{\lambda_1^t p_{\cdot jt}} \left[\frac{p_{\cdot jt}}{p_{\cdot\cdot t}} \left(\frac{p_{i \cdot t}}{p_{i \cdot\cdot}} - \frac{p_{i' \cdot t}}{p_{i' \cdot\cdot}} \right) \right]^2 = \sum_t \frac{1}{\lambda_1^t p_{\cdot\cdot t}} \left(\frac{p_{i \cdot t}}{p_{i \cdot\cdot}} - \frac{p_{i' \cdot t}}{p_{i' \cdot\cdot}} \right)^2$$

and

$$\sum_t \sum_{j \in J_t} \frac{1}{\lambda_1^t p_{\cdot jt}} \left(\frac{p_{ijt}}{p_{i \cdot\cdot}} - \frac{p_{i'jt}}{p_{i' \cdot\cdot}} \right) \left(\frac{p_{\cdot jt}}{p_{\cdot\cdot t}} \right) \left(\frac{p_{i \cdot t}}{p_{i \cdot\cdot}} - \frac{p_{i' \cdot t}}{p_{i' \cdot\cdot}} \right) = \sum_t \frac{1}{\lambda_1^t p_{\cdot\cdot t}} \left(\frac{p_{i \cdot t}}{p_{i \cdot\cdot}} - \frac{p_{i' \cdot t}}{p_{i' \cdot\cdot}} \right)^2$$

this squared distance can also be written as:

$$d^2(i,i') = \left[\sum_t \frac{1}{\lambda_1^t} \sum_{j \in J_t} \left(\frac{p_{ijt}}{p_{i \cdot\cdot}} - \frac{p_{i'jt}}{p_{i' \cdot\cdot}} \right)^2 \frac{1}{p_{\cdot jt}} \right] - \left[\sum_t \frac{1}{\lambda_1^t p_{\cdot\cdot t}} \left(\frac{p_{i \cdot t}}{p_{i \cdot\cdot}} - \frac{p_{i' \cdot t}}{p_{i' \cdot\cdot}} \right)^2 \right] \quad (13.5)$$

Equation 13.4 shows how, for the two groups i and i', the deviation for each word of table P_t is relativized by the word counts used (by these groups) in table P_t. In Equation 13.5, disregarding the weighting by the reverse of the first eigenvalue:

The first term is the distance (between the profiles i and i') in the CA of the complete table **P**.
The second term is the distance (between the profiles i and i') in the CA of the $(I \times T)$ table containing the word counts by group and city (that is, the table whose general term i,t is the count of all the words employed by group i in the city t). This term shows how this last table is neutralized in ICA.

Moreover, the overweighting by $1/\lambda_1^t$ balances the influence of the different tables. Thus, in the example (Figure 13.5), females and males are closer to each other over 50 than under 30; indeed, in the three cities, their lexical profiles are globally more similar over 50. In these proximities, MFACT has eliminated the differences between the word counts of the same groups in the various cities.

Distance between words (columns)

The distances between words can be interpreted as a resemblance between their users (Figure 13.6). More precisely, the squared distance between word j (belonging to table $\mathbf{P}_t$) and word j' (belonging to table $\mathbf{P}_{t'}$), calculated from coordinates given in Equation 13.3 (the row i has the weight $p_{i\cdot\cdot}$), is

$$d^2(j,j') = \sum_i \frac{1}{p_{i\cdot\cdot}} \left[\left(\frac{p_{ijt}}{p_{\cdot jt}} - \frac{p_{i\cdot t}}{p_{\cdot\cdot t}} \right) - \left(\frac{p_{ij't'}}{p_{\cdot j't'}} - \frac{p_{i\cdot t'}}{p_{\cdot\cdot t'}} \right) \right]^2 \tag{13.6}$$

$$d^2(j,j') = \sum_i \frac{1}{p_{i\cdot\cdot}} \left[\left(\frac{p_{ijt}}{p_{\cdot jt}} - \frac{p_{ij't'}}{p_{\cdot j't'}} \right) - \left(\frac{p_{i\cdot t}}{p_{\cdot\cdot t}} - \frac{p_{i\cdot t'}}{p_{\cdot\cdot t'}} \right) \right]^2 \tag{13.7}$$

Case 1: the words belong to a same table ($t = t'$). The proximity between two words measures the resemblance between profiles exactly as in the usual CA.

Case 2: the words j and j′ belong to different tables ($t \neq t'$).

Equation 13.6 shows that the profile of a word intervenes by its deviation from the average profile of its table. Equation 13.7 shows how the differences between word profiles are "relativized" by the differences between average profiles. In such a way, the distance between words used by different samples is interpretable. This important property gives sense to the global representation of the word columns.

Thus, in the example, according to Figure 13.6, "legumes," "vegetables," and "yasai" are rather quoted by the same respondent groups. As a matter of fact, consulting the data, we can see that "legumes," "vegetables," and "yasai" are used, respectively, 101, 71, and 65 times in female answers and only 54, 52, and 30 times in male answers (see Table 13.4). These differences between frequencies are amplified by the fact that, in the three cities and for each age, males are less prolix than females. Table 13.4 shows the detailed counts of these words.

From counts and weights presented in Table 13.4 to Table 13.6, it is easy to calculate the distances between two words. For example, from Equation 13.7, the squared distance between "légumes" and "vegetables" is

$$\begin{aligned}
&d^2(\text{légumes, vegetables})\\
&= \frac{1}{0.134}\left[\left(\frac{11}{155}-\frac{11}{123}\right)-\left(\frac{1305}{8331}-\frac{525}{5221}\right)\right]^2 + \frac{1}{0.180}\left[\left(\frac{11}{155}-\frac{17}{123}\right)-\left(\frac{1213}{8331}-\frac{1145}{5221}\right)\right]^2\\
&+ \frac{1}{0.144}\left[\left(\frac{32}{155}-\frac{24}{123}\right)-\left(\frac{1287}{8331}-\frac{786}{5221}\right)\right]^2 + \frac{1}{0.148}\left[\left(\frac{22}{155}-\frac{26}{123}\right)-\left(\frac{1328}{8331}-\frac{738}{5221}\right)\right]^2\\
&+ \frac{1}{0.208}\left[\left(\frac{32}{155}-\frac{25}{123}\right)-\left(\frac{1508}{8331}-\frac{1182}{5221}\right)\right]^2 + \frac{1}{0.185}\left[\left(\frac{47}{155}-\frac{20}{123}\right)-\left(\frac{1690}{8331}-\frac{845}{5221}\right)\right]^2\\
&= 0.1201
\end{aligned}$$

13.6.3 Transition formulas and interpretation rules

In MFACT, as in PCA, the representations of rows and columns must be jointly analyzed due to the transition formulae that relate the coordinates of rows and columns. We examine hereinafter the transition formulae in MFACT.

Table 13.4 Some word counts in the answers for males and females.

	Pizza			Hamburgers [a]		Vegetables		
Groups by Sex and Age	Paris pizza	NY pizza	Tokyo piza	NY hamburgers	Tokyo hambagu	Paris légumes	NY vegetables	Tokyo yasai
Males <30	14	12	2	14	17	11	11	5
Males 30–50	7	13	1	12	4	11	17	8
Males >50	1	5	0	4	2	32	24	17
Males total	22	30	3	47	23	54	52	30
Females <30	8	13	7	4	9	22	26	5
Females 30–50	3	14	1	1	8	32	25	22
Females >50	0	7	0	3	2	47	20	38
Females total	11	34	8	19	19	101	71	65
Total counts	33	64	11	66	42	155	123	95

[a] "Hamburger" is not cited with sufficient frequency in Paris to be retained.

Let us recall that in PCA of the table having general term x_{ij} with row weights r_i and column weights c_j, the relations (along axis s) between the coordinates $\{f_{is};\ i = 1,\dots,I\}$ of rows and the coordinates $\{g_{js};\ j = 1,\dots,J\}$ of columns, named transition formulae, are

$$f_{is} = \frac{1}{\sqrt{\lambda_s}} \sum_{j \in J} x_{ij} c_j g_{js}$$

$$g_{js} = \frac{1}{\sqrt{\lambda_s}} \sum_{i \in I} x_{ij} r_i f_{is}$$

Transition of words to groups

The relation giving (along axis s) the coordinate f_{is} of respondent group i from the coordinates $\{g^t_{js}; j = 1,\dots,J_t;\ t = 1,\dots,T\}$ of words is

$$f_{is} = \frac{1}{\sqrt{\lambda_s}} \sum_t \sum_{j \in J_t} \frac{p_{\cdot jt}}{\lambda_1^t} \left[\frac{1}{p_{i\cdot\cdot}} \left[\frac{p_{ijt}}{p_{\cdot jt}} - \frac{p_{i\cdot t}}{p_{\cdot\cdot t}} \right] \right] g^t_{js}$$

The column profiles belonging to J_t being centered (see Section 13.3.3), we now have

$$\sum_{j \in J_t} p_{\cdot jt} g^t_{js} = 0$$

Using this property, the first transition formula can be written as:

$$f_{is} = \frac{1}{\sqrt{\lambda_s}} \sum_t \frac{1}{\lambda_1^t} \frac{p_{i\cdot t}}{p_{i\cdot\cdot}} \left[\sum_{j \in J_t} \frac{p_{ijt}}{p_{i\cdot t}} g^t_{js} \right] \qquad (13.8)$$

Except for λ_1^t, this transition formula is similar to the one of usual CA. Disregarding a constant, each group lies in the centroid of the words used by this group. In other words, a group is attracted by the words it uses and rejected by the words it does not use. In the example, females quoted more often "vegetables" (and less often "hamburgers") than males (Table 13.4).

The distinction between the two summations in Equation 13.8 points out the two steps of the calculus of the location of group i along axis s:

1. Sum over j: For each cloud $N(J_t)$, the centroid of its words is calculated, giving to each of them a weight in accordance with the relative importance of this word for this group i in table $\mathbf{P}_t$. Except for a coefficient, this part of the formula is the one of the CA of table $\mathbf{P}_t$. This is the intra-table aspect of the analysis.
2. Sum over t: The centroid of the previous centroids is calculated, giving to each of them a weight proportional to the frequency of group i in table $\mathbf{P}_t$ (relative to the importance of group i in the whole set of tables); here again, the balancing of the different sets intervenes (overweighting by $1/\lambda_1^t$).

Transition from groups to words

The relation giving (along axis s) the coordinate g_{js}^t of word j,t from the coordinates $\{f_{is}, i = 1,\ldots,I\}$ of the groups is

$$g_{js}^t = \frac{1}{\sqrt{\lambda_s}}\left[\sum_i \left(\frac{p_{ijt}}{p_{\cdot jt}} - \frac{p_{i\cdot t}}{p_{\cdot\cdot t}}\right) f_{is}\right] \tag{13.9}$$

As the coefficient of f_{is} can be negative, words are not in the centroid (strictly speaking) of the groups, except when the row weights are the same in all tables. This coefficient of f_{is} measures the discrepancy between the profile of word j,t and the column margin of table $\mathbf{P}_t$. A word is attracted by (respectively rejected by) the groups that use it more (respectively less) than if there were independence between rows and columns within the table $\mathbf{P}_t$.

In the example, the distribution of the dish "pizza" (written "pizza" in Paris and New York, but "piza" in Tokyo) among the six groups is quite different in three cities (Table 13.4), as shown by their different positions on the principal plane (Figure 13.6).

13.6.4 Superimposed representation of the group clouds

The structures of the groups according to the different cities are compared as usual in MFA by superimposing the representations of the groups:

- Associated with each table $\mathbf{P}_t$ (called *partial groups*)
- Associated with the global table $\mathbf{P}$ (called *global* or *mean groups*)

The aim of these superimposed representations is similar to that of generalized Procrustes analysis (Gower 1975). In MFA, the superimposed representations derive from the word representation by means of a "restricted" transition formula. We denote by i^t the group i considered only from table $\mathbf{P}_t$; the coordinate of i^t along axis s (f_{is}^t) is obtained by

$$f_{is}^t = \frac{1}{\sqrt{\lambda_s}} \frac{p_{i\cdot t}}{p_{i\cdot\cdot}} \left[\sum_{j \in J_t} \frac{p_{ijt}}{p_{i\cdot t}} \frac{g_{js}^t}{\lambda_1^t} \right] \tag{13.10}$$

This relation is the restriction of Equation 13.8 to table $\mathbf{P}_t$: i^t is attracted by the words used by the group i in city t and rejected by the words it does not use in city t.

According to Equation 13.8 and Equation 13.10, $f_{is} = \sum_t f_{is}^t$. In practice, in the graph superimposing partial representations, these partial representations are dilated by the coefficient T (number of tables). Thus, a (mean) group is located in the centroid of the corresponding partial groups (points corresponding to a same group are linked to their centroid by lines).

Figure 13.7 shows a very strong resemblance between the three partial representations of each group. This resemblance does exist in the data, as it was previously seen in the representations from the separate analyses (see Figure 13.4). Relationships between axes of the global and separate analyses are studied in the next section.

Beyond the global resemblance between the three partial representations, some differences can be observed. Thus, in the example (Figure 13.7), these representations suggest that males under 30 years and males between 30 and 50 years are (relative to the other groups of the same city) more different in Tokyo than in New York (from a word-use point of view).

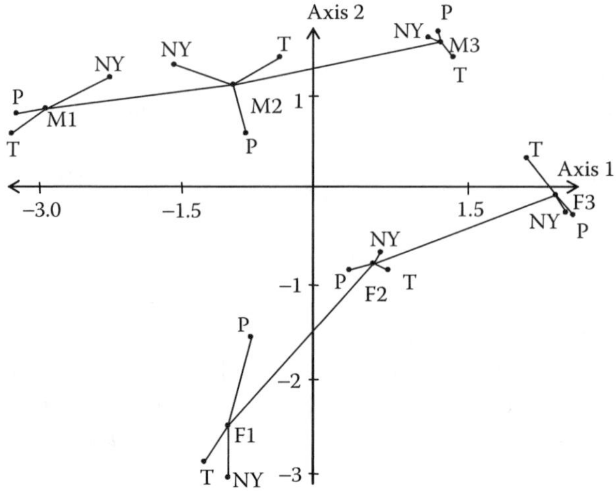

Figure 13.7 Partial-groups representation on the first plane of MFACT. Here Figure 13.5 is completed by the representation of groups as described by each of the tables $\mathbf{P}_t$ (called partial groups).

13.6.5 Relationships between global analysis and pseudoseparate analyses

To study the relationships between the global analysis (MFACT) and the separate analyses, we could think of calculating the correlations between the factors for I (vectors of the coordinates of the groups) of the global analysis and those of separate analyses. However, these two kinds of factors have not been calculated with the same row weights ($p_{i\cdot\cdot}$ for the global analysis; $p_{i\cdot t}/p_{\cdot\cdot t}$ for the separate analysis of table $\mathbf{P}_t$; concretely, they are not centered for the same weights).

A first approach might consist of calculating these correlation coefficients using one of the two sets of row weights, preferably $p_{i\cdot\cdot}$, in order to use the same weights for each coefficient. Instead, we calculate the correlations between the factors of the global analysis and those of the pseudoseparate analyses. This way is more consistent with MFACT, which depends on pseudoseparate analyses (in particular, via the reweighting). These coefficients are truly interesting when the separate and pseudoseparate analyses do not differ too much. (This is generally the case and, in particular, applies to the example presented; see Table 13.7, which is introduced later in this chapter in Section 13.7.2.)

Table 13.5 Correlations among factors of pseudoseparate analyses and of the global analysis. Correlations among the three axes of the global analysis and the corresponding axes of the pseudoseparate analyses are shown in boldface.

Pseudoseparate Analyses		Global Analysis Axis 1	Axis 2	Axis 3
Paris	Axis 1	**.979**	–.058	.061
	Axis 2	.054	**.735**	–.364
	Axis 3	.003	.541	**.337**
New York	Axis 1	**–.883**	.194	.401
	Axis 2	–.159	**–.976**	.077
	Axis 3	–.224	–.008	**–.383**
Tokyo	Axis 1	**.942**	.250	.202
	Axis 2	–.253	**.946**	.021
	Axis 3	.211	.068	**–.778**

Table 13.5 shows that the first factors of pseudoseparate and global analyses are closely correlated: the structures common to the three tables are close to the main structure of each table. The second factor of the global analysis is close to the second factors of the pseudoseparate analyses of New York and Tokyo, but not so close to the one of Paris. It would be interesting to study in what respect Paris is different. The third global factor is close only to the third pseudoseparate factor of Tokyo.

13.7 Rules for studying the suitability of MFACT for a data set

The suitability of MFACT must be examined for each application from several points of view: the centering of columns' subclouds, the overweighting, and the deformation induced by using the same row weights in each subtable. We illustrate these checks from the example.

13.7.1 Suitability of removing "between inertia"

Table 13.6 gathers the row margins of the three tables. From one city to the other, the greatest difference between the weights for corresponding groups in the different cities is for the group "males <30,"

Table 13.6 Word counts and weights for every group in the separate CA and in the CA of table **P**.

	Group Words Counts and Weights							
	Paris		New York		Tokyo		Whole	
Rows	Words	Weight	Words	Weight	Words	Weight	Words	Weight
Males <30	1305	0.157	525	0.101	620	0.132	2450	0.134
Males 30–50	1213	0.145	1145	0.219	931	0.199	3289	0.180
Males >50	1287	0.154	786	0.151	550	0.117	2623	0.144
Females <30	1328	0.159	738	0.141	635	0.135	2701	0.148
Females 30–50	1508	0.180	1182	0.226	1104	0.235	3794	0.208
Females >50	1690	0.205	845	0.162	849	0.181	3384	0.185
Total	8331	1	5221	1	4689	1	18241	1

whose weight is 0.157 in Paris and only 0.101 in New York. These differences between weights induce between-subcloud inertia in columns clouds. In our example, the ratio of inertia (between subclouds) and total inertia is 9.2%. This "between inertia" must be removed.

As a matter of fact, CA directly applied to the concatenated tables (not described here) provides a second factor that is strongly influenced by between-clouds inertia (34% of the inertia associated with this second axis is due to between-clouds inertia). It expresses an opposition between groups more numerous (or more prolix) in the New York and Tokyo samples (males and females between 30 and 50) and groups more numerous in the Paris sample. Therefore, removing between-clouds inertia is useful in this example. But the variation of row weights may induce difficulties in the comparison of tables, as discussed below.

Table 13.7 Inertia of the three column clouds in separate and pseudoseparate CA.

	Separate CA		Pseudoseparate CA	
	Global inertia (ϕ^2)	Inertia of axis 1	Global inertia (ϕ_c^2)	Inertia of axis 1 (λ_1^t)
Paris	0.149	0.0545	0.153	0.0493
New York	0.153	0.0479	0.147	0.0507
Tokyo	0.221	0.0877	0.215	0.0932

Note: Separate CA is the usual CA. Pseudoseparate CA differs from separate CA according to row weights (cf. Table 13.6), which induce the reweighting of the columns in MFACT.

13.7.2 Suitability of the overweighting

Table 13.7 gathers some inertias of separate and pseudoseparate analysis. According to Table 13.7, without balancing tables by λ_1^t, the third table (having the highest first eigenvalue) has an *a priori* influence higher than the other ones. In this case, reweighting looks to be necessary, but we have to keep in mind that the relationship between groups and words is strongest in Tokyo.

13.7.3 Separate and pseudoseparate analyses

To compare the representations provided by separate CAs, it is inevitable to distort them (distortions due to the changing of row weights). Then we evaluate these distortions, i.e., the differences between the separate CA and the pseudoseparate CA. The weights ($p_{i\cdot\cdot}$), used in the pseudoseparate analysis, and the weights ($p_{i\cdot t}/p_{\cdot\cdot t}$), used in the separate analysis, are listed in Table 13.6.

When performing the separate CA, that is to say the usual CA, global inertia is equal to ϕ^2. When performing the pseudoseparate CA, the inertia of table $\mathbf{P}_t$ can be considered as a "corrected" ϕ_c^2, whose expression is given by Equation 13.11:

$$\phi_c^2 = \sum_i \sum_j \frac{\left(p_{ijt} - \frac{p_{i\cdot t}}{p_{\cdot\cdot t}} p_{\cdot jt}\right)^2}{p_{i\cdot\cdot}\, p_{\cdot jt}} \tag{13.11}$$

To obtain an indication of distortion between separate and pseudoseparate CA, we compare ϕ^2 and ϕ_c^2 (Table 13.7). In this example, ϕ^2 and ϕ_c^2 are close to each other; in particular, the stronger structure of Tokyo appears in both cases. On the other hand, the first eigenvalues for the separate and pseudoseparate CA are very similar. The stronger structure of Tokyo appears again in both cases.

Finally, for each table $\mathbf{P}_t$, correlation coefficients between factors (of the same rank) of separate CA and pseudoseparate analyses are calculated using $p_{i\cdot\cdot}$ as row i weights (Table 13.8). For each table $\mathbf{P}_t$, separate CA and pseudoseparate CA are similar in the first-plane solutions.

We note that the structure of each table $\mathbf{P}_t$ is little modified by the change of row weights. Therefore, in this example, it can be considered that MFACT compares the structures of the three tables with almost the same point of view as that of separate CA.

Table 13.8 Correlations between the factors of the separate CAs and those of the pseudoseparate CAs.

Paris		New York		Tokyo	
$r(F_{1,ps},F_{1,sep})$	$r(F_{2,ps},F_{2,sep})$	$r(F_{1,ps},F_{1,sep})$	$r(F_{2,ps},F_{2,sep})$	$r(F_{1,ps},F_{1,sep})$	$r(F_{2,ps},F_{2,sep})$
0.999	0.927	−0.986	0.963	0.998	−0.991

Note: Correlations are calculated using $p_{i\cdot\cdot}$ as row weights.

13.8 Conclusion

MFACT proposes a solution to the two difficult problems that arise during the simultaneous analysis of several contingency tables having a common dimension.

1. The "intra-table aspect" of MFACT solves the problem of the distinct margins (corresponding to the common dimension) by centering each table with respect to its row margin.
2. The "MFA aspect" of MFACT solves the problem of balancing the influence of the various subtables in a global analysis by adopting the solution already introduced to deal with T tables of individuals × variables (quantitative or qualitative). This solution is now extended to the case of contingency tables.

Let us recall that the usual methodologies (CA of the concatenated tables; PCA of the concatenated tables transformed into percentages calculated separately for each subtable) do not take into account any of these difficulties. From this viewpoint, MFACT expands the methodological tools available for studying a set of contingency tables having the same rows.

Software note

MFACT described in this work can be performed using the step MFA of SPAD 6.0 (2005) software, provided that the matrix and weights are suitably adapted (SPAD-Groupe Test & Go, Paris, France (http://www.spadsoft. com).

Acknowledgments

This work has been partially supported by the Catalan DGR, grant 1998BEAI400100. We thank Ludovic Lebart for putting the data set used here at our disposal.

CHAPTER 14

Simultaneous Analysis: A Joint Study of Several Contingency Tables with Different Margins

Amaya Zárraga and Beatriz Goitisolo

CONTENTS

14.1 Introduction

The joint study of several data tables has given rise to an extensive list of factorial methods, some of which have been gathered by Cazes (2004), for both quantitative and categorical data tables. In the correspondence analysis (CA) approach, Cazes shows the similarity between some methods in the case of proportional row margins and shows the problem that arises in a joint analysis when the row margins are different or not proportional. In this chapter, simultaneous analysis (SA) is presented as a factorial method developed for the joint treatment of a set of several data tables, especially frequency tables whose row margins are different, for example when the tables are from different samples or different time points.

The results of applying classical methods such as CA to the concatenation of the tables can be affected by

1. The differences between the grand totals of the tables. All else being equal, the larger the grand total of a table, the greater is the table's influence in the overall analysis.
2. The differences between the inertias of the tables, especially the inertia on the first factorial axis of each table. All else being equal, the higher a table's inertia, the greater is its influence in the overall analysis.
3. The different masses of the rows in each table, i.e., the differences between the marginals.

SA allows us to overcome these disadvantages, thus providing a joint description of the different structures contained within each table as well as a comparison of them. Furthermore, SA can be applied to the joint analysis of more than two data tables in which rows refer to the same entities, but whose columns may be different.

To solve these problems, the SA allows us:

1. To balance the influence of each table in the overall analysis by transforming the grand total of each table and then expressing each contingency table relative to its total.

2. To balance the influence of the tables according to the differences in inertia between them by reweighting each as in multiple factor analysis (MFA) (Escofier and Pagès 1998).
3. To preserve the different masses of each row in an overall factorial analysis.

Multiple factor analysis for contingency tables (MFACT; Bécue-Bertaut and Pagès 2004; Chapter 13 of this volume) also seeks to provide a solution to the three problems mentioned. In MFACT, the marginal of the concatenated table is imposed on the overall analysis and is, thus, suitable when the marginal relative frequencies of the tables are similar.

After introducing SA, we shall examine the differences between SA and MFACT. We will apply SA to a set of four data tables with the same columns, different overall totals, and different row marginals. These tables represent the active population coded by gender and training level in Spain's autonomous communities. The information is taken from the Active Population Survey conducted by the Statistical Office of Spain (INE) in 2001.

14.2 Simultaneous analysis

Let $\mathbf{T} = \{1,\ldots,t,\ldots,T\}$ be the set of contingency tables to be analyzed (Figure 14.1). Each of them classifies the answers of $n_{\cdot\cdot t}$ individuals with respect to two categorical variables. All the tables have one of the variables in common, in this case the row variable with categories $\mathbf{I} = \{1,\ldots,i,\ldots,I\}$. The other variable of each contingency table can be different or the same variable observed at different time points or in different subsamples. Upon concatenating all of these contingency

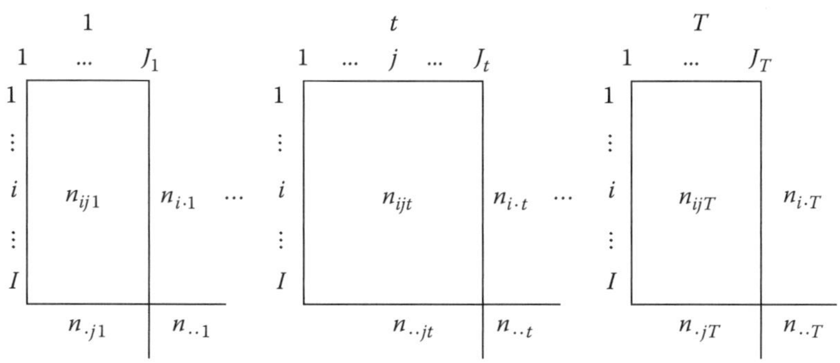

Figure 14.1 Set of contingency tables.

tables, a joint set of columns $\mathbf{J} = \{1,\ldots, j,\ldots,J\}$ is obtained. The element n_{ijt} corresponds to the total number of individuals who choose simultaneously the categories $i \in \mathbf{I}$ of the first variable and $j \in \mathbf{J}_t$ of the second variable, for table $t \in \mathbf{T}$. Sums are denoted in the usual way, for example, $n_{i \cdot t} = \sum_{j \in \mathbf{J}_{\cdot t}} n_{ijt}$, and n denotes the grand total of all T tables.

To maintain the internal structure of each table t, SA begins by obtaining the relative frequencies of each table as is usually done in CA:

$$p_{ij}^{t} = \frac{n_{ijt}}{n_{\cdot\cdot t}}$$

so that $\sum_{i \in \mathbf{I}} \sum_{j \in \mathbf{J}_t} p_{ij}^{t} = 1$for each table t. It is important to keep in mind that these relative frequencies are different from those obtained when calculating the relative frequency for the whole matrix: $p_{ijt} = n_{ijt}/n$. Notice the following relationship between our notation and that of Chapter 13: $p_{ij}^{t} = p_{ijt} \big/ p_{\cdot\cdot t}$.

SA is carried out basically in two stages.

14.2.1 *Stage one: CA of each contingency table*

Because in SA it is important for each table to maintain its own structure, the first stage carries out a classical CA of each of the T contingency tables. These separate analyses also allow us to check for the existence of structures common to the different tables. From these analyses, it is possible to obtain the weighting used in the next stage.

CA on the tth contingency table can be carried out by calculating the singular-value decomposition (SVD) of the matrix $\mathbf{X}^t$, whose general term is

$$\sqrt{p_{i\cdot}^{t}} \left(\frac{p_{ij}^{t} - p_{i\cdot}^{t} p_{\cdot j}^{t}}{p_{i\cdot}^{t} p_{\cdot j}^{t}} \right) \sqrt{p_{\cdot j}^{t}}$$

Let $\mathbf{D}_r^t$ and $\mathbf{D}_c^t$ be the diagonal matrices whose diagonal entries are, respectively, the marginal row frequencies $p_{i\cdot}^{t}$ and column frequencies $p_{\cdot j}^{t}$. From the SVD of each table X^t we retain the first squared singular value (or eigenvalue, or principal inertia), denoted by λ_1^t. Notice that the difference our formulation and that of Chapter 13 is that we perform a weighted PCA of

$$\frac{1}{p_{i \cdot t}} \left(\frac{p_{ijt}}{p_{\cdot jt}} - \frac{p_{i \cdot t}}{p_{\cdot\cdot t}} \right)$$

with weights $P_{i \cdot t}$ on the rows. A comparison with formula (13.3) of Chapter 13 shows that the difference is in the row weighting.

14.2.2 Stage two: analysis of intrastructure

In the second stage, in order to balance the influence of each table in the joint analysis, as measured by the inertia, and to prevent this joint analysis from being dominated by a particular table, SA will include a weighting on each table, α_t. The choice of α_t is discussed in Section 14.2.3.

As a result, SA proceeds by performing a principal component analysis (PCA) of the matrix:

$$\mathbf{X} = \left[\sqrt{\alpha_1}\mathbf{X}^1 \ldots \sqrt{\alpha_t}\mathbf{X}^T \ldots \sqrt{\alpha_T}\mathbf{X}^T\right]$$

The PCA results are also obtained using the SVD of X, giving singular values $\sqrt{\lambda_s}$ on the sth dimension and corresponding left and right singular vectors $\mathbf{u}_s$ and $\mathbf{v}_s$.

We calculate projections on the sth axis of the columns as principal coordinates:

$$\mathbf{g}_s = \lambda_s \mathbf{D}_c^{-1/2} \mathbf{v}_s$$

where $\mathbf{D}_c$ ($J \times J$) is a diagonal matrix of all the column masses, that is, all the $\mathbf{D}_c^t$.

One of the aims of the joint analysis of several data tables is to compare them through the points corresponding to the same row in the different tables. These points, as in MFA of quantitative variable tables, will be called *partial rows* and denoted by i^t.

The projection on the sth axis of each partial row is denoted by f_{is}^t, and the vector of projections of all the partial rows for table t is denoted by $\mathbf{f}_s^t$:

$$\mathbf{f}_s^t = \left(\mathbf{D}_r^t\right)^{-1/2}\left[0 \ldots \sqrt{\alpha_t}\mathbf{X}^t \ldots 0\right]\mathbf{v}_s$$

(Notice that our notation f_{is}^t is not the same as that of Chapter 13 (see (13.10)) because of the different row weighting mentioned above.) Especially when the number of tables is large, comparison of partial rows is complicated. Therefore, each partial row will be compared with the (overall) row, projected as:

$$\mathbf{f}_s = \left(\mathbf{D}_w\right)^{-1}\left[\sqrt{\alpha_1}\mathbf{X}^1 \ldots \sqrt{\alpha_t}\mathbf{X}^T \ldots \sqrt{\alpha_T}\mathbf{X}^T\right]\mathbf{v}_s$$

$$= \left(\mathbf{D}_w\right)^{-1}\mathbf{X}\mathbf{v}_s$$

where $\mathbf{D}_w$ is the diagonal matrix whose general term is $\Sigma_{t\in\mathrm{T}}\sqrt{p_{i\cdot}^t}$. The choice of this matrix $\mathbf{D}_w$, which allows us to expand the projections of the (overall) rows to keep them inside the corresponding set of projections of partial rows, is appropriate when the partial rows have different weights in the tables. With this weighting, the projections of the overall and partial rows are related as follows:

$$f_{is} = \sum_{t\in\mathbf{T}} \frac{\sqrt{p_{i\cdot}^t}}{\sum_{t\in\mathbf{T}}\sqrt{p_{i\cdot}^t}} f_{is}^t$$

Thus, the projection of a row is a weighted average of the projections of partial rows. It is closer to those partial rows that are more similar to the overall row in terms of the relation expressed by the axis and that have a greater weight than the rest of the partial rows. The dispersal of the projections of the partial rows with regard to the projection of their (overall) row indicates discrepancies between the same row in the different tables.

Notice that if $p_{i\cdot}^t$ is equal in all the tables, then $\mathbf{f}_s = \sum_{t\in\mathbf{T}} 1/T\mathbf{f}_s^t$, that is, the overall row is projected as the average of the projections of the partial rows.

14.2.3 Choice of the weighting of the tables

The choice of the weighting α_t depends on the aims of the analysis and on the initial structure of the information. Different values can be used, depending on the circumstances:

1. $\alpha_t = 1$ if no modification of the initial influence of each table is sought.
2. α_t is the inverse of the first eigenvalue of the separate CA of each table (stage one), as in MFA. This is also the weight

chosen for SA, $\alpha_t = 1/\lambda_1^t$, where λ_1^t denotes the first eigenvalue (square of first singular value) of table t.

3. α_t is the inverse of the total inertia of the separate CA of each table (stage one) if tables with small inertias are to be favored.

14.2.4 *Reconstructing original information*

In CA, the use of projections on all the dimensions allows to reconstruct exactly the χ^2 distances between two points (rows or columns). The proposed SA also allows us to restore the χ^2 distances between rows and between columns of a particular table, since in SA each table maintains its own weights and metrics.

As in standard factorial analyses, SA allows us to reconstruct original data as follows:

$$p_{ij}^t = p_{i\cdot}^t p_{\cdot j}^t \left(1 + \frac{1}{\sqrt{\alpha_t}} \frac{\sum_{t \in \mathbf{T}} \sqrt{p_{i\cdot}^t}}{\sqrt{p_{i\cdot}^t}} \sum_s \lambda_s^{-1/2} f_{is} g_{js} \right)$$

14.3 Interpretation rules for simultaneous analysis

In SA, the transition relations between projections of different points create a simultaneous representation that provides more detailed knowledge of the matter being studied.

14.3.1 *Relation between f_{is}^t and g_{js}*

The projection of a partial row on axis s depends on the projections of the columns:

$$f_{is}^t = \frac{\sqrt{\alpha_t}}{\sqrt{\lambda_s}} \sum_{j \in \mathbf{J}_t} \frac{p_{ij}^t}{p_{i\cdot}^t} g_{js}$$

Except for the factor $\sqrt{\alpha_t / \lambda_s}$, the projection of a partial row on axis s is, as in CA, the centroid of the projections of the columns of table t.

14.3.2 Relation between f_{is} and g_{js}

The projection of an overall row on axis s can be expressed in terms of the projections of the columns as follows:

$$f_{is} = \sum_{t \in \mathbf{T}} \sqrt{\alpha_t} \frac{\sqrt{p_{i\cdot}^t}}{\sum_{t \in \mathbf{T}} \sqrt{p_{i\cdot}^t}} \left(\frac{1}{\sqrt{\lambda_s}} \sum_{j \in \mathbf{J}_t} \frac{p_{ij}^t}{p_{i\cdot}^t} g_{js} \right)$$

Therefore, the projection of the row is, except for the coefficients $\sqrt{\alpha_t / \lambda_s}$, the weighted average of the centroids of the projections of the columns for each table.

14.3.3 Relation between g_{js} and f_{is} or f_{is}^t

The projection on the axis s of the column j for table t can be expressed in the following way:

$$g_{js} = \sqrt{\frac{\alpha_t}{\lambda_s}} \left(\sum_{i \in \mathbf{I}} \left(\sum_{t \in \mathbf{T}} \sqrt{p_{i\cdot}^t} \right) \sqrt{p_{i\cdot}^t} \left(\frac{p_{ij}^t - p_{i\cdot}^t p_{\cdot j}^t}{p_{i\cdot}^t p_{\cdot j}^t} \right) f_{is} \right)$$

This expression shows that the projection of a column is placed on the side of the projections of the rows with which it is highly associated, and on the opposite side of the projections of those with which it is less associated.

This projection is, according to partial rows:

$$g_{js} = \sqrt{\frac{\alpha_t}{\lambda_s}} \left(\sum_{i \in \mathbf{I}} \sqrt{p_{i\cdot}^t} \left(\frac{p_{ij}^t - p_{i\cdot}^t p_{\cdot j}^t}{p_{i\cdot}^t p_{\cdot j}^t} \right) \left(\sum_{t \in \mathbf{T}} \sqrt{p_{i\cdot}^t} f_{is}^t \right) \right)$$

The same aids to interpretation are available in SA as in standard factorial analysis as regards the contribution of points to principal axes and the quality of display of a point on axis s. It is also possible to calculate the contribution of each table to the principal axes as the sum of the contributions of the columns of the corresponding table. To compare the different tables, SA provides measurements of the relation between the factors of the different analyses.

14.3.4 Relation between factors of the individual analyses

The correlation coefficient can be used to measure the degree of similarity between the factors of the separate CA of different tables. This is possible when the marginals $p_{i\cdot}^{t}$ are equal.

The relation between the factors s and s' of the tables t and t', respectively, would be calculated with the correlation coefficient:

$$r\left(\mathbf{f}_{st},\mathbf{f}_{s't'}\right)=\sum_{i\in \mathbf{I}}\frac{f_{ist}}{\sqrt{\lambda_s^t}}\,p_{i\cdot}^{t}\,\frac{f_{is't'}}{\sqrt{\lambda_{s'}^{t'}}}$$

where f_{ist} and $f_{is't'}$ are the projections on the axes s and s' of the separate CA of the tables t and t', respectively, and where λ_s^t and $\lambda_{s'}^{t'}$ are the inertias associated with these axes.

When the marginals $p_{i\cdot}^{t}$ are not equal, Cazes (1982) proposes calculating the correlation coefficient between factors, assigning weight to the rows corresponding to the margins of one of the tables. Therefore, these weights, and the correlation coefficient as well, depend on the choice of this reference table. In consequence, we propose to solve this problem of the weight by extending the concept of generalized covariance (Méot and Leclerc 1997) to that of generalized correlation (Zárraga and Goitisolo 2003). The relation between the factors s and s' of the tables t and t', respectively, would be calculated as:

$$r\left(\mathbf{f}_{st},\mathbf{f}_{s't'}\right)=\sum_{i\in \mathbf{I}}\frac{f_{ist}}{\sqrt{\lambda_s^t}}\sqrt{p_{i\cdot}^{t}}\sqrt{p_{i\cdot}^{t'}}\,\frac{f_{is't'}}{\sqrt{\lambda_{s'}^{t'}}}$$

This measurement allows us to verify whether the factors of the separate analyses are similar and check the possible rotations that occur.

14.3.5 Relation between factors of the SA and factors of the separate analyses

Likewise, it is possible to calculate for each factor s of the SA the relation with each of the factors s' of the separate analyses of the different tables:

$$r\left(\mathbf{f}_{s't},\mathbf{f}_{s}\right)=\sum_{i\in\mathbf{I}}\frac{f_{is't}}{\sqrt{\lambda_{s'}^{t}}}\sqrt{p_{i\cdot}^{t}}\left(\sum_{t\in\mathbf{T}}\sqrt{p_{i\cdot}^{t}}\right)\frac{f_{is}}{\sqrt{\lambda_{s}}}$$

If all the tables of frequencies analyzed have the same row weights, this measurement is reduced to:

$$r\left(\mathbf{f}_{s't},\mathbf{f}_{s}\right)=\sum_{i\in\mathbf{I}}\frac{p_{i\cdot}^{t}f_{is't}f_{is}}{\sqrt{\sum_{i\in\mathbf{I}}p_{i\cdot}^{t}\left(f_{is't}\right)^{2}}\sqrt{\sum_{i\in\mathbf{I}}p_{i\cdot}^{t}\left(f_{is}\right)^{2}}}$$

that is, the classical correlation coefficient between the factors of the separate analyses and the factors of SA.

14.4 Comments on the appropriateness of the method

This section presents three cases that can be found when jointly analyzing several contingency tables. In these cases, different methods give equivalent or different results, depending on the row margins.

14.4.1 Case 1: equal row margins between the tables

The case of equal row margins, that is, $n_{i\cdot t}=n_{i\cdot t'}$, might be found when one wants to analyze data tables based on the same sample of individuals, for example, at different moments in time. In the concatenated table, there is no inter-inertia and therefore the results of CA of this table are equivalent to those of the intra-analysis. If each table is weighted, SA is equivalent to CA of the weighted concatenated table and to MFACT (Chapter 13).

14.4.2 Case 2: proportional row margins between the tables

The case of proportional row margins, that is, $p_{i\cdot}^{t} = p_{i\cdot}^{t'}$, would be found, for example, if surveys with different numbers of variables were observed on the same sample of individuals. In this case there is no interinertia, but the masses of the tables are different and, in consequence, SA is equivalent to a weighted CA of the concatenation of the frequency tables (Zárraga and Goitisolo 2002). In MFACT, if the entries in each table are initially expressed relative to their totals, then MFACT and SA similarly balance the effects of the different tables.

14.4.3 Case 3: row margins not proportional between the tables

In this most general case, where the row margins, $n_{i\cdot t}$, are not proportional and therefore the row weights, $p_{i\cdot}^{t}$, differ from one table to another, we find differences between MFACT and SA.

In their first stages, these methods perform a "pseudoseparate" CA and a "separate" CA of each table. The different treatment of the same contingency table according to one method or the other can cause different results that, in turn, can be carried over in the second stage to the overall analysis. The larger the differences in the margins of the rows of the tables, the larger these differences will be, as shown in the following example.

Let us consider the two contingency tables in Table 14.1. They have equal grand totals but different row margins and, hence, different row weights.

Separate CA and pseudoseparate CA of data table A

Figure 14.2 and Figure 14.3 show the first factorial planes of the two analyses. The first observation is that there is a change of axes

Table 14.1 Two small data tables.

	Data Table A				Data Table B			
	A1	A2	A3	Total	B1	B2	B3	Total
R1	10	0	0	10	58	6	6	70
R2	0	42	3	45	6	7	2	15
R3	0	3	42	45	6	2	7	15
Total	10	45	45	100	70	15	15	100

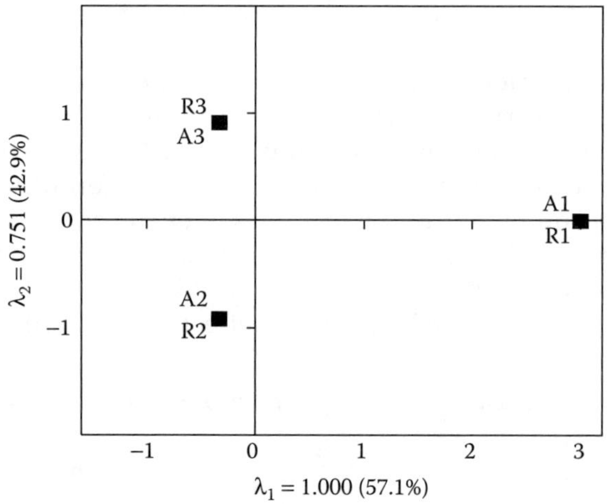

Figure 14.2 Separate CA of data table A.

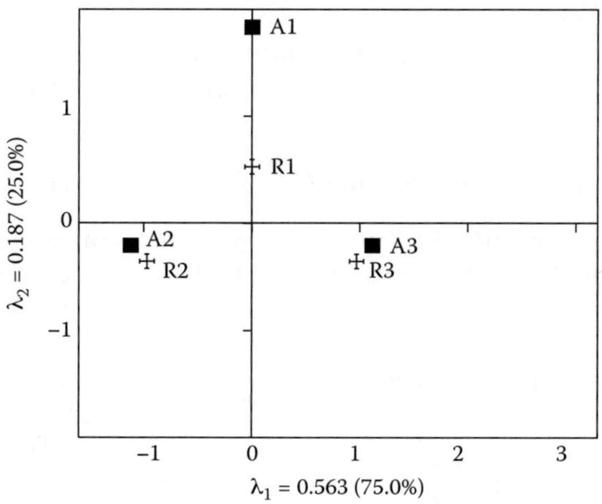

Figure 14.3 Pseudoseparate CA of data table A.

(associated with a change of projected inertias) between separate CA and pseudoseparate CA.

Another difference between the two figures is that the first eigenvalue of CA (Figure 14.2) is 1, due to the perfect association observed in the table between R1 and A1. This association is not reflected in the pseudoseparate CA (Figure 14.3). Except in this case of an eigenvalue equal to one, in this type of representation it is "not possible to deduce from the closeness of a row and column point the fact that the corresponding row and column are highly associated in the data" (Greenacre 1993a). It is the transition equations that allow one row (column) to be interpreted with respect to the set of all the columns (rows).

Another point worth making is that since table A is a symmetric table, CA provides the expected results, displaying the row and column in the same point. In pseudoseparate CA, however, each row is not displayed in the same point as the column.

Because table A has only three rows (and three columns), all the information in the table is displayed on this plane. Therefore, the plotted row-to-row distances are the inter-row χ^2 distances and the plotted column-to-column distances are the intercolumn χ^2 distances. So in CA, the greatest distance is that between row R1 and row R2 (or between R1 and R3), this distance (squared) being approximately three times the distance between R2 and R3. The same applies to columns A1, A2, and A3 due to the symmetry of the table. Nevertheless, in Figure 14.3, which represents the pseudoseparate CA of the same table, it is observed how the use of the concatenated table, both in the calculation of the relative frequencies and in the weights assigned to the rows, results in a distortion of the relations between the rows. In this pseudoseparate CA, the distance between rows R2 and R3 is more than twice the distance between either of these points and R1. In this case, despite the symmetry in table A, these distances are not maintained between the columns. This alteration in the relative positions of points and the change in the weights lies at the origin of the change observed in the relative importance of the axes.

Separate CA and pseudoseparate CA of data table B

Figure 14.4 and Figure 14.5 show the only factorial planes of the two analyses. In this case, the configuration of the two planes is similar, though it is also possible to observe how the relations between rows are altered. Whereas CA shows that the distance between points R2 and R3 is about the same as between R1 and R2 or R1 and R3, the pseudoseparate CA shows R2 and R3 to be closer. The same applies to columns. This alteration in the relative distances between the points and in the weights of the rows causes a modification in the percentages

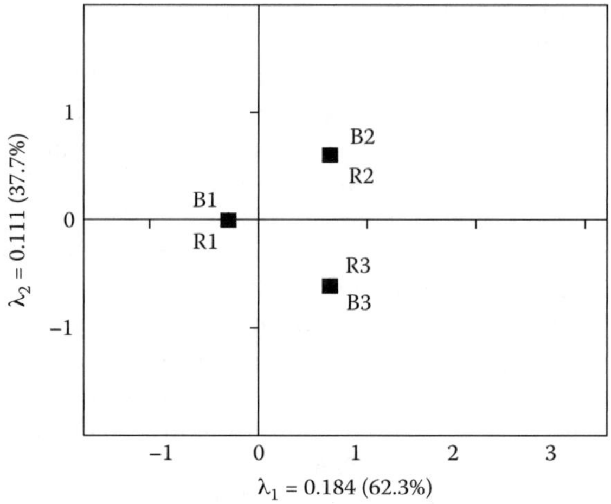

Figure 14.4 Separate CA of data table B.

of inertia of each factor. Whereas in CA symmetry between rows and columns is observed, as was the case in table A, in the pseudoseparate CA the symmetry present in the data is not reflected on the plane because the distances between points (row to row and column to column) are modified.

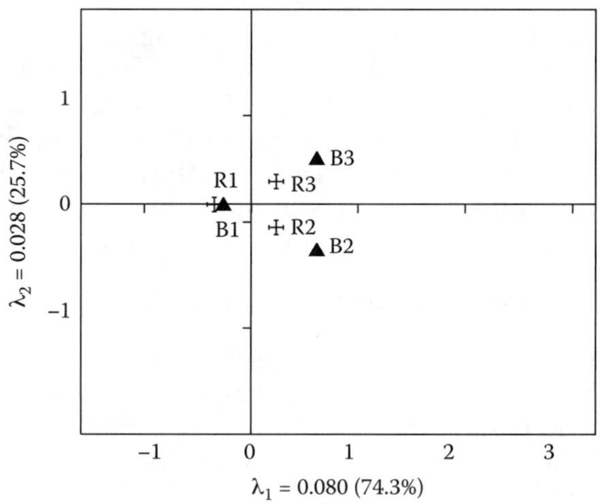

Figure 14.5 Pseudoseparate CA of data table B.

SA and MFACT of the two tables

The first factorial planes of the two analyses appear in Figure 14.6 (SA) and Figure 14.7 (MFACT). Both analyses try to jointly study tables A and B. In both SA and MFACT, each global row depends on

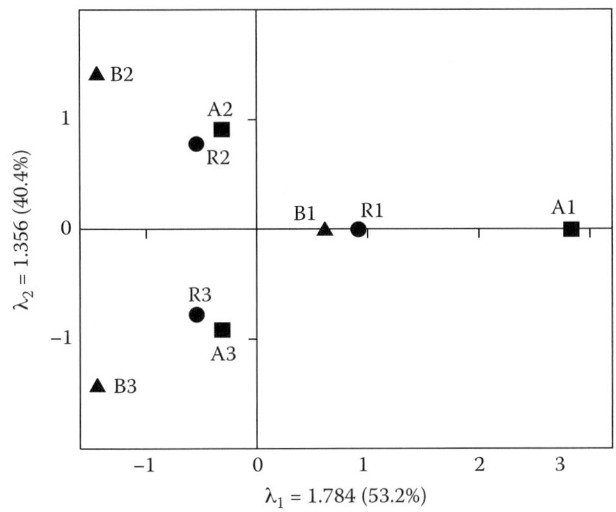

Figure 14.6 SA of data tables A and B.

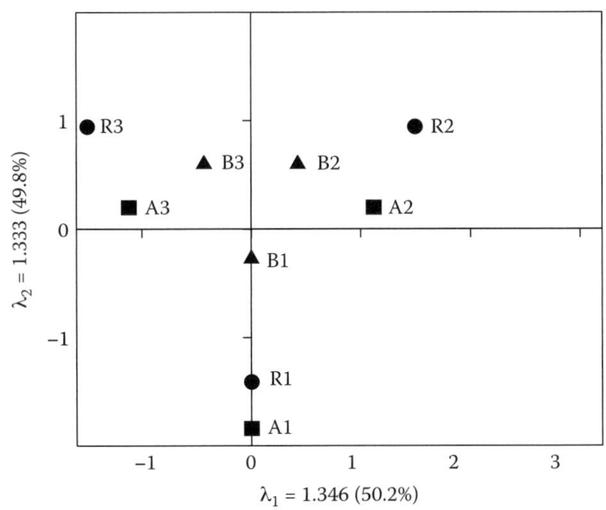

Figure 14.7 MFACT of data tables A and B.

both tables, so neither the symmetry between rows and columns nor the perfect association between a row and a column that was observed in CA of table A exists any longer.

In relation to SA (Figure 14.6), it is worth mentioning that the projected inertia on this first plane is not the total inertia. In this joint analysis there is a third axis with an inertia of less than 6.5% due to the fact that the global rows are not centered.

The axis of greater inertia of the CA of tables A and B shows the position of the first row (column) with regard to the second and third rows (columns), while the second axis reflects the opposition between R2 and R3 (equivalently for the columns). It is this structure that is observed in the SA (Figure 14.6). Moreover, we said in the CA of table A that the distance squared from A1 to A2 was approximately three times the distance between A2 and A3 (and between A1 and A3), and in the CA of table B we said that the distance between B2 and B3 was greater than the distance between B1 and B2 (and between B1 and B3), so these relative distances can also be deduced from SA (Figure 14.6).

MFACT results depend on those obtained in pseudoseparate CA, as can be seen in Figure 14.7 with regard to Figure 14.3 and Figure 14.5.

14.5 Application: study of levels of employment and unemployment according to autonomous community, gender, and level of education

In this example, SA is applied to the joint study of four data tables. The information, taken from the Active Population Survey drawn up by the INE (Statistical Office of Spain), corresponds to 2001 and represents the employed and unemployed population by gender, autonomous community, and level of instruction. The aim is to summarize and display the geographical distribution of the active population (employed and unemployed) in Spain's autonomous communities according to gender and training level.

The tables to be analyzed jointly by means of SA display the number (in thousands) of employed and unemployed according to gender and level of education for each autonomous community. There are, therefore, four data tables, with 17 rows (the autonomous communities) and 16 columns showing the levels of instruction achieved by employed men and women and by unemployed men and women. The full data set is available at http://www.ine.es.

In Table 14.2, row margins and the grand total of each of the four tables are detailed. The greatest differences are those observed

Table 14.2 Tables to be analyzed by SA, showing only the marginal percentages and the grand total ($n_{..t}$) for each table, collapsed over the four levels of instruction, P, 1S, 2S, and H.

	Employed		Unemployed	
	Men (EM)	Women (EW)	Men (UM)	Women (UW)
Andalusia	16.1	13.6	32.3	28.0
Aragon	3.0	2.9	1.2	1.4
Asturias	2.3	2.2	1.5	1.8
Balearic Islands	2.2	2.4	1.2	1.2
Canary Islands	4.5	4.4	4.8	4.3
Cantabria	1.3	1.3	1.0	1.1
Castile Leon	6.0	5.4	4.7	6.0
Castile La Mancha	4.3	3.4	3.1	3.7
Catalonia	16.6	18.7	13.8	13.5
Valencia	10.8	11.0	9.1	9.7
Extremadura	2.4	1.9	3.5	3.1
Galicia	6.4	7.2	6.7	7.4
Madrid	13.6	15.2	9.2	9.6
Murcia	2.9	2.5	2.6	2.9
Navarre	1.5	1.5	0.6	0.6
Basque Country	5.4	5.7	4.4	5.5
Rioja	0.7	0.7	0.3	0.2
Total (in 1000s)	9995.5	5902.5	806.9	1060.1

P = primary education and illiterates; 1S = first stage of secondary education and the corresponding training and job placement; 2S = second stage of secondary education and the corresponding training and job placement; H = higher education, including Ph.D.

between the employed and the unemployed, and those between men and women are smaller.

14.5.1 *Results of the four separate analyses*

Table 14.3 shows the inertias from the separate CAs, from which we deduce that the data structure is two-dimensional.

14.5.2 *Results of the CA of the concatenated table*

In the CA of the concatenated table (the map is not given here), both the between-inertia and the within-inertia of the four tables are involved. In particular, in the example presented, 33% of the total

Table 14.3 Inertias and percentages of inertia on the first two axes of the separate analyses.

	Employed		Unemployed	
	Men	Women	Men	Women
Inertia	0.044	0.035	0.051	0.054
Inertia (%)				
F1	67.1%	63.5%	63.8%	69.4%
F2	30.4%	29.7%	29.5%	26.9%
F1 + F2	97.5%	93.2%	93.3%	96.3%

inertia is between-tables inertia. Of this between-tables inertia, 67.1% is accounted for by the first factorial axis of the analysis, which represents mainly the opposition between employment and unemployment. This axis shows the association of Andalusia and Extremadura with unemployed people, whereas the rest of the communities are characterized mainly by employed women and men with the different levels of instruction achieved.

This concatenated analysis may be useful as a first stage for jointly analyzing tables. The inertias accounted for by the factorial axes are a mixture of between-tables and within-tables inertia, and their interpretation is more complex because both types of dispersion have to be taken into consideration. If we want to study the between-tables inertia in greater depth, it is necessary to resort to CA of the table of marginal frequencies. SA analyzes the internal structure of each table. In SA, a between-tables effect, which is trivial in many occasions, as indeed it is in this example, has been eliminated by centering each table internally, and what we are looking at is the dispersion within-tables across the autonomous communities.

14.5.3 Results of the simultaneous analysis

With the weighting chosen in SA, i.e., the inverse of the first eigenvalue of each one of the separate analyses, it is observed how the influence of each of the four tables is balanced in the formation of the first factor of SA (Table 14.4). There is a slightly greater contribution to the formation of the second factor by the tables related to employed and unemployed men, whereas in the formation of the third factor, the data tables related to employed and unemployed women have more influence.

Table 14.4 Table of inertias and contributions from each table to SA.

	F1	F2	F3
Eigenvalue	3.325	1.588	0.588
Percentage	54.7	26.1	9.7
Cumulative percentage	54.7	80.8	90.5
Employed men	0.257	0.261	0.185
Employed women	0.257	0.226	0.296
Unemployed men	0.251	0.288	0.200
Unemployed women	0.235	0.225	0.320

The first eigenvalue of the overall analysis is equal to 3.325 (Table 14.4), close to its maximum value of 4, which is the number of tables to be analyzed. This value indicates that the first axis of the SA is an axis of major inertia in the four data tables. This can also be seen in Table 14.5, which shows how the first axis of SA is closely related to the first axes of each of the four CA. The same can be said about the second axis of SA with regard to the second axes of each CA.

From Table 14.4 it is also deduced that the first two factors account for 80.8% of the total inertia, so a detailed analysis will be made of only the first factorial plane resulting from the SA.

The first factorial plane

The negative side of the first factor displays the categories that represent the lowest levels of instruction achieved by the active population of men

Table 14.5 Relation between the CA's first axes and SA.

		Simultaneous Analysis	
		F1	F2
CA Employed men	F1	0.925	0.006
	F2	0.009	0.955
CA Employed women	F1	0.922	0.058
	F2	0.102	0.869
CA Unemployed men	F1	0.881	−0.315
	F2	0.346	0.880
CA Unemployed women	F1	0.883	0.037
	F2	−0.003	0.956

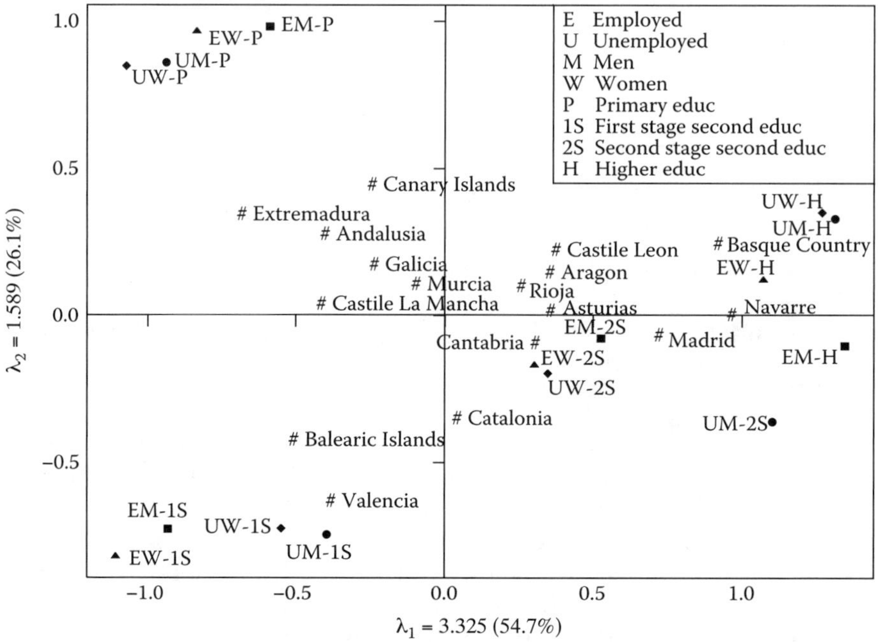

Figure 14.8 Projection of rows and columns.

and women (employed and unemployed), namely, primary and lower education (P) and the first stage of secondary education (1S) (Figure 14.8).

The positive side of the factor, with the greater contributions, displays the categories related to higher education, including Ph.D. (H). The intermediate level of instruction, the second stage of secondary education (2S), is also displayed on the positive side of the axis, although with low contributions.

Thus, this first factor sets those autonomous communities in which levels of employment and unemployment for men and women are high among the least well-educated levels of the population (primary education and first stage of secondary education) against those in which both employment and unemployment affect men and women with the highest levels of education (second stage of secondary education and higher education). Among the former are the Canary Islands, Extremadura, Andalusia, Valencia, the Balearic Islands and, to a lesser degree, Castile-La Mancha. Among the latter are the Basque Country, Navarre, and Madrid.

The second axis distinguishes those communities in this first group in which there is a prevalence of employment categories for men and women associated with primary education (Extremadura, Canary Islands, Andalusia) in contrast to those that have an active population with the subsequent level, first stage of secondary education (Balearic Islands, Valencia, and Catalonia).

On the factorial plane it is also observed that, for a given level of instruction, the columns corresponding to each of the four tables are displayed close to one another. This seems to suggest that the geographical distribution of the training levels is similar between both occupational categories, employed and unemployed, and both genders. Those autonomous communities in which the active population is characterized predominantly by a certain level of instruction can therefore be identified.

The categories related to the second stage of secondary education and to higher education are slightly more disperse than those of both prior levels, for example, UM-2S and EM-H, indicating differences in the communities in which the active population possesses these levels of instruction, as will be seen in the next subsection, where we discuss projection of the partial rows.

Projection of the partial rows

The projection of the partial rows allows the distribution of the training level reached by the four tables formed by both occupational categories and gender to be specified for each community, in other words, for employed men and women and for unemployed men and women. On the factorial plane, each community is represented by five points: four related to each of the four tables and the overall point that represents the community. Due to the number of rows in the table, the joint display of rows and partial rows makes interpretation on the same graph nearly impossible. For that reason, only the most relevant rows are displayed in Figure 14.9a, Figure 14.9b, and Figure 14.9c.

It can be observed (Figure 14.9a) that certain communities, such as Madrid and Andalusia, present partial rows that are closer to one another and close to their global row, indicating that the distribution of the four levels of instruction is similar in the four tables considered. The first of these communities is largely characterized by having an active population with higher education, and the second one by having an active population with primary education and illiterates.

The proximity of partial rows, for example, EM-Vale and EW-Vale and UW-Bale, shows how levels of instruction are distributed in a similar way between employed men and women in Valencia and unemployed women in the Balearic Islands. In all of these groups, the dominant level of instruction is the first stage of secondary education. Moreover, it is noteworthy that the lowest training level,

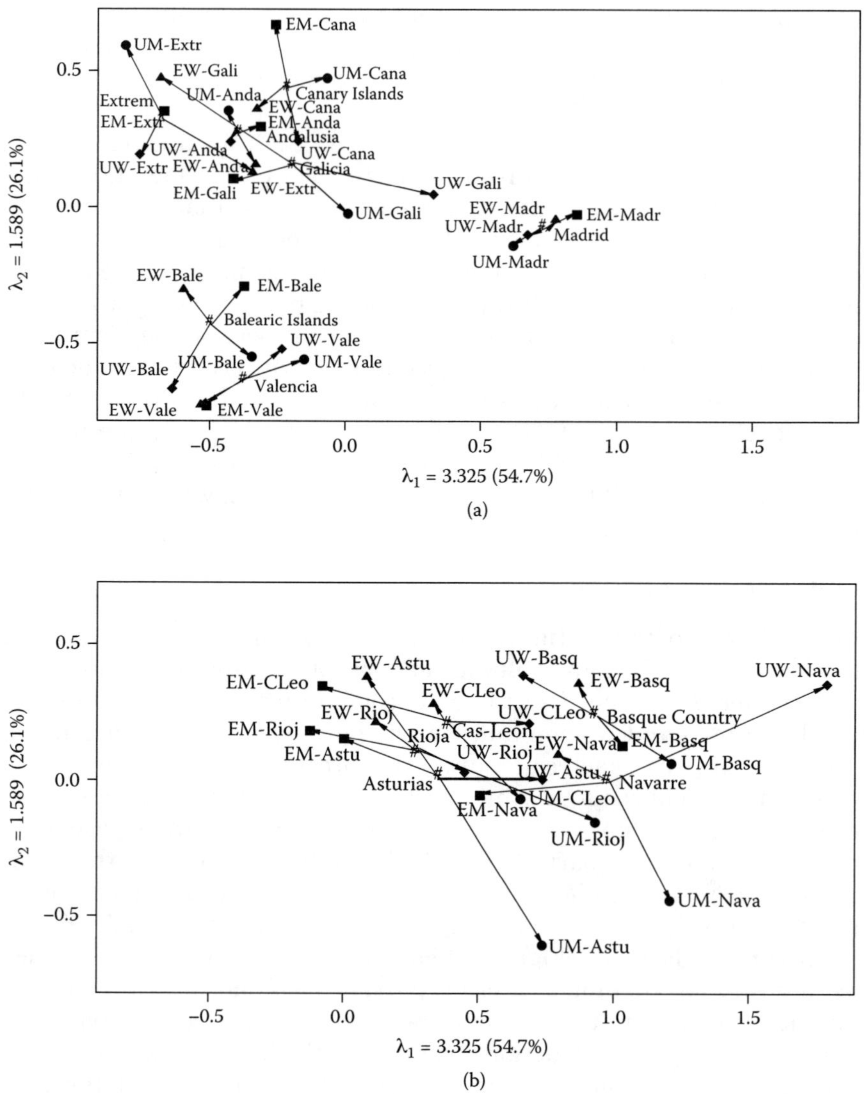

Figure 14.9 (a) Partial and overall rows; (b) partial and overall rows; (c) partial and overall rows.

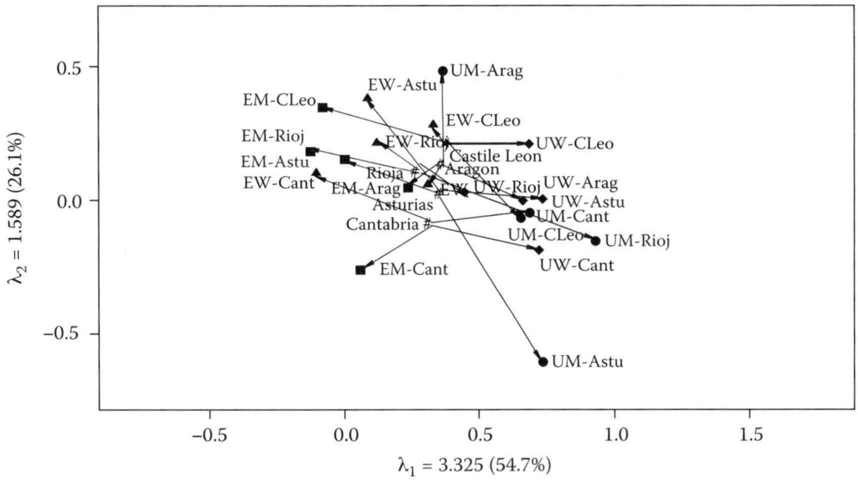

(c)

Figure 14.9 (Continued)

primary and lower education, is the most prevalent among unemployed men in Extremadura (UM-Extr), among employed women in Galicia (EW-Gali), and among employed men in the Canary Islands (EM-Cana).

There are, furthermore, certain communities, some of whose partial rows noticeably deviate from their overall row (see Figure 14.9b). This dispersal shows which group is most affected by a certain level of instruction. Thus, Navarre is the autonomous community that is most affected by unemployment among women with higher education (UW-Nava) and that presents the highest levels of unemployment among the male population with second stage of secondary education and with higher education (UM-Nava). The latter also applies to Rioja (UM-Rioj), the Basque Country (UM-Basq), Asturias (UM-Astu), and Castile-Leon (UM-CLeo) and causes the projection of the category UM-2S that is farthest from its counterparts on the factorial plane.

To conclude the interpretation provided by the partial rows, we wish to highlight (Figure 14.9c) how they allow an explanation of the structure of the active population, induced by the level of instruction achieved, of certain autonomous communities for which their position on the first factorial plane does not allow a clear interpretation, as is the case for Rioja, Castile-Leon, Aragon, Asturias, and Cantabria.

In all of these cases, the partial point corresponding to the first table, that of men in employment, deviates from its global row to the left-hand side of the level, that is, to the lowest training levels (EM-CLeo), (EM-Rioj), etc., whereas the partial points corresponding to the table of unemployed women move toward the right of the first factor, in other words, toward the highest level of instruction (UW-Rioj), (UW-CLeo), etc. These are, therefore, communities in which the level of employment is high among males with the lowest levels of instruction and unemployment mainly affects women with better academic qualifications, which explains their position on the factorial plane.

14.6 Conclusions

SA, like MFACT, treats a set of concatenated contingency tables. Both methods are equivalent in particular cases when the row margins of the tables are equal or proportional. The major difference between the methods is the way the rows are weighted and the way the partial rows are defined and analyzed. In SA, differences between the grand totals of tables are eliminated by expressing each table relative to its total. When row margins of the tables are different, SA maintains the structure of each table in the overall analysis by centering each table internally with its margins, as is done in CA. SA further admits the possibility of jointly analyzing tables of variables with different measurement scales (continuous, categorical, or frequency).

Software notes

Software for performing simultaneous analysis, written in S-Plus 2000, can be found in Goitisolo (2002). The AnSimult package for R can be obtained from the authors.

Acknowledgments

This work was supported by Basque Country University (UPV/EHU) under research grant 9/UPV 00038.321-13631/2001.

CHAPTER 15

Multiple Factor Analysis of Mixed Tables of Metric and Categorical Data

Elena Abascal, Ignacio García Lautre, and M. Isabel Landaluce

CONTENTS

15.1 Introduction

Principal component analysis (PCA) is widely recognized as a technique for analyzing tables of metric data, and multiple correspondence analysis (MCA) is the comparable method for use with categorical data. However, there are occasions when both types of data are present at once, thus giving rise to what is called a "mixed table." In such cases, multiple factor analysis (MFA) is a suitable solution, because it is the equivalent of using weighted PCA for the metric part of the data and MCA for the categorical part.

In this chapter, we focus our attention on the types of mixed tables that arise when working with metric data where some of the variables need to be recoded as categorical variables before being included in the analysis. For example, the distribution of the variables might be very skewed, including many zero or missing values, or it might have nonlinear correlation between the variables.

This chapter has a twofold aim. First, we propose a new way of analyzing the partially recoded metric data table by means of MFA, which provides us with a way of making a simultaneous analysis of categorical and metric data. We then compare our proposal with two factorial solutions (PCA of the table of metric data and MCA of the fully recoded table). Finally, we show how MFA can provide interpretable factorial planes by retaining both the features of PCA for the metric data and those of MCA for the categorical data.

We illustrate the procedure with an empirical study of voting percentages (by provinces) in the 2004 Spanish general elections. Because the study involves metric variables, a table of voting percentages could be analyzed by means of PCA. However, the large number of structural zeros present on some of the variables creates a problem that we propose to solve by MFA.

15.2 Multiple factor analysis

MFA, as proposed by Escofier and Pagès (1982, 1990), is a type of factorial analysis adapted to deal with tables of data in which a set of cases is described by several groups of variables. In various circumstances, such as when treating a set of mixed numerical and categorical variables or when handling the same set of variables measured at different points of time, application of MFA can cause groups of variables to emerge. The fact that the raw data can be structured into groups of variables enhances their interpretation. In the analysis presented in this chapter, the aim is to obtain not merely a typology of the cases described by a whole set of variables, but also to discover any possible relationships between the patterns that emerge within each of the groups.

MFA allows us to find factorial planes common to several groups of variables, which enables us to balance their contributions on the first factor. In this respect, MFA provides the typical results of such classic factorial methods as PCA, simple correspondence analysis, and MCA. In other words, MFA proceeds axis by axis to obtain the coordinates, contributions, and squared cosines of the cases; the coefficients of correlation between the continuous variables and the factors; the associated test values for the categories of the nominal variables; the centers of gravity or centroids of the cases that feature each category; and the graphical displays. Consequently, MFA can be used to consider relationships among variables, among cases, between variables and cases, as well as among groups.

MFA is based on PCA methodology and works in two stages. (Henceforth, we discuss only the case of normalized PCA, which, in practice, is the most common.) The two stages are

1. Partial analysis: Each group t of variables (where $t = 1,\dots, T$) is linked to a cloud of cases, known as the partial cloud, and represented by a rectangular table of data, $\mathbf{X}_t$ (the cases are the rows and the variables are the columns). Each $\mathbf{X}_t$ table captures the magnitude of the variables of group t in the set of cases. At this stage, a normalized PCA is performed on each of the $\mathbf{X}_t$ tables. The first eigenvalue, $\lambda_{1(t)}$, in each analysis is retained. Factors obtained in each separate analysis are referred to as partial factors.
2. Global analysis: A weighted PCA is performed on the global table $\mathbf{X}$ formed by concatenating the T tables $\mathbf{X}_t$. Each $\mathbf{X}_t$ is weighted by the inverse of the first eigenvalue retained in the partial analysis, $1/\lambda_{1(t)}$. The factors obtained in the global analysis are referred to as global factors. Two main properties are associated with the use of this weight:
   - The structure of each $\mathbf{X}_t$ table is maintained by giving the same weighting to all of the variables contained within the table.
   - The influence of the groups is balanced because the maximum inertia of any of the *partial clouds* defined by the different groups is 1, in any direction. In addition, the influence of every group in the first global factor is balanced. This is appreciated if we express the distance between any two cases (i,i') in the global cloud:

$$d^2\left(i,i'\right)=\sum_{t=1}^{T}\frac{1}{\lambda_{1(t)}}d^2\left(i_{(t)},i'_{(t)}\right)$$

where $i_{(t)}$ and $i'_{(t)}$ are the individuals i and i' characterized only by the variables of group t.

Escofier and Pagès (1986, 1990) show that MFA is suitable for analyzing tables of mixed data, including groups of both categorical and metric variables, as long as all of the variables in each group are of the same type.

Categorical variables are coded as complete disjunctive tables (or indicator matrices). Each categorical variable, with J_t categories, is represented by a set of (binary) indicator variables, $\delta_{j(t)}$ as columns of $\mathbf{X}_t$, where $\delta_{j(t)}$ ($j = 1,\ldots, J_t$) takes a value of 1 if the case i features category j or otherwise is 0. Each indicator variable $\delta_{j(t)}$ is weighted by $1 - c_{j(t)}$, where $c_{j(t)}$ is the mass associated with the category j, i.e., $c_{j(t)} = n_{j(t)}/n$ is the proportion of cases that feature category j, where n is the total number of cases and $n_{j(t)}$ is the number of cases that feature this category.

Escofier and Pagès (1990) show that weighting a normalized PCA of indicator variables in this way leads to the same factors as in MCA when the inertias of the factors of both analyses are the same up to a coefficient Q, where Q is the number of categorical variables. They also demonstrated that MFA acts as (a) a weighted PCA when the analysis involves only groups of continuous variables and (b) an MCA when it involves only nominal variables. Landaluce (1997) showed that the inertias of a matrix analyzed by MCA and MFA—with the latter properly weighted—are similar, except for the coefficient $1/Q$, and that the distances between cases, though different, are still closely related.

15.3 MFA of a mixed table: an alternative to PCA and MCA

A table of metric variables can be partially recoded, leaving some variables unaltered while the rest are transformed as categorical variables. This type of information can be analyzed by means of MFA on a mixed table where each variable (whether recoded or not) is taken as a group. This is a good alternative to classical PCA of the metric data, where none of the variables are recoded, or to MCA of the recoded data.

Variables with irregular distributions, especially those with structural zeros, a large number of zeros, or missing values, are quite common, for example, variables such as holiday spending or income from certain sources, which can have various specific magnitudes or may even be absent (no holidays, no income from a certain source). In such cases, the zero magnitude is of inherent relevance, a reality that goes unheeded if the variable in question is analyzed as a metric variable by PCA.

A similar problem arises when there is a nonlinear relationship between variables. This often occurs in the presence of threshold effects, for example, "a pathological condition may be characterized by a value that is too low or too high" (Escofier and Pagès 1990). A classic example is the relationship between the amount of sugar in a cup of coffee and the degree of satisfaction that it produces in the consumer.

When dealing with this type of variable, the usual solution (Escofier and Pagès 1990; Aluja and Morineau 1999) is to transform all of them as categorical by recoding the continuous variable on a scale. Then, a factorial method suited to this type of data, often MCA, will be used to analyze all the variables simultaneously. This solution involves recoding both the regular and irregular metric variables as categorical, with resultant loss of information from the regular variables.

Another option, however, is to convert only the irregular variables into categories, while leaving the rest metric, thus keeping as much information as possible. This process results in a mixed table of metric and categorical variables, which can also include true categorical variables such as social status or religious denomination. The mixed table thus obtained can be analyzed by MFA, where we will take each metric variable as a single group and the categories of each categorical variable (resulting from the conversion of the irregular variable) as further groups.

15.3.1 Constructing the mixed table

Suppose that we have a set of metric variables on a set of n cases, and suppose that the first H variables are regular and the remainders, from $H + 1$ onward, are irregular. The procedure is as follows:

1. The regular metric variables remain unaltered. Each variable forms a single group, represented by $\mathbf{X}_t$, $(t = 1,\ldots, H)$ giving rise to H tables of a single column and n rows.
2. Each irregular variable is recoded into a categorical variable, with J_t categories, that is, the variable is represented by the matrix $\mathbf{X}_t$ $(t = H + 1,\ldots, T)$ of indicator variables $\delta_{j(t)}$ $(j = 1,\ldots, J_t)$ which, once appropriately weighted (see Section 15.2), are incorporated into the analysis as a group. Here each $\mathbf{X}_t$ is formed by J_t columns and n rows.
3. The global table $\mathbf{X}$ is formed by concatenating tables $\mathbf{X}_t$ $(t = 1,\ldots, T)$.
4. MFA is performed on the global table $\mathbf{X}$ pursuing the same objectives as the initial classic PCA of the unconverted table. The metric variables perform in the same way as in PCA,

while the categorical variables perform as in MCA; thus, it is possible to retain the advantages of PCA for the study of the metric variables while providing a graphical display that also incorporates the categorical variables.

Note that in our example, since all the tables $\mathbf{X}_t$ are formed by a single variable (metric or categorical), the MFA weighting $(1/\lambda_{1(t)})$ for each table is 1. Thus, if the variable is metric, its variance is 1 and the first eigenvalue of the separate PCA is equal to 1. If the variable is categorical, the weight $(1 - c_{j(t)})$ assigned to indicator variables keeps the properties with respect to its inertia. Thus, every indicator variable of the table $\mathbf{X}_t$ has the same inertia with respect to the origin (Escofier and Pagès 1990: Sections 7.6.1 and 7.6.2).

$$\left(1-c_{j(t)}\right)\frac{1}{n}\sum_{i=1}^{n}\frac{\delta^2_{ij(t)}}{c_{j(t)}\left(1-c_{j(t)}\right)}=1$$

Furthermore, as all the indicator variables are orthogonal in pairs, the eigenvalues (including the first) of the weighted PCA of the indicator matrix are equal to 1. MFA weighting is, therefore, irrelevant for the analysis of our data. However, if the irregular variables are recoded into too many categories, the factors obtained may retain a small percentage of inertia and the rare categories will contribute too highly to the factors, as when MCA is applied.

15.3.2 Characteristics of MFA for the proposed mixed table

In this section we will show how MFA of the mixed table, proposed in the previous subsection, retains both the properties of PCA for the metric variables (unconverted table) and the properties of MCA for the categorical variables (fully recoded table). We also provide the guidelines needed for interpreting the simultaneous graphical display of both the metric variables and the categories of the nominal variables.

Distances

The (squared) distance *between cases* in MFA with a mixed table is the sum of the weighted Euclidean (squared) distances:

$$d^2(i,i')=\sum_{t=1}^{H}\left(\frac{x_{it}-x_{i't}}{s_t}\right)^2+\sum_{t=H+1}^{T}\sum_{j=1}^{J_t}\frac{1}{c_{j(t)}}(\delta_{ij(t)}-\delta_{i'j(t)})^2$$

This distance grows with increases in the differences in their magnitudes on the metric variables and in the number of categories in which the cases differ. It has one PCA-type component, due to the metric variables, and another MCA-type component, as a result of the categorical variables. It retains the general formula, which states that the square of the distance between cases is the weighted mean of the squares of their distances in the partial clouds formed by each of the groups (Escofier and Pagès 1986).

The distance *between columns* in MFA depends on the type of column:

- Between two metric variables, it is the same as in PCA and is interpreted in the same way.
- Between a metric variable t ($t = 1,\ldots, H$) and a category j ($j = 1,\ldots, J_{t'}$) of a categorical variable t' ($t' = H + 1,\ldots, T$), the distance is given as

$$d^2(t,j)=\sum_{i=1}^{n}\frac{1}{n}\left[z_{it}-\frac{\left(\delta_{ij(t')}-c_{j(t')}\right)}{\sqrt{c_{j(t')}\left(1-c_{j(t')}\right)}}\right]^2=2\left[1-\frac{\sum_{i=1}^{n}z_{it}\delta_{ij(t')}}{n\sqrt{c_{j(t')}(1-c_{j(t')})}}\right]=2[1-\mathrm{corr}(t,\delta_{j(t')})]$$

  where z_t is the variable t standardized and $\mathrm{corr}(t,\delta_{j(t')})$ is the linear correlation coefficient (or point biserial correlation coefficient) between variable t and the indicator variable of category j of variable t'. This distance decreases as the correlation between t and category j of variable t' gets closer to 1. The correlation increases when the cases scoring above the mean for variable t feature category j and also when category j is rare ($c_{j(t')}$ is close to 0) or very frequent ($c_{j(t')}$ is close to 1).
- Between category j of the categorical variable t and category j' of the categorical variable t', the distance is given as

$$d^2(j,j')=\sum_{i=1}^{n}\frac{1}{n}\left[\frac{(\delta_{ij(t)}-c_{j(t)})}{\sqrt{c_{j(t)}(1-c_{j(t)})}}-\frac{(\delta_{ij'(t')}-c_{j'(t')})}{\sqrt{c_{j'(t')}(1-c_{j'(t')})}}\right]^2$$

$$=2\left[1-\frac{\frac{n_{jj'}}{n}-c_{j(t)}c_{j'(t')}}{\sqrt{c_{j(t)}(1-c_{j(t)})}\sqrt{c_{j'(t')}(1-c_{j'(t')})}}\right]$$

As in MCA, the distance decreases as the number of cases in which they coincide ($n_{jj'}$) increases, especially if they are either rare or very frequent categories. This formula is simplified when the categories are of the same variable.

$$d^2(j,j') = 2\left[1 + \left(\frac{n_j}{n-n_j}\right)^{1/2}\left(\frac{n_{j'}}{n-n_{j'}}\right)^{1/2}\right] = 2\left[1 + \left(\frac{c_j}{1-c_j}\right)^{1/2}\left(\frac{c_{j'}}{1-c_{j'}}\right)^{1/2}\right]$$

In this case, the distance depends only on the marginal frequencies, just as in MCA. It increases when both categories are frequent and decreases when they are rare.

Projections onto factors

In MFA the projection of a case is

$$f_{is} = \frac{1}{\sqrt{\lambda_s}}\left[\sum_{t=1}^{H} z_{it}g_{ts} + \sum_{t=H+1}^{T}\sum_{j=1}^{J_t}\sqrt{\frac{1-c_{j(t)}}{c_{j(t)}}}(\delta_{ij(t)} - c_{j(t)})g_{js}\right]$$

$$= \frac{1}{\sqrt{\lambda_s}}\left[\sum_{t=1}^{H} z_{it}g_{ts} + \sum_{t=H+1}^{T}\sum_{j=1}^{J_t}\sqrt{\frac{1-c_{j(t)}}{c_{j(t)}}}\delta_{ij(t)}g_{js} - M\right]$$

where λ_s is the eigenvalue associated with global factor s, and M is a constant for all cases.

This expression includes both PCA-type and MCA-type components. The PCA-type component corresponds to the metric variables. Cases depart from the origin toward the metric variables t ($t = 1,\ldots,H$) on which they score highest and toward the qualitative characteristics that they feature. The MCA-type component corresponds to the categorical variables t ($t = H + 1,\ldots,T$), that is, a linear combination of the factor scores of the categories featured in case i. Thus, the rarer category j is, the closer $c_{j(t)}$ is to 0 and the greater is the coefficient of the linear combination. Inversely, if category j is more frequent, then the coefficient is smaller. Therefore, cases that feature rare categories are often separated from the rest, as occurs in MCA.

In MFA, not only does the projection of a metric variable coincide with its projection in PCA, but also the projection of a category j of a categorical variable t is defined as:

$$g_{sj} = \frac{1}{\sqrt{\lambda_s}} \sqrt{\frac{c_j}{1 - c_j}} \bar{f}_{is}$$

where $\bar{f}_{is}$ is the mean principal coordinate on dimension s of the cases where category j is present. Apart from a scaling factor, a category will be located at the centroid of the cases in which it is featured, as would be the case in MCA.

As in PCA, the projection of a variable is a measure of its correlation with the factor. In MFA the projection of a category measures the point biserial correlation between the factor and the indicator variable (Escofier and Pagès 1986).

Comparison between categorical and metric variables

When several groups of variables are analyzed simultaneously, MFA provides a display in which each group is represented by only a single point. To do this, each group is represented by a matrix. The space of these matrices is provided with the classical scalar product, which is interpreted as a measure of relationship between groups. In this context, two types of coefficients have been proposed (Escofier and Pagès 1982, 1990; Escoufier 1973):

1. The L_g coefficient is defined as the scalar product between the matrices associated with each group. These coefficients are displayed in a matrix with an interpretation analogous to that of a covariance matrix. It takes the value 0 when there is no relationship between groups, and it increases with the strength of the relationship. Its value also increases with the dimensionality of the groups.
2. The RV coefficient is defined as the the quotient of the L_g coefficient and the product of the norms of the matrices associated with each group. These coefficients are not affected by the dimensionality of the groups, their interpretation being analogous to that of a linear correlation coefficient. The coefficients are displayed in a matrix, similar to a correlation matrix, where the main diagonal consists of 1's, and the remaining elements take values between 0 and 1.

In this study, each group of the mixed table is composed of a single variable, metric, or categorical. In consequence, the L_g coefficients can be interpreted as measures of relationship between variables. In the following, we give a brief explanation of the three possible cases depending on the kind of the variables:

1. *Metric variables*: The measure of the relationship between two metric variables t and t':

$$L_g(t,t') = \text{ inertia of projection } t \text{ onto } t' = [\text{corr}(t,t')]^2$$

2. *Categorical and metric variables*: The measure of the relationship between a metric variable t and a group formed by the indicator variables $\delta_{j(t')}$ ($j = 1,\ldots, J_{t'}$) of a categorical variable t' can be considered as a measure of the relationship between the metric variable and the categorical variable. This is given by

$$L_g(t,t') = \sum_{j=1}^{J_{t'}} (\text{inertia of projection of } \delta_{j(t')} \text{ onto variable } t)$$

$$= \sum_{j=1}^{J_{t'}} (1 - c_{j(t')})(\text{corr}(t,j))^2$$

This coefficient is the sum of the point biserial correlation coefficients between the metric variable t and the indicator variables of the categorical variable t' squared, weighted by $(1 - c_{j(t')})$ (used in MFA, see Section 15.2)

3. *Categorical variables*: The measure of the relationship between two groups of indicator variables $\delta_{j(t)}$ ($j = 1,\ldots, J_t$) and $\delta_{j(t')} =$ ($j = 1,\ldots, J_{t'}$) can be considered as a measure of the relationship between the two categorical variables t and t'. That is,

$$L_g(t,t') = \sum_{j=1}^{J_t} (1 - c_{j(t)}) L_g(\delta_{j(t)}, t')$$

$$= \sum_{j=1}^{J_t} \sum_{j'=1}^{J_{t'}} (1 - c_{j(t)})(1 - c_{j'(t')})[\text{corr}(j,j')]^2$$

where corr(j, j') is the linear correlation coefficient between the indicator variables $\delta_{j(t)}$ and $\delta_{j'(t')}$.

15.4 Analysis of voting patterns across provinces in Spain's 2004 general election

To illustrate the procedure proposed above, we analyze the percentage of votes cast for the main parties in the last general election held in Spain on March 14, 2004, focusing on the results across the 50 provinces. The political parties considered are the Socialist Party (PSOE), the People's Party (PP), the United Left (IU), the main left-wing regional or autonomous parties (A. Left), and the main right-wing regional or autonomous parties (A. Right). It is should be mentioned that autonomous parties attract a strongly nationalist vote, irrespective of their political orientation, across the Basque country, Catalonia, Galicia, and Navarra. The distribution of frequencies for these two variables, A. Left and A. Right, deviates widely from the normal distribution because, in about 60% of the provinces, these parties receive no votes because they have no candidates.

Correlations between the percentage of votes for the PP and the percentage of votes for the A. Right and IU are A. Left, is negative and statistically significant, which is an indication of strongly contrasting voter attitudes toward these parties across the different provinces. A similar contrast in attitudes arises toward the PSOE and the autonomous parties (Table 15.1). However, the scatterplots (for example, Figure 15.1) suggest that the correlation between these parties is affected by the considerable number of zero values of the autonomous parties. Therefore, if we were to use PCA to study voting patterns in the Spanish provinces, the results would not give a true picture of the reality underlying the data.

PCA is applied to the table that displays the percentage of votes for the five political parties considered in each of Spain's 50 provinces. The first factor explains 52.7% of the total inertia and reveals that there are two types of provinces. The first group of provinces is made up of those in which the majority vote goes to the national parties (the PP and the PSOE), situated in the left-hand quadrants (Figure 15.2).

Table 15.1 Correlation matrix.

	PSOE	PP	IU	A. Left
PP	0.155	—	—	—
IU	0.098	−0.441[a]	—	—
A. Left	−0.541[a]	−0.579[a]	−0.091	—
A. Right	−0.539[a]	−0.831[a]	0.258	0.470[a]

[a] Correlation significant at the 0.01 level.

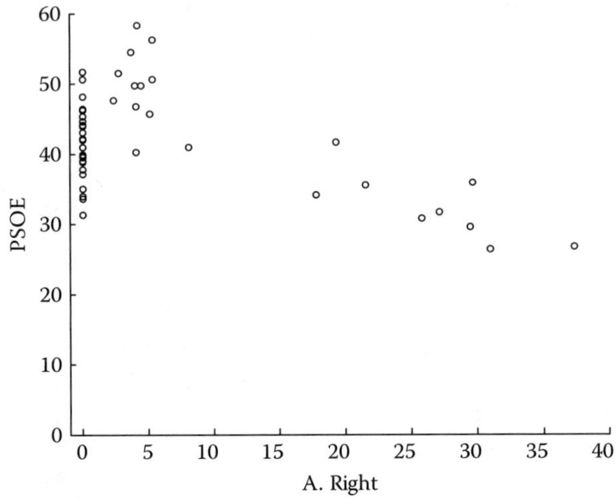

Figure 15.1 Dispersion diagram showing voting percentages for Socialist Party (PSOE) and autonomous right-wing parties (A. Right).

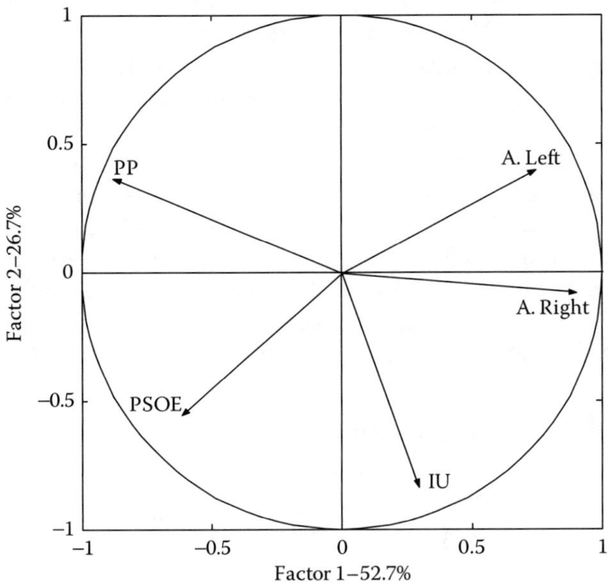

Figure 15.2 Graphical display of variables on the principal plane of PCA.

The second includes those provinces that vote mainly for the autonomous parties (both left and right wing), situated in the right-hand quadrants. The second factor, which accounts for 26.7% of the total inertia, basically represents the left-wing (A.Left), national parties opposing furthest the United Left (IU). The exploratory analysis of the votes in Spain using PCA outlines a very general political map, with some provinces displaying a "national ideology" and others a "nationalistic ideology," reflected in the significant negative correlation that exists between the variables representing the percentage of votes going to the parties that represent each of these two political ideologies. However, the inclusion of two parties with a large number of structural zeros (A. Left and A. Right) could hide some nuances.

To solve the problem arising from the structural zeros, we propose analyzing the votes of the Spanish provinces in 2004 by MFA. The two irregular variables are therefore recoded to transform them into categorical variables with four categories each (zero, low, medium, and high in each case). The first and last four rows of the concatenated data matrix used in the MFA are shown in Table 15.2.

Note that both the continuous variables and the two newly created categorical variables are treated as an MFA group (Group 1 = A. Left recoded, Group 2 = A. Right recoded, Group 3 = PSOE, Group 4 = PP, and Group 5 = IU). The **X** table to be analyzed, therefore, is made up

Table 15.2 Concatenated data matrix **X** analyzed by MFA (the first and last four rows are shown).

		A. Left (recoded)				A. Right (recoded)						
Row No.		Zero	Low	Med.	High	Zero	Low	Med.	High	PSOE	PP	IU
1	Almería	1	0	0	0	0	1	0	0	47.6	44.4	3.1
2	Cádiz	0	1	0	0	0	0	1	0	50.7	33.6	6.0
3	Córdoba	1	0	0	0	0	1	0	0	49.8	33.8	9.7
4	Granada	1	0	0	0	0	1	0	0	51.4	37.1	5.9
.	.	.	.	.	.	.	.	.	.	.	.	.
.	.	.	.	.	.	.	.	.	.	.	.	.
.	.	.	.	.	.	.	.	.	.	.	.	.
47	Álava	0	1	0	0	0	0	0	1	30.8	26.8	7.8
48	Guipúzcoa	0	0	1	0	0	0	0	1	26.4	15.1	7.7
49	Vizcaya	0	1	0	0	0	0	0	1	26.8	18.6	8.6
50	La Rioja	1	0	0	0	1	0	0	0	44.1	49.8	2.8
		$\mathbf{X}_1$				$\mathbf{X}_2$				$\mathbf{X}_3$	$\mathbf{X}_4$	$\mathbf{X}_5$

Table 15.3 Eigenvalues of global analysis of the MFA.

Factor	Eigenvalue	Eigenvalue (%)	Accumulated (%)
1	2.6685	29.7	29.7
2	1.8357	20.4	50.1
3	1.2991	14.4	64.5
4	1.2040	13.4	77.9

of 50 rows (provinces) and five $\mathbf{X}_t$ tables: the first two (A. Left and A. Right, both recoded) each have four columns; the other three (PSOE, PP, and IU) have only one column each.

Among the numerous readings that can be taken from MFA, the following are worth noting. The eigenvalues (Table 15.3) clearly show the situation reflected in this study to be more plural than would appear from the PCA of the untransformed table. Note that, as in MCA, the percentage of inertia associated with the first global factor is low due to the nature of the recoded variables. Nevertheless, these low percentages do not play a role in our interpretation of the MFA results, where we describe several typologies within the votes of the Spanish provinces.

The relationship between groups is measured by RV coefficients (Table 15.4). These show that the strongest link is between PP and right-wing autonomous parties (RV = 0.438), followed by the relationships between PSOE and right-wing autonomous parties. The L_g coefficients (Table 15.5) show the strongest links to be the same as RV, but it must be noted that these coefficients are influenced by the dimensionality of the groups.

The factorial plane of the first two global axes (Figure 15.3) gives a fuller and clearer picture of the Spanish votes at the provincial level than that given by PCA. Note that the display provides the projections of the untransformed metric variables and the centers of gravity of the categories of the new categorical variables (recoded metric variables). The coordinates of the categories are therefore greater than one.

It can be seen that in MFA, as in PCA, the principal plane of variables (Figure 15.3) reflects a strong autonomous tendency for some provinces (bottom left), while others on the right show a strong national tendency. However, the description in MFA is more detailed than in PCA, since it shows (in the upper part of the plane) how a marked left-wing national ideology (PSOE and IU) tends to appear alongside a certain nationalistic tendency (both left and right wing),

Table 15.4 RV coefficients.

Variables	A. Left[a]	A. Right[a]	PSOE	PP	IU
A. Left[a]	1.000				
A. Right[a]	0.126	1.000			
PSOE	0.170	0.315	1.000		
PP	0.203	0.438	0.024	1.000	
IU	0.122	0.062	0.010	0.194	1.000

[a] Indicates recoded variable.

Table 15.5 L_g coefficients.

Variables	A. Left[a]	A. Right[a]	PSOE	PP	IU
A. Left[a]	3.000				
A. Right[a]	0.379	3.000			
PSOE	0.295	0.546	1.000		
PP	0.352	0.759	0.024	1.000	
IU	0.211	0.107	0.010	0.194	1.000

[a] Indicates recoded variable.

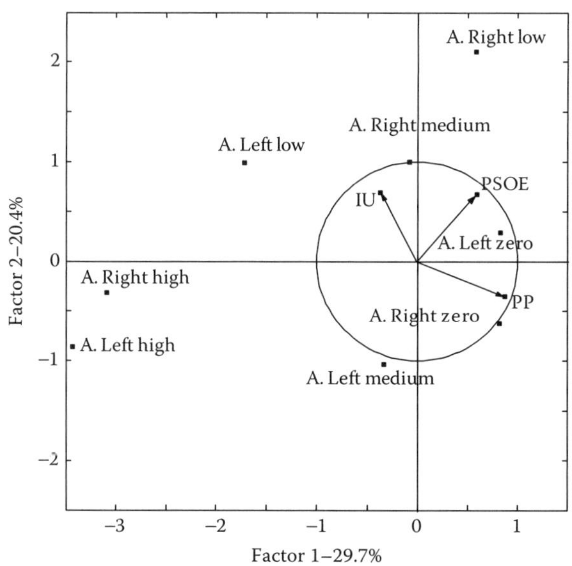

Figure 15.3 Graphical display of the metric variables and centers of gravity of the categorical variables on the principal plane of MFA.

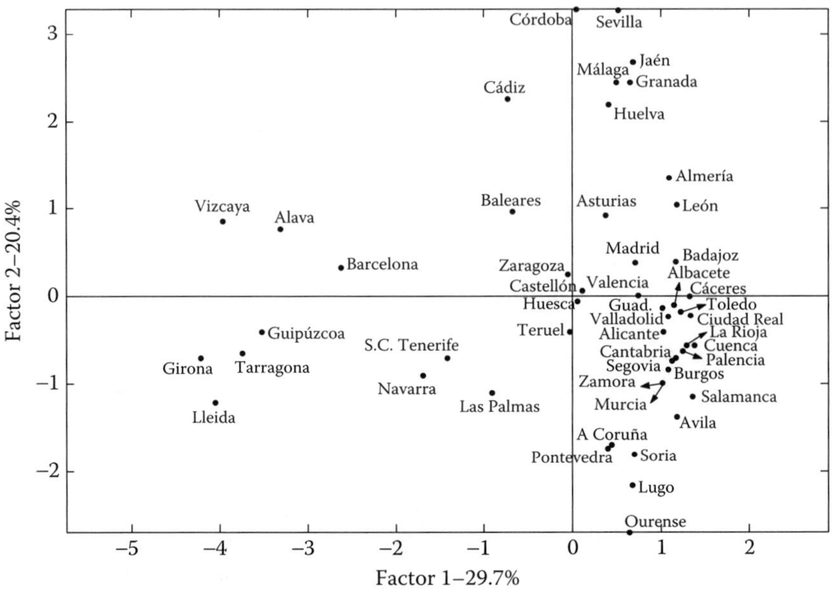

Figure 15.4 Graphical display of the Spanish provinces on the principal plane of MFA.

while a marked right-wing national ideology (PP) tends to appear unaccompanied by any other ideological inclinations.

The displays of the provinces on the main factorial plane in MFA (Figure 15.4) and PCA (not shown) are similar, although some provinces switch their positions with respect to the PCA plane. Nevertheless, the added richness of interpretation of the MFA factors permits a more detailed analysis of the positions of the various provinces than is possible with PCA. The MFA display of the provinces provides a truer representation of the voting patterns.

Finally, it is worth noting that MFA enables us to uncover some features that PCA leaves undetected, while loss of information is prevented by making all of the variables categorical.

15.5 Conclusions

MFA is useful in dealing with mixed-data tables. Using MFA, it is possible to obtain interpretable graphical displays while maintaining all of the characteristics of PCA for the metric variables and those of MCA for the categorical variables. When the data to be analyzed require the active intervention of several types of metric and categorical variables—or when some of the variables are irregular or show nonlinear correlations—there is no need to recode all variables to make them categorical (which is common practice, despite the drawbacks of recoding). MFA makes it possible to deal with both sets of data simultaneously while retaining the properties best suited to each type of data.

Software notes

The software used to obtain the results is SPAD: www.spadsoft.com.

SECTION IV

MCA and Classification

CHAPTER 16

Correspondence Analysis and Classification

Gilbert Saporta and Ndèye Niang

CONTENTS

Table 16.1 Eye and hair color of Scottish children.

	Hair Color					
Eye Color	Fair	Red	Medium	Dark	Black	Total
Blue	326	38	241	110	3	719
Light	688	116	584	188	4	1580
Medium	343	84	909	412	26	1774
Dark	98	48	403	681	85	1315
Total	1455	286	2137	1391	118	5387

16.1 Introduction

The use of correspondence analysis (CA) for discrimination purposes goes back to the "prehistory" of data analysis. In a famous paper, R.A. Fisher (1940) derived the equations of correspondence analysis while analyzing the data cross-classifying two categorical variables: hair and eye colors of 5387 Scottish children. The problem addressed by Fisher was to derive a linear combination of the indicator variables for the eye colors giving the best discrimination between the five classes of hair color (Table 16.1).

A linear combination of indicator variables leads to assigning scores to the categories, and Fisher's paper is the beginning of a long series of papers on optimal scaling (Young 1981). In other words, Fisher performs a canonical discriminant analysis between two sets of variables: the five indicators of hair color and the four indicators of eye color. Actually, Fisher used only the indicator variables of the last three eye colors (light, medium, dark), leaving out the first indicator of eye color (blue) to avoid the trivial solution.

It is well known that the optimal solution is given by the first factor of a CA of the contingency table: the optimal scores are the coordinates of the categories along the first axis. In his solution, Fisher standardized the scores to have zero mean and unit variance when weighted by the marginal frequencies, i.e., the standard coordinates in CA (Table 16.2).

The algorithm of successive averages given by Fisher (1940: 426) —"starting with arbitrarily chosen scores for eye color, determining from these average scores for hair color, and using these latter to find new scores for eye color"—is an alternating least-squares procedure and can be viewed as the ancestor of Gifi's (1990) homogeneity analysis and of Nishisato's (1980) dual scaling. Was Fisher the father of CA? Despite the fact that he derived the eigenequation of CA, one cannot

Table 16.2 Eye and hair color of Scottish children, standardized scores.

Eye Color	x	Hair Color	y
Blue	–0.8968	Fair	–1.2187
Light	–0.9873	Red	–0.5226
Medium	0.0753	Medium	–0.0941
Dark	1.5743	Dark	1.3189
		Black	2.4518

say so, for he used only the first eigenvector, preventing the use of the graphical displays that characterize CA as an exploratory data analysis technique.

When there are several categorical predictors, a commonly used technique consists of a two-step analysis: multiple correspondence analysis (MCA) on the predictors set, followed by a discriminant analysis using factor coordinates as numerical predictors (Bouroche et al. 1977). However, in economic and social science applications, logistic regression seems to be used more often instead of discriminant analysis when predictors are categorical. This tendency is also due to the flexibility of logistic-regression software. It can be easily proved, however, that discarding minor eigenvectors gives more robust results than direct logistic regression, for it is a regularization technique similar to principal components regression (Hastie et al. 2001).

Because factor coordinates are derived without taking into account the response variable, we propose to use partial least-squares (PLS) regression, which is related to barycentric discrimination (Celeux and Nakache 1994) and to nonsymmetric correspondence analysis (Verde and Palumbo 1996).

All of these methods are compared using a data set coming from a Belgian insurance company. This application is strictly operational (without any substantive interpretation) since the aim is to maximize the rate of good classifications into "good" or "bad" insurees.

16.2 Linear methods for classification

Let us consider the case of two classes and p numerical predictors. The two main techniques are Fisher's linear discriminant analysis (LDA) and logistic regression.

16.2.1 Fisher's linear discriminant analysis

The linear combination $\mathbf{u}$ of the p variables that maximizes the between- to within-variance ratio $\mathbf{u}^\top\mathbf{B}\mathbf{u}/\mathbf{u}^\top\mathbf{W}\mathbf{u}$ is given by:

$$\mathbf{u} = \mathbf{W}^{-1}(\mathbf{g}_1 - \mathbf{g}_2) \tag{16.1}$$

where $\mathbf{B}$ and $\mathbf{W}$ are, respectively, the between- and within-group covariance matrices and $\mathbf{g}_1$ and $\mathbf{g}_2$ are the centroids of the two groups. Let $\mathbf{T}$ be the total variance matrix, $\mathbf{T} = \mathbf{W} + \mathbf{B}$; it is known that $\mathbf{W}^{-1}(\mathbf{g}_1 - \mathbf{g}_2)$ is proportional to $\mathbf{T}^{-1}(\mathbf{g}_1 - \mathbf{g}_2)$. It corresponds to a linear frontier in the unit space (according to Mahalanobis metric $\mathbf{W}^{-1}$) given by the mediator hyperplane separating the two centroids. Apart from a multiplicative constant, Fisher's LDA is the ordinary least-squares estimate of $\boldsymbol{\beta}$ in the model $\mathbf{y} = \mathbf{X}\boldsymbol{\beta} + \mathbf{e}$, where $\mathbf{y}$ takes only two different values, one for each group. Fisher's score of unit $\mathbf{x}$ can be defined by:

$$S(\mathbf{x}) = (\mathbf{g}_1 - \mathbf{g}_2)^\top\mathbf{W}^{-1}\mathbf{x} - 1/2(\mathbf{g}_1^\top\mathbf{W}^{-1}\mathbf{g}_1 - \mathbf{g}_2^\top\mathbf{W}^{-1}\mathbf{g}_2) \tag{16.2}$$

There is a probabilistic model leading to this result: if the conditional distributions of the p variables in each group are normally distributed with the same covariance matrix and equal prior probabilities, then the posterior probability for an observation $\mathbf{x}$ coming from group 1 is (Hand 1981):

$$P(G_1 \mid \mathbf{x}) = \frac{\exp(S(\mathbf{x}))}{1+\exp(S(\mathbf{x}))} = \frac{e^{\beta_0+\beta_1x_1+\cdots+\beta_px_p}}{1+e^{\beta_0+\beta_1x_1+\cdots+\beta_px_p}} \tag{16.3}$$

Modifying priors changes only the constant term in the score function.

16.2.2 Logistic regression

In logistic regression, one uses Equation 16.3 as the model and not as a consequence. The coefficients β_j are estimated by conditional maximum likelihood, while in discriminant analysis they are estimated by least squares (which are also the unconditional maximum-likelihood estimates in the normal case with equal covariance matrices). Logistic regression is very popular because the β_j are related to odds ratios. The exponent $\beta_0 + \beta_1x_1 + \cdots + \beta_px_p$ is also used as a score function.

The probabilistic assumptions of logistic regression seem less restrictive than those of discriminant analysis (normal distributions with equal covariance matrices), but discriminant analysis also has a strong nonprobabilistic background, being defined as the least-squares separating hyperplane between classes. In many cases it has been observed that both solutions are very close (see Hastie et al. 2001).

16.2.3 About the notion of score

In some applications where a classification is required, there is a strict decision rule with a threshold s_0 that says that if $S(\mathbf{x}) > s_0$, then $\mathbf{x}$ is classified in group 1. But in many other applications one just uses a score as a rating of the risk to be a member of one group, and any monotonic increasing transformation of S is also a score. Let us remark that in this sense the probability $P(G_1 \mid \mathbf{x})$ is also a score ranging from 0 to 1.

16.3 The "Disqual" methodology

16.3.1 Categorical discriminant analysis

Classifying observations described by categorical predictors into one out of k classes has long been done by using models derived from the multinomial or log-linear model (Goldstein and Dillon 1978; Celeux and Nakache 1994). However, these models suffer from some curse of dimensionality and are difficult to apply when the number of category combinations is large.

Another way of dealing with categorical predictors consists of transforming them into numerical predictors by assigning values to the categories in an "optimal" way. "Optimal" means that the discrimination that will be done afterward will maximize some criterion, such as the Mahalanobis distance between the centroids. Since transforming qualitative variables into discrete numerical variables comes down to defining linear combinations of the indicator variables of the categories, linear discriminant analysis for categorical predictors is a discriminant analysis where predictors are indicator variables. The scores assigned to the categories, that is, the coefficients of indicator variables, define what is called a "scorecard" in credit-scoring applications (Thomas et al. 2002).

Let $\mathbf{Z}_1, \mathbf{Z}_2, \ldots, \mathbf{Z}_Q$ be the indicator matrices for the predictors with $J_1, J_2, \ldots, J_Q$ categories. Then one has to do a canonical analysis between the matrix **Y** of the indicator variables of groups and the super-indicator matrix **Z** of the Q predictors. However, the within-group covariance matrix **W** is not of full rank because the sum of indicator variables for each predictor is equal to 1, and thus, cannot be inverted to obtain a solution. As in the general linear model, one way of getting a solution consists of leaving out one dummy variable for each predictor, or in an equivalent way to impose a zero score for this category (usually the last one for most statistical programs). This is also the solution used in logistic regression for categorical predictors.

16.3.2 Discrimination with MCA factors

Another way of determining a scorecard, named Disqual, has been proposed by Saporta (1976) and has been widely used (at least in France) for credit scoring (Bouroche and Saporta 1988). It consists of two main steps:

1. A multiple correspondence analysis (MCA) is performed on the array of predictors **Z** with the class variable as a supplementary one.
2. A linear discriminant analysis is done using factor coordinates as predictors.

Because MCA is closely related to principal component analysis, Disqual is very close to principal component regression.

The vector **s** of overall scores given by Fisher's LDA is a linear combination of the coordinates $\mathbf{f}_j$: $\mathbf{s} = \sum_{j=1}^{J-Q} d_j \mathbf{f}_j$, which is not easy to use for new observations. However, the transition formulas of MCA ($\mathbf{f}_j = \mathbf{Z}\mathbf{v}_j$) allow us to write **s** as a sum of the partial scores (the scorecard) of all categories:

$$\mathbf{s} = \sum_{j=1}^{J-Q} d_j \mathbf{Z}\mathbf{v}_j = \mathbf{Z}\underbrace{\sum_{j=1}^{J-Q} d_j \mathbf{v}_j}_{\text{scorecard}} \tag{16.4}$$

where $\mathbf{v}_j$ is the jth principal coordinate of columns. The scorecard is a linear combination of the coordinates of the categories along the axes of MCA, where the coefficients d_j are given by Fisher's formula (Equation 16.1). We use here $\mathbf{T}^{-1}$ instead of $\mathbf{W}^{-1}$ because **T** (the covariance matrix of factor components of MCA) is the diagonal matrix of the eigenvalues (components are uncorrelated, which makes computations easier):

$$\begin{pmatrix} \cdot \\ d_j \\ \cdot \end{pmatrix} = \mathbf{T}^{-1}\left(\mathbf{g}_1 - \mathbf{g}_2\right) = \begin{pmatrix} \cdot \\ \dfrac{\bar{\mathbf{g}}_j^1 - \bar{\mathbf{g}}_j^2}{\lambda_j} \\ \cdot \end{pmatrix} \tag{16.5}$$

where $\bar{\mathbf{g}}_j^1$ and $\bar{\mathbf{g}}_j^2$ are, respectively, the mean values of both groups on the jth axis.

16.3.3 Factor selection

Using all the $J - Q$ factors is equivalent to discarding one category for each predictor. But one of the main interests of the method lies in the possibility of discarding irrelevant factors: factors are computed irrespective of the class variable, even though some may not be relevant for classification purposes. Since they are uncorrelated, one can just use univariate tests of comparison of means; if the class means do not differ significantly on an axis, one can discard that factor. Using fewer factors than $J - Q$ gives more robust and reliable prediction for new observations. The degrees of freedom are equal to the number of selected factors, which differs from the number of coefficients in the scorecard.

Statistical learning theory (Vapnik 1998) gives a rationale for using fewer factors: let h be the Vapnik–Cervonenkis (VC) dimension, which is the maximum number of points in a two-class problem that can always be perfectly separated by a classifier; h is a measure of the separability power of a classifier. For instance, for linear classifiers in $\mathbb{R}^2$, $h = 3$. In the case of perfect separation, let C be the distance of the closest point to the boundary ($2C$ is called the margin, see Figure 16.1). Then it can be proved that $h \leq \frac{\rho^2}{C^2}$, where ρ is the radius of the smallest sphere containing all the observations.

Vapnik's inequality (Equation 16.6) states that the true error risk R (for new observations from the same distribution) is, with probability $1 - q$, less than the empirical risk R_{emp} (misclassification rate on the learning sample, or resubstitution error rate) plus a quantity depending on the VC dimension h:

$$R < R_{\text{emp}} + \sqrt{\frac{h(\ln(2n/h) + 1) - \ln(q/4)}{n}} \tag{16.6}$$

where n is the size of the learning sample.

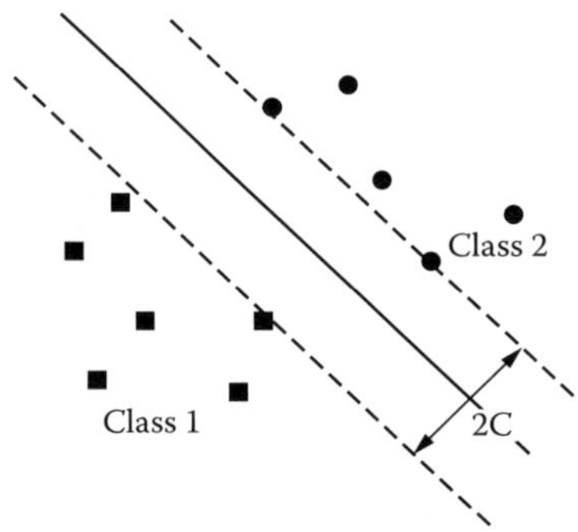

Figure 16.1 2C is the margin between the two classes.

If we select only a few factors, we work on a projection of the original data onto a subspace, which lowers ρ. Hence, if the discarded factors are irrelevant for discrimination purposes, R_{emp} does not change, and if the margin remains unchanged, h decreases and hence the bound for the true error risk R also decreases.

16.4 Alternative methods

16.4.1 Logistic regression for categorical predictors

Logistic regression is now used extensively in credit scoring applications and is used more often than discriminant analysis. Our belief is that this is due not only to its specific properties, but also to the improvement of specific statistical software for logistic regression. Categorical predictors are now easily used in most statistical software packages, since these automatically create indicator variables for each category. One simply declares predictors as "class variables," and interactions are easily included as well. Moreover, the stepwise selection is a true variable-selection procedure and not a selection of indicator variables. All of these features could be added to discriminant analysis software, but in practice that is not yet the case.

One theoretical drawback of logistic regression is that it uses the full space spanned by the indicator variables, which can be of high dimensionality and might lead to overfitting. On the other hand, logistic regression performs better than linear discriminant analysis when the conditional distributions are not normal or have different covariances. This is why we propose the following compromise: as in Disqual, a first step of MCA is performed, followed by a logistic regression with factor selection. The scorecard is then obtained by the same formula (Equation 16.4) presented earlier, the only difference being that the d_j are estimated through a (conditional) maximum-likelihood procedure instead of a least-squares one.

16.4.2 Partial least-squares regression

Partial least-squares regression (PLS) is an alternative technique to ordinary least-squares regression when strong collinearities between predictors are present in the model $\mathbf{y} = \mathbf{X}\boldsymbol{\beta} + \mathbf{e}$. PLS looks for a set of uncorrelated linear combinations of the predictors of the form $\mathbf{t} = \mathbf{Xa}$, but unlike principal component regression, PLS components $\mathbf{t}$ are computed to be related to the response $\mathbf{y}$. PLS regression has been proposed as an algorithm (see Wold et al. 1983), but in our opinion, its rationale is better understood in terms of maximization of covariance (Tenenhaus 1998), which is obtained using Tucker's criterion (1958):

$$\text{maximize } Cov^2(\mathbf{y}, \mathbf{Xa})$$

Since $Cov^2(\mathbf{y}, \mathbf{Xa}) = Cor^2(\mathbf{y}, \mathbf{Xa})\ Var(\mathbf{Xa})\ Var(\mathbf{y})$, maximizing the covariance is a compromise between the explained variance of $\mathbf{X}$ and the correlation with $\mathbf{y}$.

In a classification problem with two classes, one can use PLS regression instead of Fisher's LDA. However, a more symmetric way of dealing with indicator variables, which can be generalized to k-groups discrimination, is PLS2 (for multivariate regression), which maximizes $Cov^2(\mathbf{Yb}, \mathbf{Xa})$, where $\mathbf{Y}$ is the $n \times k$ indicator matrix of the groups. If predictors are categorical, let $\mathbf{Z}$ be the indicator matrix of all categories. The first PLS component is given by the first eigenvector $\mathbf{a}$ of $\mathbf{Z}^\mathsf{T}\mathbf{YY}^\mathsf{T}\mathbf{Z}$:

$$\mathbf{Z}^\mathsf{T}\mathbf{YY}^\mathsf{T}\mathbf{Za} = \lambda\mathbf{a} \tag{16.7}$$

$\mathbf{Y}^\mathsf{T}\mathbf{Z}$ is the concatenated matrix of all tables cross-tabulating the response variable with the predictors. Successive components are obtained by optimizing Tucker's criterion for residuals after orthogonalization. Usually the number of useful PLS components is chosen by cross-validation.

16.4.3 Barycentric discrimination

This technique (see Celeux and Nakache 1994) consists of a correspondence analysis of the table $\mathbf{Y}^\mathsf{T}\mathbf{Z}$. If $\mathbf{D}_y$ and $\mathbf{D}_z$ are the diagonal matrices of column frequencies for $\mathbf{Y}$ and $\mathbf{Z}$, the scores for the categories of the predictors are given by the first (and unique if $k = 2$) eigenvector of

$\frac{1}{Q}\mathbf{D}_z^{-1}\mathbf{Z}^\mathsf{T}\mathbf{Y}(\mathbf{Y}^\mathsf{T}\mathbf{Y})^{-1}\mathbf{Y}^\mathsf{T}\mathbf{Z} = \frac{1}{Q}\mathbf{D}_z^{-1}\mathbf{Z}^\mathsf{T}\mathbf{Y}\mathbf{D}_y^{-1}\mathbf{Y}^\mathsf{T}\mathbf{Z}$, since the row-margin diagonal matrix of $\mathbf{Y}^\mathsf{T}\mathbf{Z}$ is $Q\mathbf{D}_y$ and its column-margin diagonal is $\mathbf{D}_z$

$$\frac{1}{Q}\mathbf{D}_z^{-1}\mathbf{Z}^\mathsf{T}\mathbf{Y}\mathbf{D}_y^{-1}\mathbf{Y}^\mathsf{T}\mathbf{Z}\mathbf{a} = \lambda\mathbf{a} \tag{16.8}$$

Note that Equation 16.8 is almost the same as Equation 16.7, apart from weighting.

For a classification into two groups, $\mathbf{Y}^\mathsf{T}\mathbf{Z}$ is a matrix with two rows, and computations can be done by hand since there is only one axis. The score of a category j is given by the barycenter of $\mathbf{g}_1$ and $\mathbf{g}_2$, weighted by n_{1j} and n_{2j}. Usually $\mathbf{g}_1$ and $\mathbf{g}_2$ are put at the extremities of the interval [0,1], and the score of a unit is equal to the sum of the conditional probabilities (of being a member of group 2) of its Q categories.

Barycentric discrimination is similar to the "naive Bayes classifier," which uses a multiplicative score equal to the product of the conditional probabilities. Barycentric discrimination is equivalent to Disqual only if the predictors are pairwise independent.

16.4.4 Nonsymmetric correspondence analysis

Nonsymmetric correspondence analysis, proposed by Lauro and d'Ambra (1984) for one predictor and used by Verde and Palumbo (1996) for discrimination purpose with p predictors, is equivalent to redundancy analysis (Van den Wollenberg 1977) or PCA with instrumental variables (Rao 1964). When the matrix $\mathbf{X}$ of predictors is nonsingular (which is not the case for the disjunctive table), the linear combinations of the columns of $\mathbf{X}$ are the eigenvectors of $\mathbf{V}_{11}^{-1}\mathbf{V}_{12}\mathbf{V}_{21}$, where $\mathbf{V}_{11}$ and $\mathbf{V}_{12}$ are covariance matrices defined by: $\mathbf{V}_{11} = 1/n\mathbf{X}^\mathsf{T}\mathbf{X}$, $\mathbf{V}_{12} = 1/n\mathbf{X}^\mathsf{T}\mathbf{Y}$, $\mathbf{V}_{21} = \mathbf{V}_{12}^\mathsf{T}$.

For categorical predictors we have:

$$\mathbf{D}_z^{-1}\mathbf{Z}^\mathsf{T}\mathbf{Y}\mathbf{Y}^\mathsf{T}\mathbf{Z}\mathbf{a} = \lambda\mathbf{a}$$

When both groups have equal frequencies, this comes down to barycentric discrimination and also to the first component of PLS regression. Following Bougeard et al. (see Chapter 17), one can derive a continuous set of solutions from Disqual or MCA (α = 0) to redundancy analysis (α = 1) by maximizing

$$\alpha\, Cor^2(\mathbf{Yb},\mathbf{Xa}) + (1-\alpha)\sum_{j=1}^{p} Cor^2(\mathbf{Xa},\mathbf{X}_j\,\mathbf{a}_j)$$

16.5 A case study

16.5.1 Data description

The sample consists of 1106 automobile insurees from Belgium observed in 1992 belonging to one of two groups:

1. Those without claim $n_1 = 556$ (the "good" ones).
2. Those with one or more claims (the "bad" ones) $n_2 = 550$.

We use here nine categorical predictors with a total of 20 categories:

1. Use type (2): professional, private
2. Insuree type (3): male, female, companies
3. Language (2): French, Flemish
4. Birth cohort (3): 1890–1949, 1950–1973, unknown
5. Region (2): Brussels, other regions
6. Level of bonus-malus (2): B-M+, other B-M (–1)
7. Horsepower (2): 10–39, 40–349
8. Year of subscription (2): <86 , others
9. Year of vehicle construction (2): 1933–1989, 1990–1991

16.5.2 Multiple correspondence analysis

The class variable with categories "0 claim" or "1 or more claims" (good or bad) is a supplementary one. MCA gives 11 = 20 – 9 factors (see Table 16.3).

Table 16.4 gives the frequencies of the categories, followed by the test values (corresponding normal deviates) and coordinates along the

Table 16.3 Eigenvalues of MCA.

Number	Eigenvalue	Proportion	Cumulative
1	0.2438	19.95	19.95
2	0.1893	15.49	35.44
3	0.1457	11.92	47.36
4	0.1201	9.82	57.18
5	0.1091	8.92	66.11
6	0.0999	8.17	74.28
7	0.0855	7.00	81.28
8	0.0732	5.99	87.26
9	0.0573	4.68	91.95
10	0.0511	4.18	96.13
11	0.0473	3.87	100.00

Table 16.4 Factor coordinates and test values for all categories.

		Test Values					Coordinates				
Categories	Freq.	1	2	3	4	5	1	2	3	4	5
Use type:											
Professional	185	11.1	24.5	1.5	–2.1	–3.7	0.74	1.64	0.10	–0.14	–0.25
Private	921	–11.1	–24.5	–1.5	2.1	3.7	–0.15	–0.33	–0.02	0.03	0.05
Insuree type:											
Male	787	–9.5	–3.7	–17.4	9.6	16.6	–0.18	–0.07	–0.33	0.18	0.32
Female	249	5.7	–10.8	15.6	–8.5	–16.2	0.32	–0.61	0.87	–0.47	–0.90
Companies	70	8.0	25.5	5.7	–3.4	–3.1	0.93	2.95	0.66	–0.39	–0.36
Language:											
French	824	11.6	–7.4	14.5	17.9	–1.1	0.20	–0.13	0.26	0.31	–0.02
Flemish	282	–11.6	7.4	–14.5	–17.9	1.1	–0.60	0.38	–0.75	–0.92	0.06
Birth cohort:											
1890–1949	301	1.3	–6.9	–13.9	16.7	–15.1	0.06	–0.34	–0.69	0.82	–0.74
1950–1973	309	17.7	–13.2	0.2	–13.8	13.2	0.86	–0.64	0.01	–0.67	0.64
Unknown	496	–17.1	18.2	12.3	–2.5	1.6	–0.57	0.61	0.41	–0.08	0.05
Region:											
Brussels	367	15.3	0.1	15.8	14.7	7.1	0.65	0.00	0.67	0.63	0.30
Others regions	739	–15.3	–0.1	–15.8	–14.7	–7.1	–0.32	0.00	–0.33	–0.31	–0.15
Level of bonus-malus:											
B-M +	549	–26.9	–2.3	1.1	3.4	–5.9	–0.82	–0.07	0.03	0.10	–0.18
B-M (–1)	557	26.9	2.3	–1.1	–3.4	5.9	0.80	0.07	–0.03	–0.10	0.18

Year of subscription:											
<86 contracts	629	−23.2	5.9	10.8	7.0	0.9	−0.61	0.15	0.28	0.18	0.02
Others	477	23.2	−5.9	−10.8	−7.0	−0.9	0.80	−0.20	−0.37	−0.24	−0.03
Horsepower:											
10–39 HP	217	−3.6	−11.5	15.5	−12.3	−9.1	−0.22	−0.70	0.94	−0.75	−0.55
40–349 HP	889	3.6	11.5	−15.5	12.3	9.1	0.05	0.17	−0.23	0.18	0.13
Year of vehicle construction:											
1933–1989	823	−12.5	−3.0	9.9	−2.0	17.9	−0.22	−0.05	0.17	−0.04	0.32
1990–1991	283	12.5	3.0	−9.9	2.0	−17.9	0.46	0.15	−0.51	0.10	−0.92
Claim:											
0 claim (“good”)	556	−23.1	−1.5	2.3	1.5	−2.8	−0.69	−0.05	0.07	0.04	−0.08
≥1 claim (“bad”)	550	23.1	1.5	−2.3	−1.5	2.8	0.70	0.05	−0.07	−0.04	0.08

first five axes of MCA. The results show that the first factor is very discriminating, with a test value of 23 for the "good/bad" class variable. When all of the predictors are highly related to the group variable, this is often the case. One might remark that the centroid of "0 claim" insurees is too close to categories "Flemish," "private," and "other regions." Figure 16.2 shows the principal plane of MCA. (The label "???BD" in Figure 16.2 means "unknown birth cohort.")

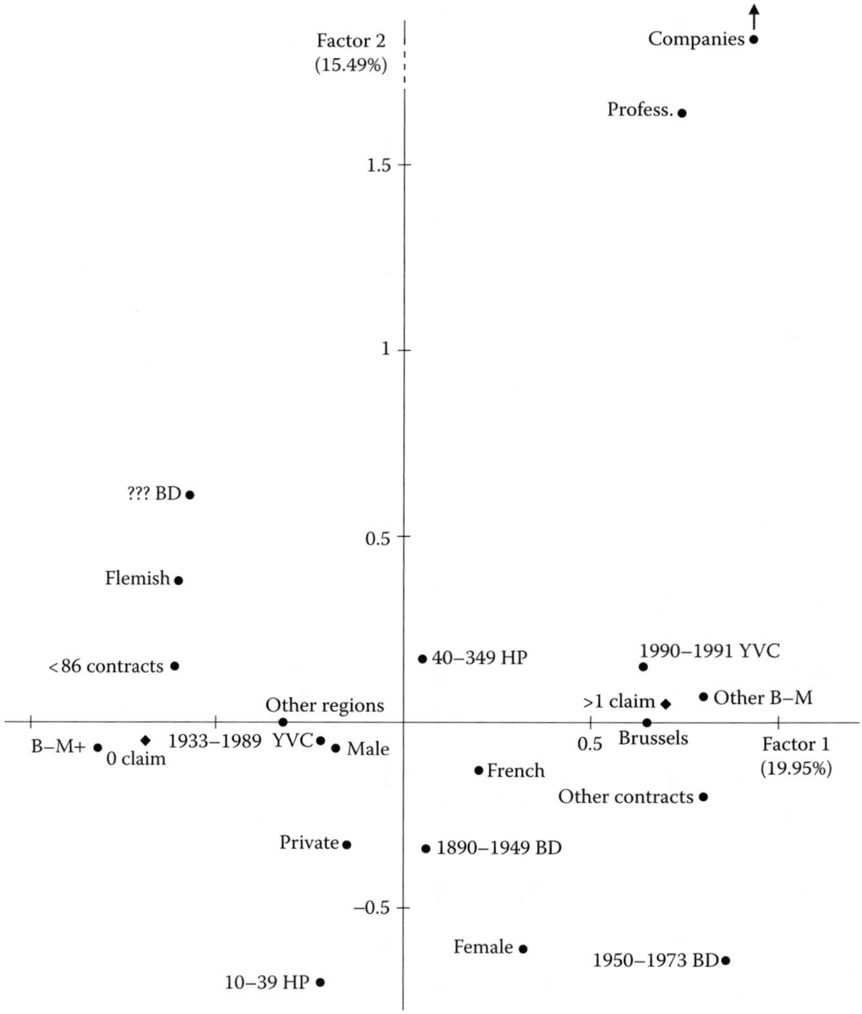

Figure 16.2 Principal plane of MCA.

The coordinate on the first axis can be taken as a discriminant score with a very simple rule: positive values for "0 claim" and negative values for "one claim or more." However, the use of other factors will improve the decision rule.

16.5.3 Disqual

The next step consists of performing Fisher's LDA using factor coordinates as predictors. Because Fisher's LDA is defined up to its sign, we have chosen an orientation such that a high value means a "good" driver.

Table 16.5 shows that the ranking according to eigenvalues is not the same as the ranking according to the prediction of the class variable and that factors F2, F4, and F7 are not discriminant. We thus decide to discard them. Because factors are uncorrelated, the coefficients do not change when some factors are discarded, as they would be in a regression with orthogonal predictors. Thus the score of a statistical unit is given by $-0.695F1 + 0.068F3 + \cdots + 0.062F11$.

Table 16.6 gives the scorecard with two options: the raw coefficients coming from the direct application of LDA to indicator variables, and the transformed coefficients standardized by a linear transformation such that the score range is [0, 1000], which is used most often in practice.

Table 16.5 Fisher's linear discriminant function (LDF) as a combination of factor coordinates.

Factors	Correlations with LDF	Coefficients Discriminant Function
F1	–0.695	–6.0525
F2	*0.046*	*0.4548*
F3	0.068	0.7639
F4	*0.045*	*0.5530*
F5	–0.084	–1.0876
F6	–0.084	–1.1369
F7	*–0.009*	*–0.1270*
F8	–0.063	1.0064
F9	0.079	1.4208
F10	0.129	2.4594
F11	0.062	1.2324

Note: Italicized entries are not discriminant and thus are discarded from the analysis.

Table 16.6 Scorecard with raw and transformed coefficients.

Categories	Coefficients Discriminant Function	Transformed Coefficients (score)
Use type:		
Professional	−4.577	0.00
Private	0.919	53.93
Insuree type:		
Male	0.220	24.10
Female	−0.065	21.30
Companies	−2.236	0.00
Language:		
French	−0.955	0.00
Flemish	2.789	36.73
Birth cohort:		
1890–1949	0.285	116.78
1950–1973	−11.616	0.00
Unknown	7.064	183.30
Region:		
Brussels	−6.785	0.00
Other regions	3.369	99.64
Level of bonus-malus:		
B-M +	17.522	341.41
Other B-M (−1)	−17.271	0.00
Year of subscription:		
<86	2.209	50.27
Others	−2.913	0.00
Horsepower:		
10–39	6.211	75.83
40–349	−1.516	0.00
Year of vehicle construction:		
1933–1989	3.515	134.80
1990–1991	−10.222	0.00

The transformed coefficients are obtained in the following way. To get a score equal to zero for the worst case, we add 4.577 to both scores of categories "use type," 2.236 to the three scores of "gender," etc. The maximum value ("best" case) is now equal to the sum of the maximal modified scores for each variable [(0.919 + 4.577) + (0.22 + 2.236) + (2.789 + 0.955) + ⋯ + (3.515 + 10.222) = 101.909]. Each score is then multiplied by a constant equal to 1000 divided by the maximum value (= 1000/101.909 = 9.8125) so that the maximal value of the sum of

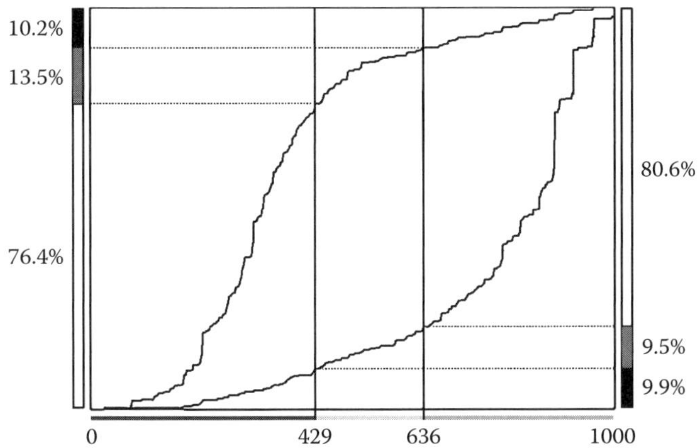

Figure 16.3 CDF of the score for the two groups.

partial scores equals 1000. The final score is obtained by adding the scores corresponding to the categories; for example, an insuree with a private use of his vehicle, male, French-speaking, etc. will get a score of 53.93 + 24.1 + 0 + ...

Figure 16.3 gives the cumulative distribution function (CDF) of the score for both groups. It is commonly used to derive classification rules, taking into account the two kinds of error risks. Here both risks have been taken approximately equal to 10%: an insuree with a score lower than 429 will be predicted as a "bad" one: 76.4% of the "bad" ones are detected and 9.9% of the "good" ones are wrongly considered as "bad." Conversely 80.6% of the "good" insurees have a score higher than 636, and only 10.2% of the "bad." The interval [429; 636] is an uncertainty domain.

16.5.4 Comparison of Disqual with logistic regression

We have applied logistic regression to the same data. Table 16.7 gives the coefficients of the logistic score. We see that the constraint used here is that one category of each predictor has a zero score. At first glance, it is not easy to compare both sets of coefficients coming from Disqual and from logistic regression, but the scatter plot (Figure 16.4) of both individual scores shows that they agree very well; the correlation coefficient between both scores is $r = 0.975$.

Table 16.7 Scorecard by logistic regression.

Categories	Coefficients Logistic Regression
Use type:	
Professional	0.00
Private	0.7060
Insuree type:	
Male	0.4797
Female	0.4868
Companies	0.00
Language:	
French	−0.1236
Flemish	0.00
Birth cohort:	
1890–1949	−0.3596
1950–1973	−1.6155
Unknown	0.00
Region:	
Brussels	−0.8585
Other regions	0.00
Level of bonus-malus:	
B-M +	0.00
Other B-M (−1)	−2.4313
Year of subscription:	
<86 contracts	0.4932
Others	0.00
Horsepower:	
10–39 HP	0.7305
40–349 HP	0.00
Year of vehicle construction:	
1933–1989	1.3362
1990–1991	0.00
Intercept	−0.2498

Another way of comparing scores is to compare their ROC (receiver operating characteristics) curves and AUC (area under the ROC curve): the ROC curve (see Bamber 1975) synthesizes the performance of a score for any threshold s. Using s as a parameter, the ROC curve links the true positive rate to the false positive rate. The true positive rate (or specificity) is the probability of being classified in G_1 for a member of $G_1 = P(S > s \mid G_1)$. The false positive rate (or 1–sensitivity) is the probability of being wrongly classified to $G_1 = P(S > s \mid G_2)$.

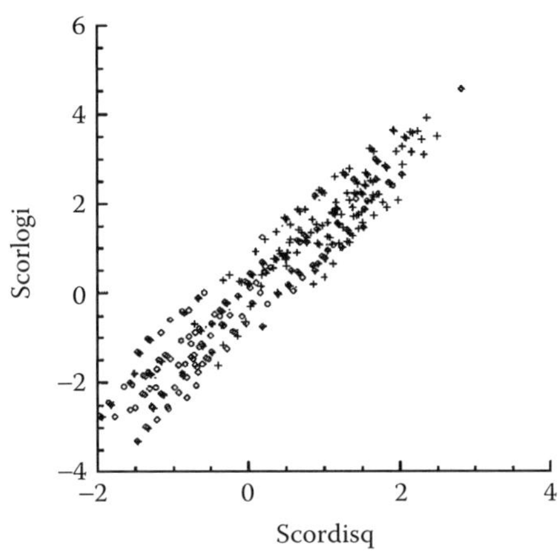

Figure 16.4 Scatterplot of scores obtained by logistic regression and by Disqual.

One of the main properties of the ROC curve is that it is invariant with respect to increasing (not only linear) transformations of S. Because the ideal curve is the one that sticks to the edges of the unit square, the favorite measure is given by the area under the ROC curve (AUC), which allows us to compare several curves, as long as there is no crossing. Theoretical AUC is equal to the probability of "concordance": AUC = $P(X_1 > X_2)$ when one draws at random two observations independently from both groups. For two samples of n_1 and n_2 observations, AUC comes down to Mann-Whitney's U statistic.

Figure 16.5 shows very close results: logistic regression gives a slightly greater AUC than Disqual: 0.908 instead of 0.904, but with a standard error of 0.01, the difference is not significant.

The above comparison was done with the total sample and may suffer from the so-called resubstitution bias since the same data set is used twice: for estimating score and for prediction. If we want to compare predicting capabilities of several methods, it is necessary to do so with an independent sample. One has to divide randomly the total sample into two parts: the training set and the test set. To avoid a pattern that is too specific, we did this random split 50 times using a stratified sampling (the strata are the two groups) without replacement of 70% for the training sample and 30% for the test sample.

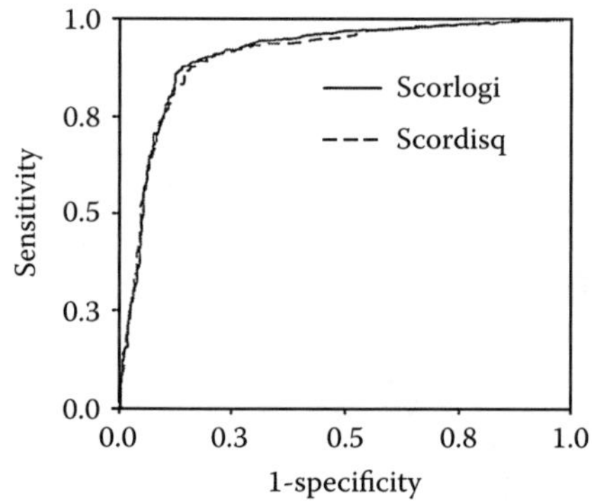

Figure 16.5 ROC curves.

We used the following five methods:

1. Disqual with an automatic selection of relevant factors with a probability level of 5%
2. Logistic regression on raw data at probability
3. Logistic regression on MCA factors with automatic selection (probability level 5%)
4. PLS regression with cross-validation factor selection
5. Barycentric discrimination

Table 16.8 give the results of factor selection for the first 15 replications. Factors F1, F5, F6, F9, and F10 are selected 15 times by Disqual; factor F7 is never selected. Logistic regression retains fewer factors than Disqual.

The performance of each of the five methods was measured by the AUC, computed 50 times on the test samples. One may remark from Table 16.9 that the methods based on a selection of MCA factors are more precise (i.e., with a lower standard deviation), even if the average is slightly lower. PLS regression was performed with a cross-validation choice for the numbers of factors: four factors were selected in 42 cases, three factors for the other eight simulations.

Actually, the five methods give very similar performances, and it is not possible to state the superiority of any one for this particular application. Barycentric discrimination has very good performance despite its simplicity. An explanation of this surprising fact might be that the nine variables were already the result of an expert selection and have low intercorrelations.

Table 16.8 Factor selection for discriminant analysis and logistic regression in 15 (out of 50) test samples.

	1	2	3	4	5	6	7	8	9	10	11	12	13	14	15
Disqual	F1	F1	F1	F1	F1	F1	F1	F1	F1	F1	F1	F1	F1	F1	F1
	F2				F2	F2		F2			F2	F2			
	F3	F3	F3	F3		F3	F3	F3	F3	F3	F3	F3	F3	F3	F3
	F4	F4		F4					F4		F4				
	F5	F5	F5	F5	F5	F5	F5	F5	F5	F5	F5	F5	F5	F5	F5
	F6	F6	F6	F6	F6	F6	F6	F6	F6	F6	F6	F6	F6	F6	F6
	F8	F8	F8	F8	F8		F8	F8	F8		F8	F8	F8	F8	F8
	F9	F9	F9	F9	F9	F9	F9	F9	F9	F9	F9	F9	F9	F9	F9
	F10	F10	F10	F10	F10	F10	F10	F10	F10	F10	F10	F10	F10	F10	F10
	F11	F11	F11	F11	F11	F11	F11	F11	F11	F11			F11		F11
No. of factors	10	9	8	9	8	8	8	9	9	7	9	8	8	7	8
Logifact	F1	F1	F1	F1	F1	F1	F1	F1	F1	F1	F1	F1	F1	F1	F1
											F2				
	F3	F3	F3	F3		F3	F3	F3	F3	F3		F3	F3	F3	F3
		F5		F5			F5	F5		F5	F5	F5	F5	F5	F5
	F6	F6	F6	F6	F6	F6	F6	F6	F6	F6	F6	F6	F6	F6	F6
	F8	F8		F8			F8	F8	F8		F8	F8		F8	
	F9	F9	F9	F9	F9	F9		F9	F9	F9		F9	F9		F9
	F10	F10	F10	F10	F10	F10	F10	F10	F10	F10	F10	F10	F10	F10	F10
				F11											
No. of factors	6	7	5	8	4	5	6	7	6	6	6	7	6	6	6

Table 16.9 AUC on 50 test samples.

	Disqual	Logistic	Logistic Factor	PLS	Barycentric Discrimination
Mean	.9024	.9044	.9023	.9035	.9039
Std. Dev.	.0152	.0156	.0146	.0157	.0155
Min.	.863	.857	.861	.856	.860
Max.	.928	.932	.928	.930	.933

16.6 Conclusion

We have advocated MCA as an intermediate step to derive numerical predictors before applying a linear discriminant analysis. Factor selection is a kind of regularization that lowers the VC dimension and brings better generalization by avoiding overfitting.

The ability of MCA to recover the data structure explains its efficiency, despite the fact that factors are computed without taking into account the response (group) variable. Fewer factors are necessary if one uses a nonsymmetrical analysis such as PLS. The comparison with logistic regression has not shown any systematic superiority of this technique, which can be combined with a selection of MCA factors.

Moreover, the factor space can also be used as a basis for nonlinear analysis, which is not possible with barycentric discrimination, and for optimizing criteria other than the Mahalanobis distance, such as AUC. The use of factor coordinates also provides a key for applying methodologies designed for numerical predictors, such as support vector machines or neural networks, to categorical data (Hastie et al. 2001).

Software notes

Multiple correspondence analysis and score functions have been performed with SPAD v5.6 (http://www.spadsoft.com) (the insurance data set is provided with this software). Logistic regression was performed with Proc Logistic from SAS v8.2. ROC curves and AUC were computed with SPSS v11.5.

CHAPTER 17

Multiblock Canonical Correlation Analysis for Categorical Variables: Application to Epidemiological Data

Stéphanie Bougeard, Mohamed Hanafi, Hicham Noçairi, and El-Mostafa Qannari

CONTENTS

17.1 Introduction: epidemiological data and statistical issues

Animal health has become an issue of paramount interest since the outbreak of major diseases in cattle and poultry. Research in epidemiology is concerned with detecting, identifying, and preventing animal diseases. Very often, the purpose of the study is to predict one or more categorical variables related to animal health or the spread of an infection based on categorical variables related to the breeding environment, alimentary factors, and farm management, among others. Disease is the result of interactions between the infectious agent, the affected animal, the environment, and management factors. Therefore, the investigation of the relationships between the variables related to the disease, on the one hand, and the variables that are deemed to have an impact on this disease, on the other hand, is of paramount importance, as it is a source of opportunities to reduce disease at multiple points in the transmission cycle. From a statistical standpoint, disease is manifested either as a binary variable (uninfected/infected) or as a categorical variable with more than two categories that cover a spectrum that can range from unapparent to fatal. The explanatory variables (risk factors) may be directly observed on the animals or measured by means of questionnaires administered to animal breeders inquiring about management and feeding habits.

We address the problem of predicting a categorical variable related to the animal disease from several categorical variables (risk factors) by means of a variant of generalized canonical analysis adapted to the case of categorical variables. For the investigation of the relationships among several categorical variables, multiple correspondence analysis (MCA) is often advocated. MCA is an exploratory multivariate technique, and the variables under study have the same role. In contrast, in the situations we consider, there is a categorical variable Y that we predict from the other categorical variables. In other words, this problem is related to discriminant analysis on categorical data.

Several procedures have been proposed in the literature to address this issue. Multiple logistic regression (Hosmer and Lemeshow 1989) is one approach that requires setting up a model that can optionally include interaction terms. However, it is well known that this technique is very sensitive to the quality of data at hand, for example, sample size and multicollinearity.

Another approach that is simpler than logistic regression and that leads to satisfactory results, either in terms of prediction or in terms of interpretation, consists of performing MCA on the predictor variables. In a subsequent stage, linear discriminant analysis is applied on the categorical response Y using the principal axes derived from MCA as predictor variables. To avoid instability in the results through multicollinearity, the principal axes associated with small eigenvalues can be discarded. This method is usually referred to as *Disqual* as an abbreviation for discrimination on qualitative variables (Saporta 1990a). This approach bears a striking similarity to regression on principal components applied within the context of multiple linear regression (Draper and Smith 1998), and the same criticism of this latter method can be applied to *Disqual*, namely that the major principal axes may not be related to the response variable.

The aim of the method of analysis discussed here is to circumvent this problem by seeking principal axes to be used for discrimination, but unlike *Disqual*, these principal axes are computed in such a way that they take account of the variable Y to be predicted. The principal axes are determined sequentially by maximizing at each step a criterion that is related to generalized canonical analysis. We start by introducing the method within a setting involving quantitative variables, and in a subsequent stage, we apply the strategy of analysis to the case of categorical variables by considering the indicator matrices associated with the categorical variables. We discuss how the method is related to other statistical techniques and give illustrations using a data set from an epidemiological context.

17.2 Multiblock canonical correlation analysis

Suppose that we have a data set $\mathbf{Y}$ consisting of P quantitative variables and Q data sets $\mathbf{X}_q$ ($q = 1, 2,\ldots,Q$), where $\mathbf{X}_q$ consists of J_q quantitative variables. All of these variables are measured on the same N individuals and centered. We aim at predicting $\mathbf{Y}$ from $\mathbf{X}_q$ ($q = 1, 2,\ldots,Q$). We denote by $\mathbf{X} = [\mathbf{X}_1, \mathbf{X}_2,\ldots,\mathbf{X}_Q]$ the data matrix formed by horizontally concatenating the data matrices $\mathbf{X}_1, \mathbf{X}_2,\ldots, \mathbf{X}_Q$.

For this purpose, we seek, in a first step, latent variables that are linear combinations of the variables in $\mathbf{Y}$ and $\mathbf{X}$:

$$\mathbf{u} = \mathbf{Yc}$$

$$\mathbf{t} = \mathbf{Xw}$$

where $\mathbf{t}$ can be partitioned according to the Q sets as:

$$\mathbf{t}_q = \mathbf{X}_q\mathbf{w}_q \ (q = 1, 2,\ldots,Q)$$

These latent variables are sought in such a way as to maximize the following criterion:

$$\alpha \operatorname{cor}^2(\mathbf{u},\mathbf{t}) + (1-\alpha)\sum_{q=1}^{Q}\operatorname{cor}^2(\mathbf{t}_q,\mathbf{t})$$

where α is a fixed scalar between 0 and 1 and cor stands for the coefficient of correlation. The overall latent variable $\mathbf{t} = \mathbf{Xw}$ plays a central role in the analysis and in the prediction procedure, as described below. The partial latent variables $\mathbf{t}_q = \mathbf{X}_q\mathbf{w}_q$ mainly serve an interpretation purpose, as they reflect the information contained in the overall latent variable $\mathbf{t}$ in terms of the variables in each data set.

By varying the tuning parameter α, various methods of analysis can be retrieved. The case $\alpha = 1$ amounts to the first step of canonical correlation analysis of $\mathbf{Y}$ and $\mathbf{X}$ (Hotelling 1936), whereas the case $\alpha = 0$ leads to generalized canonical analysis of data sets $\mathbf{X}_1, \mathbf{X}_2,\ldots,\mathbf{X}_Q$ (Carroll 1968). Of particular interest is the case $\alpha = 0.5$, where the criterion to be maximized amounts to $Cor^2(\mathbf{Yv},\mathbf{Xw}) + \sum_{q=1}^{Q} Cor^2(\mathbf{Xw},\mathbf{X}_q\mathbf{w}_q)$. Therefore, it can be seen that this choice leads to the first step of generalized canonical correlation analysis of $\mathbf{X}_1, \mathbf{X}_2,\ldots,\mathbf{X}_Q$ and $\mathbf{Y}$. This strategy of analysis ($\alpha = 0.5$) bears some similarity to latent root regression analysis (Webster et al. 1974), which, in the usual multiple-regression setting with a single response variable, starts by performing principal component analysis on the data matrix formed by the variable to be predicted and the predictors.

The solution $\mathbf{w}$ (vector of coefficients associated with the overall latent variable) is the eigenvector of the matrix

$$\left(\mathbf{X}^\mathsf{T}\mathbf{X}\right)^{-1/2}\mathbf{X}^\mathsf{T}\left(\alpha\mathbf{Y}\left(\mathbf{Y}^\mathsf{T}\mathbf{Y}\right)^{-1}\mathbf{Y}^\mathsf{T} + (1-\alpha)\sum_{q=1}^{Q}\mathbf{X}_q\left(\mathbf{X}_q^\mathsf{T}\mathbf{X}_q\right)^{-1}\mathbf{X}_q^\mathsf{T}\right)\mathbf{X}\left(\mathbf{X}^\mathsf{T}\mathbf{X}\right)^{-1/2}$$

associated with the largest eigenvalue λ (Chessel and Hanafi 1996; Lafosse and Hanafi 1997).

The vector of coefficients $\mathbf{c}$ is given by

$$\mathbf{c} = \frac{1}{\sqrt{\mu}}\left(\mathbf{Y}^\mathsf{T}\mathbf{Y}\right)^{-1}\mathbf{Y}^\mathsf{T}\mathbf{Xw}$$

where $\mu = \mathbf{w}^\mathsf{T}\mathbf{X}^\mathsf{T}\mathbf{Y}\left(\mathbf{Y}^\mathsf{T}\mathbf{Y}\right)^{-1}\mathbf{Y}^\mathsf{T}\mathbf{Xw}$.

The vectors of coefficients $\mathbf{w}_q$ are given by

$$\mathbf{w}_q = \frac{1}{\sqrt{\theta_q}}\left(\mathbf{X}_q^\mathsf{T}\mathbf{X}_q\right)^{-1}\mathbf{X}_q^\mathsf{T}\mathbf{Xw}$$

where $\theta_q = \mathbf{w}^\mathsf{T}\mathbf{X}^\mathsf{T}\mathbf{X}_q\left(\mathbf{X}_q^\mathsf{T}\mathbf{X}_q\right)^{-1}\mathbf{X}_q^\mathsf{T}\mathbf{Xw}$.

Moreover, we have the relationship:

$$\lambda = \alpha\mu + \left(1-\alpha\right)\sum_{q=1}^{Q}\theta_q$$

Once the first overall latent variable $\mathbf{t}$ and its associated partial latent variables are determined, the variables in $\mathbf{Y}$ are regressed upon $\mathbf{t}$, giving a first prediction model. But variables in the data sets $\mathbf{X}_q$ potentially contain further information that might be useful for predicting **Y**. To capture this information, we adopt the strategy commonly advocated within the PLS (partial least squares) context (Wold 1966; Garthwaite 1994). This consists of considering the residuals from the regression of **Y** variables on **t** and the residuals from the regression of the variables of the various data sets $\mathbf{X}_q$ on **t**. Thereafter, the same procedure is performed again on the residual matrices, leading to a second set of latent variables. This process can be repeated to derive subsequent latent variables. A final model for the prediction of **Y** is given by regressing **Y** on the overall latent variables that are eventually retained. A stopping strategy that can be used to choose the appropriate number of variables to be introduced in the prediction model and the appropriate parameter α can be set up on the basis of a cross-validation procedure (Stone 1974).

If instead of quantitative variables, we dispose of a categorical variable **Y** that we wish to predict from Q categorical variables $\mathbf{X}_q$ ($q = 1, 2,\ldots,Q$), we adopt the coding of all of these variables using the indicator matrices. Suppose that **Y** has K categories and that each $\mathbf{X}_q$ variable has J_q categories. The matrix $\mathbf{X} = [\mathbf{X}_1, \mathbf{X}_2,\ldots,\mathbf{X}_Q]$ is (N, J), where $J = \sum_{q=1}^{Q} J_q$ is the total number of categories. We adopt a strategy of analysis based on the determination of latent variables using a criterion similar to the criterion given above. The first case leads to a set of latent variables that depends upon the tuning parameter α. The case $\alpha = 1$ amounts to (simple) correspondence analysis of **Y** and **X**. The case $\alpha = 0$ consists of performing MCA of **X** and the case $\alpha = 0.5$ amounts to performing MCA on **Y** and **X**. It should be pointed out that the determination of the subsequent latent variables take account of the fact that the variable **Y** on the one hand and the variables $\mathbf{X}_1, \mathbf{X}_2,\ldots,\mathbf{X}_Q$ do not play a symmetrical role.

Subsequent latent variables can be determined by adopting an approach based on successive regressions of the data sets upon preceding latent variables $\mathbf{t}_1 = \mathbf{X}\mathbf{w}_1$, $\mathbf{t}_2 = \mathbf{X}\mathbf{w}_2,\ldots,\mathbf{t}_s = \mathbf{X}\mathbf{w}_s$ and considering the residuals as described above. Once a set of latent variables is determined, we advocate to predict variable **Y** using Fisher's discriminant analysis (Fisher 1936; Saporta 1990a), where $\mathbf{t}_1, \mathbf{t}_2,\ldots,\mathbf{t}_s$ become the predictor variables. When α increases, the method of analysis makes it possible to take account of the variable **Y** in the derivation of the latent variables, which ensures that these latent variables are relevant for the prediction of **Y**. The appropriate number of latent variables to be introduced in the discrimination model and the appropriate tuning parameter α can again be chosen by cross-validation.

17.3 Application

Epizootic enterocolitis of the rabbit (EER) is a serious gastrointestinal syndrome that appeared in France in late 1996 and very quickly spread throughout Europe. The risk factors that might have an effect on the development of EER are believed to be related to breeding environment and to farming management. To better understand these factors, an epidemiological survey was carried out that involved 96 rabbit breeding colonies (Klein 2002). The explanatory variables were collected through questionnaires administered to farmers. The variable of interest related to the contamination by EER is expressed by three categories: $Y = 1$ corresponding to noninfected farms ($n_1 = 28$), $Y = 3$ corresponding to infected farms ($n_3 = 37$), and $Y = 2$ corresponding to intermediate cases ($n_2 = 31$) that have shown sporadic presence of EER in the five generations of rabbit colonies preceding the survey. The explanatory variables are given in Table 17.1.

For simplicity's sake, we performed the methods of analysis by choosing $\alpha = 0.5$. In a first stage, we applied multiblock canonical correlation analysis for the prediction of variable EER from the explanatory variables.

In Table 17.2, we give the squared coefficients of correlations of the first four overall latent variables and their corresponding partial latent variables associated with the various categorical variables. Of particular interest are the correlations associated with the variable EER. It can be seen that EER is more strongly related to axes 1 and 3 than axes 2 and 4. It is worth noting in passing that a choice of a larger value for α than 0.5 is likely to result in the switching of axes 2 and 3, owing to the fact that as the weight assigned to variable $\boldsymbol{Y}$ increases, the latent variables that are related to $\boldsymbol{Y}$ will emerge as the more important components. It can also be seen that the first overall latent variable is related to variables CAGE, TRANSF, ELIMIN, MORT, and AGE. The third overall latent variable, besides being linked to EER, is also linked to variables HAND, REPLACE, and WEIGHT. Variables such as NBFEM and DIET are related to axes 2 and 4, respectively, but are not important for predicting EER.

Figure 17.1, which depicts the coefficients on the basis of axis 1 and axis 3, highlights the association among the various categories and makes it possible to identify some risk factors associated with EER. Correspondingly, Figure 17.2 gives the position of the individuals (farms) on the basis of axes 1 and 3, with these individuals being labeled by their associated codes with respect to the variable EER (1, 2, and 3 for the uninfected, the intermediate, and the infected groups,

Table 17.1 Explanatory variables.

Variables	Variable Descriptions and Categories		
WATER	Origin of the drinking water		
	(1) Sewerage	(2) Sink	(3) Well
	($n_1 = 56$)	($n_2 = 27$)	($n_3 = 13$)
CAGE	Number of cages for one feeder at fattening		
	(1) One feeder	(2) More than one feeder	
	($n_1 = 44$)	($n_2 = 52$)	
DIET	Number of dietary options during the last 6 months at fattening		
	(1) One dietary option	(2) Two dietary options	(3) Three or more dietary options
	($n_1 = 33$)	($n_2 = 43$)	($n_3 = 20$)
STRAIN	Main strain used for parents		
	(1) A	(2) B	(3) Other
	($n_1 = 62$)	($n_2 = 14$)	($n_3 = 20$)
HAND	Hand washing between maternity and weaning		
	(1) Yes	(2) No	
	($n_1 = 43$)	($n_2 = 53$)	
TRANSF	Transfer at weaning		
	(1) Of rabbit doe	(2) Of young rabbit and rabbit doe	
	($n_1 = 24$)	($n_2 = 72$)	
REPLACE	Annual replacement rate in farming		
	(1) ≥125%	(2) 112 to 125%	(3) ≤112%
	($n_1 = 28$)	($n_2 = 31$)	($n_3 = 37$)
COLIBAC	Chronic colibacillosis in herd (evolution during 4 months)		
	(1) Yes	(2) No	
	($n_1 = 29$)	($n_2 = 67$)	
NBFEM	Number of productive females in farming		
	(1) ≥600	(2) 350 to 600	(3) ≤350
	($n_1 = 28$)	($n_2 = 33$)	($n_3 = 35$)
FERT	Fertility rate		
	(1) ≥82%	(2) 78 to 82%	(3) ≤78%
	($n_1 = 28$)	($n_2 = 37$)	($n_3 = 31$)
ELIMIN	Elimination rate of alive young rabbits		
	(1) ≥4	(2) 0.5 to 4	(3) ≤0.5
	($n_1 = 33$)	($n_2 = 33$)	($n_3 = 30$)
MORT	Mortality between birth and weaning (except elimination)		
	(1) ≥12	(2) 8.5 to 12	(3) ≤8.5
	($n_1 = 30$)	($n_2 = 36$)	($n_3 = 30$)
AGE	Average weaning age		
	(1) >35 days	(2) =35 days	(3) <35 days
	($n_1 = 19$)	($n_2 = 55$)	($n_3 = 22$)
WEIGHT	Average weight at sale		
	(1) ≥2.45 kg	(2) 2.39 to 2.45 kg	(3) ≤2.39 kg
	($n_1 = 38$)	($n_2 = 26$)	($n_3 = 32$)

Table 17.2 Multiblock canonical correlation analysis: squared coefficients of correlation of the first four overall latent variables and their corresponding partial latent variables.

Variables	Axis 1	Axis 2	Axis 3	Axis 4
EER (Y)	**0.239**	0.036	**0.201**	0.066
WATER	0.059	0.183	0.084	0.046
CAGE	**0.305**	0.235	0.011	0.008
DIET	0.198	0.022	0.081	0.395
STRAIN	0.053	0.213	0.131	0.030
HAND	0.002	0.110	**0.220**	0.096
TRANSF	**0.265**	0.010	0.179	0.053
REPLACE	0.004	0.033	**0.369**	0.384
COLIBAC	0.092	0.093	0.078	0.000
NBFEM	0.072	0.516	0.088	0.228
FERT	0.188	0.144	0.033	0.197
ELIMIN	**0.462**	0.204	0.064	0.065
MORT	**0.297**	0.018	0.005	0.120
AGE	**0.280**	0.301	0.014	0.041
WEIGHT	0.017	0.032	**0.212**	0.026

Note: Boldface entries indicate significantly related variables for axes 1 and 3.

respectively). It can be seen that there is a good separation of the groups, particularly the two extreme groups (infected and uninfected groups). The conclusion that can be drawn from the graphical display shown in Figure 17.1 is that the category EER = 2 is intermediate between the category EER = 3 (infected farms) and the category EER = 1 (uninfected farms). This means that the intermediate cases can show symptoms quite similar to the contaminated farms or can stand uninfected during a long period.

If we now interpret variable categories that influence infected farms (EER = 3) and uninfected farms (EER = 1), we can determine, respectively, risk and protective factors. The first risk factor that can clearly be identified is the high mortality rate between the birth and the weaning (MORT-1), and it constitutes a warning that the infection is causing damage in the colony. The transfer at weaning (TRANSF-2), which causes a stress to the doe and to the young rabbits, is associated with EER = 3. Then, young rabbits that drink well water (WATER-3) are more contaminated. Bacteriological water analyses are not different between infected and uninfected farms, but the quality of well water (WATER-3) varies greatly according to the season. When the elimination rate of live rabbits is low (ELIMIN-3), the risk of contamination is high

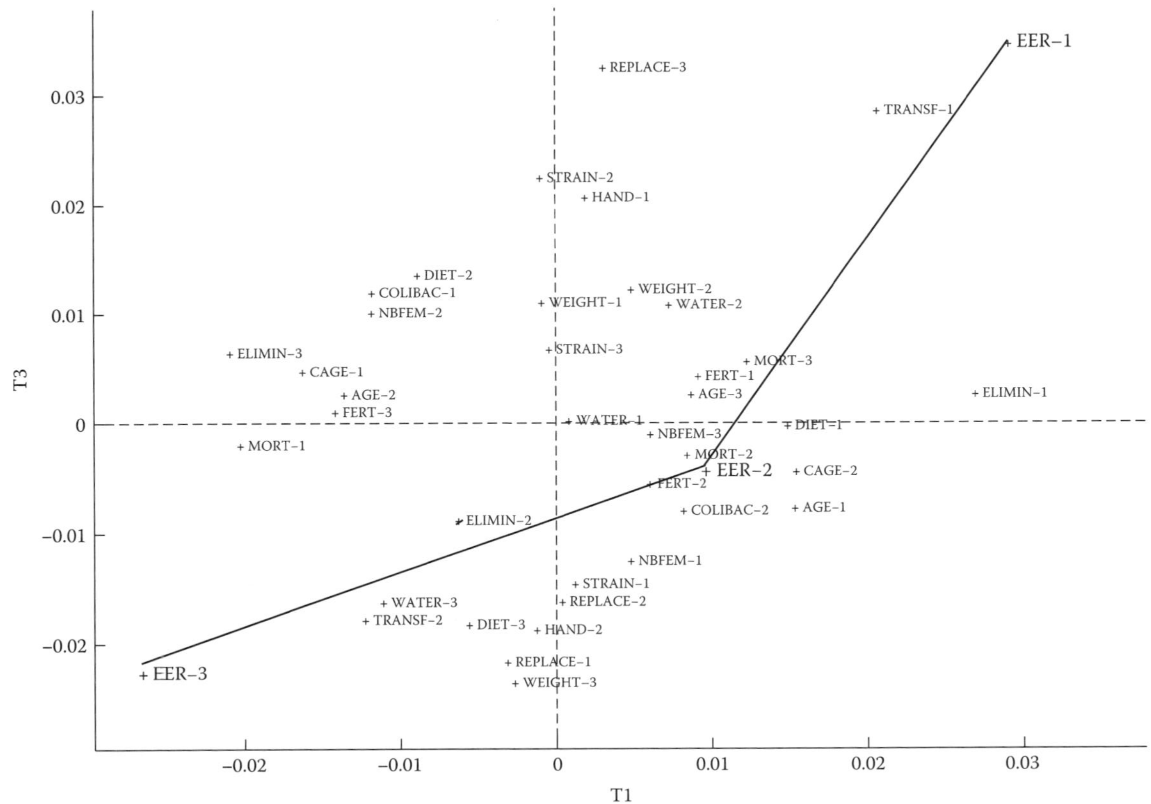

Figure 17.1 Multiblock canonical correlation analysis. Graphical display shows the association of the categories of the variables with respect to axes 1 and 3. Variables that are not judged important for interpretation are suppressed on the representation but are active for the construction (variables DIET, STRAIN, HAND, COLIBAC, NBFEM, and FERT).

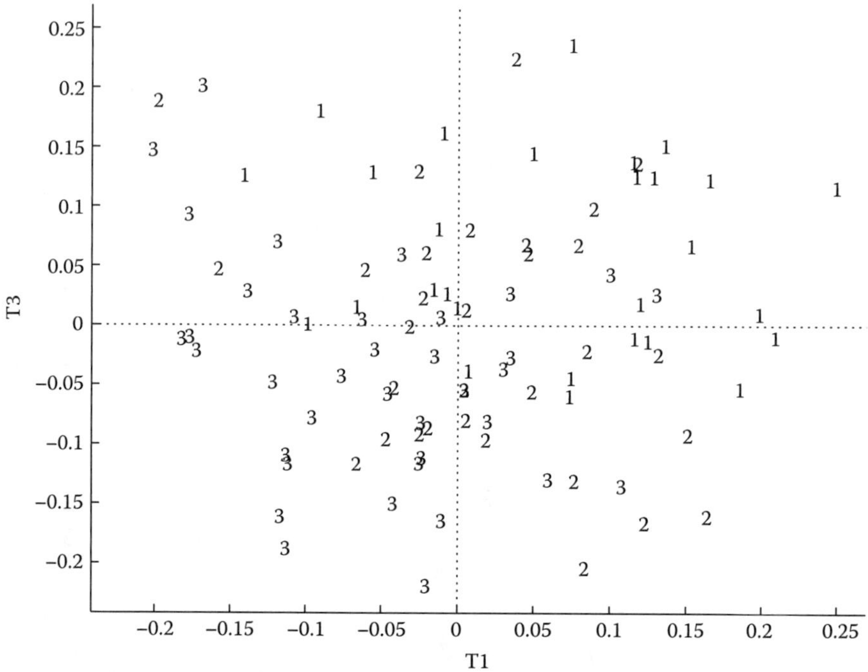

Figure 17.2 Multiblock canonical correlation analysis. Graphical display shows the individuals, coded by their EER category, with respect to axes 1 and 3.

because weak and vulnerable rabbits are kept alive, which constitutes a danger to the whole colony. Another risk factor is connected with the availability of feeders in insufficient quantities (CAGE-1), because this results in the fact that the rabbits, and particularly the weak ones, are not sufficiently fed. From this study, the protective factors that can be identified are, first, the transfer of the doe at weaning (TRANSF-1), because this results in less stress for the young rabbits, and moreover, contamination by the rabbits from other cages is curbed. Then a relatively high level of elimination (ELIMIN-1) should be undergone to keep the rabbit population under control and eradicate the weak rabbits that are likely to contract the disease. Finally, a weak annual replacement rate in farming (REPLACE-3) is important.

In a subsequent stage, we performed discriminant analysis using the overall latent variables as predictors and EER as the group variable. The outcomes in Table 17.3 are based on a cross-validation procedure and give the number of correctly classified individuals in each

Table 17.3 Percentage of correctly classified individuals obtained by cross validation.

Latent Variables	Axis 1	Axis 1 to Axis 2	Axis1 to Axis 3	Axis 1 and Axis 3	Axis 1 to Axis 4
EER = 1 (uninfected)	60.7	64.3	71.4	67.9	67.8
EER = 2 (intermediate)	19.3	25.8	48.4	51.6	48.4
EER = 3 (infected)	67.6	70.3	70.3	73.0	75.7
Average	49.2	53.4	64.1	64.1	64.9

group as a function of the number of overall latent variables introduced in the discrimination model. This percentage of correctly classified individuals is a measure of the model's fit. It turns out that the first and third latent variables are sufficient and give 64.1% of correctly classified individuals, with the infected and uninfected groups being classified best.

17.4 Discussion and perspectives

We have discussed a method for the analysis of several categorical variables with the purpose of predicting some outcome. The multiblock canonical correlation analysis is similar to performing MCA on the explanatory variables and then superimposing the variable to be predicted as a supplementary variable (Greenacre 1984). However, in the method that we propose, the variable to be predicted plays an active role, ensuring that the major principal axes will be related to this variable. We have also introduced a tuning parameter that offers a continuum approach ranging from MCA performed only on the explanatory variables to MCA performed on all variables. When α increases, the role of the variable to be predicted is increasingly taken into account and, as a consequence, the first principal axes are likely to be related to this variable. Another feature of the method is that it can be extended to the case where it is desirable to predict more than one categorical variable from other categorical variables. Regarding the tuning parameter α, it is clear that, for a particular situation, it can be fixed by means of cross-validation by seeking to maximize the number of correctly classified individuals. However, further research is needed to investigate its impact on the prediction ability. This research is currently underway in the form of a simulation study.

Further research is needed to adapt the strategy of analysis to the prediction of one or more categorical variables from a mixture of qualitative and quantitative variables. A starting point for this investigation is the canonical correspondence analysis (CCA) proposed by ter Braak (1986) and Lebreton et al. (1988). Indeed, CCA seems relevant to our purpose, as it makes it possible to explain categorical variables from quantitative variables.

CHAPTER 18

Projection-Pursuit Approach for Categorical Data

Henri Caussinus and Anne Ruiz-Gazen

CONTENTS

18.1 Introduction

Exploratory projection pursuit aims to find low-dimensional projections displaying "interesting features" in the structure of the case distribution in a cases × variables array, for example, the presence of outliers or a partition into groups. The potential features of interest are discussed more generally in basic papers dealing with this approach: see, for example, Friedman and Tukey (1974), who coined the term "projection pursuit," Huber (1985) for a synthetic presentation, and Caussinus and Ruiz-Gazen (2003) for a recent review. Biology, among other areas, provides many applications with continuous variables, and this chapter presents an example involving morphometric measurements. Microarray data also provide examples where interest focuses on finding a structure (outliers, clusters, etc.) within a large set of gene expressions. Applications are also found in the social sciences, although data sets in these fields often consist of categorical variables rather than continuous ones.

Principal component analysis (PCA) and related methods, such as correspondence analysis, produce graphical displays for users whose interest focuses primarily on preserving dispersion. Since maximum dispersion is not necessarily equivalent to maximum interest (see the previously mentioned papers), PCA may fail to disclose some of the interesting features of the data set. However, suitably generalized principal component analyses are likely to reveal several kinds of special structures in the data, thus meeting the aims of the exploratory projection-pursuit approach. A generalized PCA is to be understood as a PCA using a special metric on the case space. Proposals for such metrics can be found in Caussinus and Ruiz-Gazen (1995). In that chapter, the authors investigate the properties of their methods for continuous data. They rely basically on a mixture model

$$\int \mathcal{N}_p(x, \mathbf{W})dP(x) \tag{18.1}$$

where the "uninteresting noise" is a p-variate normal distribution $\mathcal{N}_p$, while the (nonnormal) mixing distribution P is the "structure" of interest. P is assumed to be concentrated on a subspace E of dimension m ($m < p$); the aim of the analysis is then to estimate E and project the cases onto this subspace to visualize P, or rather an empirical approximation of P. Roughly speaking, the methods we have considered look like a discriminant analysis where the classes would not be known. If $\mathbf{B}$ denotes the variance of P, then the variance of mixture (Equation 18.1) is

$$\mathbf{V} = \mathbf{B} + \mathbf{W} \tag{18.2}$$

which accounts for the decomposition of the total variance $\mathbf{V}$ into two terms, the "between" part $\mathbf{B}$ (its image belongs to E) and the "within" part $\mathbf{W}$ (variance of the uninteresting noise). Recall that a discriminant analysis is a generalized PCA with metric $\mathbf{W}^{-1}$, but the problem here is that neither P nor $\mathbf{W}$ are known (actually, the aim is to get information about P); moreover, in these conditions, $\mathbf{W}$ is fairly difficult to estimate. The techniques we have considered rely on various estimates of $\mathbf{W}$ according to different classes of models for P.

In the case of categorical variables the same methods can be formally applied to indicator matrices, but their properties are far from clear. In particular, the model expressed by Equation 18.1, with a normal distribution for the uninteresting noise, makes no more sense. The following questions then arise:

- What is an uninteresting distribution?
- How can a model be formulated to distinguish between the interesting and the uninteresting parts of the distribution?
- How can we obtain a projection displaying the interesting part?

From the viewpoint of projection pursuit, emphasis is put on displaying the cases. However, a joint display of cases and variables is advisable because it provides information about the variables that account for most of the structure revealed by the plot of the cases. The simultaneous representation of cases and variables makes sense for any kind of generalized PCA; Gabriel (2002b) gives examples with the metrics we shall use (see also Caussinus et al. 2003b). Conversely, in the case of categorical variables, users of multiple correspondence analysis (MCA) often emphasize the representation of variables. However, the simultaneous representation of cases and variables is possible and often advisable, and its interpretation is discussed in most presentations of MCA (see Le Roux and Rouanet 1998). We shall use biplots in our displays without discussing their concrete interpretation at length. Beyond the obvious scientific utility of biplots, this is indeed an opportunity to pay further homage to Gabriel's work from his pioneering paper (1971) to the most recent ones (2002a, 2002b).

18.2 Continuous variables

The data set is an $n \times p$ matrix $\mathbf{X}$ of real numbers, where n is the number of cases and p is the number of numerical variables. Let us focus on the distribution of the cases, where the aim of exploratory projection pursuit is to reveal hidden "structures of interest" within this distribution. The first step of the approach is to define "interestingness." It is assumed that:

1. Homogeneity is uninteresting; thus, the most interesting structure is the one that maximizes some criterion of nonhomogeneity,
2. Homogeneity can be measured by entropy; thus, maximum entropy corresponds to maximum homogeneity.

With continuous data, as far as the structure of the case distribution is concerned, the center and the scale are irrelevant; more generally, any (regular) affine transformation of the cases (rows of $\mathbf{X}$) does not change the structure. (For example, if the cases are divided into clusters, the clusters are preserved by such a transformation.) Hence, the mean and the variance of the distribution are irrelevant and can be chosen arbitrarily. Thus, we have

3. For given mean and variance, entropy is maximum for the normal distribution.

Finally, if there are many techniques to look for interesting projections, we have shown in previous papers that they include suitably generalized PCAs (Caussinus and Ruiz-Gazen 1995; Caussinus et al. 2003b). These techniques can be introduced as follows. Let $\mathbf{x}_i$, $i = 1,\ldots,n$, denote the transpose of the ith row of $\mathbf{X}$, that is, the vector corresponding to the ith case. Let us keep the same notation for the observations and the underlying random vectors. We assume that the random vectors $\mathbf{x}_i$ are independent with the same covariance matrix $\mathbf{W}$ and that the expectations $\mathbb{E}(\mathrm{x}_i)$ belong to an (unknown) m-dimensional subspace of $\mathbb{R}^p$ (fixed-effect model; see Caussinus 1986).

The expected values $\mathbb{E}(\mathrm{x}_i)$ can be estimated using a generalized PCA with metric $\mathbf{M}$ where, under fairly mild conditions, the best choice of $\mathbf{M}$ is $\mathbf{M} = \mathbf{W}^{-1}$ (Besse et al. 1988). In practice, the estimates of these vectors are obtained by projecting the $\mathbf{x}_i$'s onto the subspace spanned by the m eigenvectors of $\widehat{\mathbf{V}}\,\widehat{\mathbf{W}}^{-1}$ associated with the m largest eigenvalues, where $\widehat{\mathbf{V}}$ is the empirical variance of the data set and $\widehat{\mathbf{W}}$ is an estimate of $\mathbf{W}$.

Note that the fixed-effect model above is implicitly assumed when PCA is used as an exploratory technique to display the main aspects of the cases: actually, it is a model *à la* Tukey "data = structure + noise," where the means $\mathbb{E}(\mathrm{x}_i)$ are the structural part and the noise is represented by the variance $\mathbf{W}$ (Caussinus 1986). It will be replaced by the model (Equation 18.1) if, instead of being fixed, the means are random, independent, with probability distribution P—which does not basically change the point of view—and the noise has a normal distribution, since it is representative of the uninteresting part of the distribution (see points 1 to 3 above).

Let $\bar{\mathbf{x}}$ and $\widehat{\mathbf{V}}$ be the empirical mean and variance of the $\mathbf{x}_i$'s and

$$
\mathbf{S}(\beta) = \frac{\sum_{i=1}^{n} w_i(\beta)(\mathbf{x}_i - \bar{\mathbf{x}})(\mathbf{x}_i - \bar{\mathbf{x}})^{\mathsf{T}}}{\sum_{i=1}^{n} w_i(\beta)}
$$

$$
\mathbf{T}(\beta) = \frac{\sum_{i=1}^{n-1}\sum_{j=i+1}^{n} w_{ij}(\beta)(\mathbf{x}_i - \mathbf{x}_j)(\mathbf{x}_i - \mathbf{x}_j)^{\mathsf{T}}}{\sum_{i=1}^{n-1}\sum_{j=i+1}^{n} w_{ij}(\beta)} \tag{18.3}
$$

with $w_i(\beta) = \exp(-\frac{\beta}{2} \| \mathbf{x}_i - \bar{\mathbf{x}} \|^2_{\hat{V}^{-1}})$, $w_{ij}(\beta) = \exp(-\frac{\beta}{2} \| \mathbf{x}_i - \mathbf{x}_j \|^2_{\hat{V}^{-1}})$ and β a positive tuning parameter to be chosen by the user, in practice small for **S** and close to 2 for **T**. For the first (respectively second) metric, it turns out that the displays obtained with various choices of β between 0.01 and 0.10 (resp. 1.75 and 2.25) are very similar; thus, we advise taking $\beta = 0.05$ (resp. $\beta = 2$).

Properties of generalized PCAs with metrics $\widehat{\mathbf{W}}^{-1} = \mathbf{S}^{-1}(\beta)$ or $\mathbf{T}^{-1}(\beta)$ have been discussed in our above-mentioned papers. Roughly speaking, since $w_i(\beta)$ weighs down the large deviations from the mean, $\mathbf{S}(\beta)$ looks like a robust estimate of **V** (see Ruiz-Gazen 1996), which is an estimate of **W** in the presence of outlying units. Hence, PCA with $\mathbf{S}^{-1}(\beta)$ works well to display outlying cases. On the other hand, $w_{ij}(\beta)$ weighs down the large differences between units in a classical alternative to the empirical variance formula (note that $\mathbf{T}(0) = 2\hat{\mathbf{V}}$), so that $\mathbf{T}(\beta)$ is expected to approach **W** up to a multiplicative scalar in the presence of clusters (or even more general structures). Hence, PCA with $\mathbf{T}^{-1}(\beta)$ is able to reveal clusters. It is worth noting that the eigenvalues of these generalized PCAs cannot be interpreted in terms of percent of explained inertia, as in ordinary PCA. For this reason, they will not be provided in the examples below. However, the eigenvalues can be used to make inferences concerning the "good" dimensionality, that is, the displays corresponding to the largest eigenvalues that may exhibit some "structure" of interest, against the displays that are likely to show only noise and should not be interpreted (for these testing procedures, see Caussinus et al. 2002, 2003a, 2003b).

Let us briefly consider a small example to illustrate the main aspects of the proposed methods. The data (Jambu 1977) consist of six measures (variables) on 43 skulls (cases): 12 wolves and 31 dogs of various breeds. Figure 18.1 shows the two first dimensions of an ordinary (standardized) PCA. Figure 18.2 and Figure 18.3 result from a generalized PCA with metric $\mathbf{T}^{-1}(2)$ (in fact a slight variant to increase the robustness of the estimate of **V**). Figure 18.2 shows that the first axis separates the wolves from the dogs with only one exception (a St. Bernard). Actually, a histogram of the first coordinate is clearly bimodal (Caussinus et al. 2003a). It is worth noting that ordinary PCA does not provide such a clear-cut separation. Figure 18.3 suggests that (a) two or three dogs (top of the display) are somewhat different from the others according to the third axis and (b) they are characterized by small values of variables 1 and 2. Actually, these three dogs are the two boxers and the bulldog, while variables 1 and 2 are those most highly related to the length of the jaw (note that these variables distinguish more clearly different breeds of dogs than

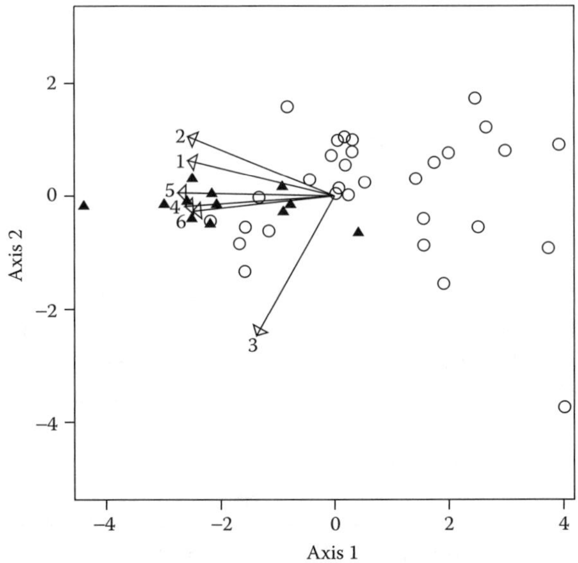

Figure 18.1 Data for dogs (circles) and wolves (triangles): standardized PCA. Axes 1 and 2: $\lambda_1 = 4.0$ (67%), $\lambda_2 = 0.9$ (15%). The variables are total length of the skull (1), upper jaw length (2), jaw width (3), upper carnassial length (4), first upper molar length (5), first upper molar width (6).

dogs from wolves). More details can be found in Gabriel (2002b) and Caussinus et al. (2003a), especially concerning the choice of the relevant dimensionality, which seems to be 3.

18.3 Categorical variables

In this section, the data matrix is a multivariate indicator matrix **Z**. There are still n rows corresponding to the n cases. The number of variables (questions) is Q, J_q denotes the number of categories for the qth variable, and the total number of columns is $J = \Sigma J_q$. Since MCA is basically a generalized PCA of the data matrix **Z**, the question arises as to whether the corresponding choice of the metric can disclose an "interesting" structure in this data set—as opposed to mere "dispersion"—or whether another choice of the metric would provide more useful displays. Of course, from an empirical point of view, the techniques of Section 18.2 can still be used heuristically,

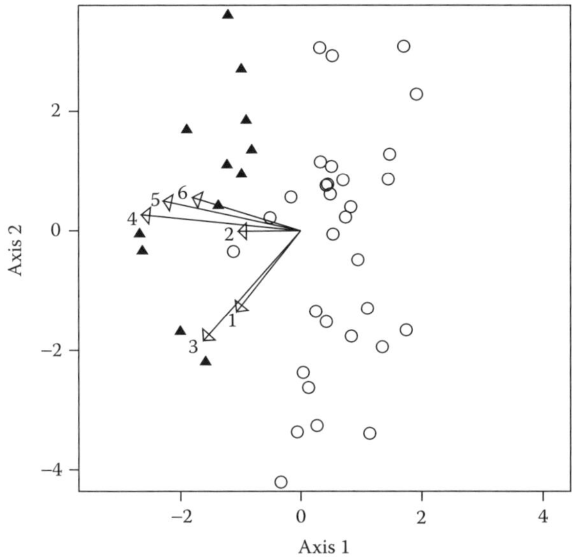

Figure 18.2 Data for dogs (circles) and wolves (triangles): biplot for the generalized (robust) PCA with metric $\mathbf{T}^{-1}(2)$. Axes 1 and 2.

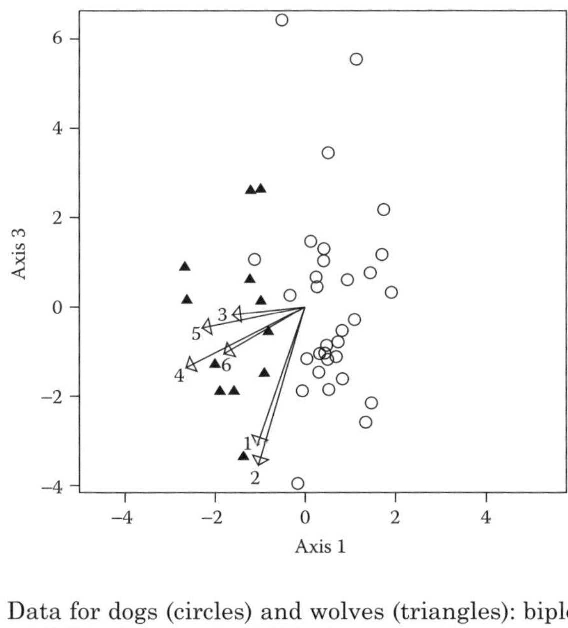

Figure 18.3 Data for dogs (circles) and wolves (triangles): biplot for the generalized (robust) PCA with metric $\mathbf{T}^{-1}(2)$. Axes 1 and 3.

but their properties are not clear at all because they rest on the normal distribution of the noise, an assumption that is no longer realistic. Moreover, the aim of an exploratory technique is to find structure as opposed to noise, so new definitions are first required for structure and noise.

18.3.1 Structure versus noise

As in Section 18.2, we first define the kind of distribution that is characteristic of the noise. Going on with the projection-pursuit paradigm, let us measure the homogeneity of a probability distribution Π by its entropy (see point 2 in Section 18.2), that is

$$\varepsilon = -\int f \log(f) d\mu$$

where f is the density of Π with respect to a measure μ. In this section, a case has a discrete distribution, so that μ is a counting measure and f is a vector in a $(J_1 \times \cdots \times J_Q)$-dimensional space, namely, the vector whose elements are the probabilities of the $J_1 \times \cdots \times J_Q$ categories of the associated Q-dimensional table. Since the margins of the distribution, that is, the probability distribution of each variable, are not related to the structural part of the data distribution, the maximum of ε is to be found under the constraint of fixed margins. For given marginal distributions, ε is a maximum when Π is the probability distribution defined by its margins and by the independence of the Q variables.

Hence, from arguments similar to those of Section 18.2, homogeneity is now characterized by independence of the Q variables in place of normality, and the mixture model (Equation 18.1) becomes

$$\int I(\pi) dP(\pi) \tag{18.4}$$

where the noise is I, a distribution assuming the independence of the variables with marginal probabilities π ($\pi \in \mathbb{R}^j$).

As in Section 18.2, it will be assumed that P is concentrated on an m-dimensional subspace to be consistent with the search of m-dimensional projections. For example, if P is discrete, concentrated on $(m + 1)$ points, Equation 18.4 turns out to be the latent class model. This is a special case where P is concentrated on an affine m-dimensional variety; if the data are centered to their column means, P is then concentrated on an m-dimensional subspace of $\mathbb{R}^j$. More generally, if P is any distribution concentrated on an affine m-dimensional variety, then Equation 18.4 is a latent-variable model where the response probabilities are J-vectors belonging to this variety. Their distribution P characterizes the structural part of the model. (Here again, the variety becomes a subspace if the data are centered, which will generally be the case.) It is worth mentioning that the model we obtain is not a "classical" latent-variable model (see, for example, Everitt 1984 for more usual models, where the probabilities π are not linear functions of the latent variables), but (a) it is consistent with the search of the structural part in a subspace of $\mathbb{R}^J$, that is, with the projection-pursuit approach and (b) it includes the usual latent class model. Now, the variance of a case can still be decomposed as in (Equation 18.2), where $\mathbf{V}$ is estimated by the empirical variance $\hat{\mathbf{V}}$ and $\mathbf{W}$ is a block-diagonal matrix due to the independence of the variables for a given π.

18.3.2 Multiple correspondence analysis

For an approach similar to the one of Section 18.2, one should use a generalized PCA of $\mathbf{Z}$ with metric equal to (an estimate of) $\mathbf{W}^-$. (It is necessary to consider a generalized inverse, since $\mathbf{W}$ is singular, but $\mathbf{V}$ and $\mathbf{W}$ generally have the same kernel, so there is no serious difficulty with this point.) Let the various covariance matrices be decomposed into blocks corresponding to the different pairs of variables; the block indexed by qq' is a $J_q \times J_{q'}$ matrix. Let $\mathbf{d}$ be the m-vector whose elements are the column sums of $\mathbf{Z}$ divided by n, that is, the relative frequencies of the J categories, and $\mathbf{D} = \text{diag}(\mathbf{d})$.

MCA can be considered as the generalized PCA of $\mathbf{Z}$ with metric $\mathbf{D}^{-1}$. Now, we have seen that $\mathbf{W}_{qq'} = 0$ for $q \neq q'$, hence $\mathbf{W}^-$ differs from $\mathbf{D}^-$ only with respect to the diagonal blocks. Roughly speaking, the use of $\mathbf{D}^-$ is equivalent to the use of the metric whose matrix is a block diagonal with diagonal blocks equal to $\mathbf{V}_{qq}^-$ instead of W_{qq}^- for $q = 1,\ldots,J$. This suggests that MCA can be fairly efficient in displaying the latent structure. In that case, MCA could be considered as a projection-pursuit technique related to the model in Equation 18.4. Let us examine this question in the light of a concrete example.

The data come from Rajadell Puiggros (1990), who studied the attitude of children toward reading. As in Bécue-Bertaut and Pagès (2001), we consider $Q = 8$ questions and $n = 804$ children (cases) who answered all these questions (see Table 18.1). The set of the J_q's is $\{3,3,3,3,2,3,3,3\}$ and $J = 23$.

The data have been processed by means of MCA. A latent class model with three classes has also been fitted. Figure 18.4 compares the two

Table 18.1 The eight questions of the reading attitudes example, with categories and marginal frequencies.

Questions	Labels	Categories	Frequencies
How much do you read at school?	1	Read little	82
	2	Read moderately	465
	3	Read a lot	257
Have you many books at home?	4	Have little	31
	5	Have moderately	229
	6	Have a lot	544
How much do you read?	7	Read little	118
	8	Read moderately	458
	9	Read a lot	228
How do you read?	10	With great difficulties	24
	11	With some difficulties	311
Do you like the books given by the teacher?	12	Easily	469
	13	Like	687
	14	Don't like	117
When do you read?	15	For work	170
	16	For pleasure	415
	17	For both	219
How do you prefer to read?	18	Low voice	588
	19	Loud voice	180
	20	Both	36
Do you like to read at school?	21	Like	133
	22	Don't like	639
	23	Like sometimes	32

approaches in the first two-dimensional subspace of MCA; the cases are labeled by the number of the most probable class conditional on the response vector, except if the largest probability is lower than .6, in which case they are labeled 4 (41 cases close to the origin, as could be expected). MCA tends to separate the classes, even if they overlap and if the cases in a given class are fairly scattered. Moreover, the biplot allows us to determine which responses are the most characteristic of the classes. Actually the negative ("low") answers point to the left (class number 3 of "nonreaders"), but the most interesting discussion might concern the characterization of class 2 versus class 1, or the fact that the three classes do not seem to be ordered along one axis.

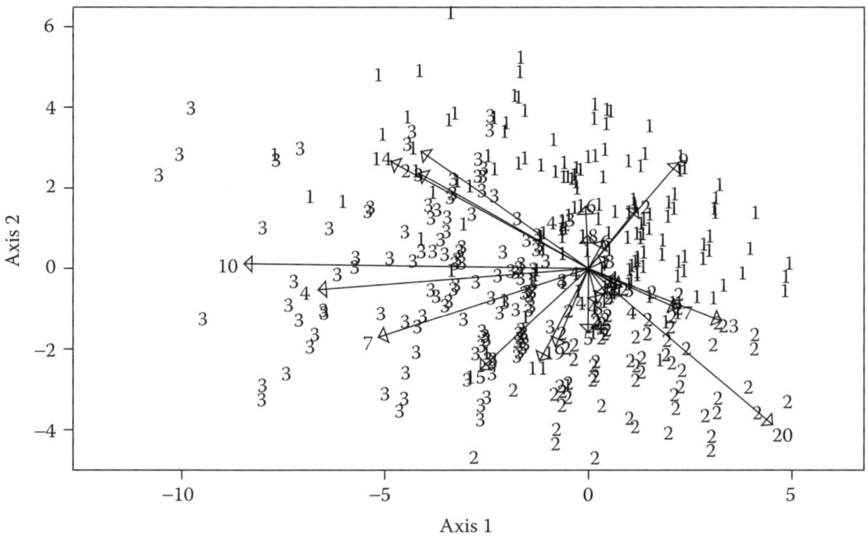

Figure 18.4 Data for children reading: MCA. Axes 1 and 2: $\lambda_1 = 0.22$ (12%), $\lambda_2 = 0.18$ (10%). Biplot of the variables (arrows for the 23 categories) and the cases (labeled with the number of the most probable class).

18.3.3 Other metrics

From the previous subsection, we can conclude that MCA works rather satisfactorily as the projection-pursuit technique for which we are looking. However, it remains that the metric $\mathbf{D}^{-1}$ is not exactly $\mathbf{W}^{-}$, which suggests that we should look for other choices of metric.

Although this approach is very heuristic, the metrics derived from Equation 18.3 for the analysis of continuous variables can still be used. This provides interesting displays with some examples, but not often enough not to be strongly advisable without further research (for example, with categorical variables, new guidelines are needed for choosing the tuning parameter β). Another possibility is to estimate $\mathbf{W}$ from a latent-variable model. The resulting analysis may provide a display of the data likely to assess or complement the model.

For example, consider the model in Equation 18.4 with P discrete, that is, a latent class model. Denote by C the number of classes, p_c ($c = 1, \ldots, C$) the probability of class c, and set π_c as the column vector of the marginal response probabilities for class c; π_c can be divided into subvectors related to each question: the J_q-vector π_{cq} is associated

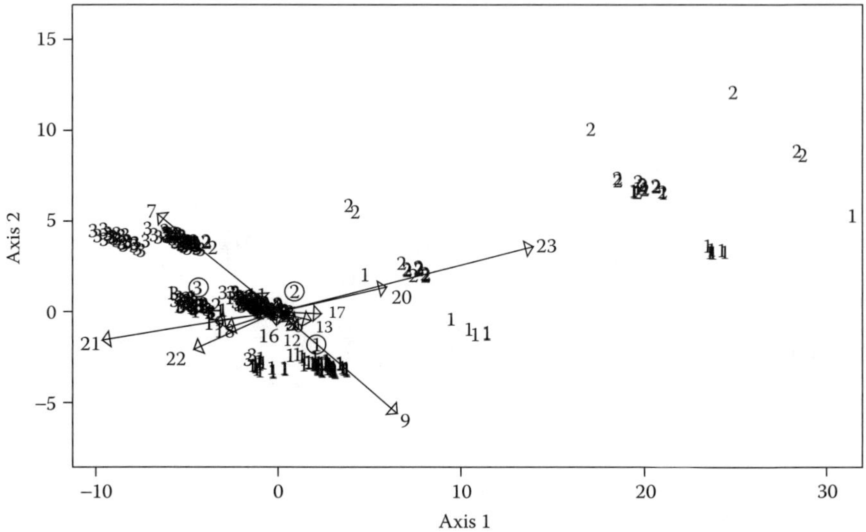

Figure 18.5 Data for children reading: PCA with metric $\widehat{\mathbf{W}}^{-}$. Axes 1 and 2. Biplot of the variables (arrows for the 23 categories) and the cases (labeled with the number of the most probable class). The circled digits correspond to the centroids of the classes.

with question q; its elements are the response probabilities to the different categories of question q for a case belonging to class c. Now, if $\mathbf{W}$ is divided into blocks associated with each pair of questions (see above), the nondiagonal blocks vanish and

$$\mathbf{W}_{qq} = \mathrm{diag}(\bar{\pi}_q) - \sum_c P_c \pi_{cq} \pi_{cq}^{\mathsf{T}}$$

with $\bar{\pi}_q = \Sigma_c p_c \pi_{cq}$.

$\widehat{\mathbf{W}}_{qq}$ is obtained by setting the parameters to their estimated values, and $\widehat{\mathbf{W}}^{-}$ is the Moore–Penrose generalized inverse of $\widehat{\mathbf{W}}$. (Actually, in practice we use a "pseudo" generalized inverse where the very small eigenvalues are set to 0 to improve stability.)

Generalized PCA of $\mathbf{Z}$ with metric $\widehat{\mathbf{W}}$ has been performed: the first two principal components provide the display of Figure 18.5, very different from the one of Figure 18.4. (We restrict ourselves to the first principal plane for the sake of simplicity; note also that this is sufficient if there are *really* only three classes.) The centroids of the classes are fairly well separated, but the classes themselves are not as well separated as might be expected. On the other hand, each class is split into "subclasses" in particular according to responses 23 and 20, the "medium" categories of the eighth and seventh questions, which are rarely observed. Now response categories 7 and 9 (respectively "low" and "high" of the third question) seem to be prevalent to distinguish the classes. However, if we consider for example the third class, it breaks up into two parts according to the answer of "low" or "medium" to this question. The new analysis depends heavily on a few variables, while more variables are involved in the characterization of the first principal plane of MCA. As a consequence, the scattering of the cases is more "homogeneous" with MCA. The new analysis stresses the discriminating power of question 3 as the price of neglecting other response categories that could be worth considering (for example, categories 1, 14, 15, and 21, which appear clearly with MCA). This can be interpreted as a less exploratory capability resulting from choosing a model prior to the analysis. The strong influence of a few categories can also make some classes overlap, for example the cluster of "3" closest to the origin hides a large number of children from cluster "1" who give the same answer of "medium" to the third question. On the other hand, it is possible that MCA gives too much importance to some rare response categories that could be included in noise rather than in substantive structure (for example, categories 4, 10, and 20). To summarize, (a) MCA can be more adequate in providing numerical scores since the data are more continuously spread out, in other words, MCA seems to "forget" more clearly the discrete nature of the data; and (b) the new analysis can provide an interesting complement to MCA rather than serving as a substitute.

18.4 Conclusion

It is well known that dimension-reducing techniques and latent-variable models are strongly related. Bartholomew and Knott (1999) point out the strong relationship between their latent-variable model and MCA and give a striking example. In fact, the relationship between latent-variable models and MCA has been recognized for a long time; the first paper that discusses this point at length is perhaps Aitkin et al. (1987). Thus, we do not claim that our approach is technically new, but the presentation in terms of the projection-pursuit paradigm can be enlightening. In particular, (a) it illustrates the relationship between modeling and graphical displays, (b) it emphasizes some questions of consistency, and (c) it is likely to provide a starting point for further research.

Let us finally give an example concerning the consistency of a particular exploratory analysis. The analysis of mixed data (numerical and categorical variables) is a challenge for the statistician (see, for example, Pagès 2004). Saporta (1990b) gives a review of the various possible approaches. Let $\mathbf{Y} = [\mathbf{X} \mid \mathbf{Z}]$, where $\mathbf{X}$ is the $n \times p$ matrix of the numerical data (centered to their mean) and $\mathbf{Z}$ is the $n \times J$ matrix of categorical data. One of the most popular approaches consists of performing a generalized PCA of $\mathbf{Y}$ with a block-diagonal metric where the $p \times p$ block related to $\mathbf{X}$ is the inverse of the diagonal matrix of the p empirical variances—equivalent to standardization—and the $J \times J$ block related to $\mathbf{Z}$ is $\mathbf{D}^{-1}$, corresponding to MCA. From the discussion above, it turns out that MCA resembles more the generalized PCA of Section 18.2 than a standardized PCA. Hence, our discussion would suggest changing the first block of the metric into $\mathbf{S}^{-1}(\beta)$ or $\mathbf{T}^{-1}(\beta)$ for the sake of consistency. Incidentally, this would provide a natural answer to the weighting of the two kinds of variables.

Acknowledgments

The authors wish to thank Mónica Bécue-Bertaut, Jérôme Pagès, and Gilbert Saporta for providing helpful material.

SECTION V

Related Methods

CHAPTER 19

Correspondence Analysis and Categorical Conjoint Measurement

Anna Torres-Lacomba

CONTENTS

19.1 Introduction

Conjoint analysis, one of the most popular methods in marketing research (see, for example, Green and Wind 1973), consists of modeling the preferences for a product as a function of the attributes constituting the product. For example, suppose that a perfume manufacturer wants to create a new perfume and that the perfume possesses attributes such as type of fragrance, intensity of fragrance, and size of bottle. Each attribute has certain levels: for example, four fragrances are to be tested (floral, oriental, citric, and leather) at two levels of intensity (high and low) in three different sizes of bottle (small

30 ml, medium 50 ml, and large 100 ml). In this simple example, there are $4 \times 2 \times 3 = 24$ combinations of perfume possible, for example, a citric perfume at low intensity in a small bottle.

Presented with this potential product (and subsequently with all the other combinations as well), a potential consumer has to express her preference for the product. There are several ways to express preference, or intention to buy: this could be a rating, say from 1 to 10 for each product, or a rank ordering of the different perfumes, or different products could be compared pairwise (e.g., a citric fragrance of low intensity in small bottle compared with a floral fragrance of low intensity in medium-sized bottle) and the preferred perfume of a pair selected. The process of evaluating preference for a product made up of a combination of attributes is called conjoint measurement. Rather than asking consumers directly how important the attributes are to them, it is believed that more valid measures of attribute importance are obtained through the indirect process of conjoint measurement, where different combinations of products are compared by the consumer, thus forcing trade-offs between the attributes. The analytical step of determining the attribute utilities, that is, the effect on preference of the levels of each attribute, is achieved by various types of regression analysis adapted to the case of categorical predictor variables (the attributes) and the response variable (preference), depending on the measurement level of the response (see, for example, Green and Srinivasan 1990; Green and Wind 1973; Rao 1977).

In this chapter we consider a specific variant of conjoint measurement called categorical conjoint measurement (CCM), where preference is measured on an ordinal discrete scale. In an often cited but unpublished paper, Carroll (1969) considered CCM and proposed an analytical procedure based on canonical correlation analysis of the attributes, with the responses all coded as dummy variables (see also Lattin et al. 2003; Rao 1977). This chapter clarifies the relationship between the analysis of CCM data and correspondence analysis (CA). Since CA can be defined as a canonical correlation analysis of dummy variables, it will be shown how CA can be used to achieve exactly Carroll's procedure for CCM. We shall also show how CCM can be extended to include interactions between attributes, how this would be achieved using canonical correlation analysis, and how the data can be coded so that CA obtains equivalent results. The advantage of the CA approach is the visualization of the attribute levels, either as "main effects" or "interaction effects," depending on the coding, along with the preference levels in a joint map. It is also interesting to display different preference structures coming from different market segments. In this chapter, most technical demonstrations have been omitted, but they can be found in Torres-Lacomba (2001).

19.2 Categorical conjoint measurement

We shall take the perfume context as our example and suppose that perfumes defined by all 24 combinations of the three attributes described in Section 19.1 are presented to a consumer, who evaluates each product on the following ordinal scale:

A: very high worth
B: just high worth
C: just low worth
D: very low worth

The data for a single consumer are shown in Table 19.1 in different forms. Table 19.1(a) contains the original response patterns, the last three columns indicating the attribute levels in what is known as a full-factorial design (that is, all combinations), and the first column indicating the preference response category. In the jargon of conjoint analysis, using a full-factorial design is called the *full profile* method of data collection. Other designs are possible when the number of combinations becomes too large, called fractional factorial designs or the *restricted profile* method. The methodology presented here applies equally to restricted designs.

Table 19.1(b) is the same data expressed in the form of an indicator matrix of dummy variables, with as many dummies as there are attribute and preference levels: 4 + 4 + 2 + 3 = 13. We use the following notation: K = number of levels of preference (indexed by k), Q = number of attributes (indexed by q), with number of levels $J_1,\ldots,J_Q$ (indexed respectively by $j_1,\ldots,j_Q$). In our example $K = 4$, $Q = 3$, $J_1 = 4$, $J_2 = 2$, $J_3 = 3$. To denote the indicator matrices that compose Table 19.1(b), we shall use the notation $\mathbf{Z}_1,\ldots,\mathbf{Z}_Q$ to denote the matrices for the Q attributes, and $\mathbf{Z}_0$ to denote the indicator matrix for the preference response. Thus, Table 19.1(b) is the concatenated matrix $[\mathbf{Z}_0\ \mathbf{Z}_1\ \mathbf{Z}_2\ \mathbf{Z}_3]$.

Carroll (1969) proposed the use of canonical correlation analysis on the matrices $\mathbf{Z}_0$ and $[\mathbf{Z}_1\ldots\mathbf{Z}_Q]$ to estimate the attribute utilities. This can be achieved, albeit inefficiently, using a regular canonical correlation algorithm where, to avoid singularity of covariance matrices, one dummy variable from each variable is dropped and serves as

Table 19.1 Data for categorical conjoint analysis on 24 perfumes for a single case, showing (a) the original response data, where the first variable is the categorical preference response (P) and the other three variables are the attributes fragrance (F), intensity (I), and size (S) in full profile form; (b) the indicator matrix of all 13 dummy variables; (c) stacked cross-tabulations of the categories of the three attributes with the preferences.

(a)				(b)													(c)				
				P[a]				F[b]				I		S				P			
P	F	I	S	A	B	C	D	1	2	3	4	1	2	1	2	3		A	B	C	D
1	1	1	1	1	0	0	0	1	0	0	0	1	0	1	0	0	F1 Flo	2	2	2	0
1	1	1	2	1	0	0	0	1	0	0	0	1	0	0	1	0	F2 Ori	1	2	2	1
2	1	1	3	0	1	0	0	1	0	0	0	1	0	0	0	1	F3 Cit	0	1	3	2
2	1	2	1	0	1	0	0	1	0	0	0	0	1	1	0	0	F4 Lea	0	0	1	5
3	1	2	2	0	0	1	0	1	0	0	0	0	1	0	1	0	I1 Lo	1	3	3	5
3	1	2	3	0	0	1	0	1	0	0	0	0	1	0	0	1	I2 Hi	2	2	5	3
3	2	1	1	0	0	1	0	0	1	0	0	1	0	1	0	0	S1 Small	2	2	3	1
3	2	1	2	0	0	1	0	0	1	0	0	1	0	0	1	0	S2 Medium	1	1	3	3
4	2	1	3	0	0	0	1	0	1	0	0	1	0	0	0	1	S3 Large	0	2	2	4
1	2	2	1	1	0	0	0	0	1	0	0	0	1	1	0	0					
2	2	2	2	0	1	0	0	0	1	0	0	0	1	0	1	0					
2	2	2	3	0	1	0	0	0	1	0	0	0	1	0	0	1					
2	3	1	1	0	1	0	0	0	0	1	0	1	0	1	0	0					
3	3	1	2	0	0	1	0	0	0	1	0	1	0	0	1	0					
3	3	1	3	0	0	1	0	0	0	1	0	1	0	0	0	1					
3	3	2	1	0	0	1	0	0	0	1	0	0	1	1	0	0					
4	3	2	2	0	0	0	1	0	0	1	0	0	1	0	1	0					
4	3	2	3	0	0	0	1	0	0	1	0	0	1	0	0	1					
3	4	1	1	0	0	1	0	0	0	0	1	1	0	1	0	0					
4	4	1	2	0	0	0	1	0	0	0	1	1	0	0	1	0					
4	4	1	3	0	0	0	1	0	0	0	1	1	0	0	0	1					
4	4	2	1	0	0	0	1	0	0	0	1	0	1	1	0	0					
4	4	2	2	0	0	0	1	0	0	0	1	0	1	0	1	0					
4	4	2	3	0	0	0	1	0	0	0	1	0	1	0	0	1					

[a] Preference (P) at four levels: A = very high worth, B = just high worth, C = just low worth, D = very low worth.

[b] Fragrance (F) at four levels: Flo = floral, Ori = oriental, Cit = citric, Lea = leather; intensity (I) at two levels: Lo = low, Hi = high; and size (S) at three levels: Small = 30 ml, Medium = 50 ml, Large = 100 ml).

the reference category. Carroll noticed that the particular form of the data led to a great simplification of the canonical correlation calculations, which could be performed by initially calculating the eigensolution of the $K \times K$ matrix

$$\mathbf{R} = (1/Q)\sum_{q=1}^{Q}\mathbf{S}_q^{\mathsf{T}}\mathbf{S}_q \tag{19.1}$$

where $\mathbf{S}_q$ is the $K \times J_q$ matrix of standardized residuals (see Chapter 1 of this volume) obtained after cross-tabulating the preference variable with the qth attribute and centering and normalizing the table with respect to the expected frequencies under the independence model. The cross-tables are shown in Table 19.1(c) stacked one atop another.

We now show how the equivalent solution can be obtained using a regular CA algorithm.

19.3 Correspondence analysis and canonical correlation analysis

Simple correspondence analysis (CA) of a two-way cross-table is equivalent to the canonical correlation analysis of the two groups of dummy variables (see, for example, Greenacre 1984: section 4.4). It can be shown that this equivalence extends in a straightforward manner to the situation where the two groups consist of multiple sets of dummy variables, in particular the present case, where there are multiple sets of dummy variables for the attributes.

Suppose that we have two sets of categorical variables, the first with Q variables, with respective number of categories per variable of $J_1, \ldots, J_Q$, and a second set with S variables, with respective number of categories per variable of $K_1, \ldots, K_S$. We code all of the data into indicator matrices of dummy variables, $J = J_1 + \cdots + J_Q$ dummy variables for the first set and $K = K_1 + \cdots + K_S$ dummy variables for the second set. Then we apply canonical correlation analysis to the two sets of variables, dropping the last dummy variable for each categorical variable to obtain the solution. Suppose the canonical coefficients obtained are

$$\tilde{x}_1, \ldots, \tilde{x}_{J_1-1}; \ldots; \tilde{x}_1, \ldots, \tilde{x}_{J_Q-1} \quad \text{for the first set}$$

$$\tilde{y}_1, \ldots, \tilde{y}_{K_1-1}; \ldots; \tilde{y}_1, \ldots, \tilde{y}_{K_S-1} \quad \text{for the second set}$$

Table 19.2 Solution when applying canonical correlation analysis to the dummy variables (column CC, see Section 19.3); also showing equivalence between this solution and the one obtained by CA of the stacked tables.

	CC	CC–c	CA
A	2.825	1.547	1.580
B	1.917	0.639	0.653
C	1.576	0.298	0.305
D	—	−1.278	−1.306
Flo	2.260	1.016	1.797
Ori	1.683	0.438	0.755
Cit	1.036	−0.209	−0.369
Lea	—	−1.245	−2.203
Hi	0.415	−0.208	−0.367
Lo	—	0.208	0.367
Small	1.108	0.612	1.082
Medium	0.381	−0.115	−0.204
Large	—	−0.496	−0.878

Note: Column CC−c has centered coefficients and column CA has the standard coordinates on the first dimension. The preference coordinates differ from the centered coefficients by the scale factor $\sqrt{24/23} = 1.022$, while the attribute coordinates differ by the scale factor $\sqrt{24 \times 3/23} = 1.769$; see Section 19.4.

Using the canonical correlation program in STATA (Statcorp 2003), the canonical coefficients are given in the first column, headed "CC," of Table 19.2. (In Section 19.4 we shall explain how to pass from these results to those from CA.) We then reparameterize the solution by calculating the last coefficient for each variable and recentering the other coefficients as follows:

$$x_{J_q} = -(r_1\tilde{x}_1 + \cdots + r_{J_q-1}\tilde{x}_{J_q-1})/r_{J_q}\,;\; x_1 = \tilde{x}_1 + x_{J_q}\;;\;\cdots\;;\; x_{J_q-1} = \tilde{x}_{J_q-1} + x_{J_q}$$

where $q = 1, \ldots, Q$

$$y_{K_s} = -(c_1\tilde{y}_1 + \cdots + c_{K_s-1}\tilde{y}_{K_s-1})/c_{K_s}\,;\; y_1 = \tilde{y}_1 + y_{K_s}\;;\;\cdots\;;\; y_{K_s-1} = \tilde{y}_{K_s-1} + y_{K_s}$$

where $s = 1, \ldots, S$

Then it can be shown that—up to simple scaling factors—the solution is identical to the standard coordinates on the first principal axis of the CA of the supermatrix of all two-way contingency tables crossing the Q variables as rows and the S variables as columns:

$$\mathbf{N} = \begin{bmatrix} \mathbf{N}_{11} & \mathbf{N}_{12} & \cdots & \mathbf{N}_{1S} \\ \mathbf{N}_{21} & \mathbf{N}_{22} & \cdots & \mathbf{N}_{2S} \\ \vdots & \vdots & \ddots & \vdots \\ \mathbf{N}_{Q1} & \mathbf{N}_{Q2} & \cdots & \mathbf{N}_{QS} \end{bmatrix}$$

This supermatrix, called a concatenated or stacked table, has the particular property that the row totals of each subtable across a row of subtables are the same; similarly, the column totals of each subtable in a column of subtables are the same. With this property, the total inertia in the CA of **N** is equal to the average of the inertias in the separate CAs of each subtable (Greenacre 1994).

19.4 Correspondence analysis and categorical conjoint analysis

Given the relationship already described in Equation (19.1) to estimate a simple additive relationship between the response and the attributes using Carroll's canonical correlation proposal, the solution can be obtained from the first dimension of simple CA applied to the stacked tables in Table 19.1(c). The coordinates of the attributes on the first principal axis estimate the utilities of the attribute levels, while the coordinates of the preference levels give an optimal scaling of preference. The second dimension of the CA summarizes the second-most-important relationship, independent of the first one, and so on. CA can visualize these results, in this case the estimated utilities and optimal positions of the preference levels, in the form of a two-dimensional map with a well-known distance or scalar product (biplot) interpretation, depending on the choice of coordinate scaling. Figure 19.1 shows the symmetric CA map, accounting for 91% of the inertia (or weighted variance) in the table. The first dimension (81% of inertia) recovers the order of the nominal preference categories, from very low worth (D) on the left to very high worth (A) on the right, which partially supports the validity of the data. The two intermediate categories of worth are projected at more or less the same position on the horizontal axis, showing that, for the respondent in this example, the scale is

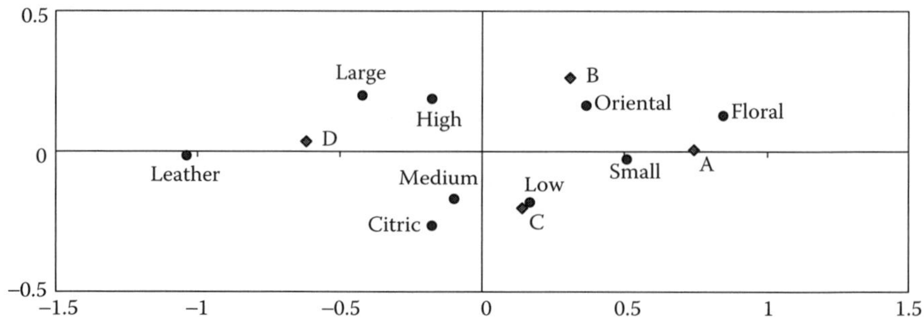

Figure 19.1 Symmetric CA map of Table 19.1(c), both rows and columns in principal coordinates (total inertia = 0.275, percentages of inertia recovered on the first two dimensions are 81% and 10%, respectively).

effectively three-point. Floral fragrance is preferred, small bottle is also toward the right, and there is a slight overall preference for low-intensity perfumes.

To illustrate how to pass from the canonical correlation results directly to the CA results, as described in Section 19.3, suppose that we denote all canonical correlation coefficients (after dropping the last category for each variable) by $\tilde{p}_1, \tilde{p}_2, \tilde{p}_3, \tilde{f}_1, \tilde{f}_2, \tilde{f}_3, \tilde{i}_1, \tilde{s}_1$, and $\tilde{s}_2$ (column CC of Table 19.2). Then, to satisfy the CA centering conditions, the weighted averages of coefficients are zero, with weights proportional to the marginal frequencies. In the case of the attributes, given the full factorial design, the weights are equal: for example, variable S (size) has weights equal to 1/3, 1/3, and 1/3. To calculate the three centered coefficients s_1, s_2, s_3, calculate the omitted category coefficient as $s_3 = -(\tilde{s}_1 + \tilde{s}_2)/3$, and then $s_1 = \tilde{s}_1 + s_3$ and $s_2 = \tilde{s}_2 + s_3$. It is readily verified that $(s_1 + s_2 + s_3)/3 = 0$, which is the CA centering condition. The case of the preference response is an example of the general case with differential weights, 3/24, 5/24, 8/24, and 8/24, respectively, for the four categories. To obtain the centered CA coefficients from the canonical coefficients $\tilde{p}_1$, $\tilde{p}_2$, and $\tilde{p}_3$, calculate the omitted category as $p_4 = -(3\tilde{p}_1 + 5\tilde{p}_2 + 8\tilde{p}_3)/8$ and recenter the remaining categories as: $p_j = \tilde{p}_j + p_4$ ($j = 1, 2, 3$). Again, the CA (weighted) centering condition can be verified: $(3p_1 + 5p_2 + 8p_3 + 8p_4)/24 = 0$. The recentered coefficients are given in column CC–c of Table 19.2. The coefficients recovered by recentering as just described are the standard coordinates in CA (column CA of Table 19.2), in this case differing by simple scaling factors of $\sqrt{n/(n-1)} = \sqrt{24/23}$ for the preference categories and $\sqrt{Qn/(n-1)} = \sqrt{3 \times 24/23}$ for the attribute levels (see Torres-Lacomba 2001):

for example, in Table 19.2 for preference category A, 1.580 = 1.547 × 1.022, and for fragrance floral, 1.797 = 1.016 × 1.769. To obtain the principal coordinates used in the maps, multiply the standard coordinates by the corresponding canonical correlation coefficient, equal to 0.471, identical to the square root of the first principal inertia of the CA.

19.5 Incorporating interactions

Green (1973) recognized the importance of interaction effects between attributes (see also Green and Wind 1973), and the literature (see Vriens 1995) includes interaction effects in their models. Carmone and Green (1981) and Vriens (1995) differentiate between crossover and noncrossover interaction effects. A crossover interaction is observed when the preference ordering for levels of an attribute, say fragrance, changes in the presence of another attribute, say intensity; for example, it might be that floral is preferred to citric at low intensity, but at high intensity citric is preferred to floral. A noncrossover interaction effect is observed when the preference ordering does not change but the strength of preference does: for example, at low intensity a citric fragrance is preferred to floral, but at high intensity citric is more preferred. In the following discussion, we are interested in introducing and diagnosing both types of interaction effects into CCM and showing how the solution can be achieved using CA.

Interactions can be incorporated into the analysis in a simple and natural way. Suppose we wish to incorporate the interaction between fragrance and intensity. The cross-tabulations in Table 19.1(c) corresponding to these two attributes are replaced by a single cross-tabulation in Table 19.3 between the response and all combinations of these two attributes, of which there are 4 × 2 = 8 (this is called interactive coding, see Chapter 1 of this volume). The two "marginal" cross-tables for fragrance and intensity can still be included in the CA, but only as so-called supplementary (or passive) points to see the positions of the individual attributes as "main effects" among the points representing the attribute combinations. Figure 19.2 shows the symmetric CA map of Table 19.3, accounting for 90% of the inertia. Note the difference in the total inertias between Figure 19.1 and Figure 19.2, increasing from 0.275 to 0.769. The inertia is greatly increased when the interaction effect is introduced, and the difference can be seen in the maps.

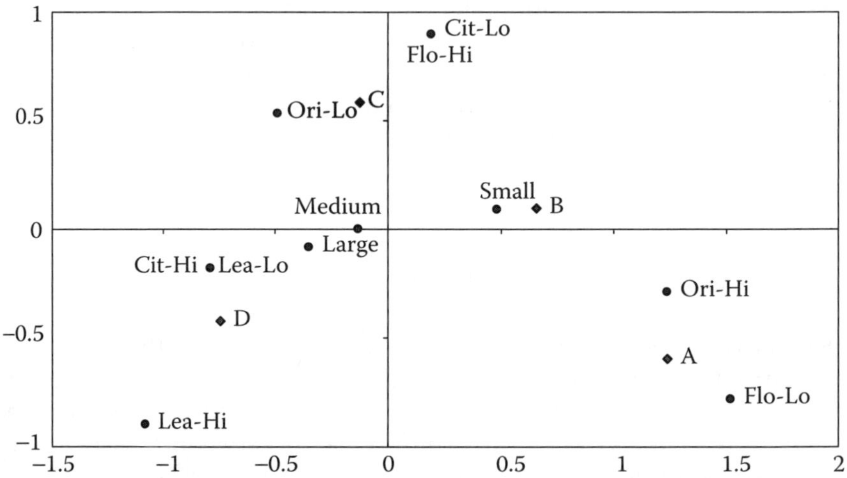

Figure 19.2 Symmetric CA map of Table 19.3, including interactively coded variable composed of fragrance and size, both rows and columns in principal coordinates (total inertia = 0.769, percentages of inertia recovered on the first two dimensions are 61% and 29%, respectively).

Table 19.3 Table for categorical conjoint analysis, with attribute interactions: attributes F (fragrance) and I (intensity) are coded interactively and cross-tabulated with preference, stacked on the table crossing the third attribute S (size) with preference.

		A	B	C	D
F1–I1	Flo–Lo	2	1	0	0
F1–I2	Flo–Hi	0	1	2	0
F2–I1	Ori–Lo	0	0	2	1
F2–I2	Ori–Hi	1	2	0	0
F3–I1	Cit–Lo	0	1	2	0
F3–I2	Cit–Hi	0	0	1	2
F4–I1	Lea–Lo	0	0	1	2
F4–I2	Lea–Hi	0	0	0	3
S1	Small	2	2	3	1
S2	Medium	1	1	3	3
S3	Large	0	2	2	4

From Figure 19.1, it appears that fragrance is more important than intensity; thus, the expected preferences based on main effects, in the absence of interaction, would be

Flo-Lo > Flo-Hi > Ori-Lo > Ori-Hi > Cit-Lo > Cit-Hi > Lea-Lo > Lea-Hi

On the first axis of Figure 19.2, however, we obtain the following ordering:

Flo-Lo > Ori-Hi > Flo-Hi & Cit-Lo > Ori-Lo > Lea-Lo & Cit-Hi > Lea-Hi

indicating an interaction between fragrance and strength. For example, the "oriental-high" combination moves up the preference scale and crosses over the "oriental-low" position. This is called a *crossover interaction*, as opposed to an interaction where the order of the levels is maintained but there is differential strengthening and weakening of the utilities across the interactive levels.

Finally, the way CA handles interactions enlightens us as to how we would analyze interactions using canonical correlation analysis software. There are two possibilities. The first is to drop one category from each variable, as before, and then define the interaction as the product of all pairs of dummy variables retained for the pair of attributes. The second is to first form a new set of dummy variables by coding all pairs of attribute categories, as we have effectively done in our CA implementation, and then dropping just one of these, treating the interaction variable as a single variable. The manner in which we can recover exactly the results of a CA with interactions using canonical correlation software, in either of the two forms described above, is fully explained and illustrated by Torres-Lacomba (2001).

19.6 Discussion and conclusions

Canonical correlation analysis is the "mother" of all classical multivariate statistical techniques, having regression analysis, principal component analysis, discriminant analysis, and correspondence analysis as special cases. Consequently, it is expected that we can recover the results of categorical conjoint analysis using CA, since categorical conjoint analysis is essentially a regression analysis with the categorical preference variable set as a multivariable response. However, we feel that it is important—from both a pedagogical and a technical point of view—to show and realize these equivalences in an effort to enrich our understanding of the methods. Such knowledge will also help in avoiding duplication and rediscovery of the methods in new and apparently different guises.

We have illustrated the equivalence using data from a single respondent. For a sample of respondents, we would simply stack each respondent's cross-tabulation, as shown in Table 19.1(c) or Table 19.3, one atop another, and then analyze the stacked table by CA. Each attribute level would thus be represented by as many points as respondents. The average attribute level, as well as its dispersion, can be summarized either by univariate intervals along principal axes or by bivariate regions (for example, confidence ellipses) in the two-dimensional map.

Acknowledgments

I am grateful to Michael Greenacre, James Nelson, and Tammo Bijmolt for their useful comments. Financial support is acknowledged from the Spanish Ministry of Education grant SEC2001-1169 as well as from the Universidad Carlos III de Madrid and Universitat Pompeu Fabra in Barcelona.

CHAPTER 20

A Three-Step Approach to Assessing the Behavior of Survey Items in Cross-National Research

Jörg Blasius and Victor Thiessen

CONTENTS

20.1 Introduction

Cross-national research is increasingly prevalent in the social sciences. Several survey programs have been designed expressly for cross-national comparisons, and the most renowned of these are the World Values Surveys and the International Social Survey Program (ISSP). Much is already known and many advances have been made to improve the standards for conducting cross-national research. This is particularly evident in two respects: (a) requirements for translating survey items and their response categories between languages so as to make them formally equivalent, and (b) the design of surveys and data collection methods to increase the quality and comparability of the data captured (Harkness et al. 2003; Hoffmeyer-Zlotnik and Wolf 2004).

The methodological requirements for valid cross-national comparisons are particularly stringent. Even if survey items are formally equivalent and the survey design and its implementation are conducted to an acceptable standard, cross-national comparison may not be valid or warranted. Respondents in different countries may not understand or interpret the questions in the same way due to different historical, political, and cultural specificities. Hence, even under ideal circumstances it is necessary to determine whether a set of items on any issue has comparable meanings in the countries that are to be compared.

Cross-national comparisons require that the structures of responses to the items in the participating countries be sufficiently equivalent to permit substantive interpretation. Our analysis is an example of a "structure-oriented" examination of cross-national comparisons, since it squarely "addresses the question as to whether an instrument measures the same construct across cultures" (Van de Vijver 2003: 143).

In this chapter we empirically assess whether all countries participating in the ISSP were able to capture data exhibiting equivalent structures and of acceptable quality to warrant cross-national comparisons. We employ a three-step approach to assessing the comparability of survey items in cross-national research. In the first step, classical principal component analysis (PCA) is used, which makes rather stringent assumptions about the distributions and the level of measurement of the survey items. In the second step, the results of PCA are compared with those obtained using categorical principal component analysis (CatPCA), also known as nonlinear principal component analysis (NLPCA). Divergences in the results of these two types of analyses indicate the existence of measurement problems. These are then explored more systematically using the biplot methodology in the third step.

Applying CatPCA in combination with the biplot methodology, we show that one can detect whether respondents from different countries have a similar understanding of the survey items and whether their answers are substantively consistent, rather than manifesting various types of methodologically induced variation. The procedure illustrated in this chapter is proposed as a prior screening device for selecting countries and items for which cross-national comparisons make sense. This methodology helps to identify both differences in the underlying structure of the survey items and violations of metric properties in the individual items. The method can also be used to screen any set of ordered categorical items, for example, trend data or subsets of groups within a certain country.

20.2 Data

20.2.1 The International Social Survey Program

To illustrate our approach, we use data from the ISSP, a program that is reputed to maintain high standards (see Scheuch 2000). The ISSP is an ongoing annual program of international collaboration on surveys covering important social science topics. It brings together preexisting social science projects and coordinates research goals. The self-administered questionnaires are originally drafted in British English and then translated into other languages. The questions are designed to be relevant in all countries and expressed in an equivalent manner in all languages (for more details, including the availability of the data, sampling procedures, and the questionnaires, see www.issp.org). We use the 1994 data, where the focus was on "Family and Changing Gender Roles" (ISSP 1994). In that year, 22 countries participated: Australia, Germany, Great Britain, U.S., Austria, Hungary, Italy, Ireland, the Netherlands, Norway, Sweden, Czech Republic, Slovenia, Poland, Bulgaria, Russia, New Zealand, Canada, the Philippines, Israel, Japan, and Spain. Because both Germany and Great Britain are expected to be quite heterogeneous with respect to views on family and gender roles, Germany is divided into East and West, and Northern Ireland is separated from Great Britain. The sample sizes vary between 647 (Northern Ireland) and 2494 (Spain).

20.2.2 Support for single- and dual-earner family forms

To illustrate our approach without undue complications arising from certain methodological shortcomings, we searched for a data set that adequately met the following criteria:

- The identical questions (or more accurately, the equivalent questions) with identical response categories were asked in a number of countries. This is a precondition for any survey-based comparative analysis.
- The reputed quality of the data is high. This should minimize the proportion of between-country variance that is due to methodological artifacts, such as unacceptably low response rates.
- The issues being addressed by the questions were salient in the participating countries. This makes it reasonable to assume that most of the respondents in all of the countries held opinions on the issues being queried.

- Multiple potential indicators for a common construct were available. This is a formal necessity for implementing the procedures we propose. Additionally, as Smith (2003) notes, without multiple indicators, substantive cross-national differences are hopelessly confounded with possible methodological artifacts.
- The available response alternatives to the items are expected to constitute an ordered categorical level of measurement. This is a formal requirement of the procedures we propose for assessing both the quality and the comparability of cross-national surveys.
- The wording of the items adheres to the basic principles of sound item construction, i.e., items of short, simple wording, with an avoidance of negatives and double-barreled construction. This is, of course, a desideratum for all survey research, but especially so when the focus is on assessing whether the structures of the responses are comparable. Nonequivalent structures can arise for both methodological (uneven quality) and substantive (societal contextual factors that are not common to all countries being compared) reasons. Blasius and Thiessen (2001b) demonstrated that less-educated respondents seemed to be particularly vulnerable to complex item construction and to negatively worded items, resulting in nonequivalent response structures.

We selected a set of 11 statements endorsing either single- or dual-earner family structures. The issue of whether one or both partners should be in the labor force is arguably a salient one in all participating countries. The wording of the items (see Table 20.1 for the English language wording) is relatively simple, avoiding negatives and double-barreled construction (with two possible exceptions), and the items have an average length of fewer than 15 words. The level of measurement is ordered categorical, with response categories "strongly agree," "agree somewhat," "neither agree nor disagree," "disagree somewhat," and "strongly disagree." Agreement with some of the statements implies a preference for a single-earner family structure, while agreement with others endorses a dual-earner household structure. For example, respondents who support a dual-earner household structure should agree with statements such as, "A working mother can establish just as warm and secure a relationship with her children as a mother who does not work" (item A), whereas those who favor a single-earner family structure should agree with statements such as, "A pre-school child is likely to suffer if his or her mother works" (item B).

Table 20.1 Items measuring support for single/dual-earner family structure.

A	A working mother can establish just as warm and secure a relationship with her children as a mother who does not work.
B	A preschool child is likely to suffer if his or her mother works.
C	All in all, family life suffers when the woman has a full-time job.
D	A job is all right, but what most women really want is a home and children.
E	Being a housewife is just as fulfilling as working for pay.
F	Having a job is the best way for a woman to be an independent person.
G	Most women have to work these days to support their families.
H	Both the man and woman should contribute to the household income.
I	A man's job is to earn money; a woman's job is to look after the home and family.
J	It is not good if the man stays at home and cares for the children and the woman goes out to work.
K	Family life often suffers because men concentrate too much on their work.

We judged three items (G, J, and K) to be ambiguous and not *a priori* classifiable as favoring either a single-or a dual-earner household. Responses to "Most women have to work these days to support their families" (item G) are potentially double-barreled, since this item may contain both an attitudinal and a factual component. For example, respondents might agree with this statement because of their poor economic situation, even though they are opposed to a dual-earner family structure. A somewhat similar problem characterizes item K ("Family life often suffers because men concentrate too much on their work"), because it is only tangentially related to support for a single- or dual-earner household; independent of their attitudes, respondents might agree or disagree with the statement. Finally, item J ("It is not good if the man stays at home and cares for the children, and the woman goes out to work") is ambiguous in that it is formulated as a reversal of traditional family roles.

Nevertheless, five of the 11 items are constructed so as to endorse a single-earner family structure (items B, C, D, E, and I). For the sake of brevity, these will be referred to simply as single-earner items. Additionally, three statements endorse a dual-earner family structure (items A, F, and H) and will correspondingly be referred to as dual-earner items.

All the variables have the same Likert-type response categories; the set of variables is constructed to be identical for all countries; the statements have a relatively simple formulation; and the issues being addressed in these statements should be relevant in all countries. Further, the surveys are reputed to have been conducted to a high standard in all countries (for example, the field research institutes are competent, the research groups are experienced), and response variation should be caused primarily by substantive differences between the countries. Hence, these items should be appropriate for cross-national comparisons. However, we will demonstrate that certain countries should be excluded from the analysis of the given data set, that some items have to be excluded from other countries, and that the remaining comparisons must take into account different response structures. Overall, our analyses reveal that the quality of data for the countries ranges from high to almost noninterpretable.

20.3 Method

PCA is a popular technique for describing and summarizing the underlying relationships among multiple variables. PCA is not appropriate for categorical data although ordered categorical variables are often treated (incorrectly) as though they were numerical. In such cases CatPCA, in which the category values (here 1 to 5) are replaced by optimal scores on each dimension, is more appropriate (for details, see Gifi 1990; Heiser and Meulman 1994; Chapter 4, this volume). The optimal scoring process allows order constraints to be imposed so that ordered categorical variables get increasing, or at least nondecreasing, quantifications within the r-dimensional space (with $r = 1, 2, \ldots$; often $r = 2$) as the category levels increase. After replacing the original scores with their optimal scores, one can use the PCA algorithm to perform CatPCA.

Responses not consistent with the implied ordering in r dimensions manifest themselves through tied optimal quantifications for two or more successive categories. Unlike PCA, the number of dimensions (r) for the fit must be specified in advance, and the solutions for r and $r + 1$ dimensions are not nested. Once the optimal scores and the quantifications in the r-dimensional space have been found, the quantifications can replace the category codes of the items, and the remainder of the analysis can be regarded as a classical PCA. Since the solution is optimized in r dimensions, applying orthogonal rotation procedures does not change the location of the axes in the r-dimensional space.

In addition to the optimal scores for the variables on each axis and the quantifications in the r-dimensional space, CatPCA (as in PCA) provides factor loadings for the variables, eigenvalues, and explained variances for the axes. In effect, CatPCA can be understood as PCA applied to ordered categorical data. On the one hand, if the raw data are of high quality (i.e., the respondents understood the question, they answered on the whole five-point scale, and the successive response categories discriminate between increasing levels or amounts of endorsements), then the solution obtained by CatPCA should be similar to the solution obtained by PCA. On the other hand, for any r-dimensional solution, if the quality of the data is poor, the explained variance in CatPCA should be substantially higher than the one in PCA. Although there is no statistical test for the differences in explained variances, they can be used as indicators of the quality of data: the greater the difference, the lower the quality. For further details on CatPCA, see de Leeuw (Chapter 4, this volume).

As mentioned above, CatPCA provides quantifications of the variable categories within r dimensions (usually $r = 2$). The process of finding coordinates of points in a lower-dimensional space is the concern of biplots. Biplots can express quantitative variables as vectors, and the approximation is often interpreted through singular-value decomposition (Gabriel 1971, 1981). Gower and Hand (1996) stress that biplot axes can be interpreted just like other coordinate axes, e.g., by projection and reading off a scale. The biplots can be drawn using the information of the variable quantifications in the r-dimensional space. Together with the distances between adjacent response options on each item, the angles between the various items within the r-dimensional space represent the response structure. For further details on biplots, see Gower and Hand (1996), as well as Chapters 2 and 3 of this volume.

20.4 Solutions

In cross-national comparisons, PCA or other techniques requiring metric data are often used. Showing the factor loadings and explained variances of the first r dimensions provides basic information on the structure of variable associations in each country. To make the structures comparable, and to allow for cross-national comparisons, some form of Procrustes rotation is often applied. Since our interest is with the quality and comparability of the data, we will not interpret the data in terms of substantive meanings.

As mentioned previously, five-point scales are often treated as metric data, i.e., the distances between the categories are treated as equal, for example, 1, 2, 3, 4, and 5 (the distance between subsequent categories is 1). In survey research, this assumption is seldom tested and unlikely to be fulfilled. However, it is quite reasonable to restrict the data to be ordered in such a way that $a1 \leq a2 \leq a3 \leq a4 \leq a5$ with $a1 \neq a5$ (and for variables in reverse order, $b5 \leq b4 \leq b3 \leq b2 \leq b1$, with $b5 \neq b1$); i.e., for example, a "strongly agree" reflects a higher agreement or at least not a smaller agreement than an "agree" (and "strongly disagree" reflects a stronger disagreement or at least not a smaller disagreement than "disagree"). If the data are of high quality, the deviations from equidistances of the optimal quantifications in the latent space should be small, i.e., the distances between successive categories are close to a constant. In general, the closer the data are to equal distances between the categories within each item, the smaller the differences between the PCA and CatPCA solutions. We will use these differences as an indicator of the quality of data. The argument in this chapter is that support for single/dual-earner family structures must be found within the first two dimensions; where that is not the case, we argue that this is sufficient evidence of low-quality data.

Imagine a one-dimensional solution with all variables measuring the same construct, say, a dimension that reflects the attitudes toward supporting single- versus dual-earner households. In such a solution, high support for single-earner households (and no support for dual-earner households) would be on one part of the dimension, while high support for dual-earner households (and no support on single-earner households) would be on the other part. This dimension should be associated with each of the single items in such a way that the higher the support for dual-earner families and the lower the support for single-earner families, the higher the respondent score (or vice versa). Because of the restriction to ordered categorical data, projecting the single categories on the latent dimension should retain their original order. For example, for an item measuring the support for dual-earner families, a "1" might become "–1.2," a "2" a "–0.3," a "3" a "0.2," a "4" a "0.7," and a "5" a "1.4." Likewise, for an item measuring support for a single-earner household, a "1" might become "1.5," a 2 a "0.8," a "3" a "–0.2," a "4" a "–0.8," and a "5" a "–1.5." In this constructed case, the distances between values are different, but the order of the categories remains the same. If the order in the latent dimension is not fulfilled, say $a1 > a2$ or $b1 < b2$, the values get tied in the latent dimension, and they are located at the same point ($a1 = a2$, $b1 = b2$).

In the case above, only one dimension was considered, on the assumption that only one underlying dimension was necessary for sufficient interpretation of the data. This might be the situation for some countries, but there might also be other countries in which attitudes toward single-earner households is relatively independent of attitudes toward dual-earner households. If this is true, two dimensions for a meaningful interpretation are needed in which the estimation of the new category values (quantifications) will be on a line that is fitted into the best two-dimensional solution. These fitted lines (each variable is represented as a line within the two-dimensional space) can be interpreted as biplots (Gower and Hand 1996), and their quantifications give a first indication of the quality of the data. Again, the distances between two successive categories reflect their similarities: the closer they are, the less difference they measure. Under the condition of ordered categorical data, the new values of the original categories retain their order, or get tied if the original order does not hold, as when an "agree" reflects a higher agreement than a "strongly agree." In other words, all variables that fail to mirror the initial order in r dimensions have ties. Ties are expected for variables that measure support for neither dual- nor single-earner families within r dimensions, as might be the given case for items G, J, and K. The fitting of distances between successive categories within the CatPCA approach improves the fit compared with a PCA of the original variables, and the differences in the explained variances between PCA and CatPCA can be used as a rough indicator for the quality of data. The greater the differences are, the lower the quality of data. Large differences are an indicator that either the respondents were not able to handle the five-point scale or the respondents did not understand the questions (which also results in an inability to handle the five-point scale). Table 20.2 shows the explained variances of the PCA and CatPCA solutions and their differences within the first two dimensions.

As mentioned previously, in contrast to PCA, the dimensions in the CatPCA-solutions are not nested. As a result, it could happen that the explained variance of a single dimension in some cases in PCA is higher than in CatPCA. For example, for Australia, in the PCA solution the first dimension explains 34.1% of the total variation, whereas the first CatPCA solution explains only 33.1%. This indicates only that there is a relatively strong second dimension in CatPCA; in the given example it explains 16.9% of the variation compared with 14.7% in PCA. It is also possible that the second dimension of the PCA solution explains more variation than the second dimension of the CatPCA solution. However, the sum of explained variances in all samples must be higher in CatPCA.

Comparing the solutions for all countries, a large range in the increase of explained variance is evident. Norway exhibits the lowest increase; in two dimensions PCA explains 47.4% of the variation and CatPCA 47.9%. From this small improvement, one can conclude that

Table 20.2 Principal component analysis and categorical principal component analysis.

	PCA			CatPCA			CatPCA – PCA	
	λ_1	λ_2	λ_{1+2}	λ_1	λ_2	λ_{1+2}	Absolute	%
Australia	.341	.147	.487	.331	.169	.500	.013	2.7
West Germany	.302	.145	.447	.304	.160	.464	.017	3.8
East Germany	.272	.140	.412	.305	.135	.440	.028	6.8
Great Britain	.333	.144	.477	.350	.139	.489	.012	2.5
Northern Ireland	.330	.134	.464	.341	.137	.478	.014	3.0
United States	.316	.143	.459	.323	.156	.478	.019	4.1
Austria	.280	.150	.430	.286	.154	.441	.011	2.6
Hungary	.234	.128	.363	.270	.132	.401	.038	10.5
Italy	.266	.140	.407	.258	.169	.428	.021	5.2
Ireland	.300	.143	.443	.287	.205	.492	.049	11.1
The Netherlands	.327	.143	.469	.332	.157	.490	.021	4.5
Norway	.340	.134	.474	.343	.137	.479	.005	1.1
Sweden	.323	.132	.456	.331	.142	.472	.016	3.5
Czech Republic	.230	.127	.357	.238	.149	.387	.030	8.4
Slovenia	.248	.141	.390	.249	.190	.440	.050	12.8
Poland	.268	.146	.413	.296	.166	.461	.048	11.6
Bulgaria	.224	.147	.371	.236	.166	.402	.031	8.4
Russia	.201	.123	.323	.271	.116	.387	.064	19.8
New Zealand	.322	.144	.466	.326	.150	.476	.010	2.1
Canada	.313	.147	.461	.323	.148	.471	.010	2.2
The Philippines	.199	.132	.330	.269	.137	.405	.075	22.7
Israel	.244	.141	.386	.240	.207	.447	.061	15.8
Japan	.201	.169	.370	.215	.182	.397	.027	7.3
Spain	.303	.141	.444	.290	.203	.493	.049	11.0

the data from Norway are approximately on a metric scale; the differences between the successive categories are almost the same within each variable. In contrast, for the Philippines, the explained variance increases from 33.0% to 40.5%. This large difference implies that the scale of the items is far from metric. The differences in explained variances for the other countries are in between. Some of them have very small differences (for example, New Zealand and Canada), while others have large differences (for example, Russia and Israel). Although the order in the differences of explained variances may not correspond to the order in the quality of data, it indicates that it is likely that there is substantial variation in the quality of the data.

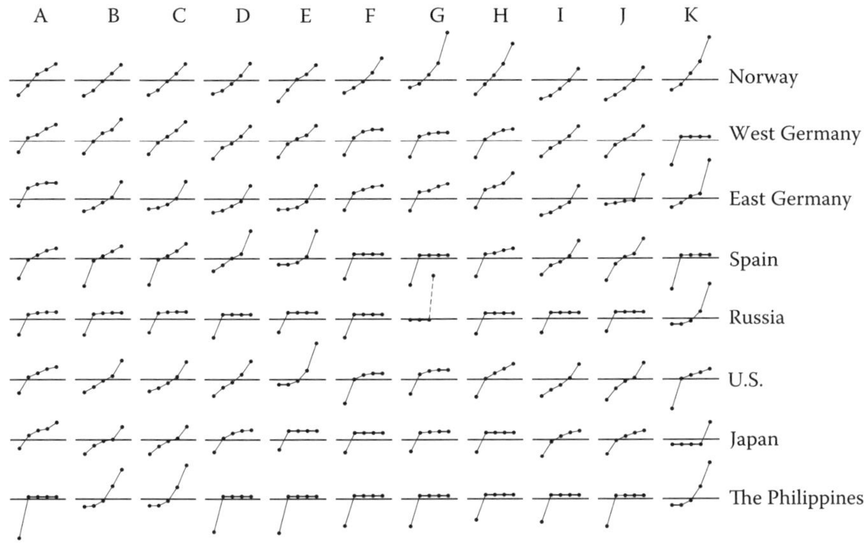

Figure 20.1 Profiles.

To inspect the country differences more closely, we show the quantifications for a subsample of eight countries (Norway, West Germany, East Germany, Spain, Russia, U.S., Japan, the Philippines) for the 11 variables (Figure 20.1).

In Figure 20.1 the distances on the horizontal axes represent the original categories, i.e., the metric scale from 1 to 5. The vertical axes reflect the quantifications in two dimensions. The straighter the lines, the more equal the distances are in the latent scale; i.e., exchanging the original categories (1, 2, 3, 4, 5) by the quantifications improves the solution to only a minor degree (for examples, see items B and C for Norway). Focusing on Norway, all items are ordinal in two dimensions, and therefore CatPCA improves the solution only slightly. In some contrast we find West Germany. Here some items are well represented with a straight line (for example, B, C, and D); others reflect mainly differences between "strongly agree" and the remaining four categories (for example items G and H). However, there is an explanation for this behavior: these items are not good indicators for either a dual- or a single-earner family structure. Item K behaves even worse; the last four categories are tied, but this item was assumed to be uncorrelated with the concept of single- and dual-earner family structure.

The patterns of quantifications for East and West Germany show some differences. In East Germany we find quite unequal distances also for variables that are assumed to measure either single- or dual-earner family structures, as for example items A, C, and E. This response pattern suggests that some respondents have strong opinions, endorsing a "strongly agree" for item A and "strongly disagree" for items C and E, while other respondents do not share these strong opinions. They further differentiate only somewhat between the remaining categories, which is reflected by the relatively small differences between them. When applying PCA to these kinds of data, the assumption of metric data is not appropriate. The data are ordered categorical, but they are not continuously scaled.

The data for Spain show relatively large differences between "strongly agree" or "strongly disagree" and the remaining categories. Respondents either hold a strong attitude toward the issue of interest or they make only minor differentiations between the categories. The items F, G, and K are tied in four categories; they do not differentiate between single- and dual-earner family structures. While this behavior was expected for items G and K, it was not for item F.

For Russia and the Philippines, nearly all items are tied in four categories. This indicates one or more of four possibilities:

1. The respondents did not understand the context of the questions.
2. The respondents could not handle five-point scales.
3. The questions were meaningless because of a different cultural background.
4. The items were poorly translated.

Regardless of the specific reason, the solutions should be handled with care.

The next step focuses on the structure of the data. Showing the biplots for the CatPCA solutions will also reveal further aspects about the quality of the data. The biplots for the 11 variables in the eight countries are shown in Figure 20.2a to Figure 20.2h.

Biplot axes represent items in a two-dimensional space. In the figures, the items are labeled at the "strongly agree" category; the first letter refers to the questions (see Table 20.1), the second characterizes the kind of support (s = single-earner, d = dual-earner, a = ambiguous). The axes are calibrated at their category-level points, i.e., their quantifications in the two-dimensional space. Empty circles are chosen to symbolize that two successive categories are tied at the same point

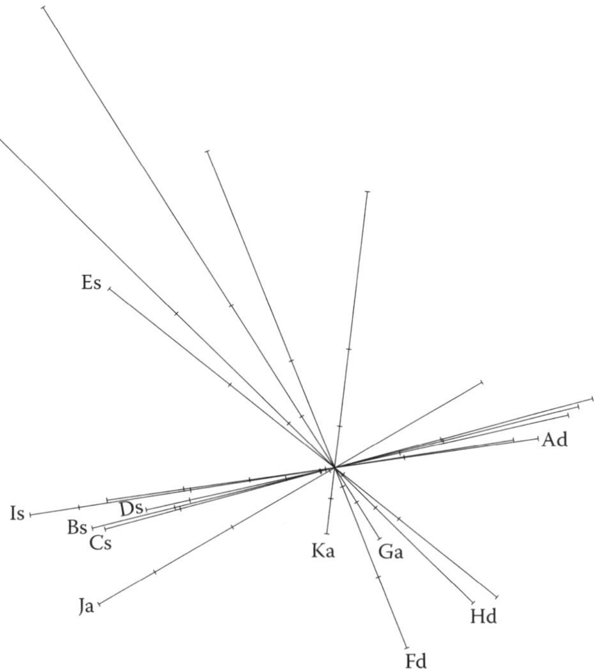

Figure 20.2a Norway.

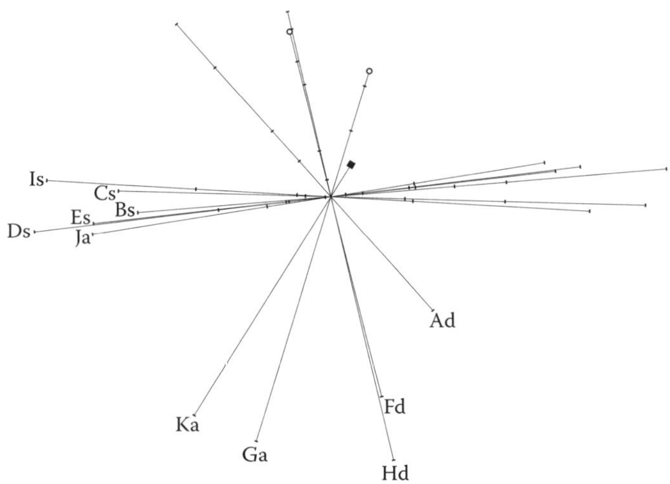

Figure 20.2b West Germany.

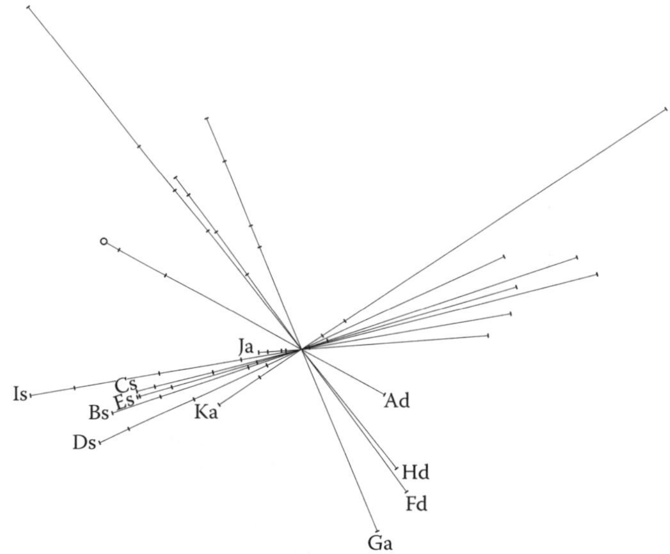

Figure 20.2c East Germany.

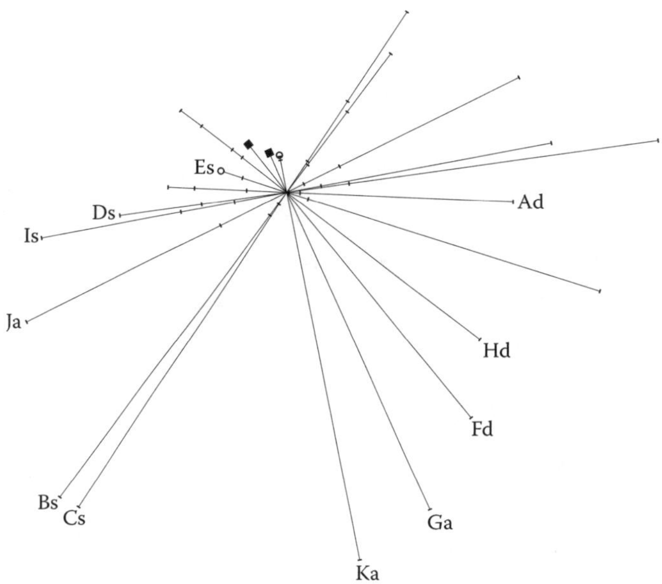

Figure 20.2d Spain.

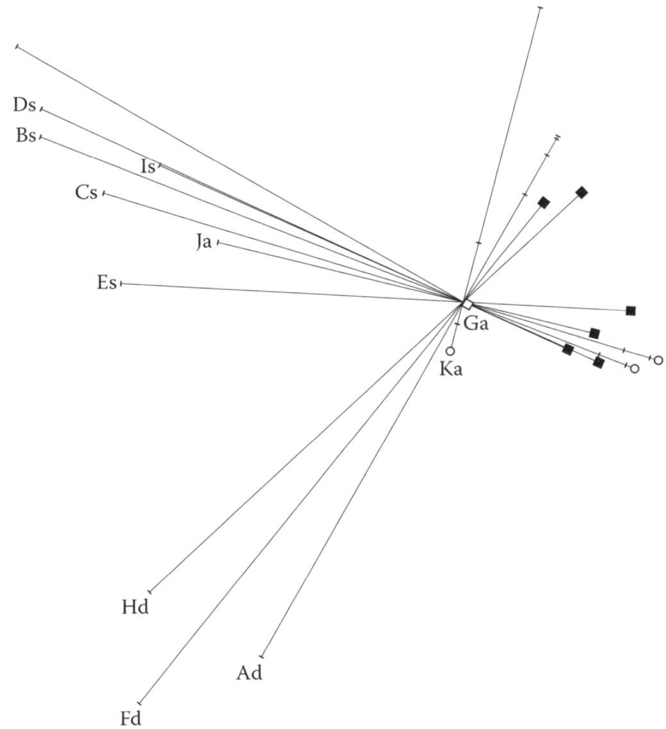

Figure 20.2e Russia.

(for example, categories "strongly disagree" and "disagree somewhat" on item G, West Germany, Figure 20.2b), empty squares symbolize three tied categories (for example, item G, Russia, Figure 20.2e), and solid squares symbolize four tied categories (for example, item K, West Germany, Figure 20.2b). The distances between two successive category-level points show their similarity. The angles between two items show their similarity: the closer they are, the higher the association between these items in the two-dimensional space. For example, in the case of Norway (Figure 20.2a), items D, B, and C are highly positively intercorrelated. In contrast, Item A is opposite to item I, i.e., high agreement with item A is associated with high disagreement with item I, and vice versa. Note that the lengths of the lines are a function of the frequencies of the categories and do not reflect the amount of explained variances of the items.

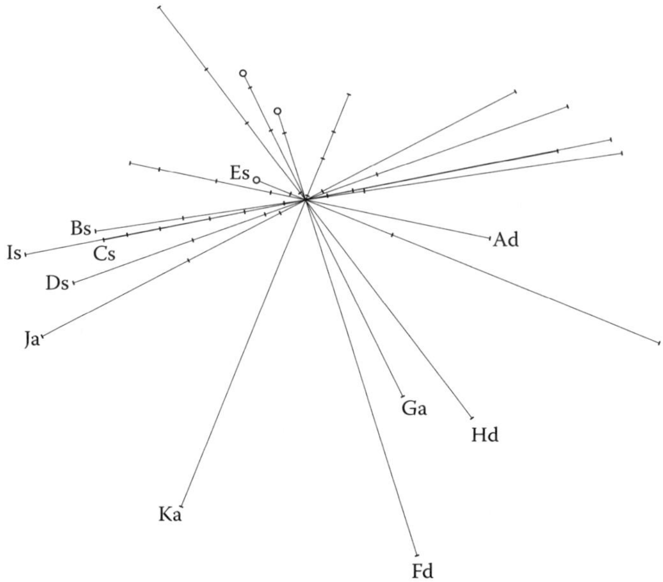

Figure 20.2f U.S.

Examining all solutions, a rough grouping of countries is possible. Dividing them into three groups, we find for Norway and the U.S. a pattern that can be described by four clusters: on the bottom right are the items H and F. Both are dual-earner items, and additionally, they all allude to economic factors (for example, most women have to work these days to support their families). Item G, which was characterized as ambiguous, is highly positively associated with items H and F, i.e., for Norway item G measures support for single-earner families. Item A, which is also a dual-earner item, is only weakly associated with these three variables. The reason might be that this variable addresses the mother–child relation rather than an economic issue. Opposite to item A are four single-earner items (I, D, B, and C). Item J is close to this cluster, and it also describes support for single-earner households. The only "economic situation" describing single-earner households (item E, comparing the status as a housewife with a person who works for pay) is opposite the items describing the economic situation in dual-earner family structures. Item K is in the middle of the two main groups; as expected, it describes support for neither single- nor dual-earner households. Although the frequencies of the two countries differ,

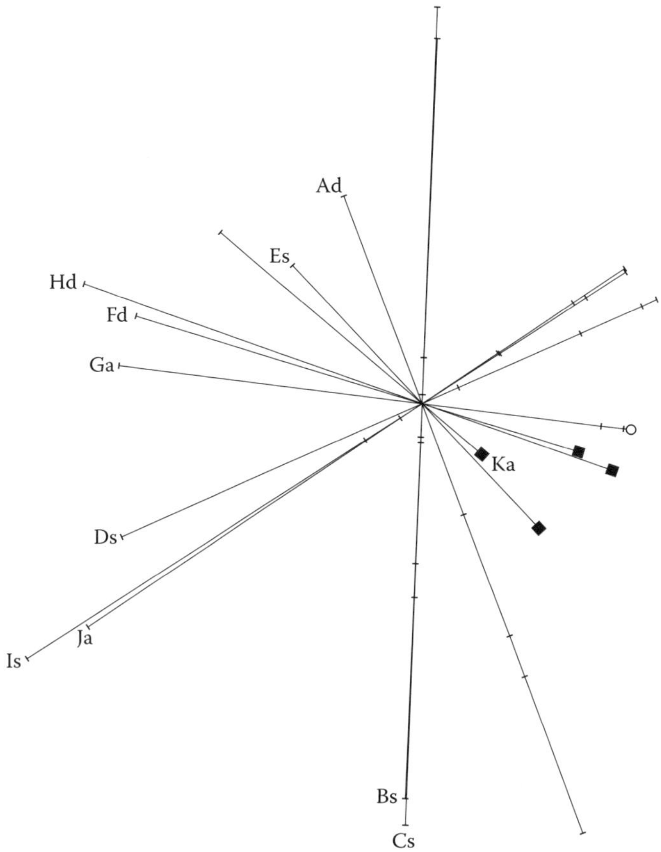

Figure 20.2g Japan.

the structures of responses are quite similar and relatively easy to interpret. Further, their items have no or only a few ties, indicating that the respondents were able to handle the five-point scales. It can be concluded that the data are of good quality and appropriate for cross-national comparison.

The second group includes East and West Germany, and to some extent Spain. All three countries are described by one cluster comprising all single-earner items (I, C, B, E, and D), regardless of whether they focus on family or economics, and again, item J belongs to this group of items. Whereas the economic item (E) is located in the middle

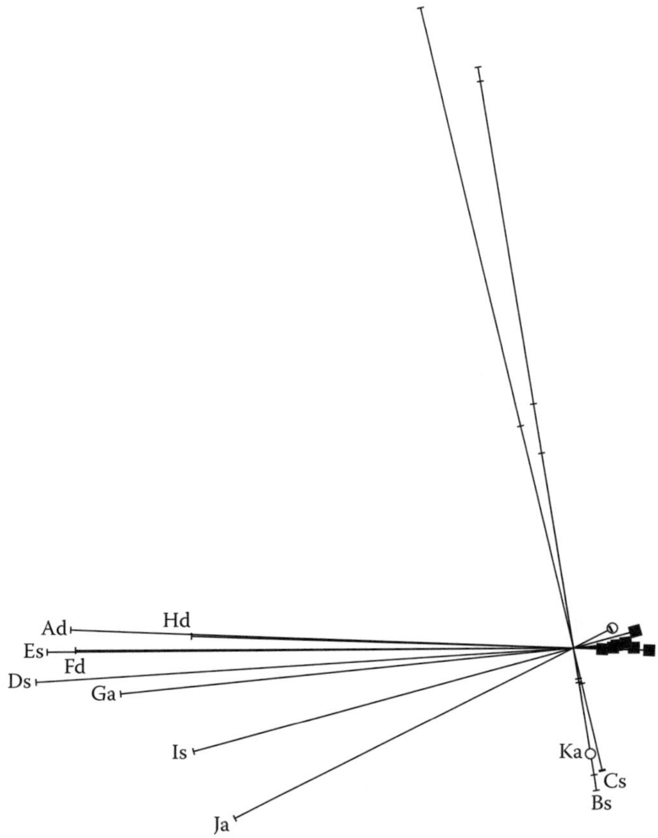

Figure 20.2h The Philippines.

of this cluster of items in both parts of Germany, in Spain this item is somewhat separate, and the items are spread out more, running from the economic issues to the family issues. This means that respondents in Spain differentiated between the various single-earner items more than respondents in both parts of Germany. The dual-earner items in all countries are independent of the single-earner items: the angles between them are approximately 90° (A, F, H, and G). Note that as was the case for the first cluster of countries, item G belongs to the dual-earner items. With respect to support for dual-earner households, respondents in all three countries differentiate between family and economic issues, but in somewhat different ways.

In East Germany, the four items are very close to each other; in West Germany they form three groups (A, which is on family; F and H, which are on economics; G, which is on economics, but not only an attitude: in many families women have to work to support their families); and in Spain they are again widely spread out. Although the structure of the data is different for the two countries discussed above, the data for both East and West Germany are of high quality (only a few ties). In Spain we find four tied categories on items F and G; these two variables have to be interpreted with care, but the remaining structure is clear.

Whereas the data for the five countries discussed above are appropriate for cross-national comparison (with some caution for Spain), we doubt the quality of responses for Japan, Russia, and the Philippines. In the case of Russia (Figure 20.2e), we find a clear subdivision between single- and dual-earner items, but many ties characterize nearly all items. Further, there is a response structure in which the respondents answered with either "strongly agree" or with any of the remaining four responses. This means the respondents did not differentiate between "somewhat agree" and "strongly disagree."

Japan (Figure 20.2g) has fewer ties than Russia but exhibits a structure without a plausible substantive interpretation. The items can be divided into four clusters, the first of which combines two single-earner items (B and C). These items and their category-level points are very close to each other, because they measure almost the same. Further, we find a cluster that combines two dual-earner items (H and F) as well as item G, which already was for other countries characterized as a dual-earner item; these items are all heavily tied (in H and F, four categories are tied; in G, two categories are tied; and the two successive categories are very close to the tied categories). The next two clusters are even more difficult to interpret, since both include single- and dual-earner items. The most plausible interpretation for these findings is that, in each cluster, one variable was asked in the wrong direction, i.e., the translation was wrong. Before making any cross-national comparison, the translation must be checked, some items probably have to be recoded, and some items probably have to be excluded.

The situation for the Philippines (Figure 20.2h) is even worse. The data are heavily tied, and one cluster combines most of the single- and most of the dual-earner items. It seems that respondents answered with either "strongly agree" or with one of the other categories, regardless of the content of the question. These data seem to be not suitable for any kind of analysis, never mind for cross-national comparison.

20.5 Discussion

The main purpose of this chapter was to investigate the comparability and quality of data in cross-national research. Comparative cross-national survey research usually proceeds either on the basis of a series of bivariate relationships (such as contingency tables or correlation coefficients) or on some form of linear multivariate data analysis. For different reasons, neither of these approaches, we contend, is well suited to the task at hand. Bivariate analyses by definition cannot reveal the underlying structure of a set of variables. Linear multivariate modeling assumes that the data are at least ordinal in nature, often in addition to other stringent assumptions, such as multivariate normal distributions.

We argued that two main criteria must be fulfilled for sound cross-national comparisons. First, the quality of the data must be sufficiently high in the countries to be compared. A simple indicator for this, we suggest, is whether the primary source of variation in the data is attributable to the substantive content of the items rather than to methodological error. Second, the underlying response structure of the items to be compared must be sufficiently compatible to warrant confidence that the same construct is tapped in the selected countries in a similar fashion. These criteria are well known and almost self-evident; our contribution consists in demonstrating how to assess the extent to which they are fulfilled.

Using the 1994 ISSP data on single- and dual-earner family structures, we could show that certain countries should be excluded from cross-national comparisons, even when common data sets are available for them, and further, that certain items in certain countries have to be interpreted cautiously, although they formally have an identical content. Stated positively, this chapter showed that meaningful comparisons are indeed possible for selected groups of countries with a selected set of items. In this chapter we showed how one can assess the fulfillment of the methodological preconditions for detecting countries that are comparable and items that can be included in such comparisons.

To overcome the inherent limitations in the previous methodological approaches, we chose CatPCA for the purpose of detecting (a) violations of ordinality in Likert-type scales (shown as tied categories), (b) presence of methodological artifacts such as response styles, and (c) national variation in the underlying structure of responses. The first two points indicate low data quality; the third restricts the comparability of countries.

On the basis of the CatPCA solutions for a subset of eight countries, we identified three subsets of countries. Norway and the U.S. exemplified one subset, characterized by a subdivision between single- and dual-earner household structures and by a further subdivision between family- and economic-focused items. The second subset of countries, exemplified by West and East Germany, also subdivided between single- and dual-earner households, but without differentiating between family- and economic-related items. However, in both cases the data are of high quality and suited for cross-national comparison. A third subset of countries was exemplified by Japan, Russia, and the Philippines. The sources of variation appear, to a high degree, to be due to incorrect translations (in the case of Japan), to questions that might be not appropriate (Russia), or to methodological artifacts, such as response styles (the Philippines). The latter data are almost not interpretable; the quality is low.

It would be unrealistic to expect the underlying structure of responses to be identical in all countries on any topic, or to expect that the data in all countries would be of the same high quality. Nevertheless, our findings indicate that, with due caution, it is possible to make substantive interpretations for subsets of countries. However, before modeling the data, the basic structure should be analyzed, and especially the quality of the data has to be evaluated.

CHAPTER 21

Additive and Multiplicative Models for Three-Way Contingency Tables: Darroch (1974) Revisited

Pieter M. Kroonenberg and Carolyn J. Anderson

CONTENTS

21.1 Introduction

In an only occasionally referenced paper, Darroch (1974) discussed the relative merits of additive and multiplicative definitions of interaction for higher-order contingency tables. In particular, he compared the following aspects: partitioning properties, closeness to independence, conditional independence as a special case, distributional equivalence, subtable invariance, and constraints on the marginal probabilities. On the basis of this investigation, he believed that multiplicative modeling is preferable to additive modeling, "but not by so wide a margin as the difference in the attention that these two definitions have received in the literature" (Darroch 1974: 213). One important aspect of modeling contingency tables did not figure in this comparison: interpretability.

The potential systematic relationships in multivariate categorical data become progressively more complex as the number of variables and/or the number categories per variable increase. In turn, interpretation becomes increasingly difficult. We consider techniques for data that can be logically formatted as three-way contingency tables. This does not limit us to three variables but, rather, it limits us to, at most, three types or modes of variables.

The focus in this chapter lies with the interpretation of the dependence present in three-way tables and how insight can be gained into complex patterns of different types of dependence. Since the major aim in most empirical sciences is to apply (statistical) models to data and to obtain a deeper insight into the subject matter, we consider it worthwhile to take up Darroch's comparison and extend his set of criteria by considering the interpretational possibilities (and impossibilities) of multiplicative and additive modeling of contingency tables. The investigation will primarily take place at the empirical level guided by a particular data set that consists of four variables but is best analyzed as a three-way table.

Similarities between additive and multiplicative modeling techniques for two-way tables have been discussed by Escoufier (1982), Goodman (1985, 1996), and Van der Heijden et al. (1994). Limited discussions of the three-way case can be found in Van der Heijden and Worsley (1988) and Green (1989). Neither of the latter two consider three-way correspondence analysis (CA) or multiple correspondence analysis (MCA). The major empirical comparisons made in this chapter are between three-way CA, which uses an additive definition of interaction (Carlier and Kroonenberg 1996, 1998), extensions of Goodman's multidimensional row–column RC(M) association model (Goodman 1979, 1985; Clogg and Shihadeh 1994; Anderson and Vermunt 2000; Anderson 2002), which use a multiplicative definition of interaction, and categorical principal component analysis (PCA), which is related to MCA, the major method discussed in this book. These methods will be compared empirically in terms of how and to what extent they succeed in bringing out the structure in a data set.

Table 21.1 Concurrent detection data.

			Confidence in High Signal (H)											
			High Signal Absent ($-\mathcal{H}$)						High Signal Present ($\mathcal{H}$)					
			1	2	3	4	5	6	1	2	3	4	5	6
Low Signal Absent ($-\mathcal{L}$)	Confidence in Low Signal Absent ($-\mathcal{L}$)	1	44	4	9	7	6	7	7	4	5	5	14	69
		2	13	30	20	8	14	7	5	7	13	15	38	37
		3	9	23	17	17	3	0	6	7	8	10	10	15
		4	16	17	10	20	2	2	4	12	5	13	6	14
		5	5	4	9	10	4	0	2	3	1	1	3	5
		6	3	3	0	1	4	1	0	0	1	1	1	3
Low Signal Present ($\mathcal{L}$)	Confidence in Low Signal Present ($\mathcal{L}$)	1	8	2	2	1	0	4	4	1	2	0	4	37
		2	5	5	5	5	5	3	0	4	0	1	8	25
		3	8	10	7	4	1	1	1	3	3	7	8	15
		4	12	17	15	13	2	2	4	4	8	17	12	21
		5	12	17	19	18	10	4	3	12	8	11	20	20
		6	31	29	25	24	12	12	11	8	12	11	12	33

Note: An (i,j) entry in the table indicates the number of times the subject expressed confidence level i in the presence/absence of the low-frequency signal and the subject's confidence level j in the presence/absence of the high-frequency signal.

Source: Wickens and Olzak (1989).

21.2 Data and design issues

We hope to show that although our attention is focused on three-way tables, the methods discussed are more general than they may appear at first glance. After describing a data set from Wickens and Olzak (1989) that guides our empirical comparison of methods for three-way tables, we briefly discuss general design issues that help to identify potential data analytic problems to which the methods presented in this chapter can be applied.

21.2.1 The Wickens and Olzak data

Wickens and Olzak (1989) report the data from a single subject on a concurrent signal-detection task (Table 21.1). In the experiment, a signal abruptly appeared on a computer screen and disappeared again after 100 milliseconds. The signal consisted of either one of two sine curves or both together. One curve had more cycles and is referred to as the high-frequency signal ($\mathcal{H}$), while the other with fewer cycles is referred to as the low-frequency signal ($\mathcal{L}$). However, in the analyses they are mostly combined into a single interactively coded factor ($\mathcal{J}$) with four categories. On each trial of the experiment, the subject was presented with either both signals, only the high-frequency signal, only the low-frequency signal, or neither; each of these was repeated 350 times. Thus, the independent variables consisted of a fully crossed 2×2 design, with a total of $4 \times 350 = 1400$ trials. However, one observation went astray, and only 1399 trials are present in the data used here.

The subject's task on each trial was to rate his confidence on a scale from 1 to 6 regarding the presence of each of the two possible stimuli. The lower the rating, the more confident the subject was that the signal was absent; the higher the rating, the more confident the subject was that the signal was present. Thus, the response variables were the two confidence ratings (H) and (L).

When a signal was present, the subject was expected to express more confidence that it was presented than when the signal was absent. This is generally confirmed by the data (Table 21.1), but there appear to be interactions as well. Wickens and Olzak's (1989) analyses and those by Anderson (2002), both of which use a multiplicative definition of interaction, confirm the presence of a three-way interaction between the presence of the two signals and the confidence rating of the low signal. However, Anderson (2002), who used a more focused test, also detected a three-way interaction between the two signals and the rating of the high signal. Since there are interactions in the data involving the two stimuli and one or the other response, in this chapter we format the data as a stimulus condition or joint signal ($\mathcal{J}$) by high rating (H) by low rating (L) cross-classification; in particular the data form an $I \times J \times K$ three-way table with $I = 4$, $J = 6$, and $K = 6$. The aim of the present analyses is to highlight and describe the nature of the interactions in the data.

21.2.2 Design issues

Several designs yield data that are naturally organized as entries in three-way contingency tables. Two aspects of design are relevant in this context: (a) the role of the variables and (b) the mutual exclusiveness of the entries in the table.

With respect to the first design aspect, variables in a study typically take on the role of response or factor (explanatory variable). In a three-way contingency table, there can be three response variables or *responses*, in which case we speak of a *response design*. When there are two responses and one factor, or when there is one response variable and two factors, we speak of *response-factor designs*. In the Wickens and Olzak (1989) experiment, the two types of ratings are the two responses and the joint signal is the factor. Especially for multiplicative modeling, it is important to consider this response–factor distinction, while for additive modeling it is primarily relevant in interpretation.

With respect to the second design aspect, in most statistical approaches to analyzing contingency tables, it is assumed that observational units are independent and that each observation's "score" is a triplet (i,j,k). The design is fully crossed if each observation falls into one and only one cell of the table. Most log-multiplicative models require fully crossed designs, while additive modeling via CA is indifferent to the design, as it does not explicitly use stochastic assumptions. In the case of the Wickens and Olzak data (1989), even though all of the observations come from a single subject, the responses over trials are considered to be independent, as is common in this type of experiment.

21.3 Multiplicative and additive modeling

Since our methods for three-way tables are characterized as being either an additive or multiplicative model, we review the defining features of the models under study in this chapter. The exposition in this section leans heavily on that by Darroch (1974).

Let π_{ijk} be the probability and p_{ijk} the proportion of observations that fall into categories i, j, and k of variables A, B, and C, respectively, where $i = 1,\ldots,I$, $j = 1,\ldots,J$, and $k = 1,\ldots, K$, and where $\sum_{ijk} \pi_{ijk} = \sum_{ijk} p_{ijk} = 1$. As usual, marginal probabilities and proportions are indicated by replacing the index over which they are summed by a dot. For example, $p_{ij.}$ is the marginal proportion for the ith category of A and the jth category of B. All summations will run from 1 to the capital letter of the index, e.g., k will run from 1 to K.

In the *multiplicative definition*, the absence of three-way interaction is defined as

$$H_m : \pi_{ijk} = \xi_{jk}\phi_{ik}\psi_{ij} \tag{21.1}$$

for all i, j, and k and for some ξ_{jk}, ϕ_{ik}, and ψ_{ij}. One way to express this is that the interaction between two variables does not depend on the values of the third variable (see Section 21.4.1). By taking logarithms of Equation 21.1, we get models that are additive in the log-scale (Roy and Kastenbaum 1956).

An equivalent expression for Equation 21.1 is

$$\frac{\pi_{ijk}\pi_{i'j'k}\pi_{i'jk'}\pi_{ij'k'}}{\pi_{i'j'k'}\pi_{i'jk}\pi_{ij'k}\pi_{ijk'}} = 1 \tag{21.2}$$

for all $i \neq i'$, $j \neq j'$, and $k \neq k'$ (Darroch 1974). Equation 21.2 is the ratio of odds ratios for two variables given the third variable. Deviations from no three-way interaction are reflected by the extent to which Equation 21.2 differs from the value of 1 (or equivalently, how much the logarithm of Equation 21.2 differs from zero).

The *additive definition* of no three-way interaction, introduced by Lancaster (1951), is

$$H_a : \frac{\pi_{ijk}}{\pi_{i\cdot\cdot}\pi_{\cdot j\cdot}\pi_{\cdot\cdot k}} = \alpha_{jk} + \beta_{ik} + \gamma_{ij} \tag{21.3}$$

for all i, j, and k and for some α_{jk}, β_{ik}, and γ_{ij}, and this is equivalent to

$$H_a : \Pi_{ijk} = \frac{\pi_{ijk}}{\pi_{i\cdot\cdot}\pi_{\cdot j\cdot}\pi_{\cdot\cdot k}} - 1 = \left(\frac{\pi_{\cdot jk}}{\pi_{\cdot j\cdot}\pi_{\cdot\cdot k}} - 1\right) + \left(\frac{\pi_{i\cdot k}}{\pi_{i\cdot\cdot}\pi_{\cdot\cdot k}} - 1\right) + \left(\frac{\pi_{ij\cdot}}{\pi_{i\cdot\cdot}\pi_{\cdot j\cdot}} - 1\right) \tag{21.4}$$

(Darroch 1974: 209). Thus, according to the additive definition of no three-way independence (i.e., according to Equation 21.4), the deviation from complete independence between A, B, and C equals the sum of the deviations from (two-way) marginal independence between A and B, A and C, and B and C. For cell (i, j, k), the three-way interaction term that is hypothesized to be zero has the form

$$\frac{\pi_{ijk} - \tilde{\pi}_{ijk}}{\pi_{i\cdot\cdot}\pi_{\cdot j\cdot}\pi_{\cdot\cdot k}} \tag{21.5}$$

where

$$\tilde{\pi}_{ijk} = \pi_{i\cdot\cdot}\pi_{\cdot j\cdot}\pi_{\cdot\cdot k} + (\pi_{ij\cdot}\pi_{\cdot\cdot k} - \pi_{i\cdot\cdot}\pi_{\cdot j\cdot}\pi_{\cdot\cdot k}) + (\pi_{i\cdot k}\pi_{\cdot j\cdot} - \pi_{i\cdot\cdot}\pi_{\cdot j\cdot}\pi_{\cdot\cdot k}) + (\pi_{\cdot jk}\pi_{i\cdot\cdot} - \pi_{i\cdot\cdot}\pi_{\cdot j\cdot}\pi_{\cdot\cdot k})$$

or

$$\tilde{\pi}_{ijk} = \pi_{ij\cdot}\pi_{\cdot\cdot k} + \pi_{i\cdot k}\pi_{\cdot j\cdot} + \pi_{\cdot jk}\pi_{i\cdot\cdot} - 2\pi_{i\cdot\cdot}\pi_{\cdot j\cdot}\pi_{\cdot\cdot k}. \tag{21.6}$$

Apart from the distinction between multiplicative and additive modeling, there is another sense in which the word "multiplicative" crops up in discussing models for contingency tables. The former distinction refers to the different ways that interactions are defined. A different question is how the interactions themselves are treated (regardless of the definition of interaction). In both types of modeling, the interactions are decomposed into multiplicative terms via two-way or three-way singular-value decompositions (SVD) to provide a lower-rank representation of the systematic patterns in the interactions. The use of the SVD allows a separation of the interaction into a systematic part and an uninterpretable remainder. In the additive framework, there is only one decomposition for the overall three-way dependence from which the decompositions of the two-way interactions are derived. However, in the multiplicative modeling definition, the decompositions are carried out either for each interaction separately or for a group of interactions jointly, but in the latter case it is not possible to separate out which part belongs to which interaction (unless restrictions are placed on the parameters).

21.4 Multiplicative models

Log-linear models, extensions of Goodman's *RC*(*M*) association model to multiway tables, and association models with latent-variable interpretations all use a multiplicative definition of dependence. Because the last two types of models are special cases of log-linear models, we start with a brief introduction and discussion of log-linear models using the Wickens and Olzak data. We use these models to show how the multiplicative definition of dependence manifests itself. Then we turn to extensions of the *RC*(*M*) association model for three-way tables.

21.4.1 Log-linear models

In this discussion, we initially treat the Wickens and Olzak (1989) data as a four-way cross-classification: presence of a high-frequency signal (i.e., "HIGH SIGNAL," $\mathcal{H}$); presence of a low-frequency signal (i.e., "low signal," $\mathcal{L}$); confidence rating in the high-frequency signal (i.e., "HIGH RATING," H); and confidence rating in the low-frequency signal (i.e., "low rating," L). Later in the chapter, we switch to a more advantageous format of a three-way table by fully crossing the two factors into a single "Joint Signal," $\mathcal{J}$, as will be justified by the results of preliminary analysis of the data.

Any table that has some form of independence (i.e., complete independence, joint independence, or conditional independence) can be expressed as the product of marginal probabilities. For our data, ideally the subject's rating of the high signal should only depend on the presence of the high signal, and the rating of the low signal should only depend on the presence of the low signal; that is, the ratings are *conditionally independent* given the stimuli. This specific hypothesis can be expressed as the product of various marginal probabilities; thus, the logarithm is an additive function of the logarithms of the marginal probabilities. Log-linear models are typically parameterized not in terms of logarithms of probabilities, but in such a way as to allow dependence structures that cannot be expressed as a product of marginal probabilities (i.e., some form of nonindependence). The parameterization of our hypothesis regarding the subject's behavior is

$$\log(\pi_{hljk}) = \lambda + \lambda_h^{\mathcal{H}} + \lambda_l^{\mathcal{L}} + \lambda_j^{H} + \lambda_k^{L} + \lambda_{hj}^{\mathcal{H}H} + \lambda_{lk}^{\mathcal{L}L} + \lambda_{hl}^{\mathcal{H}\mathcal{L}} \tag{21.7}$$

where λ is a normalization constant that ensures $\sum_{hljk} \pi_{hljk} = 1$; $\lambda_h^{\mathcal{H}}$, $\lambda_l^{\mathcal{L}}$, λ_j^{H}, and λ_k^{L} are marginal-effect terms for high signal h, low signal l, high rating j, and low rating k, respectively; and $\lambda_{lj}^{\mathcal{H}H}$, $\lambda_{lk}^{\mathcal{L}L}$, and $\lambda_{hl}^{\mathcal{H}\mathcal{L}}$ are bivariate interaction terms. The presence of the marginal-effect terms ensures that the fitted one-way marginal probabilities equal the observed margins (e.g., if $\lambda_h^{\mathcal{H}}$ is in the model, then $\pi_{h\cdots} = p_{h\cdots}$) and the presence of the interaction terms ensures that the two-way fitted probabilities equal the observed ones (e.g., if $\lambda_{hl}^{\mathcal{H}\mathcal{L}}$ is in the model, then $\pi_{hl\cdot\cdot} = p_{hl\cdot\cdot}$). In summary, the presence of a constant marginal-effect term and interaction terms guarantees that the fitted probabilities from the model reproduce the corresponding observed proportions. Since the experimenter determined the high-by-low signal margin (and it is inherently uninteresting), the term $\lambda_{hl}^{\mathcal{H}\mathcal{L}}$ should always be in the model. Location constraints are required on the log-linear model parameters to identify them (e.g., $\sum_j \lambda_j^{H} = 0$ and $\sum_h \lambda_{hj}^{\mathcal{H}H} = \sum_j \lambda_{hj}^{\mathcal{H}H} = 0$).

Besides algebraic representations, log-linear models also have schematic or graphical representations (Edwards 2000; Whittaker 1990). For example, the model in Equation 21.7 is represented by graph 1a in Figure 21.1. The boxes represent the variables; lines connecting two boxes (i.e., variables) represent possible dependence; and the absence of a line between two variables represents conditional independence. According to the model in Equation 21.7, the ratings on the two signals are conditionally independent given the stimuli, so there is no line connecting the high and low ratings in the graph. The line connecting the high signal and high rating and the one connecting the low signal and low rating indicate that dependence between these variables may exist. Whether dependence actually exists depends on whether $\lambda_{hj}^{\mathcal{H}H}$ equals zero for all h and j and $\lambda_{lk}^{\mathcal{L}L}$ equals zero for all l and k.

A slightly more complex log-linear model for the data allows for a possible dependence between the ratings themselves (e.g., perhaps a response strategy on the part of the subject). The graph representing this model is labeled 1b in Figure 21.1. The corresponding algebraic model is

$$\log(\pi_{hljk}) = \lambda + \lambda_h^{\mathcal{H}} + \lambda_l^{\mathcal{L}} + \lambda_j^{H} + \lambda_k^{L} + \lambda_{hj}^{\mathcal{H}H} + \lambda_{lk}^{\mathcal{L}L} + \lambda_{jk}^{HL} + \lambda_{hl}^{\mathcal{H}\mathcal{L}} \tag{21.8}$$

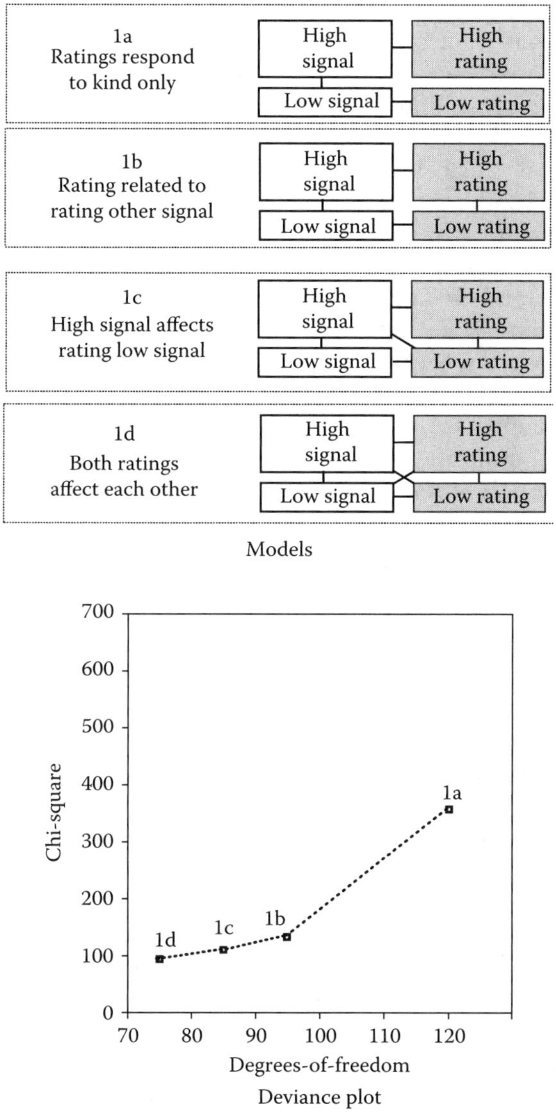

Figure 21.1 Concurrent detection data: selection of log-linear models.

The model in Equation 21.8 contains a subset of all possible bivariate interactions between pairs of variables. Deleting one or more sets of interaction terms from the model in Equation 21.8 yields a model that represents some form of independence.

When dealing with hierarchical log-linear models (i.e., models that include all lower-order terms that comprise the interaction terms), the (in)dependence between two variables, say the high signal $\mathcal{H}$ and the high rating H, given the fixed levels for the other variables, can be expressed through the odds ratios in the 2 × 2 subtables of $\mathcal{H}$ and H. For instance, if the relationship between the high signal and high rating is independent of the presence and rating of the low signal, then the odds-ratios in these subtables for high stimuli h and h' and high ratings j and j' for fixed values of k and l do not depend on the specific values of k and/or l.

To show that Equation 21.8 (and all models that only include bivariate interaction terms) use a multiplicative definition of no three-way interaction, we first examine the *conditional* or *partial* odds ratios for two variables given a fixed level of the third variable. According to the model in Equation 21.8, the odds ratio in the 2 × 2 subtables of variables $\mathcal{H}$ and H given level l and k of variables $\mathcal{L}$ and L equals

$$\theta_{hh',jj'(lk)} = \frac{\pi_{hljk}/\pi_{hlj'k}}{\pi_{h'ljk}/\pi_{h'lj'k}}$$

$$= \exp\left(\left(\lambda_{hj}^{\mathcal{H}H} - \lambda_{hj'}^{\mathcal{H}H}\right) - \left(\lambda_{h'j}^{\mathcal{H}H} - \lambda_{h'j'}^{\mathcal{H}H}\right)\right) \qquad (21.9)$$

where the subindices in parentheses indicate the categories of the conditioning variables. The odds ratios $\theta_{ll',kk'(hj)}$ and $\theta_{hh',ll'(jk)}$ are also functions of their corresponding interaction terms. The dependence between two variables in partial subtables does not depend on the value of the other variables.

The definition of no three-way interaction that is lurking behind the scenes can be seen by considering the *ratio of odds ratios* for two variables given different levels of a third variable for a fixed level of the remaining variables (if there are any). For example, the ratio of odds ratios for the high-signal conditions (i.e., h, h') and high ratings (i.e., j, j') given different low-signal conditions (i.e., l, l') and a specified low rating (i.e., k) based on the model in Equation 21.8 is

$$\frac{\theta_{hh',jj'(lk)}}{\theta_{hh',jj'(l'k)}} = \frac{\pi_{hljk}\pi_{h'lj'k}\pi_{h'l'jk}\pi_{hl'j'k}}{\pi_{h'ljk}\pi_{hlj'k}\pi_{hl'jk}\pi_{h'l'j'k}} = 1 \qquad (21.10)$$

Equation 21.10 is equivalent to the multiplicative definition given earlier in Equation 21.2. To see this, replace h in Equation 21.10 by i and note that k is constant. In other words, models that include only bivariate interactions, such as Equation 21.8, use a multiplicative definition of no three-way interaction. An implication is that a measure of three-way association is the ratio of the odds ratios and the extent to which Equation 21.2, or specifically in this case Equation 21.10, departs from 1 indicates the amount of three-way dependence in data.

A hierarchical log-linear model that includes a three-way interaction is

$$\log(\pi_{hljk}) = \lambda + \lambda_h^{\mathcal{H}} + \lambda_l^{\mathcal{L}} + \lambda_j^{H} + \lambda_k^{L} + \lambda_{hj}^{\mathcal{H}H} + \lambda_{lk}^{\mathcal{L}L} + \lambda_{jk}^{HL} + \lambda_{hl}^{\mathcal{H}\mathcal{L}} + \lambda_{hk}^{\mathcal{H}L} + \lambda_{hlk}^{\mathcal{H}\mathcal{L}L} \tag{21.11}$$

The graphical representation of this model is labeled 1c in Figure 21.1. According to this model, the ratios of odds ratios, $\theta_{ll',kk'(hj)}/\theta_{ll',kk'(h'j)}$, are functions of the three-way interaction parameters $\lambda_{hlk}^{\mathcal{H}\mathcal{L}L}$. These ratios are not equal to 1 unless $\lambda_{hlk}^{\mathcal{H}\mathcal{L}L} = 0$ for all (h,l,k).

The log-linear models given in Equation 21.7, Equation 21.8, and Equation 21.11, which correspond to graphs 1a, 1b, and 1c, respectively, in Figure 21.1, were fitted to the data, as well as a model that includes the two three-way interactions $\lambda_{hlj}^{\mathcal{H}\mathcal{L}H}$ and $\lambda_{hlk}^{\mathcal{H}\mathcal{L}L}$, which has as its graphical representation 1d in Figure 21.1. The results for these models are presented here in the form of a deviance plot (Fowlkes et al. 1988). The models in Figure 21.1 are connected to each other to show the progression of their likelihood ratio χ^2–degrees-of-freedom ratios. Ideally one wants to have a low χ^2 with as many degrees of freedom (d.f.) as possible. Based on the deviance plot, the log-linear models in graphs 1c and 1d of Figure 21.1 are candidates for selection.

There are three-way interactions between the two signals and each of the ratings; therefore, the data are reformatted as a three-way table such that there is one factor, which consists of the four possible combinations of signals (i.e., a "joint signal" or "signal condition"). In our data, we will use $\mathcal{J}$ to denote joint signal and index it as $i = 1$ (both signals absent), 2 (high present, low absent), 3 (high absent, low present), or 4 (both present). Therefore, in the next section, our starting log-linear model is

$$\log(\pi_{ijk}) = \lambda + \lambda_i^{\mathcal{J}} + \lambda_j^{H} + \lambda_k^{L} + \lambda_{ij}^{\mathcal{J}H} + \lambda_{ik}^{\mathcal{J}L} + \lambda_{jk}^{HL} \tag{21.12}$$

which has bivariate interaction terms that represent the three-way interactions between the two stimuli and each of the responses (i.e., $\lambda_{ij}^{\mathcal{J}H}$ and $\lambda_{ik}^{\mathcal{J}L}$).

21.4.2 Log-multiplicative association models

In this section, we turn to those extensions of log-linear models that provide more interpretable representations of dependence in three-way tables. In particular, we seek to separate the systematic patterns from the unsystematic parts of the interactions so as to gain insight into the nature of the association between the variables. The main tool for this will be the singular value decomposition (SVD) (for a good introduction to the SVD, see Greenacre 1984: 340).

Interaction terms in log-linear models are unstructured in the sense that they equal whatever they need to equal such that the corresponding margins of the data are fit perfectly. Goodman (1979, 1985) proposed that the two-way interaction terms in a saturated log-linear model for a two-way table be replaced by a lower rank approximation based on a SVD of the unstructured interaction terms (Clogg 1986). The resulting model, known as the multidimensional row–column or *RC*(*M*) association model, is log-multiplicative rather than log-linear because it includes multiplicative terms for the interaction. Given the success of the *RC*(*M*) model at providing interpretable representations of dependence in two-way tables, numerous extensions to three- and higher-way tables were proposed. Many of these proposals use bilinear terms (e.g., Becker 1989), trilinear terms (e.g., Anderson 1996), or both bilinear and trilinear terms (e.g., Choulakian 1988a). Wong (2001) provides an extensive summary of most of these proposals, including the required identification constraints.

We present a subset of possible log-multiplicative models for three-way tables (i.e., those that prove useful for the data at hand). Since our starting log-linear model, Equation 21.12, includes only bivariate interaction terms, we present a general log-multiplicative model that includes only SVDs of two-way interactions. Following a brief review of the essential elements of log-multiplicative association models, we present a general strategy for modeling data that connects substantive research hypotheses to models through the use of a latent-variable interpretation of log-multiplicative models.

Bivariate association models

We start with a general log-multiplicative model for three-way tables discussed by Becker (1989) that includes only bivariate interactions. In this model, each of the two-way interaction terms is replaced by a sum of bilinear terms, which can be computed via the SVD. In the case of our signal-detection data, we replace the interaction terms $\lambda_{ij}^{\mathcal{J}H}$, $\lambda_{ik}^{\mathcal{J}L}$, and λ_{jk}^{HL} in Equation 21.12 with separate bilinear terms; that is,

$$\log(\pi_{ijk}) = \lambda + \lambda_i^{\mathcal{J}} + \lambda_j^H + \lambda_k^L + \sum_{r=1}^{R} \sigma_{\mathcal{J}H(r)}^2 \omega_{ir}^{\mathcal{J}H} \nu_{jr}^{\mathcal{J}H}$$

$$+ \sum_{s=1}^{S} \sigma_{\mathcal{J}L(s)}^2 \omega_{is}^{\mathcal{J}L} \eta_{ks}^{\mathcal{J}L} + \sum_{t=1}^{T} \sigma_{HL(t)}^2 \nu_{jt}^{HL} \eta_{kt}^{HL} \tag{21.13}$$

where $\omega_{ir}^{\mathcal{J}H}$ and $\nu_{jr}^{\mathcal{J}L}$ are scale values for joint signal i and high rating j on dimension r, representing the $\mathcal{J}H$ association; $\omega_{is}^{\mathcal{J}L}$ and $\eta_{ks}^{\mathcal{J}L}$ are scale values for joint signal i and low rating k on dimension s, representing the $\mathcal{J}L$ association; ν_{jt}^{HL} and η_{kt}^{HL} are the scale values for high rating j and low rating k, representing the HL association on dimension t; and $\sigma_{\mathcal{J}H(r)}^2$, $\sigma_{\mathcal{J}L(s)}^2$, and $\sigma_{HL(t)}^2$ are association parameters that measure the strength of the $\mathcal{J}H$, $\mathcal{J}L$, and HL relationships on dimensions r, s, and t, respectively. For identification, location constraints are required on all parameters (e.g., $\sum_i \lambda_i^{\mathcal{J}} = \sum_i \omega_{ir}^{\mathcal{J}H} = 0$), and additional scaling and orthogonality constraints are required for the scale values (e.g., $\sum_i \omega_{ir}^{\mathcal{J}H} \omega_{ir'}^{\mathcal{J}H} = 1$ if $r = r'$ and 0 otherwise).

The model in Equation 21.13 is equivalent to log-linear model Equation 21.12 when $R = \min(I,J) - 1$, $S = \min(I,K) - 1$, and $T = \min(J,K) - 1$. Models where $R < \min(I,J) - 1$, where $S < \min(I,K) - 1$, or where $T < \min(J,K) - 1$ are all special cases of the log-linear model in Equation 21.12. Additional special cases can be obtained by placing equality restrictions on the scale values and association parameters across interaction terms, such as $\omega_{ir}^{\mathcal{J}H} = \omega_{is}^{\mathcal{J}L}$ and $\sigma_{\mathcal{J}H(r)}^2 = \sigma_{\mathcal{J}L(s)}^2$ for $r = s$. Based on our experience and published applications, models with one or two dimensions often fit data well (i.e., R, S, and/or $T = 1$ or 2).

Given Equation 21.13, the conditional odds ratios are functions of the scale values and association parameters. Based on the model in Equation 21.13,

$$\log(\theta_{ii',jj'(k)}) = \sum_r \sigma_{\mathcal{J}H(r)}^2 (\omega_{ir}^{\mathcal{J}H} - \omega_{i'r}^{\mathcal{J}H})(\nu_{jr}^{\mathcal{J}H} - \nu_{j'r}^{\mathcal{J}H})$$

$$\log(\theta_{ii',kk'(j)}) = \sum_s \sigma_{\mathcal{J}L(s)}^2 (\omega_{is}^{\mathcal{J}L} - \omega_{i's}^{\mathcal{J}L})(\eta_{ks}^{\mathcal{J}L} - \eta_{k's}^{\mathcal{J}L})$$

$$\log(\theta_{jj',kk'(i)}) = \sum_t \sigma_{HL(t)}^2 (\nu_{jt}^{HL} - \nu_{j't}^{HL})(\eta_{kt}^{HL} - \eta_{k't}^{HL})$$

Using the fact that odds ratios are functions of scale values and association parameters, plots of scale values provide visual displays of the dependence structure in the data as measured by a multiplicative definition of interaction (i.e., odds ratios).

With higher-way tables, not only does one have the flexibility to decide what interactions to represent by multiplicative term(s), but also what restrictions should be placed on parameters. Given the large number of possibilities, we turn to an approach that provides guidance for the construction of an appropriate model or a subset of models for a given application.

Log-multiplicative latent-variable models

The approach advocated here starts with a researcher's substantive theories about underlying processes. We describe in this section a latent-variable model from which we derive log-multiplicative association models (Anderson and Böckenholt 2000; Anderson and Vermunt 2000; Anderson 2002). The latent-variable model described here is based on statistical graphical models for discrete and continuous variables (Lauritzen and Wermuth 1989; Whittaker 1989; Wermuth and Lauritzen 1990). Just as we have graphical representations of log-linear models, adopting the latent-variable perspective provides graphical representations of log-multiplicative association models.

In the latent-variable model, the discrete variables are observed and the continuous variables are unobserved. In this chapter, the observed discrete variables, or a subset of them, are conditionally independent given the latent continuous variables. Provided that a number of assumptions are met, the model derived for the observed discrete variables is a log-multiplicative association model (for details, see Anderson and Vermunt 2000; Anderson 2002).

Examples of graphical representations for latent-variable models implying log-multiplicative models are given in Figure 21.2. In these figures, observed (discrete) variables are represented by boxes, and the latent (continuous) variables are represented by circles. Lines connecting variables indicate that the variables may be conditionally dependent, and the absence of a line between two variables indicates that the variables are conditionally independent. Note that in graphs 2a and 2d in Figure 21.2 three of the observed variables (boxes) are conditionally independent of each other given the latent variable(s). The other two graphs in Figure 21.2 (i.e., graphs 2b and 2c) permit conditional dependence between two of the observed discrete variables (i.e., the high and low ratings). Unstructured two-way interaction terms are included if

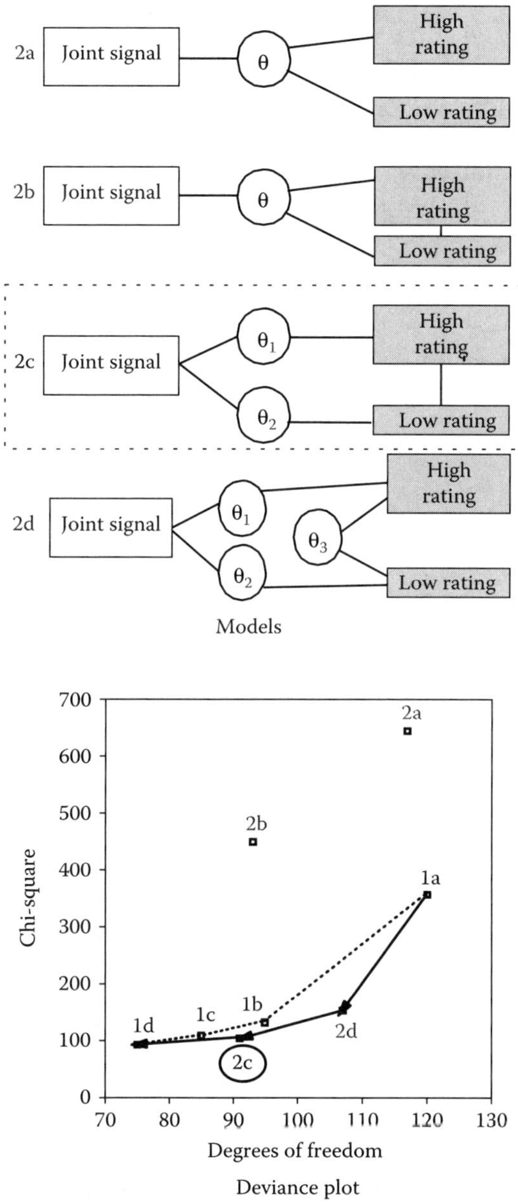

Figure 21.2 Concurrent detection data: selection of latent-association models. Boxes are observed variables; circled θ indicates latent variables. Model 2c is the preferred one.

discrete (observed) variables are connected by a line. A multiplicative term is included if there is a path between two observed discrete values that passes through one latent variable. For example, consider graph 1a in Figure 21.2, which postulates one latent variable (perhaps a subjective impression of the joint stimulus) where each of the ratings is related to this latent variable. The log-multiplicative model with this graphical representation is

$$\log(\pi_{ijk}) = \lambda + \lambda_i^{\mathcal{J}} + \lambda_j^H + \lambda_k^L + \sigma^2\omega_i^{\mathcal{J}}\nu_j^H + \sigma^2\omega_i^{\mathcal{J}}\eta_k^L \tag{21.14}$$

where σ^2 is the variance of latent variable θ and, given the additive model assumptions for the mean of the latent variable in cell (i,j,k), is

$$\mu_{ijk} = \sigma^2(\omega_i^{\mathcal{J}} + \nu_j^H + \eta_k^L)$$

The term $\sigma^2\omega_i^{\mathcal{J}}\nu_j^H$ is included in the model because there is a path between the "Joint signal" and the "HIGH RATING" that goes through the latent variable, and the term $\sigma^2\omega_i^{\mathcal{J}}\eta_k^L$ is included in the model because there is a path between the "joint signal" and the "low rating" that goes through the latent variable. A slightly more complex model is given in graphs 2b in Figure 21.2, which has a line connecting the high and low ratings. The algebraic model for graph 2b is Equation 21.14 with the (unstructured) interaction term λ_{jk}^{HL} added to represent a possible association between the high and low ratings.

In the deviance plot in Figure 21.2, the "convex hull" connecting the outer points on the lower edge is drawn to identify candidate models for selection. The models on the hull are preferable to models inside the hull because they have a more favorable deviance-d.f. ratio. In principle, one prefers to have a low χ^2 with as many degrees of freedom as possible. The models corresponding to graphs 2a and 2b in Figure 21.2 clearly do not fit very well relative to the log-linear models. The latent-variable structure for these models is too simple; therefore, we add an additional latent variable. It may be the case that there are separate internal (subjective) impressions of the signals and that each of the ratings is based on their subjective impressions. Graphs 2c and 2d in Figure 21.2 are diagrams for this conjecture. The log-multiplicative model corresponding to graphic 2c in Figure 21.2 is

$$\log(\pi_{ijk}) = \lambda + \lambda_i^{\mathcal{J}} + \lambda_j^H + \lambda_k^L + \lambda_{jk}^{HL} + \sigma_1^2\omega_i^{\mathcal{J}H}\nu_j^H + \sigma_2^2\omega_i^{\mathcal{J}L}\eta_k^L \tag{21.15}$$

where σ_1^2 and σ_2^2 are the variances of latent variables θ_1 and θ_2, respectively. Since there is a line connecting the box labeled "HIGH RATING" and "low rating," we included the term λ_{jk}^{HL}. We include the multiplicative term $\sigma_1^2\,\omega_i^{JH}\nu_j^H$ because there is a path between the box labeled "Joint Signal" and "HIGH RATING" that passes through the circle labeled "θ_1." The second multiplicative term $\sigma_2^2\,\omega_i^{JL}\eta_k^L$ was included because there is a path between the box labeled "Joint Signal" and "low rating" that passes through the circle labeled "θ_2." In the model corresponding to graph 2d in Figure 21.2, the multiplicative term $\sigma_3^2\,\nu_j^{HL}\eta_k^{HL}$ replaces λ_{jk}^{HL}.

Comparing the fit of the log-linear models from Figure 21.1 (i.e., graphs 1a to 1d) and the log-multiplicative models corresponding to graphs 2c and 2d in Figure 21.2, the latter are more parsimonious. Based on a detailed analysis, Anderson (2002) considered model 2c to be adequate for describing the patterns in the data. Note that although the fits of model 1c and model 2c are almost equal, the latter model retains more degrees of freedom. This model has two latent variables, one for each signal. We can further refine this model by placing restrictions on the parameters that correspond to specific hypotheses. For example, the way the subject "translates" the subjective impressions of a signal into a response may be the same regardless of whether rating the high or low signal. This conjecture would be represented by restricting $\nu_j^H = \eta_j^L = \nu_j$ for all $j = 1,\ldots,6$. Specific hypotheses regarding the differential effect of the design variables on ratings were tested by placing restrictions on ω_i^H and/or ω_k^L. Additional restrictions were placed on the interactively coded design variables to test, for example, whether the subjective confidence of the low signal when it is absent is the same regardless of whether the high signal is present (i.e., $\omega_1^{JL} = \omega_2^{JL}$).

Figure 21.3 represents the final model with the paths between variables labeled by parameters of the model representing the association. Table 21.2 contains the parameters' estimates of the final model that include restrictions on the parameters.

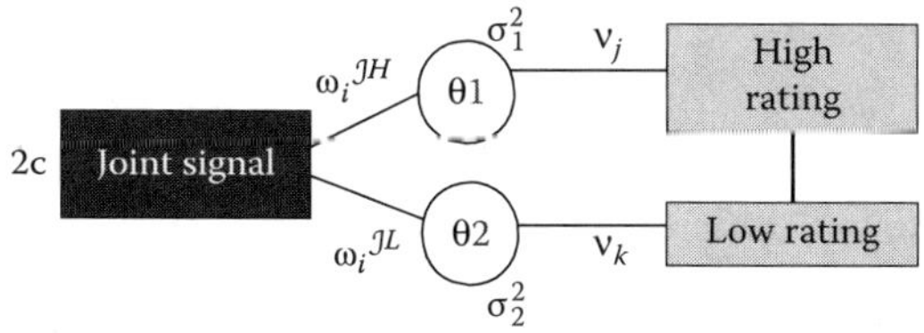

Figure 21.3 Concurrent detection data: final latent-association model with a joint-signal variable as a factor and two latent variables.

Table 21.2 Concurrent detection data: parameters of the final latent-association model and quantifications of categorical PCA (CatPCA) analysis.

Variable	Level	Parameter Estimate	CatPCA Quantifications[a] High	CatPCA Quantifications[a] Low
Ratings (high confidence = low confidence) (ν_j)	1	–0.41	–0.47	–0.44
	2	–0.37	–0.38	–0.38
	3	–0.22	–0.33	–0.21
	4	–0.05	–0.11	0.06
	5	0.31	0.28	0.41
	6	0.74	0.66	0.66
Low ratings ($\omega_i^{\mathcal{J}L}$)	HIGH – & low –	–0.50		
	HIGH + & low –	–0.50		
	HIGH – & low +	0.58		
	HIGH + & low +	0.44		
	Variance	3.40	dim. 1	dim. 2[b]
HIGH RATINGS ($\omega_i^{\mathcal{J}H}$)	HIGH – & low –	–0.58	–0.49	–0.74
	HIGH – & low +	–0.39	–0.51	–0.04
	HIGH + & low –	0.41	0.58	0.10
	HIGH + & low +	0.58	0.41	0.67
	Variance	2.69		

[a] CatPCA quantifications are scaled to unit length.
[b] 45° rotated versions of category quantifications from a CatPCA on the three-way table (see Section 21.6 with joint signal ($\mathcal{J}$) as one of the variables). These quantifications are independent of the confidence ratings.

Note that the differences between the successive pairs of scale values increase with the values of the ratings. Variances can be interpreted as measures of strength of the relationships between stimuli and responses. The estimated cell means of the subjective confidence for the high signal, θ_1, are equal to $\mu_{1(ij)} = \sigma_1^2(\omega_i^{\mathcal{J}H} + \nu_j)$, and those for the low signal, θ_2, are equal to $\mu_{2(ik)} = \sigma_2^2(\omega_i^{\mathcal{J}L} + \nu_k)$.

21.5 Additive models: three-way correspondence analysis

In Section 21.3 it was pointed out that models using the multiplicative definition of interaction decompose each group of interactions separately, while models using the additive definition of interaction decompose the complete dependence with a single multiplicative model.

The decomposition of the separate interactions is then derived from the global one. The most prominent model using the additive definition is three-way correspondence analysis (CA) and, in contrast with the association models, no distributional assumptions are made or used. In this sense, three-way CA is an exploratory technique. Its most important properties are that (1) the dependence between the variables in a three-way table is additively decomposed such that the relative size of each interaction term can be assessed with respect to each other; (2) a single multiplicative model for dependence operating at the level of proportions rather than on the log-scale is used to model global, marginal, and partial dependence, allowing for additively assessing the contributions of these interactions to the modeled dependence; and (3) graphical comparisons between all interactions can be made within a single graph. Full details can be found in Carlier and Kroonenberg (1996, 1998); earlier technical results are contained in Dequier (1974), Choulakian (1988b), and Kroonenberg (1989).

21.5.1 Measuring dependence

Measuring global dependence

The global dependence between the joint signal $\mathcal{J}$ and the confidence in the low signal L and the confidence in the high signal H is measured by the *mean squared contingency* Φ^2, defined as

$$\Phi^2 = \sum_i \sum_j \sum_k \frac{(p_{ijk} - p_{i\cdot\cdot}p_{\cdot j\cdot}p_{\cdot\cdot k})^2}{p_{i\cdot\cdot}p_{\cdot j\cdot}p_{\cdot\cdot k}} = \sum_i \sum_j \sum_k p_{i\cdot\cdot}p_{\cdot j\cdot}p_{\cdot\cdot k}\left[\frac{p_{ijk}}{p_{i\cdot\cdot}p_{\cdot j\cdot}p_{\cdot\cdot k}} - 1\right]^2$$

$$= \sum_i \sum_j \sum_k p_{i\cdot\cdot}p_{\cdot j\cdot}p_{\cdot\cdot k}P_{ijk}^2 \qquad (21.16)$$

where $P_{ijk} = p_{ijk}/p_{i\cdot\cdot}p_{\cdot j\cdot}p_{\cdot\cdot k}$ measures the global dependence of the (i,j,k)th cell (see Chapter 20). Φ^2 is based on the deviations from the three-way independence model, and it contains the global dependence due to all two-way interactions and the three-way interaction.

For instance, the two-way marginal total of the two confidence ratings is defined as the sum over the joint signal weighted by its marginal proportion. If the conditional proportions for all values of the joint signal i are equal, then $p_{jk|i} = p_{\cdot jk}$. Then $P_{ijk} = P_{\cdot jk}$, and the three-way table can be analyzed with ordinary CA between the two confidence ratings. The symmetric statement after permutation of the indices holds as well. One-way marginal totals are weighted sums over two indices, and they are zero due to the definition of P_{ijk}, and thus the overall total is zero as well.

Measuring marginal and three-way dependence

The global dependence P_{ijk} can be split into additive contributions of the two-way interactions and the three-way interaction (see also Equation 21.5).

$$P_{ijk} = \frac{p_{ij\cdot} - p_{i\cdot\cdot}p_{\cdot j\cdot}}{p_{i\cdot\cdot}p_{\cdot j\cdot}} + \frac{p_{i\cdot k} - p_{i\cdot\cdot}p_{\cdot\cdot k}}{p_{i\cdot\cdot}p_{\cdot\cdot k}} + \frac{p_{\cdot jk} - p_{\cdot j\cdot}p_{\cdot\cdot k}}{p_{\cdot j\cdot}p_{\cdot\cdot k}} + \frac{p_{ijk} - \tilde{p}_{ijk}}{p_{i\cdot\cdot}p_{\cdot j\cdot}p_{\cdot\cdot k}} \tag{21.17}$$

where $\tilde{p}_{ijk} = p_{ij\cdot}p_{\cdot\cdot k} + p_{i\cdot k}p_{\cdot j\cdot} + p_{\cdot jk}p_{i\cdot\cdot} - 2p_{i\cdot\cdot}p_{\cdot j\cdot}p_{\cdot\cdot k}$ (see Equation 21.6). The last term of Equation 21.17 measures the contribution of three-way interaction to the dependence for cell (i,j,k) (see also Section 21.3, Equation 21.5).

Using the definition of global dependence of cells, the measure of global dependence of a table is defined as the weighted sum over all cell dependencies, $\Phi^2 = \sum_{ijk} P_{i\cdot\cdot}P_{\cdot j\cdot}P_{\cdot\cdot k}P_{ijk}^2$. Due to the additive splitting of the dependence of individual cells, Φ^2 can also be partitioned additively

$$\Phi^2 = \sum_i \sum_j p_{i\cdot\cdot}p_{\cdot j\cdot}\left(\frac{p_{ij\cdot} - p_{i\cdot\cdot}p_{\cdot j\cdot}}{p_{i\cdot\cdot}p_{\cdot j\cdot}}\right)^2 + \sum_i \sum_k p_{i\cdot\cdot}p_{\cdot\cdot k}\left(\frac{p_{i\cdot k} - p_{i\cdot\cdot}p_{\cdot\cdot k}}{p_{i\cdot\cdot}p_{\cdot\cdot k}}\right)^2$$

$$+ \sum_j \sum_k p_{\cdot j\cdot}p_{\cdot\cdot k}\left(\frac{p_{\cdot jk} - p_{\cdot j\cdot}p_{\cdot\cdot k}}{p_{\cdot j\cdot}p_{\cdot\cdot k}}\right)^2 + \sum_i \sum_j \sum_k p_{i\cdot\cdot}p_{\cdot j\cdot}p_{\cdot\cdot k}\left(\frac{p_{ijk} - \tilde{p}_{ijk}}{p_{i\cdot\cdot}p_{\cdot j\cdot}p_{\cdot\cdot k}}\right)^2$$

$$= \Phi^2_{IJ} + \Phi^2_{IK} + \Phi^2_{JK} + \Phi^2_{IJK} \tag{21.18}$$

The importance of the decomposition in Equation 21.18 is that it provides measures of fit for each of the interactions and thus their contributions to the global dependence.

The left-hand panel of Table 21.3 shows this partitioning for the Wickens–Olzak data. The two-way margin of the two confidence ratings $L \times H$ is distinctly smaller (18%) than the two-way interaction $\mathcal{J}H$ of the joint signal with the high ratings (28%) and $\mathcal{J}L$ between the joint signal and the low ratings (33%), with the three-way interaction in between (21%). The $L \times H$ interaction is not really interpretable given the presence of the other interactions, because it is the sum of the four tables in Table 21.1. Summing the graphs virtually eliminates all

Table 21.3 Concurrent detection data: partitioning fit for the $2 \times 2 \times 2$ CA model.

Source	*df*	Global Dependence χ^2	Global Dependence %	Residual Dependence χ^2	Residual Dependence %	Percent of Interaction[a]
Joint signal ($\mathcal{J}$)	3	0	0	0	0	—
HIGH RATINGS (H)	5	0	0	9	2	—
Low ratings (L)	5	0	0	1	0	—
$\mathcal{J} \times H$	15	357	28	18	4	5
$\mathcal{J} \times L$	15	416	33	15	4	4
$L \times H$	25	222	18	199	46	90
Three-way interaction	75	261	21	193	44	74
Total dependence	130	1256	100	436	100	35

[a] Percent of interaction = Residual χ^2/Global $\chi^2 \times 100\%$; e.g., 90% = 199/222 × 100%.

systematic patterns present in the data, because each has its maximum in another corner of the table. The other two two-way interactions have straightforward interpretations, as they indicate to what extent there is a high (low) confidence in the presence of a particular combination of signals. Finally, the three-way interaction represents about a fifth of the dependence. Due to the presence of error, generally only a small part of this interaction contains interpretable information about the mutual influence of the ratings and the joint signal.

21.5.2 *Modeling dependence*

Up to this point, the decomposition of Equation 21.17 is the additive parallel of the log-linear model of Equation 21.11. In particular, it has separate terms for each of the interactions, but no decomposition of the interaction terms themselves has been considered yet. In ordinary CA, the SVD is used to acquire a more detailed insight into the nature of the dependence, and in three-way CA a three-way analogue of the SVD is needed.

Modeling global dependence

There are several candidates for the three-way generalization of the SVD, in particular Tucker's three-mode decomposition (Tucker 1966) and Harshman's (1970) PARAFAC. In this chapter, we will discuss only the Tucker3 model, i.e., the global dependence is decomposed as

$$P_{ijk} = \sum_r \sum_s \sum_t g_{rst} a_{ir} b_{js} c_{kt} + e_{ijk} \tag{21.19}$$

where a_{ir} are scale values for the joint signal, and they are orthogonal with respect to their weights $p_{i\cdot\cdot}$ (i.e., $\sum_i p_{i\cdot\cdot} a_{ir} a_{ir'} = 1$ if $r = r'$ and 0 otherwise). Similarly, the b_{js} are the orthogonal scale values for the confidence in the presence of the high signal, and the c_{kt} are those for the confidence in the presence of the low signal, with dimensions r, s, and t, respectively. The g_{rst} are the three-way association parameters or analogues of the singular values, and the e_{ijk} represent the errors of approximation. In three-way CA, a weighted least-squares criterion is used: the parameters g_{rst}, a_{ir}, b_{js}, and c_{kt} are those that minimize

$$\sum_i \sum_j \sum_k p_{i\cdot\cdot} p_{\cdot j\cdot} p_{\cdot\cdot k} e_{ijk}^2$$

Thus, the global measure of dependence, Φ^2, can be split into a part fitted with the three-way SVD and a residual part.

Modeling marginal dependence

The marginal dependence of the joint signal i and high ratings j is equal to

$$P_{ij\cdot} = \left(\frac{p_{ij\cdot} - p_{i\cdot\cdot} p_{\cdot j\cdot}}{p_{i\cdot\cdot} p_{\cdot j\cdot}} \right)$$

with similar expressions for the other two marginal dependencies. The elements $P_{ij\cdot}$ are derived via a weighted summation over k from the global dependence of the cells,

$$P_{ij\cdot} = \sum_k p_{\cdot\cdot k} P_{ijk} \tag{21.20}$$

Given that we have modeled the global dependence P_{ijk} with the Tucker3 model in Equation 21.19, we can use Equation 21.20 to find the model for the marginal dependence,

$$P_{ij\cdot} = \sum_r \sum_s \sum_t g_{rst} a_{ir} b_{js} c_{\cdot t} + e_{ij\cdot} \tag{21.21}$$

with $c_{\cdot t} = \sum_k p_{\cdot\cdot k} c_{kt}$ and $e_{ij\cdot} = \sum_k p_{\cdot\cdot k} e_{ijk}$. Inspecting this formula leads to the conclusion that the marginal dependence between the joint signal and the high ratings is derived from the overall model by averaging the low-ratings components.

Whereas in standard CA an optimal common dimension has to be determined for the rows and the columns, three-way CA requires a determination of the numbers of components for all three ways. For the Wickens–Olzak data, all models with dimensions ≤3 for each way were investigated and displayed in a deviance plot. The most relevant models are included in Figure 21.4 together with the relevant log-linear and log-multiplicative models. From this figure we can see that the confidence ratings each needed two dimensions, and the joint-signal mode needed either two or three dimensions. Given the desire for relative simplicity—and the fact that the χ^2 percentages of the signal dimensions for the 2 × 2 × 2 model are 39% and 26%, and for

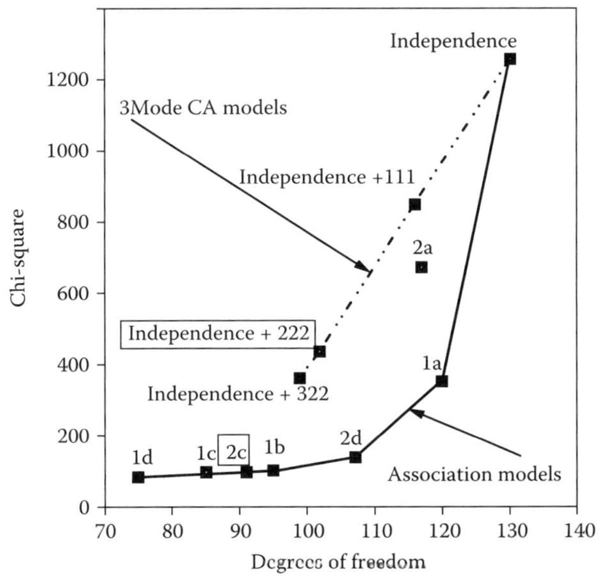

Figure 21.4 Concurrent detection data: deviance plot containing both multiplicative and additive models. "Independence" indicates the model of three-way independence; "Independence + *rst*" indicates the independence model with a decomposition of the global dependence by an $r \times s \times t$-Tucker3 model. "Independence + 222" is the preferred additive model, and Model 2c is the preferred multiplicative model.

the $3 \times 2 \times 2$ model 39%, 26%, and 7%—led us to prefer the former. Thus of the total χ^2 in the three-way table, 65% is fitted by the three-way CA.

From Figure 21.4 it can be seen that the multiplicative models clearly outperform the additive models; they combine the same degrees of freedom with a lower chi-square. It is not quite clear why the CA fits so dramatically worse. One reason might be that the additive models try to fit the complete interaction (i.e., the global dependence), while the log-multiplicative models fit only the three-way interaction.

The results for the $2 \times 2 \times 2$ three-way CA are summarized in the right-hand panel of Table 21.3. The $L \times H$ interaction has a bad fit of only 10%, but as argued above, it contains very little interesting information about the problem at hand anyway. The other two two-way interactions are fitted very well by the model, with 95% and 96% of their interaction explained by the model. Finally, only 26% of the three-way interaction is captured by the three-way model, but this is as expected. The fitted χ^2's of $\mathcal{J} \times H$ and $\mathcal{J} \times L$ together account for 90% of the total fitted χ^2 of 820 (= 1256 − 436). In other words, it is primarily these two interactions that are being fitted by the three-way CA.

21.5.3 Plotting dependence

The three-way CA model is symmetric in its three ways. However, this symmetry cannot be maintained when graphing the dependence because no spatial representations exist to portray all three ways simultaneously in one graph. A strict parallel with ordinary CA can therefore not be maintained. To display the dependence or its approximation in three-way CA, we will make use of a *nested-mode biplot*, previously called the *interactive biplot* in Carlier and Kroonenberg (1998).

Plotting global dependence: nested-mode biplot

The nested-mode biplot aims to portray all three ways in a single biplot. As a biplot has only two types of markers, two ways have to be combined into a single way. In the Wickens–Olzak data we have combined the two confidence ratings, indexed by (j,k), and represented it by a single marker ℓ. Thus, for the construction of the plot, the confidence ratings are coded interactively. The remaining mode, i.e., the joint signal, defines the plotting space and it also supplies a set of markers; it will be referred to as the *reference mode*.

The construction of the biplot for the fitted part of global dependence, designated as $\widehat{P}_{ijk}$, follows directly from three-way SVD of the global dependence.

$$\widehat{P}_{ijk} = \sum_r \left[\sum_s \sum_t g_{rst} b_{js} c_{kt} \right] a_{ir} = \sum_r d_{(jk)r} a_{ir}$$

$$\widehat{P}_{\ell i} = \sum_r d_{\ell r} a_{ir}. \tag{21.22}$$

By replacing the (jk) with a new index ℓ, we see that the coordinates of the combined response variables on component r are the $d_{\ell r}$ and that the a_{ir} are the coordinates of the joint signals. Given this construction, the combined response variables are in principal coordinates and the joint signals in standard coordinates. When plotting these coordinates in the nested-mode biplot, we will portray the combined response variables as points and the joint-signal coordinates as arrows.

The interpretation of the nested-mode biplot can be enhanced by exploiting the ordinal nature of both response variables. In particular, we can draw a grid by connecting in their natural order both the high-confidence ratings for each value of the low-confidence ratings and vice versa. The result is the grid in Figure 21.5, which shows the nested-mode biplot for the $2 \times 2 \times 2$ three-way CA. As 90% of the global dependence consists of the $\mathcal{J} \times H$ and $\mathcal{J} \times L$ dependence, these interactions primarily determine the shape of the plot.

What can one learn from such a plot? First, note that the coordinates of the joint signals ($\mathcal{L},\mathcal{H}$; $-\mathcal{L},\mathcal{H}$; $\mathcal{L},-\mathcal{H}$; and $-\mathcal{L},-\mathcal{H}$) form nearly a rectangular cross, indicating the relative independence of the high and low signals. The slight upward turn of the arrows for the condition where only one signal is present indicates that slightly more often the subject judged the not-presented signal to be present as well. A characteristic of the confidence grid is that the presence (absence) of a signal generates the corresponding appropriate confidence score. In addition, the lack of signal generally leads to low confidence scores but there is not much differentiation between the lower levels of the confidence scores, especially in the case of the absence of a low signal. The longer grid lines for the high-signal confidence, when the confidence in the presence of the low signal is at its lowest, indicate that in those circumstances the confidence in the presence of the high signal is much more marked.

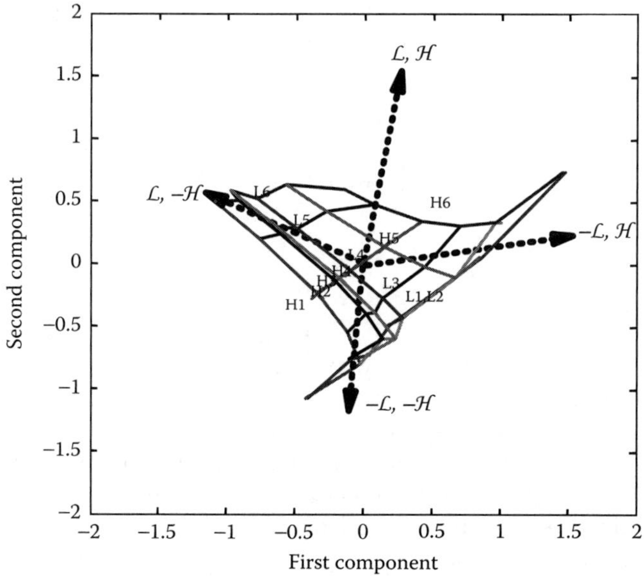

Figure 21.5 Concurrent detection data: nested-mode biplot from three-mode correspondence analysis. Hi (Li) *i*th level of confidence in presence of high (low) signal, i.e., coordinates of the HIGH RATINGS × Joint Signal (low ratings × Joint Signal) two-way margins. The script letters with associated arrows indicate the coordinates of the Joint-Signal categories.

Plotting marginal dependence

In Section 21.5.2 we saw that the marginal dependence of the joint signal i and the high ratings j was derived by summing over the mode with low ratings k

$$\hat{P}_{ij\cdot} = \sum_r \left[\sum_s \sum_t g_{rst} b_{js} c_{\cdot t} \right] a_{ir} = \sum_r d_{(j\cdot)r} a_{ir} \tag{21.23}$$

with $c_{\cdot t} = \sum_k p_{\cdot\cdot k} c_{kt}$, the weighted average of c_{kt} over k. This leads to the conclusion that the marginal coordinates for the high ratings can be derived from the overall model by averaging the appropriate components (here: the c_{kt}), and similarly for those of the low ratings by averaging the component scores d_{jkr} over j. Therefore, the $d_{j\cdot r}$ are the coordinates on the joint-signal axes for the marginal dependence of the high ratings. These marginal coordinates are indicated by $H1,\ldots,H6$ in

Figure 21.5, and similarly $L1,\ldots,L6$ are the coordinates for the marginal dependence of the low ratings.

21.6 Categorical principal component analysis

As explained in Chapter 4, a categorical principal component analysis (CatPCA) in which all variables have a multiple nominal measurement level is equivalent to multiple correspondence analysis (MCA). Meulman and Heiser (1998) investigated to what extent MCA is able to portray higher-order interactions in contingency tables using a $2 \times 2 \times 2 \times 2$ data set as an example.

The Wickens–Olzak data are a four-way data set with two ordinal six-point confidence-rating variables and two binary signal variables. If we want to apply the Meulman and Heiser conceptualization, we could carry out an MCA by performing a categorical PCA on the four variables without respecting their measurement levels. Given that in CatPCA one can do justice to the measurement levels, it seemed that the ordinal confidence ratings would be a better choice. Commonly, ordinal variables in the context of CatPCA are modeled with second-order splines with two interior knots (see, e.g., Hastie et al. 2001: chap. 5), and this is what was done here.

A CatPCA of the Wickens–Olzak data with two binary variables (presence or absence of a high or low signal) and two ordinal variables (measure of confidence in presence of high and low signal) was fitted in two dimensions, providing a fit of 75%. Also the results for the three-way table with the joint signal were examined, but these are essentially the same due to the completely balanced design. The grids shown in the biplot of the profiles and the variables (Figure 21.6) have been made in the same way as in the three-mode CA by connecting the confidence ratings in their natural orders. But here we have a separate grid for each signal presence–absence combination (cf. similar figures in Meulman and Heiser 1998). Each point in the plot represents a cell of the four-way contingency table. For instance, the topmost point is the cell with both signals present and both confidence ratings equal to 6. The lowest point is the cell with both signals absent and both confidence ratings equal to 1. In the present graph, the four grids are practically equal and translated versions of each other, illustrating additivity of the quantifications. The results suggest that the theory developed by Meulman and Heiser (1998) can be extended to ordinal variables, as these authors suggested but this subject needs further investigation.

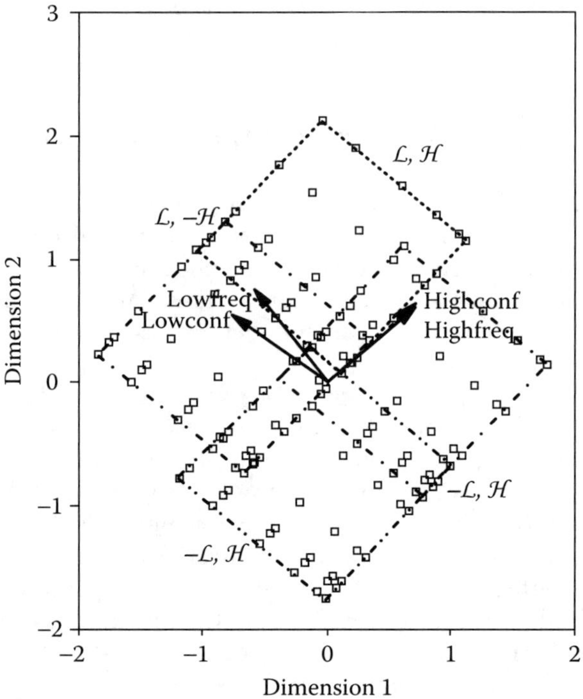

Figure 21.6 Concurrent detection data: symmetric two-dimensional biplot for categorical PCA with profile points connected per Joint Signal; $\mathcal{H}$,$\mathcal{L}$ = both low and high signal present, ⋯ , −$\mathcal{H}$,−$\mathcal{L}$ = both signals absent; arrows for two ordinal confidence variables and the two binary signal-present variables.

The optimal scaling transformation plots are given in Figure 21.7, and they show how the original categories of the ordinal confidence variables were transformed from their original numbering of 1 through 6. These graphs are based on the values reported in Table 21.2, and it can be observed that the categorical PCA parameters are very similar to the jointly estimated values of the latent-association model. The values for the joint-signal variable are also given in Table 21.2, where it can be seen that the first dimension of the categorical PCA concurs very well with the scale values from the latent-variable association model. These outcomes concur with an observation for the two-way case by Van der Heijden et al. (1994: 106), "The conclusion is that in CA the interaction is decomposed approximately in a log-multiplicative way.... This close relation between CA and models with log-bilinear terms holds also for more complex models."

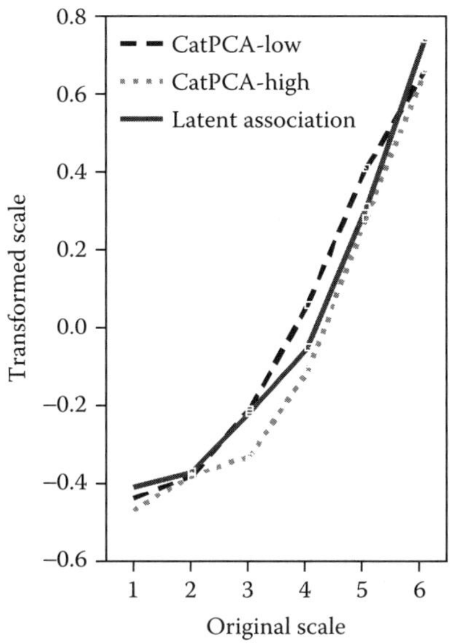

Figure 21.7 Concurrent detection data: transformed confidence scores vs. original scores for the latent-association model (joint transformation for both scores; see section 21.4.2) and for categorical PCA (separate transformations for each confidence score). The transformations for the ordinal scores in the categorical PCA are based on B-spline transformations of degree two and two interior knots (see Section 21.6).

21.7 Discussion and conclusions

The additive and multiplicative approaches to modeling are characterized by first fitting a model to the original frequencies. In the additive case, one can choose an appropriate model by examining the relative contributions of the interaction terms and then combining the chosen ones, while in the multiplicative case one has to fit series of models to assess the necessity of including each interaction term.

When dealing with variables with more than a few categories, straightforwardly fitting the models does not suffice because of the large number of parameters in the interaction terms. To model the interaction terms in both cases, multiplicative decompositions based on two-way or three-way SVD are used. Again, the two approaches differ in the way a model is "assembled." Within the additive approach, the complete dependence is decomposed and the contributions of the interactions are assessed to come to an appropriate model. In the multiplicative approach, first, a model consisting of several log-additive terms and one or more multiplicative decompositions of log-interactions is constructed, which is then assessed with respect to its goodness of fit. Several models constructed in this way are then compared to find the preferred model. There is an additional phase of sophistication for the multiplicative model in which latent variables are used to develop more parsimonious and powerful models.

Furthermore, methods have been developed to include restrictions on the estimated parameters for further simplification of the model and improved interpretation, as was shown in the Wickens–Olzak example. Such sophistication is at present not readily available for the additive approach to three-way tables, but several papers have appeared that introduce constraints in MCA (e.g., Böckenholt and Takane 1994; Hwang and Takane 2002). A sophisticated archeological example for the two-way case was presented by Groenen and Poblome (2002), but three-way extensions of this approach are not yet available.

With the multiplicative decomposition, several graphical representations are available for log-multiplicative models that can aid interpretation, such as graphs and schematic diagrams that represent the model itself as well as plots of scale values that sometimes are one-dimensional and thus easy to interpret. In the case of latent-variable models, one can fruitfully plot estimates of the means of the latent variables and use these to create insightful plots. With three-way CA, all dependence can be displayed in a single plot so that an integrated overall view of the global, marginal, and partial dependence can be obtained. However, such displays require careful inspection, and it is necessary to develop a fairly detailed understanding of what can and cannot be read in them.

The main interpretational conclusion of the paper is that both the multiplicative and additive models for three-way tables have much to offer compared with simple significance tests and numerical inspection of the interactions or modeling of them via log-linear models. When the variables have few categories, an inspection or log-linear modeling might still be an option, but for larger tables it is becomes almost impossible. If the interactions have a good structure and stochastic assumptions can be met, sophisticated multiplicative modeling can lead to parsimonious models with good interpretability. Moreover, it becomes possible to zoom in to test detailed hypotheses. When the assumptions are more difficult to meet or when the structure is messy, the additive approach has much to recommend itself, especially in supplying a grand overview of all aspects of the dependence, and as such it can even be useful when more-precise modeling is possible.

Table 21.4 Properties of multiplicative and additive models after Darroch (1974).

Property	Multiplicative Definition: Log-Linear	Multiplicative Definition: Log-Multiplicative	Additive Definition: Three-Mode CA
Partition properties	no	no	yes
Closest to independence	yes	yes	yes
Conditional independence as special case	yes	yes	no
Distributional equivalence	no	no	yes
Subtable invariance	yes	yes	no
No constraints on marginal probabilities	yes	yes	no
Statistical tests	yes	yes	no
Estimation	usually possible	can be difficult	always possible
Modeling of data	exploratory	confirmatory	exploratory
Ease of interpretation	difficult	simple given model found	good overview; can be complex

In the future, with more sophisticated machinery for the additive approach, it may be possible to model in a more refined way as well. For the present example, the log-multiplicative models outperformed the three-way CA in terms of deviance–degrees of freedom ratio, but why this is so is not yet properly understood.

CatPCA and MCA can also be used as an alternative when there are problems with log-multiplicative models, and it is expected that these will give similar results. However, a detailed theoretical analysis is required to establish differences and similarities between the models discussed in this chapter.

To summarize our attempt to introduce an interpretational element in the decision as to whether one should prefer multiplicative or additive modeling, we have extended Darroch's (1974) original table with properties of the models (Table 21.4). Its top part lists the properties of additive and multiplicative methods as given in Darroch while the lower part shows the interpretational properties of the models presented here. Unfortunately, our investigations have not led us to a more unequivocal conclusion than Darroch's. This is primarily due to the fact that interpretability is very much data dependent, and no model-based arguments can solve content-based limitations or preferences.

Software notes

The first author has written a program for three-way CA as part of his general suite of programs for three-way analysis, 3WayPack. Information about this software can be found on the Web site of The Three-Mode Company (http://three-mode.leidenuniv.nl/). An Splus program (with mainly French commentary and output) is also available. It was written by the late André Carlier and is distributed via the Laboratoire de Statistique et Probabilité (LSP), Université Paul Sabatier, Toulouse, France. Further information can be obtained from the LSP Web site (http://www.lsp.ups-tlse.fr/index.html); after reaching the site, the user should search for "multidim."

Log-multiplicative modeling (including latent variables) was carried out with *ℓEM* (Jeroen Vermunt, University of Tilburg, Tilburg, The Netherlands) (http://www.uvt.nl/faculteiten/fsw/organisatie/departementen/mto/software2.html). The output and details of the log-multiplicative models of the Wickens–Olzak data, as presented in the paper by Anderson (2002), can be found on her Web site (http://www.ed.uiuc.edu/faculty/cja/lem/index.html).

Categorical PCA was carried out with software developed by the Leiden Data Theory Group as implemented in the program module *Categories* of SPSS 11.0 (Meulman and Heiser 2001; http://www.spss.com/spssbi/categories/index.htm).

Acknowledgments

This chapter was written while the first author was a fellow-in-residence at the Netherlands Institute for Advanced Study (NIAS) in the Humanities and Social Sciences. This work was supported by grant #SES-0351175 from the National Science Foundation, awarded to the second author.

CHAPTER 22

A New Model for Visualizing Interactions in Analysis of Variance

Patrick J.F. Groenen and Alex J. Koning

CONTENTS

22.1 Introduction

In many situations of empirical research, there is the need to predict a numerical variable by one or more categorical variables. Analysis of variance (ANOVA) is often used to test whether there are differences between means of the dependent variable across the categories of the predictors.

As soon as there are two or more categorical predictor variables, interaction effects may turn up. If so, then different combinations of the categories of the predictor variables have different effects. The majority of the papers in the literature only report an ANOVA table showing which effect is significant, but they do not present or interpret the terms belonging to the effects themselves.

Each effect is characterized by a number of terms, depending on the categories involved. For example, a main effect is characterized by J_q terms, where J_q is the number of categories of the predictor variable q. An interaction effect gives the combined effect of the categories of two or more variables. For example, the two-way interaction between variables q and r consists of $J_q \times J_r$ terms. In this chapter, we argue that it is worthwhile to consider the terms that constitute the effects directly. Doing so may be difficult if the number of categories and the number of categorical predictor variables grows because the number of interaction-effect terms will grow dramatically. Therefore, we describe a new interaction-decomposition model that allows two-way interactions to be visualized in a reasonably simple manner. The interpretation of the interaction plot is similar to that of correspondence analysis. Although, in principle, the method could be used to analyze higher-way interactions, we limit ourselves to two-way interactions only.

The remainder of this chapter is organized as follows. First, we discuss the empirical data set on predicting holiday spending and apply a common ANOVA. Then, we explain the interaction-decomposition model more formally and apply our model to the holiday-spending data set.

22.2 Holiday-spending data

The data concern the holiday spending of 708 respondents, and data were gathered by students of the Erasmus University Rotterdam in 2003. The purpose of this research is to predict the amount of holiday spending (in euros) from seven categorical predictor variables: number of children, income, destination, other big expenses, accommodation, transport, and holiday length. Table 22.1 gives the frequencies of each of the categories of the predictor variables. Most categories are reasonably filled, with the exception of number of children, where almost 80% of the respondents have no children, whereas the other 20% have one to five children.

For travel agencies it is important to understand the relation between amount of money spent on a holiday and the predictor variables. With this knowledge they can provide better-suited arrangements for their clients. It can be expected *a priori* that the variable "holiday spending" is heavily skewed. The reason for this is that there will always be a few people in a sample who are able to spend much more money on a holiday than the middle 50%. Such skewness is quite common in economics for variables such as price, income, and spending in general. To make the variable less skewed and more like a normal distribution, we take the logarithm of the holiday spending. The histogram of log holiday spending is given in Figure 22.1. Indeed, the logarithm transformation has made the variable behave much more like a normal distribution. Thus, throughout this chapter, log holiday spending will be the dependent variable.

The most obvious method to investigate the relation between holiday spending and the predictor variables is analysis of variance (ANOVA).

Table 22.1 Frequencies and percentages of categorical predictor variables for the holiday-spending data.

Holiday Length	Freq.	%	**Income**	Freq.	%
<7 days	65	9.2	<400 euro	75	10.6
7–13 days	277	39.1	400–799 euro	122	17.2
14–20 days	223	31.5	800–1599 euro	124	17.5
21–27 days	88	12.4	1600–3199 euro	237	33.5
28 ≥ days	55	7.8	≥ 3200 euro	150	21.2
Destination	Freq.	%	**Other Big Expenses**	Freq.	%
Within Europe	543	76.7	No big expenses	522	73.7
Outside Europe	165	23.3	Other big expenses	186	26.3
Number of Children	Freq.	%	**Accommodation**	Freq.	%
0 children	559	79.0	Camping	162	22.9
1 child	49	6.9	Apartment	189	26.7
2 children	63	8.9	Hotel	216	30.5
3+ children	37	5.2	Other	141	19.9
Transport	Freq.	%			
By car	261	36.9			
By airplane	377	53.2			
Other transport	70	9.9			

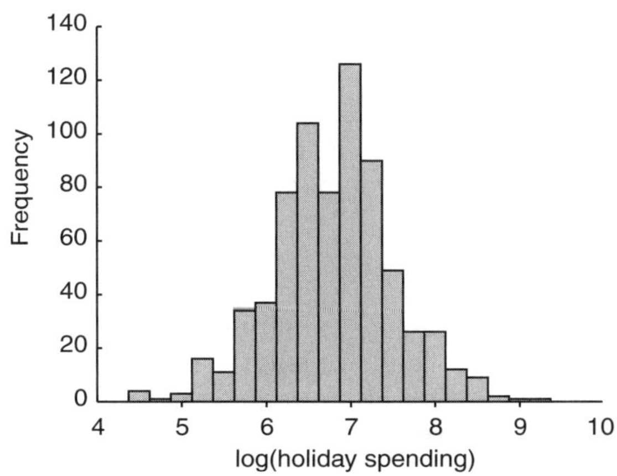

Figure 22.1 Histogram of the natural logarithm of the holiday spending.

In essence, ANOVA compares the means of the dependent variable with the groups defined by the predictor variables. ANOVA can be done in different ways. First, we can analyze the grouping for one predictor variable at a time. The disadvantage is that in our case there are seven different ANOVAs to be analyzed. Moreover, this approach imposes severe restrictions on the inferences to be drawn, especially so because the predictor variables form an unbalanced design. A better approach is to include all seven predictor variables simultaneously in the prediction, the so called main-effects model. In this manner, the predicted value of the dependent variable can be written conveniently as a linear sum of terms for each of the categories of the predictor variables.

However, joint effects of two predictor variables are not taken into account. For our data, it is possible that people with more children spend more and those who have longer holidays also spend more, but that the joint effect of having two or more children and having long holidays leads to less spending because cheaper overnight accommodations, for example, a camping site, are chosen. Such effects are called *interaction* effects. In principle, it is possible to consider interaction effects between any number of predictor variables, but we limit ourselves to two-way interactions, that is, we only consider joint effects of two predictor variables simultaneously. One important reason for doing so is that interpreting three- or higher-way interactions gets increasingly more difficult.

Let us look at an ANOVA of the holiday spending, including main effects and all two-way interactions of the predictor variables. Almost always, the results of the ANOVA are presented in an analysis of variance table that shows how the sum of squares of the dependent variable can be decomposed into contributions by the main and interaction effects and whether these effects are significant or not. In Table 22.2, these results are presented for the holiday-spending data. Note that the last column contains the partial η^2, which measures the proportion of the variance of the dependent variable accounted for by the current factor.

From Table 22.2, we see from the values of η^2 that important contributors to the prediction of holiday spending are the main effects for "income" and "holiday length" and the interaction effects of "transport" by "holiday length," "holiday length" by "children," and "holiday length" by "income." A few other main and interaction effects are significant and have reasonable effect size. The overall R^2 is equal to .606, also given as the η^2 of the total model. The column Type III sum of squares reports the reduction in residual sum of squares when the variable is entered after all other variables are modeled.

Table 22.2 ANOVA table of all main effects and all interaction effects for the holiday-spending data.

Source	Type III Sum of Squares[a]	d.f.	Mean Square	p	Partial η^2
Destination	1.756	1	1.756	.009	.012
Transport	0.256	2	0.128	.609	.002
Holiday length	4.576	4	1.144	.002	.031
Accommodation	1.357	3	0.452	.155	.009
Big expenses	0.376	1	0.376	.228	.003
Children	1.920	3	0.640	.060	.013
Income	4.778	4	1.194	.001	.032
Destination × transport	0.225	2	0.113	.646	.002
Destination × holiday length	0.770	4	0.192	.561	.005
Destination × accommodation	0.500	3	0.167	.585	.003
Destination × big expenses	0.293	1	0.293	.287	.002
Destination × children	1.763	3	0.588	.079	.012
Destination × income	1.531	4	0.383	.206	.010
Transport × holiday length	7.359	8	0.920	.000	.048
Transport × accommodation	2.561	6	0.427	.130	.017
Transport × big expenses	0.505	2	0.253	.376	.003
Transport × children	1.405	6	0.234	.489	.010
Transport × income	2.270	8	0.284	.362	.015
Holiday length × accommodation	1.888	12	0.157	.835	.013
Holiday length × big expenses	0.947	4	0.237	.453	.007
Holiday length × children	5.427	12	0.452	.053	.036
Holiday length × income	5.589	16	0.349	.159	.037
Accommodation × big expenses	0.151	3	0.050	.899	.001
Accommodation × children	3.547	9	0.394	.135	.024
Accommodation × income	3.152	12	0.263	.430	.021
Big expenses × children	0.825	3	0.275	.363	.006
Big expenses × income	3.060	4	0.765	.019	.021
Children × income	1.718	7	0.245	.466	.012
Total of model	222.224	147	1.512	.000	.606
Error	144.479	560	0.258	—	—
Dependent variable	366.703	707	—	—	—

Note: The "total model" sum of squares is not equal to the sum of all effects because not all combinations of categories were observed (i.e., the data form an unbalanced ANOVA design).

[a] The sum of squares is in deviation of the overall mean.

Because not all combinations of categories are observed here, the present data set forms an unbalanced design. As a consequence, the Type III sum of squares of all effects does not sum to 222.224, which is the variation of the prediction based on the total model. Dividing the Type III sum of squares by the degrees of freedom (d.f.) gives the mean square, which can be seen as an estimate of the variance of the effect. For example, we expect large effects for "destination," as its mean square is 1.756, for "income" with mean square 1.194, "holiday length" with mean square 1.144, and the interaction effect "transport × holiday length" with mean square .920.

However, to understand *how* a certain main or interaction effect affects the prediction, one has to inspect the estimated terms of the effect. In this chapter, we refer to a main effect as being the collection of terms belonging to the categories of a single predictor variable. A two-way interaction effect is the collection of terms belonging to all paired categories of two predictor variables. Considering our example, the number of parameters (and thus terms) to be considered depends on the number of categories per predictor variable. In total, there are 6 + 6 + 2 + 2 + 5 + 4 + 3 = 28 terms for the main effects and

$$
\begin{array}{cccccccccccccc}
6\times6 & + & 6\times2 & + & 6\times2 & + & 6\times5 & + & 6\times4 & + & 6\times3 & + & & \\
 & + & 6\times2 & + & 6\times2 & + & 6\times5 & + & 6\times4 & + & 6\times3 & + & & \\
 & & & + & 2\times2 & + & 2\times5 & + & 2\times4 & + & 2\times3 & + & & \\
 & & & & & + & 2\times5 & + & 2\times4 & + & 2\times3 & + & & \\
 & & & & & & & + & 5\times4 & + & 5\times3 & + & & \\
 & & & & & & & & & + & 4\times3 & = & 327 &
\end{array}
$$

terms for the interaction effects, summing to 28 + 327 = 355 terms to be interpreted. Clearly, this number of terms is too large to be interpreted. Certainly, one could choose to interpret only those main effects and interaction effects that are significant or have a high effect size, but even then, the number of terms to be interpreted can be quite large, especially if the number of categories of the predictor variables or the total number of predictor variables increases.

Another problem with interactions terms can occur if the predictor variables form an unbalanced design, which generally is the case for nonexperimental data. In this case, some of the interaction terms cannot be estimated due to the absence of relevant data.

Therefore, the reports in many studies are limited to an ANOVA table such as Table 22.2, thereby ignoring the most important part of the analysis, that is, the terms of the effects themselves. The main purpose of this chapter is to discuss a new approach that allows direct visualization of the interaction effects. The main advantage is that the effects are easier to interpret. The next section discusses the new model more formally.

22.3 Decomposing interactions

To express our interaction-decomposition model more formally, we need to introduce some notation. Let y_i be the value of the dependent variable, the logarithm of "holiday spending," for subject i, where $i = 1,\ldots,n$. Suppose there are Q categorical predictor variables. Each category can be presented as a dummy variable that equals 1 if subject i falls in that particular category and 0 otherwise. The collection of all dummy variables for a single categorical predictor variable q can be gathered in an $n \times J_q$ indicator matrix $\mathbf{Z}_q$, where J_q denotes the number of categories of variable q. The main-effects model in ANOVA is given by

$$y_i = c + \sum_{q=1}^{Q} \mathbf{z}_{iq}^{\mathsf{T}} \mathbf{a}_q + e_i \qquad (22.1)$$

where c is the overall constant, $\mathbf{a}_q$ is the $J_q \times 1$ vector of main effects for variable q, $\mathbf{z}_{iq}^{\mathsf{T}}$ is the ith row of $\mathbf{Z}_q$, and e_i is the error in prediction for subject i. In ANOVA, the main effects $\mathbf{a}_q$ and the constant c are estimated by minimizing the sum of squares of the errors e_i, that is, by minimizing the loss function

$$L_{\text{main}}(c, \mathbf{a}) = \sum_{i=1}^{n} \left(y_i - [c + \sum_{q=1}^{Q} \mathbf{z}_{iq}^{\mathsf{T}} \mathbf{a}_q]\right)^2 \qquad (22.2)$$

where, for notational convenience, $\mathbf{a}$ is a $J \times 1$ vector containing all main-effects terms, $J = \sum_q J_q$.

To specify an interaction effect between predictor variables q and r, consider the $J_q \times J_r$ matrix $\mathbf{B}_{qr}$ that contains all the terms of the interaction effect for all combinations of the categories of variables q and r. Suppose that subject i falls in category j of predictor variable q and in category k of predictor variable r. Then, the term of the interaction effect needed for this respondent is $b_{jk}^{(qr)}$, the element in row j and column

k of $\mathbf{B}_{qr}$. For this person i, this element can also be picked out using the appropriate indicator vectors

$$b_{jk}^{(qr)} = \mathbf{z}_{iq}^{\mathsf{T}}\mathbf{B}_{qr}\mathbf{z}_{ir}.$$

To make the notation more compact, let all interaction effects be gathered in the symmetric partitioned block matrix

$$\mathbf{B} = \begin{bmatrix} \mathbf{0} & \mathbf{B}_{12} & \cdots & \mathbf{B}_{1Q} \\ \mathbf{B}_{21} & \mathbf{0} & \cdots & \mathbf{B}_{2Q} \\ \vdots & \vdots & \ddots & \vdots \\ \mathbf{B}_{Q1} & \mathbf{B}_{Q2} & \cdots & \mathbf{0} \end{bmatrix}.$$

Note that the diagonal blocks are zero because $\mathbf{z}_{iq}^{\mathsf{T}}\mathbf{B}_{qq}\mathbf{z}_{iq}$ selects only the diagonal and thus estimates a main effect for variable q. Because main effects are already taken care of by $\mathbf{a}_q$, we set the diagonal blocks $\mathbf{B}_{qq}$ to be 0, since it does not make sense to model a main effect twice.

Now, the ANOVA model with all main effects and all two-way interaction effects minimizes

$$L_{\text{int}}(c,\mathbf{a},\mathbf{B}) = \sum_{i=1}^{n}\left(y_i - \left[c + \sum_{q=1}^{Q}\mathbf{z}_{iq}^{\mathsf{T}}\mathbf{a}_q + \sum_{q=1}^{Q}\sum_{r=q+1}^{Q}\mathbf{z}_{iq}^{\mathsf{T}}\mathbf{B}_{qr}\mathbf{z}_{ir}\right]\right)^2 \tag{22.3}$$

The number of parameters to be estimated is 1 for the constant, $\sum_{q=1}^{Q} J_q$ for the main effects in $\mathbf{a}$, and $\sum_{q=1}^{Q}\sum_{r=q+1}^{Q} J_q \times J_r$ for the interaction effects in $\mathbf{B}_{qr}$. Note that some additional constraints are necessary to prevent interaction effects from picking up main effects and the main effects from picking up the constant effect. The constraints that are often imposed include setting the sum of each of the main effects $\mathbf{a}_q$ to be equal to zero and setting the interaction effects $\mathbf{B}_{qr}$ such that the row and column sums are equal to zero. The number of free parameters to be estimated is accordingly reduced.

We now turn to the interaction-decomposition model proposed in this chapter. The key idea of this model is that the interaction terms $\mathbf{B}_{qr}$ are constrained such that an easy graphical representation is possible. The type of constrained used in the interaction-decomposition model is that of common rank reduction, that is, we require that

$$\mathbf{B}_{qr} = \mathbf{Y}_q\mathbf{Y}_r^{\mathsf{T}}, \tag{22.4}$$

where the $J_q \times S$ matrix $\mathbf{Y}_q$ has a rank not higher than S. Equivalently, we can write that, in the interaction-decomposition model, an interaction term of category j of predictor variable q and category k of predictor variable r is given by

$$b_{jk}^{(qr)} = \mathbf{y}_{qj}^{\mathsf{T}}\mathbf{y}_{rk}$$

where $\mathbf{y}_{qj}^{\mathsf{T}}$ denotes row j of $\mathbf{Y}_q$. Using this kind of constraint, the interaction term is graphically represented by a projection of the vector $\mathbf{y}_{qj}^{\mathsf{T}}$ onto $\mathbf{y}_{rk}^{\mathsf{T}}$. Thus, high projections indicate large interaction terms, and small projections indicate small interaction terms. Such a rank restriction is also used in bi-additive models such as correspondence analysis, multiple correspondence analysis, and joint correspondence analysis. The main advantage of the interaction-decomposition model is that there is only a single vector $\mathbf{y}_{qj}^{\mathsf{T}}$ to be estimated for each category of a predictor variable. In other words, the number of interaction parameters to be estimated only grows linearly with the number of categories, and not quadratically as for the unconstrained ANOVA interaction model in Equation 22.3.

Because projections do not change under rotation, the vectors in $\mathbf{Y}_q$ are determined up to a common rotation that is the same for all $\mathbf{Y}_q$. Other than the rotational indeterminacy, the $\mathbf{Y}_q$'s and thus the interactions $\mathbf{B}_{qr} = \mathbf{Y}_q\mathbf{Y}_r^{\mathsf{T}}$ can be estimated uniquely if the number of dimensions is low enough. This contrasts with standard ANOVA, which cannot estimate all interaction terms if some combinations of predictor categories are absent. For example, there are no observations that have income less then 400 euros and have one or more children. If all combinations were present, then the degrees of freedom for the interaction effect of "income" and "no. of children" would be $3 \times 4 = 12$. However, because of the absence of some combinations, Table 22.2 reports only the estimation of seven independent parameters.

Obviously, because the interaction-decomposition model imposes constraints on the interactions, it will generally not fit as well as the unconstrained ANOVA interaction model in Equation 22.3. Note that we still have to require that $\mathbf{B}_{qr}$ has zero row and column sum to avoid confounding of the main and interaction effects. This restriction implies that each $\mathbf{Y}_q$ must have column sum zero so that indeed $\mathbf{B}_{qr} = \mathbf{Y}_q\mathbf{Y}_r^{\mathsf{T}}$ will have row and column sums equal to zero. To fit the interaction-decomposition model, we have developed a prototype in MatLab that minimizes $L_{\text{int}}(c,\mathbf{a},\mathbf{B})$ subject to the constraints Equation 22.4 by alternating least squares.

22.4 Interaction-decomposition of holiday spending

Let us apply the interaction-decomposition model to the holiday-spending data. The R^2 for the main-effects-only model is .47 and for the interaction-decomposition model is .53, so it makes sense to consider the interaction effects.

First, we consider the main effects for all the categories (see Figure 22.2). The largest main effects are obtained by the variable "holiday length." Not surprisingly, we find that more money than

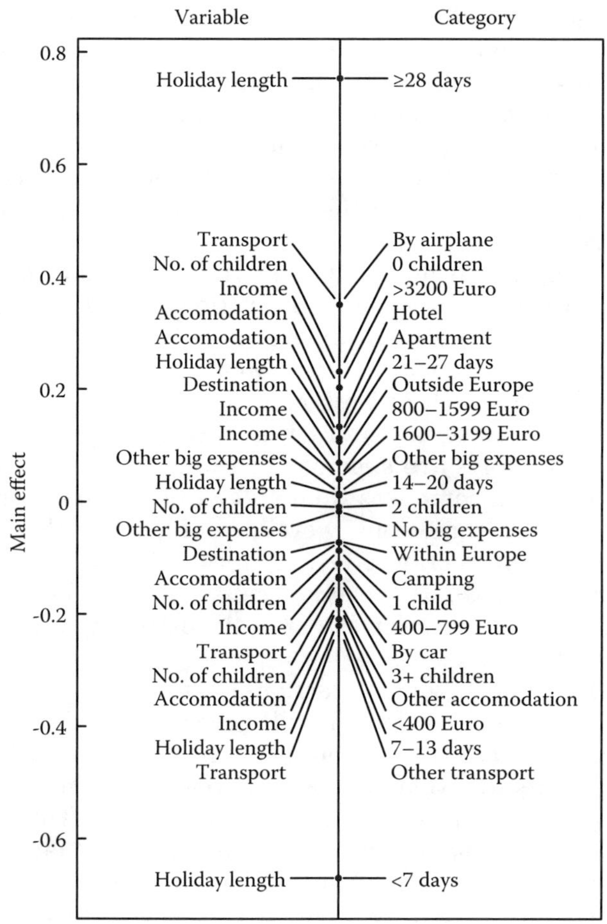

Figure 22.2 Main effects with respect to the overal mean c.

average is spent for holidays of 28 days or longer, whereas much less money is spent on holidays shorter than 7 days. Since the dependent variable is the logarithm of holiday spending, the effect of having holidays of 28 or more days implies that the holiday spending goes up by a factor exp(.75) = 2.12 and for holidays shorter than 7 days this factor is exp(−.67) = .51. For the remaining main effects, we choose to interpret only those effects larger than ±.2, as the holiday spending changes by a factor exp(.2) = 1.22 and thus increases by 22%. We see that more money is spent if the travel takes place by airplane, if there are no children, and if the income is 3200 euros or higher. On the other hand, holiday spending is reduced for holidays shorter than two weeks or made by transport other than car or airplane. The other main effects are reasonably small, suggesting that they are not very important.

The interaction effects are more interesting, because we can investigate the joint effects of two predictor variables. Figure 22.3 shows the plot of the interaction effects by the interaction-decomposition model. Panel (a) shows all effects simultaneously, and Panel (b) zooms in on the center part in the box in Panel (a). The basic way to interpret the interaction solution is as follows. First, condition on a single category of interest. Then project all categories of other variables onto this vector. High positive projections indicate a high positive interaction effect (thus, more money spent during the holiday), and high negative projections indicate a high negative interaction effect (thus, less money spent during the holiday). In addition, vectors that are exactly orthogonal have no interaction. Note that a reasonable interaction effect can still occur for vectors that are almost orthogonal if one or both of the vectors are long enough. Also, long vectors generally have larger interaction with all other categories.

Note that variables with only two categories will have equal-length vectors that are mirrored in the origin. In Figure 22.3b, we see an example of this case for the variable "other big expenses" (with categories "no big expenses" and "other big expenses"). The reason for two equal-length and opposite vectors lies in the restriction that the coordinates have zero column sum per variable.

To see how the holiday money is spent on holidays shorter than a week, we have projected the categories of other variables onto the vector of category "holiday, <7 days" in Figure 22.4. Thus, we are considering interaction terms conditioned on the category "holiday, <7 days." These conditional effects are also presented separately in Figure 22.5 summed with the main-effect term of about −.67 for the category "holiday, <7 days." The interaction-decomposition model predicts that holiday spending increases for these short holidays if the transport is

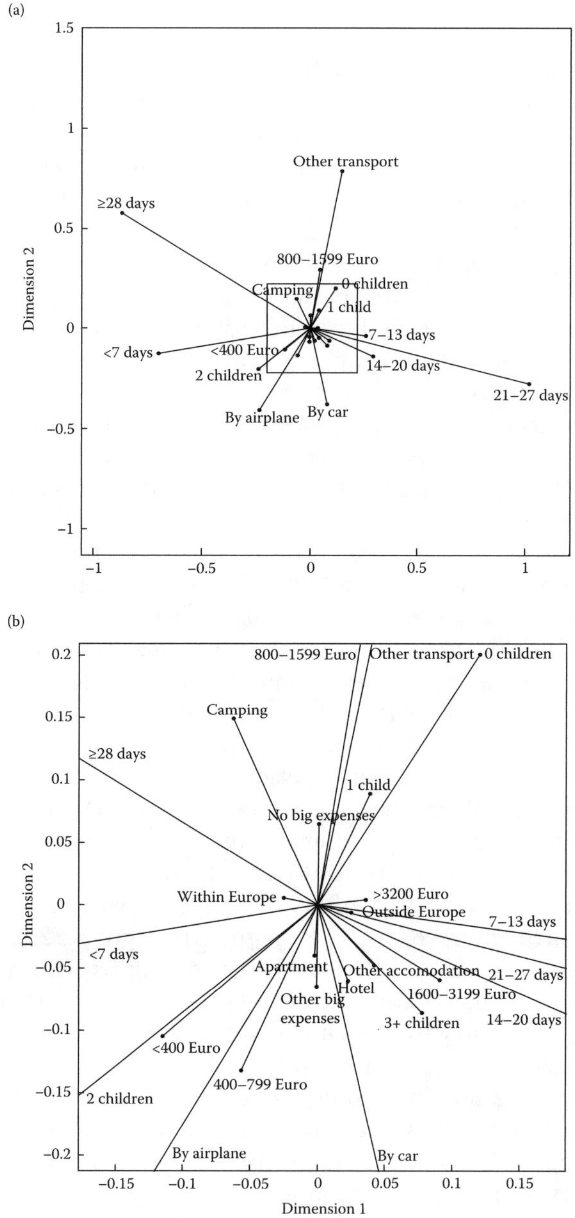

Figure 22.3 Interaction plot for the interaction-decomposition model on the holiday-spending data. Panel (b) zooms in on the box in Panel (a) to view the labels more clearly.

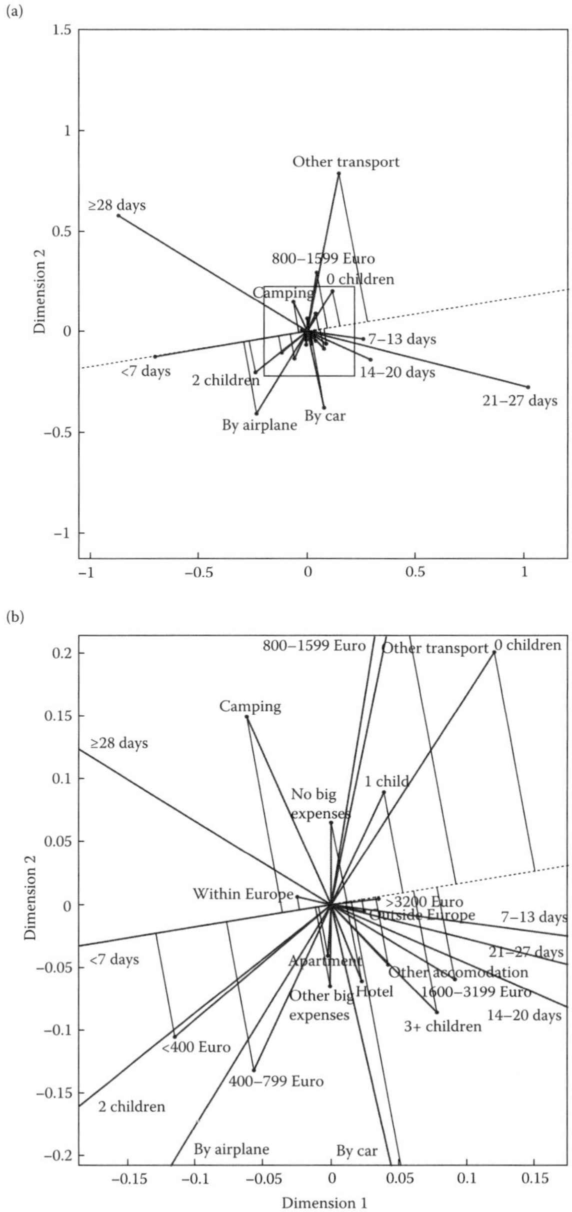

Figure 22.4 Interaction-decomposition plot with projections onto the category "holiday, <7 days" of holiday. Panel (b) zooms in on the box in Panel (a) to present the labels more clearly.

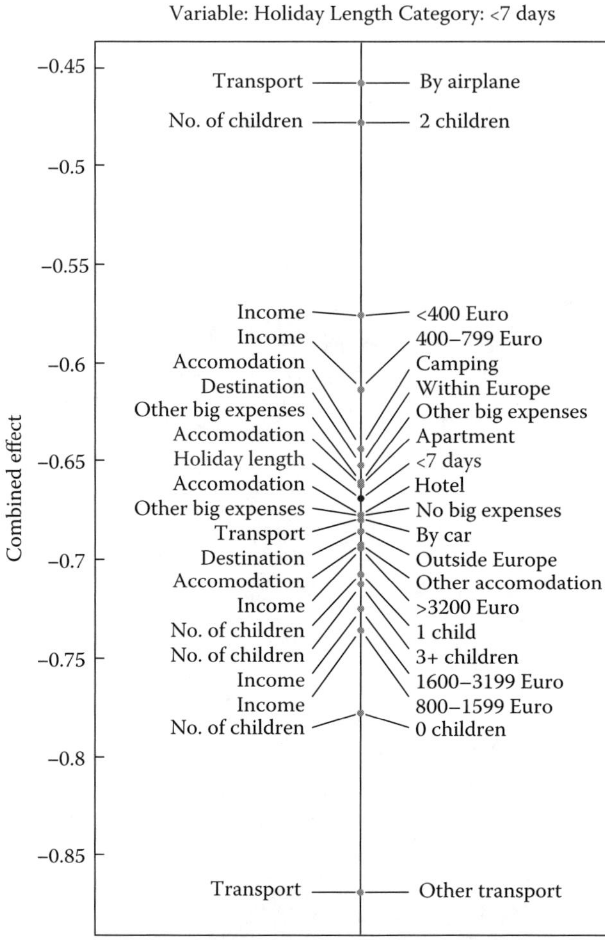

Figure 22.5 Sum of the main effect for the category "holiday <7 days" and the interaction effects of this category with the categories of all other variables.

by airplane or if there are two children. Compared with the main effect of "holiday, <7 days," the holiday spending decreases if other transport is used or if there are no children. In general, a plot as in Figure 22.5 is very helpful in interpreting the interaction terms conditioned on a certain category.

The most important interactions can also be derived directly from Figure 22.3 by looking at the largest vectors. The categories that matter most in the interactions are holidays that are shorter than 7 days, 21 to 27 days, or 28 days or more, and all three categories for transport (car, airplane, and other). For example, if other transport is used, the model predicts higher spending if the holiday is 28 days or more, there are no children, and the income is between 800 and 1599 euros (because these vectors project highly positive). Also, for other transportation, lower spending is predicted if there are two children or if the holiday is shorter than a week (because the vectors project negatively).

It is certainly possible to describe more interaction effects predicted by the interaction-decomposition model. Of course, we have to keep in mind that the solution does not fit perfectly. Therefore, it is a compromise solution that tries to capture as much information as possible about the relations present in the data.

22.5 Conclusions

To investigate interactions in the traditional ANOVA framework, most researchers limit themselves to an ANOVA table. In this chapter, we argued that it is worthwhile to study the interaction terms themselves. Because the number of interaction terms increases rapidly with the number of categorical predictor variables and the number of categories per variable, we have proposed a new model, called the interaction-decomposition model, that allows visualization of the two-way interactions.

For two categorical predictor variables, similar decomposition models have been proposed in the literature (Choulakian 1996; de Falguerolles and Francis 1992; Gabriel 1998; van Eeuwijk 1995). In psychometrics, such models have been known under the name FANOVA (Gollob 1968). A different method for modeling three-way interactions among three categorical predictors was presented by Siciliano and Mooijaart (1997). The main difference between their approach and ours is that they focus on three-way interactions, whereas we limit ourselves to two-way interactions only. We believe that two-way interactions convey the most important information while it is still reasonably easy to interpret. For three- or higher-way interactions, the interpretation becomes far more difficult. Another difference from previous models in the literature is that the interaction-decomposition model is not limited to two or three categorical predictor variables, but can handle any number of predictors.

The current model has some resemblance to joint correspondence analysis (Greenacre 1988) and multiple correspondence analysis (see, for example, Chapter 2 of this volume; Gifi 1990; Greenacre 1984).

Similar to joint correspondence analysis, the diagonal effects of $\mathbf{B}_{qq}$ are simply discarded. However, the main difference lies in the fact that the main effects in joint correspondence analysis are not linearly modeled by separate terms $\mathbf{z}_{iq}^{\mathsf{T}}\mathbf{a}_q$, but are included as weights in the loss function. A minor difference consists in the different normalization of the $\mathbf{Y}_q$.

In principle, the current model can be extended to generalized linear models (McCullagh and Nelder 1989; Nelder and Wedderburn 1972), but it remains to be seen whether the present model needs to be adapted.

Acknowledgment

We would like to thank Philip Hans Franses for kindly making the holiday-spending data available.

CHAPTER 23

Logistic Biplots

José L. Vicente-Villardón, M. Purificación Galindo-Villardón, and Antonio Blázquez-Zaballos

CONTENTS

23.1 Introduction

The biplot method (Gabriel 1971) is a simultaneous graphical representation of the rows and the columns of a given data matrix. The main uses are exploratory although it has also been used as a graphical representation for more formal models (Gabriel 1998). In practice, biplot fitting occurs either by computing the singular-value decomposition (SVD) of the data matrix or by performing an alternating regressions procedure (Gabriel and Zamir 1979). Jongman et al. (1987) fit the biplot by alternating a regression and a calibration step, essentially equivalent to the alternating regressions. Gower and Hand (1996) use the term "interpolation" rather than "calibration."

When data are binary, classical linear biplots are not suitable because the response along the dimensions is linear; in the same way, linear regression is not suitable when the response is binary. Multiple correspondence analysis (MCA) is commonly used for categorical data and can be considered as a particular form of biplot for binary (or indicator) matrices. Gabriel (1995) as well as Gower and Hand (1996) develops the complete theory for MCA in relation to biplots, the latter authors proposing what they call "prediction regions" as an extension of the usual linear projections. The prediction regions are based on distances from the individual points to the category points. The representation space is divided into regions that predict each category or combination of categories.

In this chapter, we propose a linear biplot for binary data in which response along the dimensions is logistic. Each individual is represented as a point and each character as a direction through the origin. The projection of an individual point onto a character direction predicts the probability of presence of that character. The method is related to logistic regression in the same way that biplot analysis is related to linear regression. Thus, we refer to the method as the logistic biplot.

We take here an exploratory point of view as opposed to the modeling approach in chapters by Gabriel (1998) or de Falguerolles (1998). The main aim is to analyze a data matrix (individuals by variables) rather than to model a two-way (contingency) table using a bilinear model. A preliminary version of the logistic biplot is proposed by Vicente-Villardón (2001). Schein et al. (2003) propose a generalized linear model for principal components of binary data, without the biplot point of view. Our proposal is closely related to MCA and some psychometric latent-variable procedures, such as item-response theory or latent traits. The main theoretical results are applied to a molecular classification of cancer by monitoring gene expression.

23.2 Classical biplots

Let $\mathbf{X}$ be a data matrix with I rows and J columns containing the measures of J variables (usually continuous) on I individuals. A low-dimensional biplot is a graphical representation of a data matrix $\mathbf{X}$ by means of markers $\mathbf{a}_1,\ldots,\mathbf{a}_I$ for its rows and markers $\mathbf{b}_1,\ldots,\mathbf{b}_J$ for its columns, in such a way that the product $\mathbf{a}_i^\mathsf{T}\mathbf{b}_j$ approximates x_{ij} as closely as possible. Arranging the markers as row vectors in two matrices $\mathbf{A}$ and $\mathbf{B}$, the approximation of $\mathbf{X}$ can be written as $\mathbf{X} \approx \mathbf{AB}^\mathsf{T}$. Although the classical biplot is well known, we include here a description, in terms of alternating regressions, related to our proposal. We also describe the geometry of regression biplots.

23.2.1 *Linear biplot based on alternating regressions/interpolations*

If we consider the row markers **A** as fixed, the column markers can be computed by regression

$$\mathbf{B}^{\mathsf{T}} = (\mathbf{A}^{\mathsf{T}}\mathbf{A})^{-1}\mathbf{A}^{\mathsf{T}}\mathbf{X} \tag{23.1}$$

In the same way, fixing **B**, **A** can be obtained as

$$\mathbf{A}^{\mathsf{T}} = (\mathbf{B}^{\mathsf{T}}\mathbf{B})^{-1}\mathbf{B}^{\mathsf{T}}\mathbf{X}^{\mathsf{T}} \tag{23.2}$$

By alternating the steps 1 and 2 (Equation 23.1 and Equation 23.2, respectively) and normalizing **A**, for example, after each two-step iteration, the algorithm converges to the SVD of **X**. The algorithm can then be completed with an orthogonalization step to ensure the uniqueness of its solution. The regressions in Equation 23.1 and Equation 23.2 can be separated for each row and column of the data matrix. This symmetrical process is commonly used to adjust bilinear (or biadditive) models with symmetrical roles for rows and columns. For a data matrix of individuals by variables, the roles of rows and columns are nonsymmetrical; nevertheless the algorithm is still valid and is interpreted as a two-step process, alternating a regression step and an interpolation/calibration step. The regression step adjusts a separate linear regression for each column (variable), and the interpolation step interpolates an individual using the column markers as the reference.

23.2.2 *Geometry of regression biplots*

The geometry of biplots for linear subspaces is described in Gower and Hand (1996), but the geometry of regression biplots is not. Let us suppose that the biplot is in two-dimensional space (a plane). Then we want to find the direction $\boldsymbol{\beta}_j$ in the space L spanned by the two columns of **A**, such that the projections of the markers in **A** onto that direction predict the values of variable j as closely as possible. That is, for the jth column $\mathbf{x}_j$ of **X**:

$$\mathbf{x}_j \approx \mathbf{A}\,\boldsymbol{\beta}_j \tag{23.3}$$

As we showed before, in Equation 23.1, this direction is given by the markers of the jth column. Without loss of generality, we can assume that the variables and thus the row markers are mean centered.

We add a third dimension for the jth variable and fit the usual regression plane. Let us call it H (see Figure 23.1). The set of points

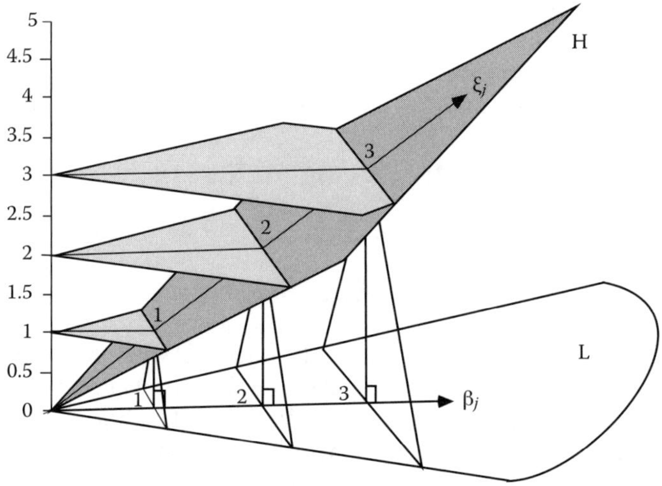

Figure 23.1 Geometry of a linear regression biplot.

in H predicting a fixed value is given by the straight line that is the intersection between the regression plane H and the plane through the fixed value parallel to L. Different fixed values lead to different parallel straight lines in H. Let $\boldsymbol{\xi}_j$ denote the line in H normal to all those straight lines (the level curves) and intersecting the third axis (at point 0 for centered data). The line $\boldsymbol{\xi}_j$ can be used as a reference for prediction, as shown in Gower and Hand (1996). The points in L predicting different values of the variable are also on parallel straight lines; the projection of $\boldsymbol{\xi}_j$ onto L is perpendicular to all these lines and it is called the biplot axis, with direction vector $\boldsymbol{\beta}_j$ (see Figure 23.1).

The projection of the row markers, onto the biplot axis $\boldsymbol{\beta}_j = (b_{j1}, b_{j2})$, gives the predictions in L. The biplot axis can be completed with scales. To find the marker on the biplot axis $\boldsymbol{\beta}_j$ that predicts a fixed value μ of the observed variable, we look for the point (x, y) that verifies

$$y = \frac{b_{j2}}{b_{j1}} x \quad \text{and} \quad \mu = b_{j0} + b_{j1}x + b_{j2}y \tag{23.4}$$

Solving for x and y, we obtain

$$x = \frac{(\mu - b_{j0})b_{j1}}{b_{j1}^2 + b_{j2}^2} \quad \text{and} \quad y = \frac{(\mu - b_{j0})b_{j2}}{b_{j1}^2 + b_{j2}^2} \tag{23.5}$$

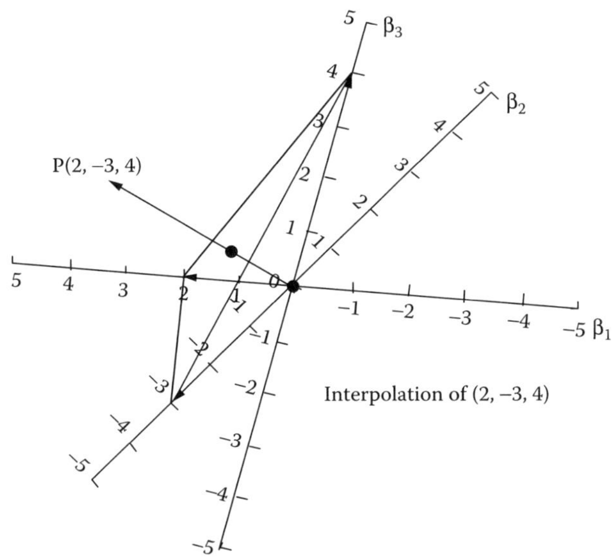

Figure 23.2 Interpolation of a point into the biplot.

The goodness of fit is measured by the squared correlation coefficients R_j^2 for the regressions. They are interpreted as measures of the "quality of the representation" in the manner commonly used in correspondence analysis.

The interpolation of an individual with an observed vector $(x_{i1},\ldots,x_{ip})$, for a fixed set of column markers **B**, is computed as the linear combination in Equation 23.2. When the columns of **B** are orthonormal, $\mathbf{B}^\mathsf{T}\mathbf{B} = \mathbf{I}$ and the combination is $\mathbf{a}_i = \mathbf{B}^\mathsf{T}\mathbf{x}_i$, i.e., the sum of vectors

$$\mathbf{a}_i = \sum_{j=1}^{J} x_{ij}\mathbf{b}_j \tag{23.6}$$

This is the geometry of the interpolation as described in Gower and Hand (1996) (see Figure 23.2).

The unit markers for interpolation are the markers $\mathbf{b}_j$, i.e., the interpolant of a general point **x** is given by the vector sum of the unit points weighted by $x_{i1},\ldots,x_{ip}$. Figure 23.2 illustrates a simple method for interpolating a point by summing the vectors for its three markers on the biplot axes $\boldsymbol{\beta}_1$, $\boldsymbol{\beta}_2$, and $\boldsymbol{\beta}_3$. We illustrate the interpolation of the point (2, −3, 4). We first select the value on each biplot axis using the graduations and then sum the resulting vectors. G is the centroid of the three points, and the interpolated point is at three (the number of biplot axes) times the vector OG from the origin to G.

Observe that the directions for interpolation are the same as for prediction, but the unit markers are different. When the columns of **B** are not orthonormal, we also have a linear combination, but the expressions of the unit markers are more complicated.

The constraint of orthogonal **B** leads to the row metric preserving biplot. The same constraint can be applied to **A**, leading to the column metric preserving biplot (Gabriel and Odoroff 1990).

23.3 Logistic biplot

23.3.1 Formulation

Let **X** be a data matrix in which the rows correspond to I individuals and the columns to J binary characters. Let $\pi_{ij} = E(x_{ij})$ be the expected probability that the character j is present at individual i, and x_{ij} the observed probability. Usually x_{ij} is either 0 or 1, resulting in a binary data matrix. The logistic biplot is formulated as

$$\pi_{ij} = \frac{e^{b_{j0}+\Sigma_s b_{js}a_{is}}}{1+e^{b_{j0}+\Sigma_s b_{js}a_{is}}} \tag{23.7}$$

where a_{is} and b_{js} ($i = 1,\dots,I$; $j = 1,\dots, J$; $s = 1,\dots, S$) are the model parameters used as row and column markers, respectively. The model in Equation 23.7 is a generalized bilinear model having the logit as a link function.

$$\text{logit}(\pi_{ij}) = \log\left(\frac{\pi_{ij}}{1-\pi_{ij}}\right) = b_{j0} + \sum_{s=1}^{S} b_{js}a_{is} = b_{j0} + \mathbf{a}_i^\mathsf{T}\mathbf{b}_j \tag{23.8}$$

where $\mathbf{a}_i = (a_{i1},\dots,a_{is})^\mathsf{T}$ and $\mathbf{b}_j = (b_{j1},\dots,b_{js})^\mathsf{T}$.
In matrix form

$$\text{logit}(\Pi) = \mathbf{1}\mathbf{b}_0^\mathsf{T} + \mathbf{A}\mathbf{B}^\mathsf{T} \tag{23.9}$$

where Π is the matrix of expected probabilities, **1** is a vector of ones, $\mathbf{b}_0$ is the vector containing the constants, and **A** and **B** are the matrices containing the markers for the rows and columns of **X**. This is a matrix generalization of the logit function (Schein et al. 2003). The constants b_{j0} have been added because it is not possible to center the data matrix in the same way as in linear biplots. The constant allows for calculating the probability π_{j0} at the point (0,0). The constant is the displacement of the gravity center in the same way as it is the trivial axis in correspondence analysis. This axis does not affect the prediction or the final representation.

The model has a close similarity to the models used in the social sciences, such as latent traits, item-response theory, or Rasch models. In fact, a biplot is implicit in many of these models. The main difference is that here the model is descriptive, and the main issue is either the ordination of individuals on a latent factor or the dimension reduction to gain a better insight into the interpretation of a complex problem. In psychometric models, the aim is to estimate the parameters of a factor model to explain the correlation between variables. Latent-trait models can be found in Bartholomew and Knott (1999) and item-response models in Baker (1992).

23.3.2 Parameter estimation

The model in Equation 23.9 is similar to the latent-trait or item-response theory models, in that ordination axes are considered as latent variables that explain the association between the observed variables. In this framework, we suppose that individuals respond independently to variables and that the variables are independent for given values of the latent traits.

With these assumptions, the likelihood function is

$$\text{Prob}(x_{ij} \mid \mathbf{b}_0, \mathbf{A}, \mathbf{B}) = \prod_{i=1}^{I} \prod_{j=1}^{J} \pi_{ij}^{x_{ij}} (1 - \pi_{ij})^{1-x_{ij}}$$

Taking the logarithm of the likelihood function yields

$$L = \log\,(\text{Prob}(x_{ij} \mid \mathbf{b}_0, \mathbf{A}, \mathbf{B}) = \sum_{i=1}^{I} \sum_{j=1}^{J} [x_{ij} \log(\pi_{ij}) + (1 - x_{ij}) \log(1 - \pi_{ij})] \tag{23.10}$$

To obtain the estimates, it is necessary to take the derivatives of L with respect to all the parameters, equate them to zero, and solve

$3J + 2I$ simultaneous equations. The Newton–Raphson method can be used to solve the system of equations, but if the number of individuals or variables is large, the computation problem becomes too large. Gabriel (1998) proposed what he called "generalized bilinear models" and proposed an estimation procedure based on segmented models. This procedure uses an alternating least-squares (crisscross) algorithm that separately adjusts the rows and the columns. The procedure is efficient when the number of rows and columns is small. While it is useful to model a contingency table, when the data matrix is large, the procedure is inefficient because of the size of the data matrices involved.

Our method for fitting the parameters of the logistic biplot is an iterative scheme that alternates between updates of **A** and **B**. Essentially, one set of parameters is updated while the other is held fixed, and this procedure is repeated until the likelihood converges to a desired degree of precision. At each step the log-likelihood function in Equation 23.10 can be separated into a part for each row or each column of the data matrix. We maximize each part separately, obtaining nondecreasing values of the likelihood. This succession is expected to converge at least to a local maximum. It can be considered as a heuristic generalization of the regression/interpolation procedure for the classical biplot; moreover, if the data are normally distributed and we use the identity (rather than logit) as a link function, the procedure converges to the solution of the classical biplot.

For **A** fixed, Equation 23.10 can be separated into J parts, one for each variable.

$$L = \sum_{J=1}^{J} L_j = \sum_{J=1}^{J} \sum_{i=1}^{I} [x_{ij} \log(\pi_{ij}) + (1 - x_{ij}) \log(1 - \pi_{ij})]$$

Maximizing each L_j is equivalent to performing a logistic regression using the jth column of **X** as a response and the columns of **A** as regressors. This is the regression step. In the same way, the probability function can be separated into several parts, one for each row of the data matrix:

$$L = \sum_{i=1}^{I} L_i = \sum_{i=1}^{I} \sum_{j=1}^{J} [x_{ij} \log(\pi_{ij}) + (1 - x_{ij}) \log(1 - \pi_{ij})]$$

To maximize each part, we use the Newton–Raphson method. The partial derivatives with respect to a_{is}, $(s = 1,\ldots, S)$ are

$$\frac{\partial L_i}{\partial a_{is}} = \sum_{j=1}^{J} x_{ij} \frac{1}{\pi_{ij}} \frac{\partial \pi_{ij}}{\partial a_{is}} + \sum_{j=1}^{J} (1 - x_{ij}) \frac{1}{(1 - \pi_{ij})} \frac{\partial (1 - \pi_{ij})}{\partial a_{ik}}$$

with

$$\frac{\partial \pi_{ij}}{\partial a_{is}} = b_{js} \pi_{ij} (1 - \pi_{ij}) \qquad \frac{\partial (1 - \pi_{ij})}{\partial a_{is}} = -b_{js} \pi_{ij} (1 - \pi_{ij})$$

Then the gradient vector **g** has elements

$$g_s = \frac{\partial L_i}{\partial a_{is}} = \sum_{j=1}^{J} b_{js} (x_{ij} - \pi_{ij})$$

The Hessian matrix of second derivatives has elements

$$h_{ss} = \frac{\partial^2 L_i}{\partial a_{is}^2} = -\sum_{j=1}^{J} b_{js}^2 \pi_{ij} (1 - \pi_{ij})$$

$$h_{ss'} = \frac{\partial^2 L_i}{\partial a_{is} a_{is'}} = -\sum_{j=1}^{J} b_{js} b_{js'} \pi_{ij} (1 - \pi_{ij})$$

The iterative Newton–Raphson method is as follows:

1. Set initial values for $[a_{i1},\ldots,a_{iS}]_0^{\mathrm{T}}$, usually the row markers in the previous biplot step, and $t = 0$.
2. Update $[a_{i1},\ldots,a_{iS}]_{t+1}^{\mathrm{T}}$ with

$$\begin{bmatrix} a_{i1} \\ \vdots \\ a_{iS} \end{bmatrix}_{t+1} = \begin{bmatrix} a_{i1} \\ \vdots \\ a_{iS} \end{bmatrix} + \mathbf{H}_g^{-1} \qquad (23.11)$$

where we estimate π_{ij} as the probability using the parameter values in the update t.
3. Increment the counter $t = t + 1$.
4. If changes in $[a_{i1},...,a_{iS}]^{\mathsf{T}}$ are small, then finish; if not, go to step 2.

Some problems with these procedures have been encountered when the response vectors are sparse or for response vectors with only 0s or 1s. The problem can be solved using slight corrections for expected probabilities equal to 0 or 1. Nevertheless, the method without any correction has proven to work in most cases.

To summarize, the general algorithm for the logistic biplot is as follows:

Step 1: Choose initial values for the parameters **A**. For example, take **A** from the principal component analysis of **X**.
Step 2: Orthonormalize **A** to avoid indeterminacies (optional).
Step 3 (regression step): Calculate $b_{j0}, b_{j1},...,b_{jS}$ using separate standard logistic regressions for each column $\mathbf{x}_j$ of **X**.
Step 4 (interpolation step): Interpolate each individual separately (calculate $a_{i1}, ..., a_{iS}$) using the Newton–Raphson method described above.
Step 5: If changes in the log-likelihood are small, then finish; if not, go to step 2.

The orthonormalization (step 2) provides a tool for the uniqueness of the parameter estimates in the same way as unit-length vectors are taken in principal components. The constraint can also be done on **B**. The step is optional, and the orthonormalization can be done *a posteriori* taking the SVD of the expected values in $\mathbf{AB}^{\mathsf{T}}$. In any case, we would obtain the same space with different rotations of the solution.

Note that steps 4 and 5 can be used to project supplementary (or illustrative) variables or individuals onto the graph, or even to produce a logistic biplot when a set of fixed coordinates has been obtained from another technique.

23.3.3 Geometry of logistic biplots

We describe the logistic biplot geometry for a two-dimensional solution: if we fix the markers in **A** and adjust the model in Equation 23.7, we obtain a logistic response surface H as in Figure 23.3. In this case, the

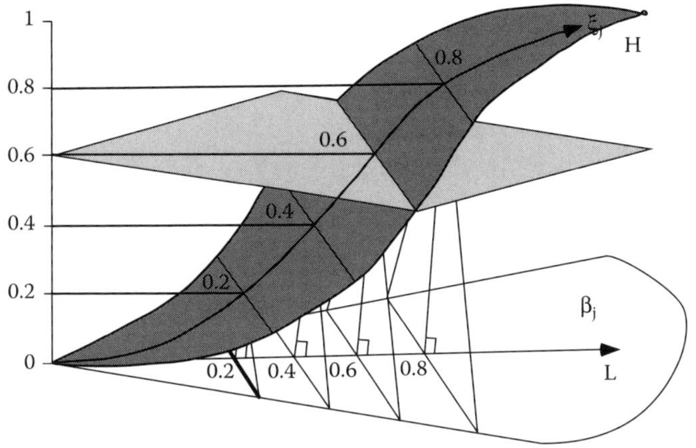

Figure 23.3 Geometry of logistic biplots.

third axis shows a scale for the expected probabilities. Although the response surface is nonlinear, the intersections of the planes normal to the probability axis and H are straight lines on H. Similar to the linear case, the lines for different probabilities are parallel. Suppose we select, on the response surface, a curve intersecting the third axis at the point $(0, 0, p_{j0})$ and whose tangent is perpendicular, at any point, to the corresponding prediction line. That curve is a nonlinear prediction axis $\boldsymbol{\xi}_j$. The projection of $\boldsymbol{\xi}_j$ onto the representation space is a straight line, the biplot axis $\boldsymbol{\beta}_j$. The direction of $\boldsymbol{\beta}_j$ is given by (b_{j1}, b_{j2}), i.e., the parameters in Equation 23.7.

The points in L predicting different probabilities are also on parallel straight lines; this means that predictions on the logistic biplot are made in the same way as on the linear biplots, i.e., projecting a row marker $\mathbf{a}_i = (a_{i1}, a_{i2})$ onto a column marker $\mathbf{b}_j = (b_{j1}, b_{j2})$. The biplot axis $\boldsymbol{\beta}_j$ is completed with marks for points predicting probabilities by projection; the main difference with the linear biplot is that equally spaced marks do not correspond to equally spaced probabilities.

To find the scale marker for a fixed probability p, we look for the point (x, y) that predicts p and is on the biplot axis, i.e., on the line joining the points $(0, 0)$ and (b_{j1}, b_{j2}):

$$y = \frac{b_{j2}}{b_{j1}} x \tag{23.12}$$

The prediction verifies

$$\text{logit}(p) = b_{j0} + b_{j1}x + b_{j2}y \tag{23.13}$$

Using Equation 23.12 in Equation 23.13, we obtain

$$x = \frac{(\text{logit}(p) - b_{j0})b_{j1}}{b_{j1}^2 + b_{j2}^2} \quad \text{and} \quad y = \frac{(\text{logit}(p) - b_{j0})b_{j2}}{b_{j1}^2 + b_{j2}^2} \tag{23.14}$$

For example, the point on axis $\boldsymbol{\beta}_j$ predicting 0.5 (logit(0.5) = 0) is

$$x = \frac{-b_{j0}b_{j1}}{b_{j1}^2 + b_{j2}^2} \quad \text{and} \quad y = \frac{-b_{j0}b_{j2}}{b_{j1}^2 + b_{j2}^2}$$

The final representation is a linear biplot interpreted by projection, even though the response surface is not linear. The result is not surprising because we are dealing with generalized linear models (the biplot is linear in the logit scale). The direct representation in the logit scale is difficult to interpret; the probability scale is simpler and easier to understand, especially for nontrained users.

23.4 Application: microarray gene expression data

The proposed method is useful for any binary data. We have chosen here an example taken from Golub et al. (1999), and it is related to the classification of two different kinds of leukemia. The logistic biplot has been used as a dimension-reduction technique, prior to the classification.

Classification of patient samples is a crucial aspect of cancer diagnosis and treatment. Although cancer classification has improved over the past years, there has been no general approach for identifying new cancer classes or for assigning tumors to known classes. Recent research uses gene-expression monitoring by DNA as a tool for classification.

For this example we use 38 bone marrow samples divided into two groups: acute lymphoblastic leukemia (ALL), with 27 members, and acute myeloid leukemia (AML), with 11 members. An additional set of 34 samples (20 ALL, 14 AML) was used to validate the classification. A more detailed description of the data can be found in Golub et al. (1999).

Distinguishing ALL from AML is critical for successful treatment. Remissions can be achieved using ALL therapy for AML (and vice versa), but cure rates are markedly diminished, and unwarranted toxicities are encountered. Although the distinction between ALL and AML has been well established, no single test is currently sufficient to establish the diagnosis. Rather, current clinical practice involves several analyses, each performed in a separate, highly specialized laboratory. Golub et al. (1999) developed a systematic approach to cancer classification based on the simultaneous monitoring of expression of thousands of genes using DNA microarrays.

Although the blueprint encoding all human genes is present in each cell, only a fraction of the proteins that they can produce is active in any particular cell. The process of transcribing a gene's DNA sequence into RNA (which serves as a template for protein production) is known as "gene expression." A gene's expression level indicates the approximate number of copies of that gene's RNA produced in a cell; this is thought to correlate with the amount of the corresponding protein made. A signal value is calculated that assigns a relative measure of the abundance of the transcript. A "detection P-value" is evaluated to determine the "detection call," which indicates whether a transcript is reliably detected (present) or not detected (absent). The binary values are noisy indicators of the presence or absence of mRNA. The detection call is provided by the Affymetrix Gene Chip software (details can be found at http://www.affymetrix.com). Expression chips (biological chips), manufactured using technologies derived from computer-chip production, can now measure the expression of thousands of genes simultaneously.

For our analysis, we use the binary matrix with the presence/absence of 6817 genes. The main difficulty with this kind of data is the extreme dimensionality of the problem. The logistic biplot is used here as a tool for dimension reduction and class identification. The most important genes for the separation of the classes are identified in the biplot. The logistic biplot is also a tool for summarizing the information of many correlated gene expressions.

The first issue is to explore whether there were genes whose expression pattern was strongly correlated with the class distinction to be predicted. The 6817 genes were sorted by their degree of correlation. After the ordering process, the 15 genes most correlated with the groups were selected. The number selected is somewhat arbitrary; we

Table 23.1 Goodness of fit for each column of the data matrix.

Sample No.	Gene Label	Training Set (n = 38) Deviance	R^2	# C.C.[a]	% C.C.[b]	Validation Set (n = 34) # C.C.[a]	% C.C.[b]
1	D21262	39.3	0.906	34	89.5	23	67.6
2	M31211	33.9	0.917	35	92.1	30	88.2
3	D49950	74.8	0.933	38	100.0	29	85.3
4	M31166	44.7	0.850	36	94.7	28	82.4
5	M84526	54.8	0.978	38	100.0	33	97.1
6	X70297	31.1	0.755	35	92.1	28	82.4
7	D88422	55.4	0.918	35	92.1	25	73.5
8	X62535	95.2	0.981	38	100.0	26	76.5
9	L47738	40.6	0.942	37	97.4	30	88.2
10	M11722	38.2	0.942	36	94.7	26	76.5
11	M77142	57.0	0.895	34	89.5	22	64.7
12	Y12670	46.3	0.932	37	97.4	30	88.2
13	U61836	45.0	0.744	34	89.5	29	85.3
14	M27783	57.9	0.984	38	100.0	28	82.4
15	M20203	57.9	0.984	38	100.0	29	85.3

Note: The deviance follows a chi-squared distribution with 1 degree of freedom.

[a] # C.C.: Number of correctly classified points.
[b] % C.C.: Percentage of correctly classified points.

use a small number for purposes of illustration. A list of the selected genes is shown in Table 23.1.

The logistic biplot was calculated using the alternative procedure proposed previously. We used the standardized principal components of the raw data matrix as a starting point for the algorithm. The algorithm converged in ten iterations. A rough measure of the goodness of fit is the percentage of correctly classified individuals on the projections onto the variables: an expected probability greater than 0.5 predicts a presence. The percentage of explained variance for these data in two dimensions was 96.3%, so the two-dimensional biplot is an adequate summary of the binary data matrix containing 15 columns. The procedure has been applied using different numbers of genes, and in all the cases a two-dimensional representation summarized the data adequately.

Figure 23.4 shows the graphical representation. The genes are represented by arrows pointing in the direction of increasing predictions of the probabilities. The start of the arrow is the point predicting 0.5, and the end is the point predicting 0.75. The marks have been placed using Equation 23.15 with the appropriate probabilities. The length of the

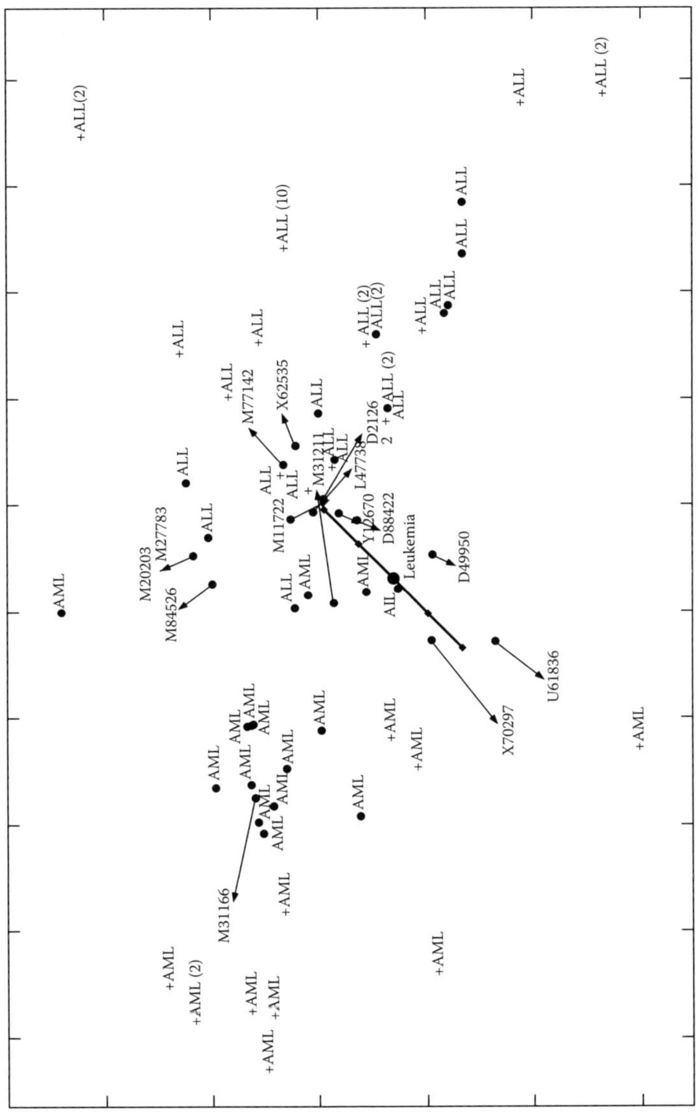

Figure 23.4 Logistic biplot for the microarray data applied to the classification of two kinds of leukemia.

segment is related to the discriminatory power of each character, i.e., to the capacity to predict the presence or absence of the gene. Short arrows correspond to the genes with greater discriminatory power. This concept is also used in item-response theory (for example, see Baker 1992), and it is related to the slope of the tangent to the logistic curve at the point predicting 0.5. Greater slopes are associated with greater discriminatory power and also with shorter vectors in the graph. Large vectors predict slowly changing probabilities, so they are not useful for interpretation. In this case, all the variables have an acceptable quality of representation as measured by the discriminatory power.

Two genes pointing in the same direction are highly correlated; two genes pointing in opposite directions are negatively correlated; and two genes forming an angle near 90° are not correlated.

Table 23.1 shows some measures of the goodness of fit of each variable: the deviance for the comparison of the whole model with the reduced model with a constant ($p < 0.0001$ for all the genes), the R^2 comparing observed and predicted probabilities, and the percent of correct classifications for the prediction of presence. All the genes have a high value in all the measures of goodness of fit.

The individuals (samples) have been represented on the biplot using points (Figure 23.4). The distance between points is interpreted as similarity/dissimilarity between individuals, although this measure needs further investigation. The training set is represented by a plus symbol (+) and the validation set by filled circles (•). The initial representation has been calculated using only the training set; the individuals belonging to the validating set have been projected using Equation 23.11.

The probability of presence of a gene in a particular sample can be approximated by projecting the sample point onto the gene direction (using the scale as a reference). The presence of the gene is predicted when such probability is greater than 0.5.

The type of leukemia has been projected (as a supplementary variable) onto the biplot using standard logistic regression and Equation 23.14. The resulting direction has been marked on the graph together with the points predicting 0.10, 0.25, 0.50, 0.75, and 0.90 (Figure 23.5).

To clarify the interpretation of the graphical representation, Figure 23.5 has been supplemented with some graphical aids. The markers of each group (ALL and AML) have been surrounded by separate polygons for the training and validating sets. The graph has been divided into three regions predicting different probabilities of presence for ALL (<0.25, between 0.25 and 0.75, >0.75) corresponding to the complementary probabilities for prediction of AML. That division has been made

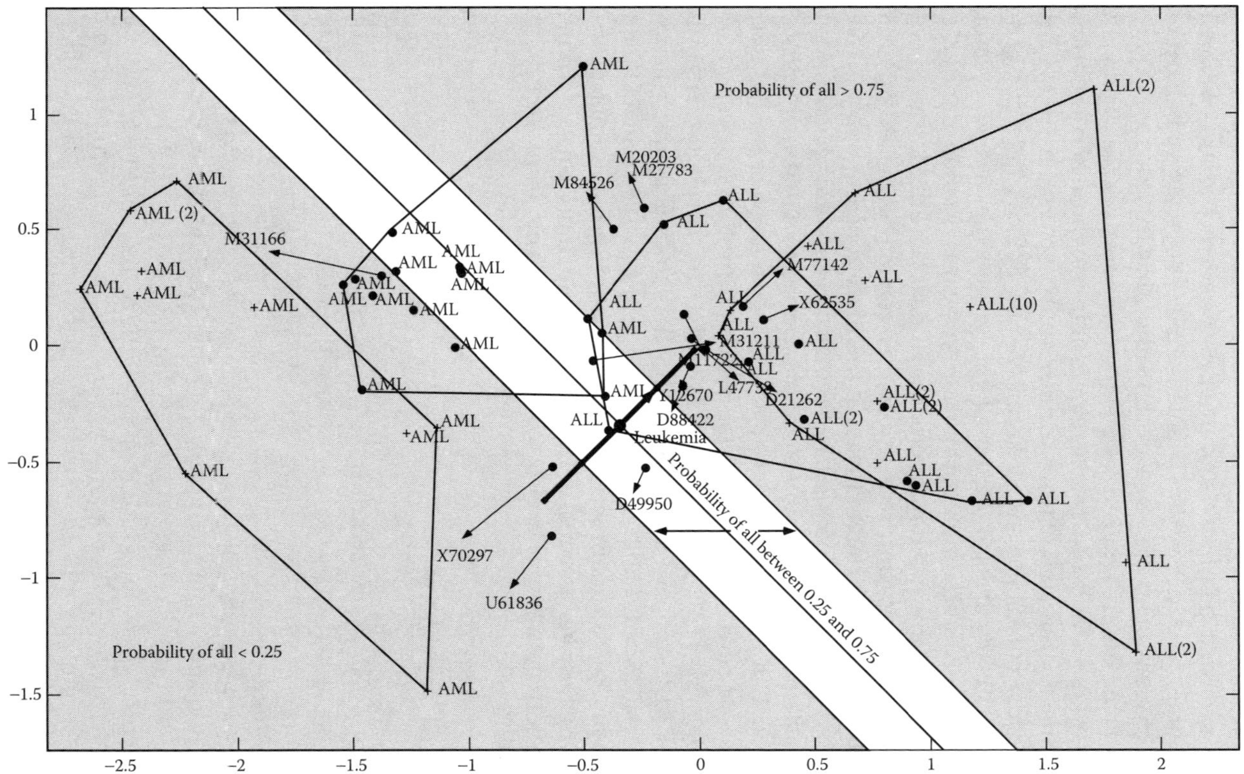

Figure 23.5 Some aids to the interpretation of the biplot. The training and the validation set has been indicated using convex hulls for both kinds of leukemia. The representation space has been divided into regions predicting different probabilities of leukemia.

according to the prediction rules, i.e., according to the projections of the points onto the direction representing the leukemia group. We achieve a complete separation of the two groups in the training set, obtaining estimated probabilities >0.75 of being in the correct group. For the validating, set most points (88.2%) predict the correct group. We have tried the same procedure with a higher number of genes and obtained 100% correct classifications, even in the validating set.

Once a clear differentiation between groups has been established, the next step is to find the genes responsible for the discrimination. The genes most correlated with the differentiation between groups will be those almost parallel to the leukemia vector. For example, the gene labeled D49950, toward the center of the graph, has a short vector, indicating a high discriminatory power and thus a good quality of representation. Moreover, its direction is almost parallel to the leukemia vector, indicating that both are highly correlated, so the gene is a good predictor of the groups. Because the directions are opposite, we conclude that the gene is present in the AML group and absent in the ALL group. The same reasoning could be applied to the genes D88422 and Y12670. The genes X70297 and U61836 are also correlated with the presence of AML, but with a smaller discriminatory power.

The genes M77142, X62535, and M31211 are correlated with the presence of ALL and are absent in AML. Some other genes, such as M31166, M84526, M20203, M27783, M11722, L47738, and D21262, are perpendicular to the direction of leukemia and thus are not useful in separating between groups. A discussion of the importance of proper discrimination between the two types of leukemia is presented in Golub et al. (1999).

23.5 Final remarks

We have obtained an important reduction of the dimension of the data, summarizing in just two latent variables the information provided by the chosen set of variables. The new latent variables are useful in discriminating between the two kinds of leukemia.

The main advantage of this approach is that we can take all the important information of the genes, thereby avoiding redundant information and taking advantage of the correlation between variables (presence of genes). Other approaches, for example, in Golub et al. (1999), consider the genes one by one, introducing redundant information that does not take into account the correlation between the variables. They consider the "between groups" correlation but not the "within groups" correlation.

The classification procedure used here is equivalent to logistic discrimination, taking the latent variables as predictors. The proposed biplot is a useful graphical tool to interpret the results. The results suggest that the procedure could be useful in detecting changes in several experimental conditions using microarray gene-expression data, although the procedure could also be applied to other disciplines where many binary variables are observed simultaneously.

Software note

For the calculations of the biplot in the example, the algorithm has been programmed using MATLAB®, a programming environment oriented to matrices. For more details, contact the first author (villardon@usal.es).

APPENDIX

Computation of Multiple Correspondence Analysis, with Code in R

Oleg Nenadić and Michael Greenacre

A.1 Introduction

Multiple correspondence analysis (MCA) is essentially the application of the simple correspondence analysis (CA) algorithm to multivariate categorical data coded in the form of an indicator matrix or a Burt matrix (see Chapter 2). Greenacre and Blasius (1994) described in detail the computations involved in CA. In this appendix we shall describe the steps involved in computing MCA solutions as well as related results such as the coordinates of supplementary points and the adjustment of principal inertias (eigenvalues). We shall also describe an algorithm for joint correspondence analysis (JCA). Our computing environment is mainly the freeware program R (www.r-project.org), but all the analyses described in this appendix have also been implemented in the XLSTAT package (www.xlstat.com) and the results have all been corroborated using XLSTAT. In Section A.8 we give the actual code in R that computed all the results described here.

The computing steps are illustrated using the same data set described in Chapter 2, the western German sample taken from the International Survey Program on Environment (ISSP 1993). Recall from Chapter 2 that there were four questions on attitudes to science, labeled A to D, with responses on a five-point scale (1 = agree strongly to 5 = disagree strongly), as well as three demographic variables: sex (two categories), age (six categories), and education (six categories).

Table A.1 Extract from the ISSP survey.

	A	B	C	D	Sex	Age	Education
1	2	3	4	3	2	2	3
2	3	4	2	3	1	3	4
3	2	3	2	4	2	3	2
4	2	2	2	2	1	2	3
5	3	3	3	3	1	5	2
⋮	⋮	⋮	⋮	⋮	⋮	⋮	⋮
871	1	2	2	2	2	3	6

Source: ISSP, International Social Survey Program, www.issp.org, 1993.

Table A.1 shows a part of the survey data for western Germany. This format of the original data is frequently called the "response pattern matrix." The columns of the response pattern matrix contain Q (= 4) questions corresponding to the active variables and Q' (= 3) questions corresponding to the supplementary variables. Each question q, active or supplementary, has a certain number J_q of response categories. In R terminology, each of the questions defines a *factor*, and each factor has a certain number of levels. In our example, $J_q = 5$ for each active factor, and the supplementary factors have 2, 6, and 6 levels, respectively. Initially we consider only the MCA of questions A, B, C, and D. The supplementary variables sex, age, and education are used at a later stage to show the computations for supplementary points in MCA.

A.2 Computations based on the indicator matrix

The most classical and standard approach to MCA is to apply a simple CA to the indicator matrix **Z**. The indicator matrix $\mathbf{Z} = \{z_{ij}\}$ corresponds to a binary coding of the factors; instead of using a factor with J_q levels, one uses J_q columns containing binary values, also called dummy variables. Table A.2 illustrates a part of the indicator matrix for the ISSP survey data.

We assume in this appendix that a (simple) CA program is available, based on the singular-value decomposition (SVD). Otherwise, if one wants to program the method from scratch, Chapter 1 defines the matrix of standardized residuals that is decomposed in CA (see Greenacre and Blasius 1994 for more details of the computations involved), while Chapter 2 gives the corresponding matrix in the case of indicator coding.

Table A.2 Extract from the indicator matrix for the first four columns of Table A.1.

	A					B					C					D				
	1	2	3	4	5	1	2	3	4	5	1	2	3	4	5	1	2	3	4	5
1	0	1	0	0	0	0	0	1	0	0	0	0	0	1	0	0	0	1	0	0
2	0	0	1	0	0	0	0	0	1	0	0	1	0	0	0	0	0	1	0	0
3	0	1	0	0	0	0	0	1	0	0	0	1	0	0	0	0	0	0	1	0
4	0	1	0	0	0	0	1	0	0	0	0	1	0	0	0	0	1	0	0	0
5	0	0	1	0	0	0	0	1	0	0	0	0	1	0	0	0	0	1	0	0
⋮	⋮	⋮	⋮	⋮	⋮	⋮	⋮	⋮	⋮	⋮	⋮	⋮	⋮	⋮	⋮	⋮	⋮	⋮	⋮	⋮
871	1	0	0	0	0	0	1	0	0	0	0	1	0	0	0	0	1	0	0	0

The number of nonzero singular values of an indicator matrix based on Q factors with a total of J levels ($J = \sum_q J_q$) is $J - Q$, which in our example is $20 - 4 = 16$. Table A.3 gives the 16 principal inertias (squares of the singular values) and the explained percentages of inertia for each dimension (again, we can show only some of these results for reasons of space).

The column standard and principal coordinates (b_{js} and g_{js}, respectively) for the first two dimensions ($s = 1, 2$) are shown in Table A.4. The results are given only for the responses to the questions A and D.

Table A.3 Some principal inertias and explained inertia for the CA of Table A.2.

s	1	2	3	4	...	16
λ_s	0.457	0.431	0.322	0.306	...	0.125
Explained inertia (%)	11.4	10.8	8.0	7.7	...	3.1

Table A.4 Some column standard and principal coordinates for the first two dimensions.

	A					...	D				
	1	2	3	4	5	...	1	2	3	4	5
b_{j1}	1.837	0.546	−0.447	−1.166	−1.995	...	1.204	−0.221	−0.385	−0.222	0.708
b_{j2}	−0.727	0.284	1.199	−0.737	−2.470	...	−1.822	0.007	1.159	0.211	−1.152
g_{j1}	1.242	0.369	−0.302	−0.788	−1.349	...	0.814	−0.150	−0.260	−0.150	0.479
g_{j2}	−0.478	0.187	0.787	−0.484	−1.622	...	−1.196	0.005	0.761	0.138	−0.756

Table A.5 Some row principal coordinates for the first two dimensions.

i	1	2	3	4	5	...	871
f_{i1}	−0.210	−0.325	0.229	0.303	−0.276	...	0.626
f_{i2}	0.443	0.807	0.513	0.387	1.092	...	0.135

A small part of the row principal coordinates f_{is} is displayed in Table A.5

Figure A.1 gives the complete result as a (symmetric) map for the first two dimensions. The four questions from the survey are coded with different symbols and in the rows (i.e., individuals) are displayed as small dots.

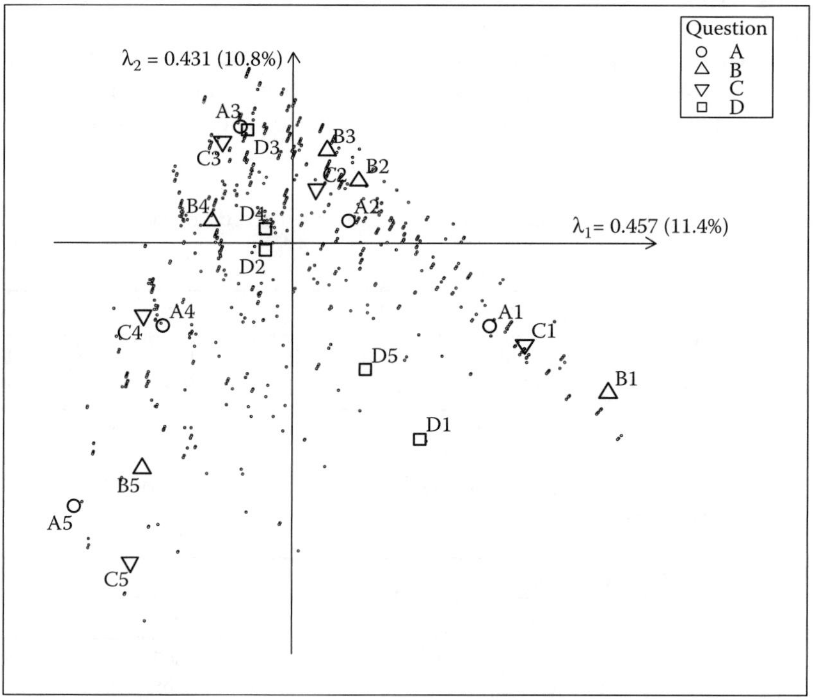

Figure A.1 Symmetric map of the ISSP data set.

Table A.6 Data of the ISSP survey in the form of a Burt matrix.

		A						D				
		1	2	3	4	5	...	1	2	3	4	5
	1	119	0	0	0	0		15	25	17	34	28
	2	0	322	0	0	0		22	102	76	68	54
A	3	0	0	204	0	0	...	10	44	68	58	24
	4	0	0	0	178	0		9	52	28	54	35
	5	0	0	0	0	48		4	9	13	12	10
	⋮			⋮			⋱			⋮		
	1	15	22	10	9	4		60	0	0	0	0
	2	25	102	44	52	9		0	232	0	0	0
D	3	17	76	68	28	13	...	0	0	202	0	0
	4	34	68	58	54	12		0	0	0	226	0
	5	28	54	24	35	10		0	0	0	0	151

A.3 Computations based on the Burt matrix

The Burt matrix **C** is obtained directly from the indicator matrix **Z**: $\mathbf{C} = \mathbf{Z}^\mathsf{T}\mathbf{Z}$. Table A.6 shows a part of the Burt matrix from the ISSP data set, corresponding to questions A and D (see Chapter 2, Table 2.3, for the complete matrix).

The computation of MCA is again the application of the (simple) CA algorithm to the Burt matrix **C**. Notice, however, the following properties of this analysis and its relation to the CA of the indicator matrix **Z**.

- Because **C** is symmetric, the solution for the rows and columns is identical.
- The analysis of **C** only gives a solution for the response categories (i.e., what were previously the columns of **Z**).
- The standard coordinates of the rows (equivalent to columns) of **C** are identical to the standard coordinates of the columns of **Z**.
- The principal inertias of **C** are the squares of those of **Z**.
- Because the matrix of standardized residuals in the analysis of **C** is symmetric (see Chapter 2, Equation 2.7), the singular values in the analysis of **C** are also eigenvalues.

This last point of terminology can cause some confusion and needs clarification. The principal inertia in a CA is often referred to as an eigenvalue since it is the square of the singular value in the SVD of the usually *rectangular* matrix **S** of standardized residuals in the SVD. In fact, CA is often computed via the eigendecomposition rather than the SVD, first calculating the cross-product matrix $\mathbf{S}^\mathsf{T}\mathbf{S}$ and then applying an eigendecomposition, in which case the eigenvalues obtained are the squares of the singular values and thus exactly the principal inertias. Let us suppose, just for the moment, that the CA algorithm does proceed this way, that is, it starts with a rectangular data matrix, forms cross-products, and calculates eigenvalues and eigenvectors and then the coordinates in the usual way. In the analysis of the indicator matrix **Z**, this square symmetric matrix of cross-products is exactly the matrix of standardized residuals for the Burt matrix **C**. That is, the analysis of **Z** involves the eigendecomposition of a matrix based on **C**. Now, to perform a CA of **C**, there would be no need to calculate a cross-product matrix since **C** is already square symmetric, but of course the application of CA to **C** would involve the calculation of a cross-product matrix a second time, which is why the principal inertias of **C** are the squares of those of **Z**. It is much clearer to consider CA as an SVD rather than an eigendecomposition, but in the case of the Burt matrix **C**, the SVD based on **C** is technically an eigendecomposition too, hence the possible confusion. Its eigenvalues are actually singular values and should be squared, as always, to give principal inertias of **C**. On the other hand, the singular values based on **C** are the squares of the singular values based on **Z** and are, thus, the principal inertias of **Z**.

It is useful here to give the steps for a minimal stand-alone computing algorithm for performing an MCA starting from the Burt matrix **C**:

1. Divide **C** by its grand total $n = \sum_i \sum_j c_{ij}$ to obtain the correspondence matrix **P**:

$$\mathbf{P} = \{p_{ij}\} = c_{ij}/n \tag{A.1}$$

and calculate the row totals (masses) r_i (equal to column masses).

2. Perform an eigenvalue–eigenvector decomposition on standardized residuals **A** (which, as we explained above, is the same as the SVD)

$$\mathbf{S} = \{s_{ij}\} = (p_{ij} - r_i r_j)/\sqrt{r_i r_j} \tag{A.2}$$

The decomposition returns the eigenvectors $\mathbf{U} = \{u_{is}\}$ and the eigenvalues δ_s from the solution of $\mathbf{S} = \mathbf{U}\boldsymbol{\Delta}\mathbf{U}^\mathsf{T}$. The eigenvalues (= singular values) are equivalent to the λ_s from Table A.3 and,

Table A.7 Some principal inertias and explained inertia for the CA of Table A.6.

s	1	2	3	4	...	16
λ_s^2	0.2092	0.1857	0.1036	0.0939	...	0.0157
Explained inertia (%)	18.6	16.5	9.2	8.3	...	1.4

hence, are the principal inertias of **Z**. If the principal inertias of **C** are required, they need to be squared. Table A.7 gives the principal inertias for the MCA based on the Burt matrix.

3. The ith row (or column) standard coordinate for the sth dimension is obtained as

$$a_{is} = u_{is} / \sqrt{r_i} \tag{A.3}$$

4. The corresponding principal coordinates are given by

$$f_{is} = a_{is}\lambda_s \tag{A.4}$$

Table A.8 contains a part of the eigenvectors for the first two dimensions (u_{i1} and u_{i2}), the (row or column) category masses (r_i), and the (row or column) standard and principal coordinates (a_{i1} and a_{i2} and f_{i1} and f_{i2}, respectively). The principal coordinates differ from those reported in Table A.4 because they are the standard coordinates scaled by λ_s, whereas in Table A.4 they are scaled by $\sqrt{\lambda_s}$.

Table A.8 The eigenvectors, masses, and standard and principal coordinates for the analysis of the Burt matrix.

	A						D				
	1	2	3	4	5	...	1	2	3	4	5
u_{i1}	0.339	0.166	−0.108	−0.264	−0.234	...	0.158	−0.057	−0.093	−0.056	0.147
u_{i2}	−0.134	0.086	0.290	−0.167	−0.290	...	−0.239	0.002	0.279	0.054	−0.240
r_i	0.034	0.092	0.059	0.051	0.014	...	0.017	0.067	0.058	0.065	0.043
a_{i1}	1.837	0.546	−0.447	−1.166	−1.995	...	1.204	−0.221	−0.385	−0.222	0.708
a_{i2}	−0.727	0.284	1.199	−0.737	−2.470	...	−1.822	0.007	1.159	0.211	−1.152
f_{i1}	0.840	0.250	−0.204	−0.533	−0.913	...	0.551	−0.101	−0.176	−0.101	0.324
f_{i2}	−0.314	0.123	0.517	−0.318	−1.064	...	−0.785	0.003	0.499	0.091	−0.496

Table A.9 Adjusted principal inertias and explained inertia for the ISSP survey.

s	1	2	3	4	5	6
$(\lambda_s^{\text{adj}})^2$	0.07646	0.05822	0.00920	0.00567	0.00117	0.00001
Explained inertia (%)	44.9	34.2	5.4	3.3	0.7	0.0

A.4 Adjustment of inertias

As described in Chapter 2, the so-called percentage of inertia problem can be partly improved by using adjusted inertias:

$$\left(\lambda_s^{\text{adj}}\right)^2 = \left(\frac{Q}{Q-1}\right)^2 \left(\lambda_s - \frac{1}{Q}\right)^2 \tag{A.5}$$

The adjusted inertias are calculated only for each singular value λ_s that satisfies the inequality $\lambda_s \geq 1/Q$. They are expressed as a percentage of the average off-diagonal inertia, which can be calculated either by direct calculation on the off-diagonal tables in the Burt matrix, or from the total inertia of **C** as follows:

$$\frac{Q}{Q-1}\left(\text{inertia}(\mathbf{C}) - \frac{J-Q}{Q^2}\right) \tag{A.6}$$

where inertia (**C**) is the sum of the principal inertias $\Sigma_s \lambda_s^2$ in Table A.7. The value of Equation A.6 in our ISSP example is 0.17024, and Table A.9 lists the adjusted inertias for the six dimensions that satisfy $\lambda_s \geq 1/4$.

A.5 Joint correspondence analysis

The main diagonal submatrices of the Burt matrix are the key issue. The JCA analysis of the ISSP example is performed here by using iteratively weighted least squares for updating the diagonal submatrices of the Burt matrix. The updating is carried out by calculating the MCA solution for the dimensions 1,…,S^* where S^* is the required dimensionality of the solution (this has to be chosen in advance since the solution is no longer nested).

The procedure is carried out in the following steps:

1. Set $\mathbf{C}^* = \{c_{ij}^*\} = c_{ij}$.
2. Perform an MCA on $\mathbf{C}^*$.
3. Reconstruct the approximation of the data from the solution:

$$\hat{\mathbf{C}} = \{\hat{c}_{ij}\} = nr_i r_j \left(1 + \sum_{s=1}^{S^*} \lambda_s a_{is} a_{js} \right) \tag{A.7}$$

   where S^* is the required dimensionality of the solution. (In the first iteration of the algorithm, one can optionally use the adjusted inertias λ_s^{adj}.)
4. Update main diagonal submatrices of $\mathbf{C}^*$ with the corresponding entries of $\hat{\mathbf{C}}$.
5. Repeat steps 2 to 4 until convergence.

One possibility to measure the convergence is given by considering the maximum absolute difference between the entries of the main diagonal matrices of $\mathbf{C}^*$ and $\hat{\mathbf{C}}$ in step 4.

Table A.10 shows the updated Burt matrix after 50 iterations based on the first two dimensions. Values in the main diagonal submatrices that were modified are typed in boldface.

When measuring the quality of the approximation, it should be remembered that (a) in the final CA of the modified Burt matrix $\mathbf{C}^*$

Table A.10 The updated (or modified) Burt matrix.

		A						D				
		1	2	3	4	5	…	1	2	3	4	5
	1	**30.72**	**53.14**	**18.59**	**13.97**	**2.58**		15	25	17	34	28
	2	**53.14**	**130.55**	**76.80**	**51.80**	**9.71**		22	102	76	68	54
A	3	**18.59**	**76.80**	**62.95**	**38.86**	**6.80**	…	10	44	68	58	24
	4	**13.97**	**51.80**	**38.86**	**53.51**	**19.85**		19	52	28	54	35
	5	**2.58**	**9.71**	**6.80**	**19.85**	**9.06**		4	9	13	12	10
	⋮			⋮			⋱			⋮		
	1	15	22	10	9	4		**9.02**	**14.67**	**5.03**	**13.27**	**18.01**
	2	25	102	44	52	9		**14.67**	**62.46**	**55.78**	**60.90**	**38.20**
D	3	17	76	68	28	13	…	**5.03**	**55.78**	**63.56**	**56.49**	**21.14**
	4	34	68	58	54	12		**13.27**	**60.90**	**56.49**	**59.74**	**35.60**
	5	28	54	24	35	10		**18.01**	**38.20**	**21.14**	**35.60**	**38.04**

Table A.11 Adjusted principal inertias and explained inertia for the JCA case.

s	1	2
$(\lambda_s^{JCA})^2$	0.09909	0.06503
Explained inertia (%)	54.3	35.6

in Table A.10, the total inertia includes contributions due to the modified diagonal blocks and (b) these are perfectly fitted by the two-dimensional solution. These must be discounted from both the total inertia and the first two principal inertias. The principal inertias and the explained inertia for the two-dimensional JCA solution are given in Table A.11, corresponding to a total inertia of the modified Burt matrix of 0.18242.

By direct calculation on Table A.10, the contributions to the total inertia due to the different submatrix blocks are given in Table A.12.

The total of Table A.12 is 0.18242, the total inertia of the modified Burt matrix (Table A.10), of which the sum of the diagonal values (0.05474) needs to be discounted from the first two principal inertias as well as the total. This gives the proportion of inertia explained by the two-dimensional JCA solution as

$$\frac{0.09909 + 0.06503 - 0.05474}{0.18242 - 0.05474} = \frac{0.10938}{0.12768} = 0.8567$$

hence 85.7% of the (off-diagonal) inertia is explained by the JCA solution. This percentage is necessarily less than the percentage explained in the whole matrix (equal to 89.9%, see Table A.11) since the latter calculation includes the modified diagonal blocks, which are fitted perfectly.

Table A.12 Contributions to total inertia of each submatrix of Table A.10.

	A	B	C	D
A	0.00745	0.01486	0.01215	0.00329
B	0.01486	0.02244	0.01858	0.00530
C	0.01215	0.01858	0.02103	0.00966
D	0.00329	0.00530	0.00966	0.00381

Notice finally that the denominator 0.12768 of this proportion (and thus the difference of 0.05474 due to the diagonal blocks) can be obtained easily from our previous results concerning the average off-diagonal inertia, Equation A.6. We had calculated the average off-diagonal inertia to be 0.17024, which is the average of 12 inertias from individual off-diagonal tables, whereas our present calculation involves the average of 16 tables. Hence, the inertia in the modified Burt matrix due to the off-diagonal tables is $(12/16) \times 0.17024 = 0.12768$. Hence, the part due to the modified tables on the diagonal is $0.18242 - 0.12768 = 0.05474$, and so it is actually not necessary to calculate the contributions on the diagonal of Table A.12 directly.

A.6 Supplementary variables

In simple CA, supplementary row or column points are commonly calculated as weighted averages of the column or row standard coordinates, respectively: for example, to position a supplementary column in principal coordinates, the profile of the supplementary column is used to calculate a weighted average of the row standard coordinates. As described in Chapter 2, Section 2.4, in the MCA of the indicator matrix there are two ways to represent supplementary variables: first, as regular supplementary columns as just described, which amounts to averaging respondent points in standard coordinates, and second, as averages of respondent points in principal coordinates, which amounts to appending to the indicator matrix, as rows, the concatenated cross-tabulations of the supplementary variables with the active variables. The latter option is preferable because it is a unified strategy across all forms of MCA, in particular the MCA with adjusted inertias, which is the form we prefer.

Before treating the second option, let us first recall the standard case in the analysis of the indicator matrix $\mathbf{Z}$, i.e., supplementary column categories whose position is obtained by averaging over row standard coordinates. Suppose that we have a supplementary variable coded in indicator form as the matrix $\mathbf{Z}^*$, with (i,j)th column element z^*_{ij} and $z^*_{\cdot j}$ its corresponding column sum. Given the standard row (respondent) coordinates a_{is}, the supplementary column principal coordinates g_{js} are given as

$$g^*_{js} = \sum_{i=1}^{I} \frac{z^*_{ij}}{z^*_{\cdot j}} a_{is} \tag{A.8}$$

which, since the values z^*_{ij} are either 0 or 1, is the average of the standard coordinates of those respondents in category j. Table A.13 shows these column coordinates.

The second (and preferable) method is based on averaging the respondent row points in principal coordinates, which is equivalent to appending the cross-tabulations $\mathbf{Z}^{*\mathsf{T}}\mathbf{Z}$ (of the supplementary variable with the active variables) as supplementary rows of $\mathbf{Z}$. As shown in Chapter 2, Section 2.4, this will give the same numerical coordinates as appending $\mathbf{Z}^{*\mathsf{T}}\mathbf{Z}$ as supplementary rows to the Burt matrix $\mathbf{C}$. Or, since the Burt matrix is symmetric, one can append the transposed cross-tabulations $\mathbf{Z}^{\mathsf{T}}\mathbf{Z}^*$ to $\mathbf{C}$ as supplementary columns. To illustrate the calculations, suppose that $\mathbf{C}^*$ denotes the latter cross-tabulations $\mathbf{Z}^{\mathsf{T}}\mathbf{Z}^*$ (stacked vertically) appended as columns to the Burt matrix, with general element c^*_{ij} and column sums $c^*_{\cdot j}$. In the analysis of $\mathbf{C}$, denote the (row) standard coordinates of the active response categories by $\tilde{a}_{is}$. (These are not the same as the a_{is} of Equation A.8, which in that case refer to standard coordinates of respondent points i in the analysis of the indicator matrix $\mathbf{Z}$; in this case, i refers to an active row category of the Burt matrix and runs from 1 to J.) Now the positions of the supplementary columns $\mathbf{C}^*$ are obtained by weighted averaging as follows:

$$\tilde{g}_{js} = \sum_{i=1}^{J} \frac{c^*_{ij}}{c^*_{\cdot j}} \tilde{a}_{is} \tag{A.9}$$

Table A.14 gives the supplementary principal coordinates for the response categories of the variables sex, age, and education in the ISSP example, which (to emphasize) can be considered either as supplementary *rows* of $\mathbf{Z}$ or as supplementary *rows or columns* of $\mathbf{C}$. Notice that, since Table A.13 contains averages over standard coordinates and Table A.14 has effectively calculated averages of row (respondent) points over principal coordinates, the values in Table A.14 are those in Table A.13 multiplied by square roots of corresponding principal inertias in the analysis of the indicator matrix $\mathbf{Z}$:

$$\tilde{g}_{js} = g_{js}\sqrt{\lambda_s}$$

For example, the coordinate $g_{11} = -0.143$ for sex1 (male) on the first dimension in Table A.13 would be multiplied by the square root of 0.457 (see Table A.3) to give: $-0.143 \times \sqrt{0.457} = -0.097$. This checks with the corresponding element $\tilde{g}_{11}$ in Table A.14.

Table A.13 Supplementary principal coordinates for the variables sex, age, and education as columns of the indicator matrix.

	Sex		Age						Education					
	1	2	1	2	3	4	5	6	1	2	3	4	5	6
g_{j1}	−0.143	0.137	−0.166	−0.087	−0.025	−0.031	0.016	0.281	0.180	0.161	−0.068	−0.227	−0.172	−0.308
g_{j2}	0.029	−0.028	−0.014	−0.081	−0.004	0.057	0.047	0.033	0.060	0.093	0.090	−0.279	−0.263	−0.291

Table A.14 Supplementary principal coordinates for the variables sex, age, and education computed as supplementary rows of Z or supplementary rows or columns of C.

	Sex		Age						Education					
	1	2	1	2	3	4	5	6	1	2	3	4	5	6
$\tilde{g}_{j1}$	−0.097	0.093	−0.112	−0.059	−0.017	−0.021	0.011	0.190	0.122	0.109	−0.046	−0.154	−0.116	−0.209
$\tilde{g}_{j2}$	0.019	−0.018	−0.009	−0.053	−0.003	0.038	0.031	0.022	0.039	0.061	0.059	−0.183	−0.172	−0.191

Table A.15 Some principal inertias and explained inertia for the subset MCA of Table A.6, excluding the categories "neither ... nor."

s	1	2	3	4	...	16
λ_s^2	0.2016	0.1489	0.0980	0.0721	...	0.0017
Explained inertia (%)	23.4	17.3	11.4	8.4	...	0.2

A.7 Subset analyses

In this section we detail briefly the adaptation needed to the basic CA algorithm to perform the subset analyses of Chapter 8. For example, suppose we wished to exclude the middle "neither agree nor disagree" responses for questions A to D from the analysis. Again, we can approach this from the viewpoint either of the analysis of the indicator matrix **Z** or of the Burt matrix **C**. The idea is to execute the same CA algorithm to the corresponding submatrix of **Z** or **C** but maintain the original row and column margins of the matrix. We would thus first calculate the complete matrix to be decomposed, as in Chapter 2, Equations 2.6 and 2.7, center them as usual with respect to their row and column margins, and then extract the submatrix of interest, excluding rows or columns to be ignored. In the case of **Z** in our example, this would give an 871 × 16 submatrix (excluding the four columns corresponding to the four "neither ... nor" categories, while in the case of **C** this would give a 16 × 16 submatrix (excluding four rows and columns). In terms of our earlier description, the latter option would mean performing an SVD on the submatrix of **S** calculated as in Section A.2. The results are given in Table A.15 and Table A.16. Note that in this case the

Table A.16 Some column standard and principal coordinates for the first two dimensions of the subset MCA of Table A.6, excluding the categories "neither ... nor."

	A				...	D			
	1	2	4	5	...	1	2	4	5
b_{j1}	1.696	0.538	−1.316	−2.449	...	0.888	−0.273	−0.222	0.482
b_{j2}	1.153	−0.517	−0.015	2.746	...	2.482	−0.462	−0.683	1.210
g_{j1}	0.761	0.242	−0.591	−1.100	...	0.399	−0.123	−0.100	0.216
g_{j2}	0.445	−0.200	−0.006	1.060	...	0.957	−0.178	−0.264	0.467

subset analysis also has 16 dimensions because there are no linear dependencies between the rows or columns of the submatrix analyzed. Notice further that the percentages calculated in Table A.15 are relative to that part of the inertia contained in the 16×16 submatrix, which is inflated by values on diagonal blocks, just as in MCA of the full Burt matrix. Whether there are simple ways in this case to adjust these principal inertias to obtain more realistic percentages of inertia still needs to be investigated.

A.8 A sample session in R for MCA and JCA

In this section we give some selected code using the programming language R (R Development Core Team 2005). The complete language and associated material such as program manuals and contributed packages from researchers all over the world are freely downloadable from www.r-project.org. The code that follows produces the results given in Table A.1 through Table A.16.

We assume that the data set is loaded into R as a data.frame named `dat`, using the function `read.table()` with the option `colClasses="factor"` so that columns are declared to be factors. Suppose that the first lines of the data file look like this:

A	B	C	D	sex	age	edu
2	3	4	3	2	2	3
3	4	2	3	1	3	4
2	3	2	4	2	3	2
2	2	2	2	1	2	3
3	3	3	3	1	5	2
.	.	.	.	.	.	.

that is, a header with the variable (column) names, followed by the data for each respondent. Further, suppose it is stored under the name `WG93.txt` in the current working directory. Then the R statement to input the data matrix would be

```
dat <- read.table("WG93.txt", header = TRUE,
                  colClasses = "factor")
```

The following R statements that are displayed with an indent are used to calculate the numerical results given in the previous tables. The results for the tables are displayed with the R commands written without indent at the bottom of each subsection.

Table A.1 (response pattern matrix)

```
  sup.ind <- 5:7
  dat.act <- dat[,-sup.ind]
  dat.sup <- dat[,sup.ind]
  I       <- dim(dat.act)[1]
  Q       <- dim(dat.act)[2]
dat[c(1:5,I),]
#     A B C D sex age edu
# 1   2 3 4 3   2   2   3
# 2   3 4 2 3   1   3   4
# 3   2 3 2 4   2   3   2
# 4   2 2 2 2   1   2   3
# 5   3 3 3 3   1   5   2
# 871 1 2 2 2   2   3   6
```

Table A.2 (indicator matrix)

```
  lev.n  <- unlist(lapply(dat, nlevels))
  n      <- cumsum(lev.n)
  J.t    <- sum(lev.n)
  Q.t    <- dim(dat)[2]
  Z      <- matrix(0, nrow = I, ncol = J.t)
  newdat <- lapply(dat, as.numeric)
  offset <- (c(0, n[-length(n)]))
  for (i in 1:Q.t)
    Z[1:I + (I * (offset[i] + newdat[[i]] - 1))] <- 1
  fn     <- rep(names(dat),
                unlist(lapply(dat,
                nlevels)))
  ln     <- unlist(lapply(dat, levels))
  dimnames(Z)[[2]] <- paste(fn, ln, sep = "")
  dimnames(Z)[[1]] <- as.character(1:I)
  ind.temp    <- range(n[sup.ind])
  Z.sup.ind <- (ind.temp[1]-1):ind.temp[2]
  Z.act       <- Z[,-Z.sup.ind]
  J           <- dim(Z.act)[2]
```

```
Z.act[c(1:5,I),]
#    A1 A2 A3 A4 A5 B1 B2 B3 B4 B5 C1 C2 C3 C4 C5 D1 D2 D3 D4 D5
#1    0  1  0  0  0  0  0  1  0  0  0  0  0  1  0  0  0  1  0  0
#2    0  0  1  0  0  0  0  0  1  0  0  1  0  0  0  0  0  1  0  0
#3    0  1  0  0  0  0  0  1  0  0  0  1  0  0  0  0  0  0  1  0
#4    0  1  0  0  0  0  1  0  0  0  0  1  0  0  0  0  1  0  0  0
#5    0  0  1  0  0  0  0  1  0  0  0  0  1  0  0  0  0  1  0  0
#871 1  0  0  0  0  0  1  0  0  0  0  1  0  0  0  0  1  0  0  0
```

 # Table A.3 (inertias of Z)

```
    P    <- Z.act / sum(Z.act)
    cm   <- apply(P, 2, sum)
    rm   <- apply(P, 1, sum)
    eP   <- rm %*% t(cm)
    S    <- (P - eP) / sqrt(eP)
    dec  <- svd(S)
    lam  <- dec$d[1:(J-Q)]^2
    expl <- 100*(lam / sum(lam))
  rbind(round(lam[c(1:4,(J-Q))], 3),
        round(expl[c(1:4, (J-Q))], 1))
  #          [,1]   [,2]  [,3]  [,4]  [,5]
  # [1,]  0.457  0.431 0.322 0.306 0.125
  # [2,] 11.400 10.800 8.000 7.700 3.100
```

 # Table A.4 (column standard/principal coordinates)

```
    b.s1   <- dec$v[,1] / sqrt(cm)
    b.s2   <- dec$v[,2] / sqrt(cm)
    g.s1   <- b.s1 * sqrt(lam[1])
    g.s2   <- b.s2 * sqrt(lam[2])
  round(rbind(b.s1,b.s2,g.s1,g.s2)[,c(1:5,16:20)], 3)
#          A1    A2     A3     A4     A5     D1     D2     D3     D4     D5
# b.s1  1.837 0.546 -0.447 -1.166 -1.995  1.204 -0.221 -0.385 -0.222  0.708
# b.s2 -0.727 0.284  1.199 -0.737 -2.470 -1.822  0.007  1.159  0.211 -1.152
# g.s1  1.242 0.369 -0.302 -0.788 -1.349  0.814 -0.150 -0.260 -0.150  0.479
# g.s2 -0.478 0.187  0.787 -0.484 -1.622 -1.196  0.005  0.761  0.138 -0.756
```

Table A.5 (row principal coordinates)

```
  f.s1 <- dec$u[,1] * sqrt(lam[1]) / sqrt(rm)
  f.s2 <- dec$u[,2] * sqrt(lam[2]) / sqrt(rm)
  a.s1 <- f.s1 / sqrt(lam[1])
  a.s2 <- f.s2 / sqrt(lam[2])
round(rbind(f.s1,f.s2)[,c(1:5,I)], 3)
#            1      2     3     4      5   871
# f.s1 -0.210 -0.325 0.229 0.303 -0.276 0.626
# f.s2  0.443  0.807 0.513 0.387  1.092 0.135
```

Table A.6 (Burt matrix)

```
  B <- t(Z.act) %*% Z.act
B[c(1:5,16:20), c(1:5,16:20)]
#      A1  A2  A3  A4  A5  D1  D2  D3  D4  D5
# A1 119   0   0   0   0  15  25  17  34  28
# A2   0 322   0   0   0  22 102  76  68  54
# A3   0   0 204   0   0  10  44  68  58  24
# A4   0   0   0 178   0   9  52  28  54  35
# A5   0   0   0   0  48   4   9  13  12  10
# D1  15  22  10   9   4  60   0   0   0   0
# D2  25 102  44  52   9   0 232   0   0   0
# D3  17  76  68  28  13   0   0 202   0   0
# D4  34  68  58  54  12   0   0   0 226   0
# D5  28  54  24  35  10   0   0   0   0 151
```

Table A.7 (principal inertias of Burt matrix)

```
  P.2     <- B / sum(B)
  cm.2    <- apply(P.2, 2, sum)
  eP.2    <- cm.2 %*% t(cm.2)
  S.2     <- (P.2 - eP.2) / sqrt(eP.2)
  dec.2   <- eigen(S.2)
  delt.2  <- dec.2$values[1:(J-Q)]
  expl.2  <- 100*(delt.2 / sum(delt.2))
  lam.2   <- delt.2^2
  expl.2b <- 100*(lam.2 / sum(lam.2))
rbind(round(lam.2, 4),round(expl.2b, 1))[,c(1:4,16)]
# [,1]   [,2]   [,3]   [,4]   [,5]
# [1,]  0.2092 0.1857 0.1036 0.0939 0.0157
```

```
# [2,] 18.6000 16.5000 9.2000 8.3000 1.4000
# Addendum: "check" that δs is equivalent to λs:
rbind(round(delt.2, 3),round(expl.2, 1),
      round(lam, 3),round(expl, 1))

#          [,1]    [,2]    [,3]   [,4]   [,5]   [,6]   [,7]   [,8]
#  [1,]   0.457   0.431   0.322  0.306  0.276  0.252  0.243  0.235
#  [2,]  11.400  10.800   8.000  7.700  6.900  6.300  6.100  5.900
#  [3,]   0.457   0.431   0.322  0.306  0.276  0.252  0.243  0.235
#  [4,]  11.400  10.800   8.000  7.700  6.900  6.300  6.100  5.900
#          [,9]   [,10]   [,11]  [,12]  [,13]  [,14]  [,15]  [,16]
#  [1,]   0.225   0.221    0.21  0.197  0.178  0.169  0.153  0.125
#  [2,]   5.600   5.550    5.20  4.900  4.400  4.200  3.800  3.100
#  [3,]   0.225   0.221    0.21  0.197  0.178  0.169  0.153  0.125
#  [4,]   5.600   5.500    5.20  4.900  4.400  4.200  3.800  3.100
```

Table A.8 (eigenvectors, column masses, column sc/pc's)

```
  u.s1  <- dec.2$vectors[,1]
  u.s2  <- dec.2$vectors[,2]
  a2.s1 <- u.s1 / sqrt(cm.2)
  a2.s2 <- u.s2 / sqrt(cm.2)
  f2.s1 <- a2.s1 * sqrt(lam.2[1])
  f2.s2 <- a2.s2 * sqrt(lam.2[2])
round(rbind(u.s1,u.s2,cm,a2.s1,
           a2.s2,f2.s1,f2.s2), 3)[,c(1:5,16:20)]
#          A1     A2     A3     A4     A5     D1     D2     D3     D4     D5
#u.s1   0.339  0.166 -0.108 -0.264 -0.234  0.158 -0.057 -0.093 -0.056  0.147
#u.s2  -0.134  0.086  0.290 -0.167 -0.290 -0.239  0.002  0.279  0.054 -0.240
#cm     0.034  0.092  0.059  0.051  0.014  0.017  0.067  0.058  0.065  0.043
#a2.s1  1.837  0.546 -0.447 -1.166 -1.995  1.204 -0.221 -0.385 -0.222  0.708
#a2.s2 -0.727  0.284  1.199 -0.737 -2.470 -1.822  0.007  1.159  0.211 -1.152
#f2.s1  0.840  0.250 -0.204 -0.533 -0.913  0.551 -0.101 -0.176 -0.101  0.324
#f2.s2 -0.314  0.123  0.517 -0.318 -1.064 -0.785  0.003  0.499  0.091 -0.496
```

Table A.9 (adjusted inertias)

```
  lam.adj  <- (Q/(Q-1))^2 * (delt.2[delt.2 >= 1/Q] - 1/Q)^2
  total.adj <- (Q/(Q-1)) * (sum(delt.2^2) - ((J - Q)/Q^2))
```

```
rbind(round(lam.adj, 5),
      100 * round(lam.adj / total.adj, 3))
#          [,1]     [,2]   [,3]    [,4]    [,5]  [,6]
# [1,]  0.07646  0.05822 0.0092 0.00567 0.00117 1e-05
# [2,] 44.90000 34.20000 5.4000 3.30000 0.70000 0e+00
```

Table A.10 (updated Burt matrix)

```
nd      <- 2
maxit   <- 1000
epsilon <- 0.0001
lev     <- lev.n[-sup.ind]
n       <- sum(B)
li      <- as.vector(c(0,cumsum(lev)))
dummy   <- matrix(0, J, J)
for (i in 1:(length(li)-1)) {
  ind.lo <- li[i]+1
  ind.up <- li[i+1]
  ind.to <- diff(li)[i]
  dummy[rep(ind.lo:ind.up, ind.to) +
        (rep(ind.lo:ind.up, each = ind.to)-1) * J] <- 1
  }
iterate <- function(obj, dummy, nd, adj = FALSE) {
  Bp      <- obj/n
  cm      <- apply(Bp, 2, sum)
  eP      <- cm %*% t(cm)
  cm.mat  <- diag(cm^(-0.5))
  S       <- cm.mat %*% (Bp - eP) %*% cm.mat
  dec     <- eigen(S)
  lam     <- dec$values
  u       <- dec$vectors
  phi     <- u[, 1:nd] / matrix(rep(sqrt(cm), nd),ncol = nd)
  if (adj)
    lam <- (Q / (Q - 1))^2 * (lam[lam >= 1 / Q] - 1 / Q) ^ 2
  for (s in 1:nd) {
    if (exists("coord")) {
      coord <- coord + lam[s] * (phi[,s] %*% t(phi[,s]))
      } else {
        coord <- lam[s] * (phi[,s] %*% t(phi[,s]))
        }
    }
```

```
    obj * (1 - dummy) + n * eP * dummy * (1 + coord)
    }
  # first iteration (adjusted lambda)
  B.star <- iterate(B, dummy, 2, adj = TRUE)
  # subsequent iterations
  k  <- 1
  it <- TRUE
  while (it) {
    temp    <- iterate(B.star, dummy, 2)
    delta.B <- max(abs(B.star - temp))
    B.star  <- temp
    if (delta.B <= epsilon | k >= maxit) it <- FALSE
    k <- k + 1
    }
round(B.star [c(1:5,16:20), c(1:5, 16:20)], 2)
#      A1     A2    A3    A4    A5    D1     D2    D3    D4    D5
# A1 30.72  53.14 18.59 13.97  2.58 15.00  25.00 17.00 34.00 28.00
# A2 53.14 130.55 76.80 51.80  9.71 22.00 102.00 76.00 68.00 54.00
# A3 18.59  76.80 62.95 38.86  6.80 10.00  44.00 68.00 58.00 24.00
# A4 13.97  51.80 38.86 53.51 19.85  9.00  52.00 28.00 54.00 35.00
# A5  2.58   9.71  6.80 19.85  9.06  4.00   9.00 13.00 12.00 10.00
# D1 15.00  22.00 10.00  9.00  4.00  9.02  14.67  5.03 13.27 18.01
# D2 25.00 102.00 44.00 52.00  9.00 14.67  62.46 55.78 60.90 38.20
# D3 17.00  76.00 68.00 28.00 13.00  5.03  55.78 63.56 56.49 21.14
# D4 34.00  68.00 58.00 54.00 12.00 13.27  60.90 56.49 59.74 35.60
# D5 28.00  54.00 24.00 35.00 10.00 18.01  38.20 21.14 35.60 38.04
```

Table A.11 (JCA inertias)

```
  P.3    <- B.star / sum(B.star)
  cm.3   <- apply(P.3, 2, sum)
  eP.3   <- cm.3 %*% t(cm.3)
  S.3    <- (P.3 - eP.3) / sqrt(eP.3)
  delt.3 <- eigen(S.3)$values
  lam.3  <- delt.3^2
  expl.3 <- 100*(lam.3 / sum(lam.3))
rbind(round(lam.3, 5),round(expl.3, 1))[,1:2]
# [,1]    [,2]
# [1,]  0.09909 0.06503
# [2,] 54.30000 35.60000
```

Table A.12 (inertia conributions of submatrices)

```
  subinr <- function(B, ind) {
   nn   <- length(ind)
   subi <- matrix(NA, nrow = nn, ncol = nn)
   ind2 <- c(0,cumsum(ind))
   for (i in 1:nn) {
     for (j in 1:nn) {
       tempmat   <- B[(ind2[i]+1):(ind2[i+1]),
                      (ind2[j]+1):(ind2[j+1])]
       tempmat   <- tempmat / sum(tempmat)
       ec        <- apply(tempmat, 2, sum)
       er        <- apply (tempmat, 1, sum)
       ex        <- er%*%t(ec)
       subi[i,j] <- sum((tempmat - ex)^2 / ex)
       }
     }
   subi / nn^2
   }
  si <- subinr(B.star, lev)
round(si, 5)
#           [,1]     [,2]     [,3]     [,4]
# [1,]  0.00745  0.01486  0.01215  0.00329
# [2,]  0.01486  0.02244  0.01858  0.00530
# [3,]  0.01215  0.01858  0.02103  0.00966
# [4,]  0.00329  0.00530  0.00966  0.00381
```

Table A.13 (supplementary principal coordinates as columns of Z)

```
  Z.star  <- Z[,Z.sup.ind]
  I.star  <- dim(Z.star)[1]
  cs.star <- apply(Z.star, 2, sum)
  base    <- Z.star / matrix(rep(cs.star, I.star), nrow =
                I.star, byrow = TRUE)
  b.star1 <- t(base) %*% cbind(a.s1, a.s2)
round(t(b.star1), 3)
#        sex1   sex2   age1   age2   age3   age4  age5  age6
# a.s1 -0.143  0.137 -0.166 -0.087 -0.025 -0.031 0.016 0.281
# a.s2  0.029 -0.028 -0.014 -0.081 -0.004  0.057 0.047 0.033
#       edu1  edu2   edu3   edu4   edu5   edu6
# a.s1  0.18 0.161 -0.068 -0.227 -0.172 -0.308
# a.s2  0.06 0.093  0.090 -0.279 -0.263 -0.291
```

Table A.14 (supplementary principal coordinates via cross-tabulation)

```
  ct.star   <- t(Z.star)%*%Z.act
  I.star2   <- dim(ct.star)[2]
  cs.star2  <- apply(ct.star, 1,sum)
  base2     <- ct.star / matrix(rep(cs.star2,
                 I.star2),ncol = I.star2)
  b.star2   <- base2 %*% cbind(a2.s1, a2.s2)
round(t(b.star2), 3)
#          sex1   sex2   age1   age2   age3   age4  age5  age6
# a2.s1 -0.097  0.093 -0.112 -0.059 -0.017 -0.021 0.011 0.190
# a2.s2  0.019 -0.018 -0.009 -0.053 -0.003  0.038 0.031 0.022
#        edu1  edu2   edu3   edu4   edu5   edu6
# a2.s1 0.122 0.109 -0.046 -0.154 -0.116 -0.209
# a2.s2 0.039 0.061  0.059 -0.183 -0.172 -0.191
```

Table A.15 (principal inertias from subset analysis)

```
  sub.ind  <- c(3,8,13,18)
  P.4      <- B / sum(B)
  cm.4     <- apply(P.4, 2, sum)
  eP.4     <- cm.4 %*% t(cm.4)
  S.sub    <- ((P.4 - eP.4) / sqrt(eP.4)) [-sub.ind,-
                sub.ind]
  dec.sub  <- eigen(S.sub)
  lam.sub  <- dec.sub$values[1:(J-Q)]^2
  expl.sub <- 100*(lam.sub / sum(lam.sub))
rbind(round(lam.sub, 4),round(expl.sub,  1))[,c(1:4,(J-Q))]
#          [,1]    [,2]   [,3]   [,4]   [,5]
# [1,]  0.2016  0.1489  0.098 0.0721 0.0017
# [2,] 23.4000 17.3000 11.400 8.4000 0.2000
```

Table A.16 (column standard and principal coordinates from subset analysis)

```
  cm.sub   <- cm.4[-sub.ind]
  u.sub.s1 <- dec.sub$vectors[,1]
  u.sub.s2 <- dec.sub$vectors[,2]
  a.sub.s1 <- u.sub.s1 / sqrt(cm.sub)
  a.sub.s2 <- u.sub.s2 / sqrt(cm.sub)
  f.sub.s1 <- a.sub.s1 * sqrt(lam.sub[1])
  f.sub.s2 <- a.sub.s2 * sqrt(lam.sub[2])
round(rbind(a.sub.s1,a.sub.s2,f.sub.s1, f.sub.s2), 3)[,
          c(1:4,13:16)]
#             A1     A2     A4     A5    D1     D2     D4    D5
# a.sub.s1 1.696  0.538 -1.316 -2.449 0.888 -0.273 -0.222 0.482
# a.sub.s2 1.153 -0.517 -0.015  2.746 2.482 -0.462 -0.683 1.210
# f.sub.s1 0.761  0.242 -0.591 -1.100 0.399 -0.123 -0.100 0.216
# f.sub.s2 0.445 -0.200 -0.006  1.060 0.957 -0.178 -0.264 0.467
```

A.9 R functions for CA, MCA, and JCA

The above code and more has been implemented in an R package `mjca`. The package comprises two core functions: `ca()` for simple correspondence analysis and `mjca()` for multiple and joint forms of correspondence analysis. Each core function has methods for printing, summarizing, and plotting (in two dimensions and three dimensions).

A short description of these functions, extracted from their help files, follows.

`ca` *Simple correspondence analysis*

Description

Computation of simple correspondence analysis.

Usage

```
ca(obj, nd = NA, suprow = NA, supcol = NA,
subsetrow = NA, subsetcol = NA)
```

Arguments

`obj`	A two-way table of nonnegative data, usually frequencies

`nd`	Number of dimensions to be included in the output; if `NA` the maximum possible dimensions are included
`suprow`	Indices of supplementary rows
`supcol`	Indices of supplementary columns
`subsetrow`	Row indices of subset
`subsetcol`	Column indices of subset

Details

The function `ca` computes a simple correspondence analysis based on the singular-value decomposition. The options `suprow` and `supcol` allow supplementary (passive) rows and columns to be specified. Using the options `subsetrow` and/or `subsetcol` result in a subset CA being performed.

Value

`sv`	Singular values
`rownames`	Row names
`rowmass`	Row masses
`rowdist`	Row chi-square distances to centroid
`rowinertia`	Row inertias
`rowcoord`	Row standard coordinates
`rowsup`	Indices of row supplementary points
`colnames`	Column names
`colmass`	Column masses
`coldist`	Column chi-square distances to centroid
`colinertia`	Column inertias
`colcoord`	Column standard coordinates
`colsup`	Indices of column supplementary points

`mjca` *Multiple and joint correspondence analysis*

Description

Computation of multiple and joint correspondence analysis.

Usage

```
mjca(obj, nd = NA, lambda = "adjusted", suprow = NA,
supcol = NA)
```

Arguments

`obj`	A response pattern matrix containing factors
`nd`	Number of dimensions to be included in the output; if `NA` the maximum possible dimensions are included
`lambda`	Gives the scaling factor for the eigenvalues; possible values include "`indicator`," "`Burt`," "`adjusted`," and "`JCA`"; using `lambda` = "`JCA`" results in a joint correspondence analysis
`suprow`	Indices of supplementary rows
`supcol`	Indices of supplementary columns

Details

The function `mjca` computes a multiple or joint correspondence analysis based on the eigenvalue decomposition of the Burt matrix.

Value

`ev`	Eigenvalues
`lambda`	Scaling method for the eigenvalues
`levelnames`	Names of the factor/level combinations
`levels.n`	Number of levels in each factor
`rownames`	Row names
`rowmass`	Row masses
`rowdist`	Row chi-square distances to centroid
`rowinertia`	Row inertias
`rowcoord`	Row standard coordinates
`rowsup`	Indices of row supplementary points
`colnames`	Column names
`colmass`	Column masses
`coldist`	Column chi-square distances to centroid
`colinertia`	Column inertias
`colcoord`	Column standard coordinates
`colsup`	Indices of column supplementary points
`Burt`	Burt matrix
`subinertia`	Inertias of sub-matrices
`call`	Return of `match.call`

A.10 XLSTAT implementation of CA and MCA

XLSTAT is a commercial statistical package for performing a range of univariate and multivariate statistical analyses in a Microsoft Excel environment. The package includes simple and multiple correspondence analysis. We have collaborated to update these two programs in XLSTAT so that they contain additional features not found in regular correspondence analysis software, such as the adjustment of principal inertias in MCA and the subset correspondence analysis option. The advantage of XLSTAT is its simple interface and the fact that all results are returned in the Excel environment. This means that it is very easy to make additional computations on the results as well as to modify maps (changing the coordinate values, labels, etc.). An example of its ease of use is the following. In the context of our present example, we can read the whole data matrix into Excel, both the four substantive variables as well as the three demographic variables, and then produce the complete Burt matrix of all cross-tabulations of these seven variables, in Excel format, using the MCA program. Then we can freely select with the mouse which part of the Burt matrix we want to analyze using the CA program. The fact that both data and results are contained in the same format in Excel worksheets means that there is a seamless interface that is very easy to use, especially for Excel users. For more information, see www.xlstat.com.

A.11 Summary of software mentioned in this book

Method	Software	Chapters	Web / E-mail / ftp
Simple and multiple correspondence analysis	`ca` and `mjca` in R	2, appendix	www.r-project.org oleg.nenadic@wi-wiss.uni-goettingen.de
	XLSTAT	2, appendix	www.xlstat.com
	ADDAD	5	ftp.math-info.univ-paris5.fr/pub/MathPsy/AGD
	EyeLID	5	www.math-info.univ-paris5.fr/~lerb
	DTM	7	www.lebart.org
	SPAD	16	www.spadsoft.com
Joint correspondence analysis	`mjca` in R	2, appendix	www.r-project.org
Nonlinear principal component analysis	`PRINQUAL` in SAS	4	www.sas.com
	`Categories` in SPSS	4, 20, 21	www.spss.com/spssbi/categories/index.htm
	`gifi` in R	4	www.stat.ucla.edu/gifi
Subset correspondence analysis	`ca` and `mjca` in R	8	www.r-project.org
	XLSTAT	8	www.xlstat.com
Regularized multiple correspondence analysis	MATLAB® code	11	takane@takane2.psych.mcgill.ca
Multiple factor analysis of contingency tables; multiple factor analysis	SPAD	13, 15	www.spadsoft.com

Simultaneous analysis	S-PLUS code, `AnSimult` in R	14	etpzacaa@bs.ehu.es
Logistic regression	`Logistic` in SAS	16	www.sas.com
Three-mode correspondence analysis	3-WayPack	21	www.three-mode.leidenuniv.nl
	`multidim` in S-PLUS	21	www.lsp.ups-tlse.fr/index.html
Log-multiplicative modeling and latent variables	ℓEM	21	www.uvt.nl/faculteiten/fsw/organisatie/departementen/mto/software2.html
Logistic biplot	MATLAB® code	23	villardon@usal.es

Note: Links to this software as well as the data sets used in this book can be found at www.carme-n.org, the web page of the CARME-network (Correspondence Analysis and Related Methods).

References

Abdessemed, L. and Escofier, B. (1996). Analyse factorielle multiple de tableaux de frequencies: comparaison avec l'analyse canonique des correspondences. *Journal de la Société de Statistique de Paris*, 137, 3–18.

Adachi, K. (2002). Homogeneity and smoothness analysis for quantifying a longitudinal categorical variable. In *Measurement and Multivariate Analysis*, S. Nishisato, Y. Baba, H. Bozdogan, and K. Kanefuji, Eds. Tokyo: Springer, pp. 47–56.

Aitchison, J.A. (1986). *Compositional Data Analysis*. London: Chapman and Hall.

Aitchison, J.A. and Greenacre, M.J. (2002). Biplots of compositional data. *Applied Statistics*, 51, 375–392.

Aitkin, M., Francis, B., and Raynal, N. (1987). Une étude comparative d'analyses des correspondances ou de classifications et des modèles de variables latentes ou de classes latentes. *Revue de Statistique Appliquée*, 35 (3), 53–82.

Aluja, T. and Morineau, A. (1999). *Aprender de los Datos: El Análisis de Componentes Principales*. Barcelona: Ediciones Universitarias de Barcelona.

Alvarez, R., Bécue-Bertaut, M., and Valencia, O. (2004). Etude de la stabilité des valeurs propres de l'AFC d'un tableau lexical au moyen de procédures de rééchantillonnage. In *Le Poids des Mots*, G. Purnelle, C. Fairon, and A. Dister, Eds. Louvain: PUL, pp. 42–51.

Anderson, C.J. (1996). The analysis of three-way contingency tables by three-mode association models. *Psychometrika*, 61, 465–483.

Anderson, C.J. (2002). Analysis of multivariate frequency data by graphical models and generalizations of the multidimensional row-column association model. *Psychological Methods*, 7, 446–467.

Anderson, C.J. and Böckenholt, U. (2000). RC regression models for polytomous variables. *Psychometrika*, 65, 479–509.

Anderson, C.J. and Vermunt, J.K. (2000). Log-multiplicative association models as latent variable models for nominal and/or ordinal data. *Sociological Methodology*, 30, 81–121.

Anderson, T.W. (1958). *An Introduction to Multivariate Statistical Analysis*. New York: Wiley.

Andrich, D. (1988). The application of an unfolding model of the PIRT type to the measurement of attitude. *Applied Psychological Measurement*, 12, 33–51.

Andrich, D. (1996). A hyperbolic cosine latent trait model for unfolding polytomous data: reconciling Thurstone and Likert methodologies. *British Journal of Mathematical Psychology*, 49, 347–365.

Andrich, D. and Luo, G. (1993). A hyperbolic cosine latent trait model for unfolding dichotomous single-stimulus responses. *Applied Psychological Measurement*, 17, 253–276.

Bailey, R.A. and Gower, J.C. (1990). Approximating a symmetric matrix. *Psychometrika*, 55, 665–675.

Baker, F.B. (1992). *Item Response Theory: Parameter Estimation Techniques*. New York: Marcel Dekker.

Bamber, D. (1975). The area above the ordinal dominance graph and the area below the receiver operating graph. *Journal of Mathematical Psychology*, 12, 387–415.

Bartholomew, D.J. and Knott, M. (1999). *Latent Variable Models and Factor Analysis*. London: Arnold.

Becker, M.P. (1989). Models for the analysis of association in multivariate contingency tables. *Journal of the American Statistical Association*, 84, 1014–1019.

Bécue-Bertaut, M. and Pagès, J. (1999). Intra-sets multiple factor analysis. Application to textual data. In *Applied Stochastic Models and Data Analysis*, H. Bacelar-Nicolau, F. Costa-Nicolau, and J. Janssen, Eds. Lisbon: INE, pp. 72–79.

Bécue-Bertaut, M. and Pagès, J. (2001). Analyse simultanée de questions ouvertes et de questions fermées — méthologie, exemples. *Journal de la Société Française de Statistique*, 142 (4), 91–104.

Bécue-Bertaut, M. and Pagès, J. (2004). A principal axes method for comparing multiple contingency tables: MFACT. *Computational Statistics and Data Analysis*, 45, 481–503.

Beh, E. (2004). *A Bibliography of the Theory and Application of Correspondence Analysis*, Vol. III (by year). http://www.uws.edu.au/download. php?file_id =7127&filename=Bibby_year.pdf&mimetype=application/pdf.

Bekker, P. and de Leeuw, J. (1988). Relation between variants of nonlinear principal component analysis. In *Component and Correspondence Analysis*, J.L.A. van Rijckevorsel and J. de Leeuw, Eds. Chichester: Wiley, pp. 1–31.

Benali, H. and Escofier, B. (1987). Stabilité de l'analyse des correspondences multiples en cas données manquantes et de modlités à faible effectif. *Revue de Statistique Appliquée*, 35, 41–51.

Benzécri, J.-P. (1979). Sur le calcul des taux d'inertie dans l'analyse d'un questionnaire: Addendum et Erratum à (Bin. Mult.). *Les Cahiers de l'Analyse des Données*, 4, 377–378.

Benzécri, J.-P. (1982a). Sur la généralisation du tableau de Burt et son analyse par bandes. *Les Cahiers de l'Analyse des Données*, 7, 33–43.

Benzécri, J.-P. (1982b). *Histoire et Préhistoire de l'Analyse des Données.* Paris: Dunod.

Benzécri, J.-P. (1983). Analyse de l'inertie intraclasse par l'analyse d'un tableau de contingence. *Les Cahiers de l'Analyse des Données*, 8, 351–358.

Benzécri, J.-P. (1992). *Correspondence Analysis Handbook.* New York: Dekker.

Benzécri, J.-P. (2003). *Qu'est-ce que l'Analyse des Données?* Text read by B. Le Roux at the Conference on Correspondence Analysis and Related Methods (CARME 2003), Barcelona; 29.06.-02.07.2003.

Benzécri, J.-P. et al. (1973). *L'Analyse des Données: L'Analyse des Correspondances.* Paris: Dunod.

Bernard, J.M., Baldy, R., and Rouanet, H. (1988). The language for interrogating data. In *Data Analysis and Informatics V*, E. Diday, Ed. Amsterdam: North Holland, pp. 461–468.

Bernard, J.M., Le Roux, B., Rouanet, H., and Schiltz, M.A. (1989). L'analyse des données multidimensionnelles par le langage d'interrogation de données: Au delà de l'analyse des correspondances. *Bulletin de Méthodologie Sociologique*, 23, 3–46.

Besse, P., Caussinus, H., Ferré, L., and Fine, J. (1988). Principal component analysis and optimization of graphical displays. *Statistics*, 19, 301–312.

Blasius, J. (1994). Correspondence analysis in social science research. In *Correspondence Analysis in the Social Sciences: Recent Developments and Applications*, M. J. Greenacre and J. Blasius, Eds. London: Academic Press, pp. 23–52.

Blasius, J. and Greenacre, M. (1994). Computation of correspondence analysis. In *Correspondence Analysis in the Social Sciences*, M. J. Greenacre and J. Blasius, Eds. London: Academic Press, pp. 53–78.

Blasius, J. and Greenacre, M. J., Eds. (1998). *Visualization of Categorical Data.* London: Academic Press.

Blasius, J. and Thiessen, V. (2001a). The use of neutral responses in survey questions: an application of multiple correspondence analysis. *Journal of Official Statistics*, 17, 351–367.

Blasius, J. and Thiessen, V. (2001b). Methodological artifacts in measures of political efficacy and trust: a multiple correspondence analysis. *Political Analysis*, 9, 1–20.

Blasius, J. and Winkler, J. (1989). Gibt es die "feinen Unterschiede"? Eine empirische Überprüfung der Bourdieuschen Theorie. *Kölner Zeitschrift für Soziologie und Sozialpsychologie*, 41, 72–94.

BMS, The (van Meter, K.M., Schlitz, M.-A., Cibois, P., and Mounier, L.) (1994). Correspondence analysis: a history and French sociological perspective. In *Correspondence Analysis in the Social Sciences,* M. J. Greenacre and J. Blasius, Eds. London: Academic Press, pp. 128–137.

Böckenholt, U. and Takane, Y. (1994). Linear constraints in correspondence analysis. In *Correspondence Analysis in the Social Sciences: Recent Developments and Applications*, M.J. Greenacre and J. Blausius, Eds., London: Academic Press, pp. 70–111.

Böhning, D. and Lindsay, B.G. (1988). Monotonicity of quadratic-approximation algorithms. *Annals of the Institute of Statistical Mathematics*, 40, 641–663.

Boik, R.J. (1996). An efficient algorithm for joint correspondence analysis. *Psychometrika*, 61, 255–269.

Bonnet, P., Le Roux, B., and Lemaine, G. (1996). Analyse géométrique des données: une enquête sur le racisme. *Mathématiques et Sciences Humaines*, 136, 5–24.

Bourdieu, P. (1979). *La Distinction: Critique Sociale du Jugement.* Paris: Editions de Minuit.

Bourdieu, P. (1984). *Distinction.* Boston: Harvard University Press.

Bourdieu, P. (1999). Une révolution conservatrice dans l'édition. *Actes de la Recherche en Sciences Sociales*, 126–127, 3–28.

Bourdieu, P. (2001). *Langage et pouvoir Symbolique.* Paris: Fayard.

Bourdieu, P. and Saint-Martin, M. (1978). Le patronat. *Actes de la Recherche en Sciences Sociales*, 20–21, 3–82.

Bouroche, J.M. and Saporta, G. (1988). Les méthodes et les applications du credit-scoring. *Atti 34° Riunione Scientifica della Sociéta Italiana di Statistica*, 19–26.

Bouroche, J.M., Saporta, G., and Tenenhaus, M. (1977). Some methods of qualitative data analysis. In *Recent Developments in Statistics*, J.R. Barra, Ed. Amsterdam: North-Holland, pp. 749–755.

Carlier, A. and Kroonenberg, P.M. (1996). Decompositions and biplots in three-way correspondence analysis. *Psychometrika*, 61, 355–373.

Carlier, A. and Kroonenberg, P.M. (1998). The case of the French cantons: an application of three-way correspondence analysis. In *Visualization of Categorical Data*, J. Blasius and M.J. Greenacre, Eds. New York: Academic Press, pp. 253–275.

Carmone, F.J. and Green, P.E. (1981). Model misspecification in multiattribute parameter estimation. *Journal of Marketing Research*, 18, 87–93.

Carroll, J.D. (1968). A generalization of canonical correlation analysis to three or more sets of variables. In *Proceedings of 76th Annual Convention of the American Psychological Association*, Washington, DC, pp. 227–228.

Carroll, J.D. (1969). Categorical Conjoint Measurement. Paper presented at the Meeting of Mathematical Psychology, Ann Arbor, MI.

Caussinus, H. (1986). Quelques réflexions sur la part des modèles probabilistes en analyse des données. In *Data Analysis and Informatics IV,* E. Diday, Y. Escoufier, L. Lebart, J. Pagès, Y. Schektman, and R. Tomassone, Eds. Amsterdam: North-Holland, pp. 151–165.

Caussinus, H., Fekri, M., Hakam, S., and Ruiz-Gazen, A. (2003a). A monitoring display of multivariate outliers. *Computational Statistics and Data Analysis*, 44, 237–252.

Caussinus, H., Hakam, S., and Ruiz-Gazen, A. (2003b). Projections révélatrices contrôlées: groupements et structures diverses. *Revue de Statistique Appliquée*, 51 (1), 37–58.

Caussinus, H., Hakam, S., and Ruiz-Gazen, A. (2002). Projections révélatrices contrôlées: recherche d'individus atypiques. *Revue de Statistique Appliquée*, 50 (4), 5–37.

Caussinus, H. and Ruiz-Gazen, A. (1995). Metrics for finding typical structures by means of principal component analysis. In *Data Science and its Applications*, Y. Escoufier and C. Hayashi, Eds. Tokyo: Academic Press, pp. 177–192.

Caussinus, H. and Ruiz-Gazen, A. (2003). Projections révélatrices exploratoires. In *Analyse des Données*, G. Govaert, Ed. Paris: Hermès, pp. 83–104.

Cazes, P. (1980). Analyse de certains tableaux rectangulaires décomposés en blocs. *Les Cahiers de l'Analyse des Données*, 5, 145–161 and 387–403.

Cazes, P. (1982). Note sur les éléments supplémentaires en analyse des correspondances. *Les Cahiers de l'Analyse des Données*, 7, 9–23 and 133–154.

Cazes, P. (2004). Quelques methodes d'analyse factorielle d'une serie de tableaux de données. *La Revue de Modulad*, 31, 1–31.

Cazes, P. and Moreau, J. (1991). Analysis of a contingency table in which the rows and the columns have a graph structure. In *Symbolic-Numeric Data Analysis and Learning*, E. Diday and Y. Lechevallier, Eds. New York: Nova Science Publishers, pp. 271–280.

Cazes, P. and Moreau, J. (2000). Analyse des correspondances d'un tableau de contingence dont les lignes et les colonnes sont munies d'une structure de graphes bistochastique. In *L'Analyse des Correspondances et les Techniques Connexes: Approches Nouvelles pour l'Analyse Statistique des Données*, J. Moreau, P.A. Doudin, and P. Cazes, Eds. Berlin–Heidelberg: Springer, pp. 87–103.

Celeux, G. and Nakache, J.P. (1994). *Discrimination sur Variables Qualitatives.* Paris: Polytechnica.

Chambers, J.M., Cleveland, W.S., Kleiner, B., and Tukey, P.A. (1983). *Graphical Methods for Data Analysis.* Pacific Grove, CA: Wadsworth and Brooks/Cole Publishing Co.

Chateau, F. and Lebart, L. (1996). Assessing sample variability in visualization techniques related to principal component analysis: bootstrap and alternative simulation methods. In *COMPSTAT 1996*, A. Prats, Ed. Heidelberg: Physica, pp. 205–210.

Chessel, D. and Hanafi, M. (1996). Analyses de la co-inertie de K nuages de points. *Revue de Statistique Appliquée*, 46, 35–60.

Cheung, K.C. and Mooi, L.C. (1994). A comparison between the rating scale model and dual scaling for Likert scales. *Applied Psychological Measurement*, 18, 1–13.

Chiche, J., Le Roux, B., Perrineau, P., and Rouanet, H. (2000). L'espace politique des électeurs français à la fin des années 1990. *Revue Française de Science Politique*, 50, 463–487.

Choulakian, V. (1988a). Exploratory analysis of contingency tables by loglinear formulations and generalizations of correspondence analysis. *Psychometrika*, 53, 235–250 and Errata 593.

Choulakian, V. (1988b). Analyse factorielle des correspondances de tableaux multiples. *Revue de Statistique Appliquée*, 36, 33–41.

Choulakian, V. (1996). Generalized bilinear models. *Psychometrika*, 61, 271–283.

Cleveland, W.S. (1994). *The Elements of Graphing Data.* Murray Hill, New Jersey: AT&T Bell Laboratories.

Clinton, J., Jackman, S., and Rivers, D. (2004). The statistical analysis of roll call data. *American Political Science Review*, 98, 355–370.

Clogg, C.C. (1986). Statistical modeling versus singular value decomposition. *International Statistical Review*, 54, 284–288.

Clogg, C.C. and Shihadeh, E.S. (1994). *Statistical Models for Ordinal Variables.* Thousand Oaks, CA: Sage.

Converse, P.E. (1964). The nature of belief systems in mass publics. In *Ideology and Discontent*, D.E. Apter, Ed. New York: Free Press, pp. 206–261.

Converse, P.E. (1970). Attitudes and non-attitudes: continuation of a dialog. In *The Quantitative Analysis of Social Problems*, E.R. Tufte, Ed. Reading: Addison-Wesley, pp. 168–189.

Coombs, C.H. (1950). Psychological scaling without a unit of measurement. *Psychological Review*, 57, 145–158.

Coombs, C.H. (1964). *A Theory of Data.* New York: Wiley.

Coombs, C.H. and Coombs, L. (1976). "Don't know": item ambiguity or respondent uncertainty. *Public Opinion Quarterly*, 40, 497–514.

Coombs, C.H. and Kao, R.C. (1955). *Nonmetric Factor Analysis*, Engineering Research Bulletin 38. Ann Arbor: University of Michigan Press.

Cramér, H. (1946). *Mathematical Methods of Statistics.* Princeton, NJ: University Press.

Cronbach, L.J. (1951). Coefficient alpha and the internal structure of tests. *Psychometrika*, 16, 297–334.

Croon, M. (1993). Ordinal latent class analysis for single-peaked items. *Kwantitatieve Methoden*, 42, 128–142.

Darroch, J.N. (1974). Multiplicative and additive interaction in contingency tables. *Biometrika*, 61, 207–214.

de Falguerolles, A. (1998). Log-bilinear biplots in action. In *Visualization of Categorical Data*, J. Blasius and M.J. Greenacre, Eds. San Diego: Academic Press, pp. 527–540.

de Falguerolles, A. and Francis, B. (1992). Algorithmic approaches for fitting bilinear models. In *COMPSTAT 1992*, Y. Dodge and J. Whittaker, Eds. Heidelberg: Physica–Verlag, pp. 77–82.

de Leeuw, J. (1973). *Canonical Analysis of Categorical Data.* Report RN 007-68, Psychological Institute, Leiden University, Netherlands; reprinted as a book and published in 1984, Leiden: DSWO Press.

de Leeuw, J. (1982). Nonlinear principal component analysis. In *COMPSTAT 1982*, H. Caussinus, Ed. Vienna: Physica–Verlag, pp. 77–86.

de Leeuw, J. (1988). Multivariate analysis with linearizable regressions. *Psychometrika*, 53, 437–454.

de Leeuw, J. (1990). Multivariate analysis with optimal scaling. In *Progress in Multivariate Analysis*, S. Das Gupta and J. Sethuraman, Eds. Calcutta: Indian Statistical Institute.

de Leeuw, J. (1994). Block relaxation methods in statistics. In *Information Systems and Data Analysis*, H.H. Bock, W. Lenski, and M.M. Richter, Eds. Berlin: Springer, pp. 308–325.

de Leeuw, J. (2005). Monotonic regression. In *Encyclopedia of Statistics in Behavioral Science*, B.S. Everitt and D.C. Howell, Eds. Chichester: Wiley, pp. 1260–1261.

de Leeuw, J. (2006). Principal component analysis of binary data by iterated singular value decomposition. *Computational Statistics and Data Analysis*, 50, 21–39.

de Leeuw, J. and Meulman, J.J. (1986). Principal component analysis and restricted multidimensional scaling. In *Classification as a Tool of Research*, W. Gaul and M. Schader, Eds. Amsterdam: North-Holland, pp. 83–96.

de Leeuw, J. and Michailidis, G. (1999). Graph layout techniques and multidimensional data analysis. In *Game Theory. Optimal Stopping, Probability and Statistics*, F.T. Bruss and L. LeCam, Eds. Beachwood, OH: Institute of Mathematical Statistics, pp. 219–248.

de Leeuw, J. and Michailidis, G. (in press) *Block Relaxation and Majorization Algorithms in Statistics.* Berlin: Springer.

de Leeuw, J., Michailidis, G., and Wang, D.Y. (1999). Correspondence analysis techniques. In *Multivariate Analysis: Design of Experiments and Survey Sampling*, S. Ghosh, Ed. New York: Marcel Dekker, pp. 523–547.

Dequier, A. (1974). Notes sur les tables de contingence entre trois caractères. Unpublished document 1033. Arcueil: CIRO.

DiPillo, P.J. (1976). The application of bias to discriminant analysis. *Communications in Statistics—Theory and Methods*, 5, 843–859.

Draper, N.R. and Smith, H. (1998). *Applied Regression Analysis*, 2nd ed. New York: Wiley.

Duncan, O.D. and Stenbeck, M. (1988). No opinion or not sure. *Public Opinion Quarterly*, 52, 513–525.

Eckart, C. and Young, G. (1936). The approximation of one matrix by another of lower rank. *Psychometrika*, 1, 211–218.

Edwards, D. (2000). *Introduction to Graphical Modelling*, 2nd ed. New York: Springer.

Efron, B. (1979). Bootstraps methods: another look at the jackknife. *Annals of Statistics*, 7, 1–26.

Efron, B. and Tibshirani, R.J. (1993). *An Introduction to the Bootstrap*. New York: Chapman and Hall.

Escofier, B. (1984). Analyse factorielle en référence à un modèle: application à l'analyse d'un tableau d'échanges. *Revue de Statistique Appliquée*, 32, 25–36.

Escofier, B. (2003). *Analyse des Correspondances: Recherches au Cœur de l'Analyse des Données*. Rennes: Presses universitaires de Rennes.

Escofier, B. and Drouet, D. (1983). Analyse des différences entre plusieurs tableaux de fréquence. *Les Cahiers de l'Analyse des Données*, 8 (4), 491–499.

Escofier, B. and Pagès, J. (1982). Comparaison de groupes de variables définies sur le même ensemble d´individus: un exemple d´applications. Working paper 165. Institut National de Recherche en Informatique et en Automatique, Le Chesnay, France.

Escofier, B. and Pagès, J. (1986). Le traitement des variables qualitatives et des tableaux mixtes par analyse factorielle multiple. In *Data Analysis and Informatics*, Vol. IV, E. Diday, Ed. Amsterdam: North-Holland, pp. 179–191.

Ecofier, B. and Pagès, J. (1990). *Analyse Factorielles Simples et Multiples.* Paris: Dunod.

Escofier, B. and Pagès, J. (1994). Multiple factor analysis (AFMULT package). *Computational Statistics and Data Analysis*, 18, 121–140.

Escofier, B. and Pagès, J. (1998). *Analyses Factorielles Simples et Multiples. Objectifs, Méthodes et Interprétation*, 2nd edition. Paris: Dunod.

Escoufier, Y. (1973). Le traitement des variables vectorielles. *Biometrika*, 29, 751–760.

Escoufier, Y. (1982). L'analyse des tableaux de contingence simples et multiples. *Metron*, 40, 53–77.

Everitt, B.S. (1984). *An Introduction to Latent Variable Models.* London: Chapman and Hall.

Fisher, R.A. (1936). The use of multiple measurements in taxonomy problem. *Annals of Eugenics*, 7, 179–188.

Fisher, R.A. (1940). The precision of discriminant functions. *Annals of Eugenics*, 10, 422–429.

Forman, A.K. (1993). Latent class models for monotone and nonmonotone dichotomous items. *Kwantitatieve Methoden*, 42, 143–160.

Fowlkes, E.B., Freeny, A.E., and Landwehr, J.M. (1988). Evaluating logistic models for large contingency tables. *Journal of the American Statistical Association*, 83, 611–622.

Friedman, J. (1989). Penalized discriminant analysis. *Journal of the American Statistical Association*, 84, 165–175.

Friedman, J. and Tukey, J. (1974). A projection pursuit algorithm for exploratory data analysis. *IEEE Transactions on Computers*, 23, 881–889.

Gabriel, K.R. (1971). The biplot-graphic display of matrices with applications to principal component analysis. *Biometrika*, 58, 453–467.

Gabriel, K.R. (1981). Biplot display of multivariate matrices for inspecting of data and diagnosis. In *Interpreting Multivariate Data*, V. Barnett, Ed. Chichester: Wiley, pp. 147–174.

Gabriel, K.R. (1995). Biplot display of multivariate categorical data, with comments on multiple correspondence analysis. In *Recent Advances in Descriptive Multivariate Analysis*, W. Krzanowsky, Ed. Oxford: Oxford Science Publications, pp. 190–226.

Gabriel, K.R. (1998). Generalised bilinear regression. *Biometrika*, 85, 689–700.

Gabriel, K.R. (2002a). Goodness of fit of biplots and correspondence analysis. *Biometrika,* 89, 423–436.

Gabriel, K.R. (2002b). Le biplot–outil d'exploration des données multidimensionnelles. *Journal de la Société Française de Statistique*, 143, 3–4 and 5–55.

Gabriel, K.R. and Odoroff, C.L. (1990). Biplots in biomedical research. *Statistics in Medicine*, 9, 469–485.

Gabriel, K.R. and Zamir, S. (1979). Lower rank approximation of matrices by least squares with any choice of weights. *Technometrics*, 21, 489–498.

García Lautre, I. (2001). Medición y Análisis de las Infraestructuras: Una Nueva Metodología Basada en el Análisis Factorial Múltiple. Unpublished Ph.D. thesis, Department of Statistics and Operations Research. Pamplona: Public University of Navarra.

García Lautre, I. and Abascal, E. (2004). A methodology for measuring latent variables based on multiple factor analysis. *Computational Statistics and Data Analysis*, 45, 505–517.

Garthwaite, P.H. (1994). An interpretation of partial least squares. *Journal of American Statistical Association*, 89, 122–127.

Gifi, A. (1990). *Nonlinear Multivariate Analysis*. Chichester: Wiley.

Gilljam, M. and Granberg, D. (1993). Should we take don't know for an answer? *Public Opinion Quarterly*, 57, 348–392.

Gleason, J.R. (1988). Algorithms for balanced bootstrap simulations. *The American Statistician*, 42, 263–266.

Goitisolo, B. (2002). *El Análisis Simultáneo. Propuesta y Aplicación de un Nuevo Método de Análisis Factorial de Tablas de Contingencia*. Bilbao: Basque Country University Press.

Goldstein, M. and Dillon, W.R. (1978). *Discrete Discriminant Analysis*. New York: Wiley.

Gollob, H.F. (1968). A statistical model which combines features of factor analytic and analysis of variance techniques. *Psychometrika*, 33, 73–116.

Golub, T.R., Slonim, D.K., Tamayo, P., Huard, C., Gaasenbeek, M., Mesirov, J.P., Coller, H., Loh, M.L., Downing, J.R., Caligiuri, M.A., Bloomfield, C.D., and Lander, E.S. (1999). Molecular classification of cancer: class discovery and class prediction by gene expression monitoring. *Science*, 286, 531–537.

Goodman, L. A. (1968). The analysis of cross-classified data: independence, quasi-independence and interactions in contingency tables with and without missing entries. *Journal of the American Statistical Association*, 63, 1091–1131.

Goodman, L.A. (1979). Simple models for the analysis of association in cross-classifications having ordered categories. *Journal of the American Statistical Association*, 74, 537–552.

Goodman, L.A. (1985). The analysis of cross-classified data having ordered and/or unordered categories: association models, correlation models, and asymmetry models for contingency tables with or without missing entries. *Annals of Statistics*, 13, 10–69.

Goodman, L.A. (1991). Measures, models, and graphical display in the analysis of cross-classified data (with discussion). *Journal of the American Statistical Association*, 86, 1085–1138.

Goodman, L.A. (1996). A single general method for the analysis of cross-classified data: reconciliation and synthesis of some methods of Pearson, Yule, and Fisher, and also some methods of correspondence analysis and association analysis. *Journal of the American Statistical Association*, 91, 408–428.

Gower, J.C. (1968). Adding a point to vector diagram in multivariate analysis. *Biometrika*, 55, 582–585.

Gower, J.C. (1975). Generalized Procrustes analysis. *Psychometrika*, 40, 33–51.

Gower, J.C. (1990). Three dimensional biplots. *Biometrika*, 77, 773–785.

Gower, J.C. (1993). The construction of neighbour-regions in two dimensions for prediction with multi-level categorical variables. In *Information and Classification: Concepts-Methods-Applications*, O. Opitz, B. Lausen, and R. Klar, Eds. Berlin: Springer, pp. 174–189.

Gower, J.C. (1998). The role of constraints in determining optimal scores. *Statistics in Medicine*, 17, 2709–2721.

Gower, J.C. (1999). Discussion of contrast between psychometric and statistical approaches to multiway data-analysis. *Bulletin of the International Statistical Institute*, Tome 58, Book 3, 101–102.

Gower, J.C. (2002). Categories and quantities. In *Measurement and Multivariate Analysis*, S. Nishisato, Y. Baba, H. Bozdogan, and K. Kanefuji, Eds. Tokyo: Springer, pp. 1–12.

Gower, J.C. (2004). The geometry of biplot scaling. *Biometrika*, 91, 705–714.

Gower, J.C. and Hand, D.J. (1996). *Biplots*. London: Chapman and Hall.

Gower, J.C. and Harding, S.A. (1998). Prediction regions for categorical variables. In *Correspondence Analysis in the Social Sciences*, M.J. Greenacre and J. Blasius, Eds. London: Academic Press, pp. 405–419.

Gower, J.C., Gardner, S., and LeRoux, N.J. (2006). A synthesis of canonical variate analysis, generalised canonical correlation and Procrustes analysis. *Computational Statistics and Data Analysis,* 50, 107–134.

Green, M. (1989). Discussion of Van der Heijden, P.G.M., de Falguerolles, A., and de Leeuw, J.: a combined approach to contingency table analysis using correspondence analysis and log-linear analysis. *Applied Statistics*, 38, 249–292.

Green, P.E. (1973). On the analysis of interactions in marketing research data. *Journal of Marketing Research*, 10, 410–420.

Green, P.E. and Srinivasan, V. (1990). Conjoint analysis in marketing: new developments with implications for research and practice. *Journal of Marketing*, 54, 3–19.

Green, P.E. and Wind, Y. (1973). *Multiattribute Decisions in Marketing: A Measurement Approach*. Hindsdale: The Dryden Press.

Greenacre, M.J. (1981). Practical correspondence analysis. In *Interpreting Multivariate Data*, V. Barnett, Ed. Chichester: Wiley, pp. 119–146.

Greenacre, M.J. (1984). *Theory and Applications of Correspondence Analysis.* London: Academic Press.
Greenacre, M.J. (1988). Correspondence analysis of multivariate categorical data by weighted least squares. *Biometrika*, 75, 457–467.
Greenacre, M.J. (1989). The Carroll-Green-Schaffer scaling in correspondence analysis: a theoretical and empirical appraisal. *Journal of Marketing Research*, 26, 358–365.
Greenacre, M.J. (1993a). *Correspondence Analysis in Practice.* San Diego: Academic Press.
Greenacre, M.J. (1993b). Biplots in correspondence analysis. *Journal of Applied Statistics*, 20, 251–269.
Greenacre, M.J. (1994). Multiple and joint correspondence analysis. In *Correspondence Analysis in the Social Sciences*, M.J. Greenacre and J. Blasius, Eds. London: Academic Press, pp. 141–161.
Greenacre, M.J. (2000). Correspondence analysis of square asymmetric matrices. *Applied Statistics*, 49, 297–310.
Greenacre, M.J. (2004). *Weighted Metric Multidimensional Scaling.* Working paper 777. Department of Economics and Business, Universitat Pompeu Fabra, Barcelona.
Greenacre, M.J. and Blasius, J., Eds. (1994). *Correspondence Analysis in the Social Sciences: Recent Developments and Applications.* London: Academic Press.
Greenacre, M.J. and Hastie, T.J. (1987). The geometric interpretation of correspondence analysis. *Journal of the American Statistical Association*, 82, 437–447.
Greenacre, M. J. and Pardo, R. (in press). Subset correspondence analysis visualizing relationships among a selected set of response categories from a questionnaire survey. *Sociological Methods and Research*, 35.
Groenen, P. and Poblome, J. (2002). Constrained correspondence analysis for seriation in archaeology applied to Sagalassos ceramic tableware. In *Exploratory Data Analysis in Empirical Research. Proceedings of the 25th Annual Conference of the Gesellschaft für Klassifikation*, O. Opitz and M. Schwaiger, Eds. Heidelberg: Springer, pp. 90–97.
Groß, J. (2003). *Linear Regression.* Berlin: Springer.
Guttman, L. (1941). The quantification of a class of attributes: a theory and method of scale construction. In *The Prediction of Personal Adjustment*, P. Horst, Ed. New York: Social Science Research Council, pp. 321–348.
Guttman, L. (1950a). The basis for scalogram analysis. In *Measurement and Prediction*, Vol. 4 of *The American Soldier*, S.A. Stouffer, L. Guttman, E.A. Suchman, P.E. Lazarsfeld, S.A. Star, and J.A. Clausen, Eds. Princeton, NJ: Princeton University Press, pp. 60–90.
Guttman, L. (1950b). The principal components of scale analysis. In *Measurement and Prediction*, Vol. 4 of *The American Soldier*, S.A. Stouffer, L. Guttman, E.A. Suchman, P.E. Lazarsfeld, S.A. Star, and J.A. Clausen, Eds. Princeton, NJ: Princeton University Press, pp. 312–361.

Guttman, L. (1953). Image theory for the structure of quantitative variables. *Psychometrika*, 18, 277–296.

Guttman, L. (1954). The principal components of scalable attitudes. In *Mathematical Thinking in the Social Sciences*, P.F. Lazarsfeld, Ed. Glencoe, U.K.: Free Press, pp. 216–257.

Hand, D.J. (1981). *Discrimination and Classification*. New York: Wiley.

Harkness, J.A., Van de Vijver, F.J.R., and Mohler, P.P. (2003). *Cross-Cultural Survey Methods*. Hoboken NJ: Wiley.

Harshman, R.A. (1970). Foundations of the PARAFAC procedure: models and conditions for an "explanatory" multi-modal factor analysis. *UCLA Working Papers in Phonetics*, 16, 1–84.

Hastie, T., Tibshirani, R., and Friedman, J. (2001). *The Elements of Statistical Learning: Data Mining, Inference, and Prediction*. New York: Springer.

Hayashi, C. (1950). On the quantification of qualitative data from the mathematico-statistical point of view. *Annals of the Institute of Statistical Mathematics*, 2, 35–47.

Hayashi, C. (1952). On the prediction of phenomena from qualitative data and the quantification of qualitative data from the mathematico-statistical point of view. *Annals of the Institute of Statistical Mathematics*, 3, 35–47.

Hayashi, C. (1954). Multidimensional quantification—with the applications to the analysis of social phenomena. *Annals of the Institute of Statistical Mathematics*, 5, 231–245.

Hayashi, C., Suzuki, T., and Sasaki, M. (1992). *Data Analysis for Social Comparative Research: An International Perspective*. Amsterdam: North-Holland.

Heiser, W.J. (1981). Unfolding Analysis of Proximity Data. Unpublished doctoral dissertation. Leiden, the Netherlands: University of Leiden.

Heiser, W.J. (1995). Convergent computing by iterative majorization: theory and applications in multidimensional data analysis. In *Recent Advantages in Descriptive Multivariate Analysis*, W.J. Krzanowski, Ed. Oxford: Clarendon Press, pp. 157–189.

Heiser, W.J. and Meulman, J.J. (1994). Homogeneity analysis: exploring the distribution of variables and their nonlinear relationships. In *Correspondence Analysis in the Social Sciences: Recent Developments and Applications*, M.J. Greenacre and J. Blasius, Eds. London: Academic Press, pp. 179–209.

Hill, M.O. (1969). On looking at large correlation matrices. *Biometrika*, 56, 249–253.

Hill, M.O. (1974). Correspondence analysis: a neglected multivariate method. *Journal of the Royal Statistical Society C (Applied Statistics)*, 23, 340–354.

Hill, M.O. and Gauch, H.G., Jr. (1980). Detrended correspondence analysis: an improved ordination technique. *Vegetatio*, 42, 47–58.

Hirschfeld, H.O (1935). A connection between correlation and contingency. *Proceedings of the Cambridge Philosophical Society*, 31, 520–524

Hjellbrekke, J., Le Roux, B., Korsnes, O., Lebaron, F., Rosenlund, L., and Rouanet, H. (in press). The Norwegian field of power anno 2000. *European Societies.*

Hochberg, Y. (1988). A sharper Bonferroni procedure for multiple tests of significance. *Biometrika*, 75, 800–803.

Hoerl, A.F. and Kennard, R.W. (1970). Ridge regression: biased estimation for nonorthogonal problems. *Technometrics*, 12, 55–67.

Hoffmeyer-Zlotnik, H.P. and Wolf, C., Eds. (2004). *Advances in Cross-National Comparison.* New York: Kluwer.

Hoijtink, H. (1990). A latent trait model for dichotomous choice data. *Psychometrika*, 55, 641–656.

Hoijtink, H., Ed. (1993a). Special issue on unidimensional unfolding analysis. *Kwantitatieve Methoden*, 42.

Hoijtink, H. (1993b). The analysis of dichotomous preference data: models based on Clyde H. Coombs' parallelogram model. *Kwantitatieve Methoden*, 42, 9–18.

Horst, P. (1935). Measuring complex attitudes. *Journal of Social Psychology*, 6, 369–374.

Hosmer, D.W. and Lemeshow, S. (1989). *Applied Logistic Regression.* New York: John Wiley and Sons.

Hotelling, H. (1933). Analysis of a complex of statistical variables into principal components. *Journal of Educational Psychology*, 24, 417–441 and 498– 520.

Hotelling, H. (1936). Relations between two sets of variates. *Biometrika*, 28, 321–377.

Householder, A.S. and Young, G. (1938). Matrix approximation and latent roots. *American Mathematical Monthly*, 45, 165–171.

Hsu, J.C. (1996). *Multiple Comparisons: Theory and Methods.* London: Chapman and Hall.

Huber, P.J. (1985). Projection pursuit (with discussion). *Annals of Statistics*, 13, 435–525.

Hwang, H. and Takane, Y. (2002). Generalized constrained multiple correspondence analysis. *Psychometrika*, 67, 211–224.

ISSP (1993). International Social Survey Program: Environment. http:// www. issp.org.

ISSP (1994). International Social Survey Program: Family and Changing Gender Roles II. http://www.issp.org.

ISSP (2003). International Social Survey Program: National Identity III. www.issp.org.

Jambu, M. (1977). Sur l'utilisation conjointe d'une classification hiérarchique et de l'analyse factorielle en composantes principales. *Revue de Statistique Appliquée*, 25 (4), 5–35.

Jolliffe, I.T. (2002). *Principal Component Analysis.* 2nd ed. Springer Series in Statistics. Berlin: Springer Verlag.

Jongman, R.H.G., ter Braak, C.J.F., and Van Tongeren, O.F.R. (1987). *Data Analysis in Community and Landscape Ecology.* Cambridge: Cambridge University Press.

Kendall, M.G. and Stuart, A. (1973). *The Advanced Theory of Statistics*, Vol. 2. London: Griffin.

Klein, M.F. (2002). *Etude des Facteurs de Risques de l'Entérocolite Épizootique du Lapin en Engraissement*. Veterinary thesis, Nantes: Ecole Nationale Vétérinaire de Nantes.

Kroonenberg, P.M. (1989). Singular value decompositions of interactions in three-way contingency tables. In *Multiway Data Analysis*, R. Coppi and S. Bolasco, Eds. Amsterdam: Elsevier, pp. 169–184.

Kruskal, J.B. and Shepard, R.N. (1974). A nonmetric variety of linear factor analysis. *Psychometrika*, 39, 123–157.

Lafosse, R. and Hanafi, M. (1997). Concordance d'un tableau avec K tableaux: définition de K+1uples synthétiques. *Revue de Statistique Appliqué*, 45, 111–126.

Lancaster, H.O. (1951). Complex contingency tables treated by the partition of chi-square. *Journal of the Royal Statistical Society B*, 13, 242–249.

Landaluce, M.I. (1995). Estudio de la Estructura de Gasto Medio de las Comunidades Autónomas Españolas: Una Aplicación del Análisis Factorial Múltiple. Unpublished Ph.D. thesis, Department of Applied Economy (Statistics and Econometry), Bilbao: Public University of Basque Country.

Landaluce, M.I. (1997). Análisis factorial de tablas mixtas: nuevas equivalencias entre PCA normado y MCA. *Qüestiió*, 21, 99–108.

Lange, K., Hunter, D.R., and Yang, I. (2000). Optimization transfer using surrogate objective functions. *Journal of Computational and Graphical Statistics*, 9, 1–20.

Lattin, J., Carroll, J.D., and Green, P.E. (2003). *Analyzing Multivariate Data*. London: Thomson Learning.

Lauritzen, S.L. and Wermuth, N. (1989). Graphical models for association between variables of which some are qualitative and some are quantitative. *Annals of Statistics*, 17, 31–57.

Lauro, N. and d'Ambra, L. (1984). L'analyse non-symétrique des correspondances. In *Data Analysis and Informatics III*, E. Diday, Ed. Amsterdam: North Holland, pp. 433–446.

Lebaron, F. (2000). *La Croyance Économique: Les Économistes Entre Science et Politique*. Paris: Seuil.

Lebaron, F. (2001). Economists and the economic order: the field of economists and the field of power in France. *European Societies*, 3, 91–110.

Lebart, L. (1975). L'orientation du dépouillement de certaines enquêtes par l'analyse des correspondances multiples. *Consommation*, 2, 73–96.

Lebart, L. (1976). The significance of eigenvalues issued from correspondence analysis: proceedings in computational statistique. In *COMPSTAT 1976*, J. Gordesch and P. Naeve, Eds. Vienna: Physica Verlag, pp. 38–45.

Lebart, L. (1995). Assessing and comparing patterns in multivariate analysis. In *Data Science and Application*, C. Hayashi et al., Eds. Tokyo: Academic Press, pp. 193–204.

Lebart, L. and Fénelon, J.P. (1971). *Statistique et Informatique Appliquées*. Paris: Dunod.

Lebart, L., Morineau, A., and Piron, N. (1995). *Statistique Exploratoire Multidimensionelle*. Paris: Dunod.
Lebart, L., Morineau, A., and Tabard, N. (1977). *Techniques de la Description Statistique: Méthodes et Logiciels pour l'Analyse des Grands Tableaux*. Paris: Dunod.
Lebart, L., Morineau, A., and Warwick, K. (1984). *Multivariate Descriptive Statistical Analysis*. New York: Wiley.
Lebart, L., Piron, M., and Steiner, J.F. (2003). *La Sémiométrie*. Paris: Dunod.
Lebart, L., Salem, A., and Berry, E. (1998). *Exploring Textual Data*. Dordrecht: Kluwer.
Lebreton, J.D., Chessel, D., Prodon, E., and Yoccoz, N. (1988). L'analyse des relations espèces-milieu par l'analyse canonique des correspondances; I: Variables de milieu quantitatives. *Acta Œcologica, Œcoligica Generalis*, 9, 53–67.
Legendre, P. and Legendre, L. (1998). *Numerical Ecology*. Amsterdam: North Holland.
Lehmann, E.L. (1966). Some concepts of dependence. *Annals of Mathematical Statistics*, 37, 1137–1153.
Le Roux, B. (1999). Analyse spécifique d'un nuage euclidien: application à l'étude des questionnaires. *Mathématiques, Informatique et Sciences Humaines*, 146, 65–83.
Le Roux, B. and Chiche, J. (2004). Specific multiple correspondence analysis. *Hellenike Tetradia Analysis Dedomenon*, 4, 30–41.
Le Roux, B. and Rouanet, H. (1984). L'analyse multidimensionnelle des données structures. *Mathématiques et Sciences Humaines*, 85, 5–18.
Le Roux, B. and Rouanet, H. (1998). Interpreting axes in multiple correspondence analysis—method of the contributions of points and deviations. In *Visualization of Categorical Data*, J. Blasius and M.J. Greenacre, Eds. San Diego: Academic Press, pp. 197–220.
Le Roux, B. and Rouanet, H. (2004a). *Geometric Data Analysis: From Correspondence Analysis to Structured Data*. Dordrecht: Kluwer.
Le Roux, B. and Rouanet, H. (2004b). *Individual Differences in Gifted Students*. http://epgy.stanford.edu/research/GeometricDataAnalysis.pdf.
Lewis, J. and de Leeuw, J. (2004). *A General Method for Fitting Spatial Models of Politics*, technical report. Los Angeles: UCLA Department of Statistics.
Likert, R. (1932). A technique for the measurement of attitudes. *Archives of Psychology*, 140, 44–53.
Lingoes, J.C. (1968). The multivariate analysis of qualitative data. *Multivariate Behavioral Research*, 3, 61–94.
Lingoes, J.C. (1973). *The Guttman–Lingoes Nonmetric Program Series*. Ann Arbor: Mathesis Press.
Lingoes, J.C. and Guttman, L. (1967). Nonmetric factor analysis: a rank reducing alternative to linear factor analysis. *Multivariate Behavioral Research*, 2, 485–505.
Markus, M.T. (1994a). Bootstrap confidence regions for homogeneity analysis: the influence of rotation on coverage percentages. In *COMPSTAT*

1994, R. Dutter and W. Grossmann, Eds. Heidelberg: Physica, pp. 337–342.

Markus, M.T. (1994b). *Bootstrap Confidence Regions in Nonlinear Multivariate Analysis.* Leiden, the Netherlands: DSWO Press.

McCullagh, P. and Nelder, J.A. (1989). *Generalized Linear Models.* London: Chapman and Hall.

McDonald, R.P. (1983). Alternative weights and invariant parameters in optimal scaling. *Psychometrika,* 48, 377–391.

Méot, A. and Leclerc, B. (1997). Voisinages a priori et analyses factorielles: illustration dans le cas de proximités géographiques. *Revue de Statistique Appliquée,* 45, 25–44.

Meulman, J.J. and Heiser, W.J. (1998). Visual display of interaction in multiway contingency tables by use of homogeneity analysis: the 2 × 2 × 2 × 2 case. In *Visualization of Categorical Data,* J. Blasius and M.J. Greenacre, Eds. San Diego: Academic Press, pp. 253–275.

Meulman, J.J. and Heiser, W.J. (1999). *SPSS Categories 10.0.* Chicago: SPSS Inc.

Meulman, J.J. and Heiser, W.J. (2001). *SPSS Categories 11.0.* Chicago: SPSS Inc.

Michailidis, G. and de Leeuw, J. (1998). The Gifi system for descriptive multivariate analysis. *Statistical Science,* 13, 307–336.

Milan, L. and Whittaker, J. (1995). Application of the parametric bootstrap to models that incorporate a singular value decomposition. *Applied Statistics,* 44, 31–49.

Mokken, R.J. (1970). *A Theory and Procedure of Scale Analysis with Applications in Political Research.* New York: De Gruyter.

Nelder, J.A. and Wedderburn, R.W.M. (1972). Generalized linear models. *Journal of the Royal Statistical Society A,* 135, 370–384.

Murtagh, F. (2005). *Correspondence Analysis and Data Coding with R and Java.* Boca Raton, FL, Chapman and Hall.

Nishisato, S. (1980). *Analysis of Categorical Data: Dual Scaling and Its Applications.* Toronto: University of Toronto Press.

Nishisato, S. (1984). Forced classification: a simple application of a quantification technique. *Psychometrika,* 49, 25–36.

Nishisato, S. (1994). *Elements of Dual Scaling: An Introduction to Practical Data Analysis.* Hillsdale, NJ: Lawrence Erlbaum.

Nishisato, S. (1996). Gleaning in the field of dual scaling. *Psychometrika,* 61, 559–599.

Nishisato, S. (2002). Differences in data structures between continuous and categorical variables from dual scaling perspectives, and a suggestion for a unified mode of analysis [in Japanese]. *Japanese Journal of Sensory Evaluation,* 6, 89–94.

Nishisato, S. (2003a). Geometric perspectives of dual scaling for assessment of information in data. In *New Developments in Psychometrics,* H. Yanai, A. Okada, K. Shigemasu, Y. Kano, and J. Meulman, Eds. Tokyo: Springer, pp. 453–462.

Nishisato, S. (2003b). Total information in multivariate data from dual scaling perspectives. *The Alberta Journal of Educational Research*, 49, 244–251.

Nishisato, S. and Baba, Y. (1999). On contingency, projection and forced classification of dual scaling. *Behaviormetrika*, 26, 207–219.

Nishisato, S. and Gaul, W. (1990). An approach to marketing data analysis: the forced classification procedure of dual scaling. *Journal of Marketing Research*, 27, 354–360.

Nishisato, S. and Hemsworth, D. (2004). Quantification of ordinal variables: a critical inquiry into polychoric and canonical correlation. In *Recent Advances in Statistical Research and Data Analysis*, Y. Baba, A.J. Hayter, K. Kanefuji, and S. Kuriki, Eds. Tokyo: Springer, pp. 49–84.

Okabe, A., Boots, B., Sugihara, K., and Chiu, S.N. (2000). *Spatial Tessellations*. 2nd ed. New York: Wiley.

Pagès, J. (2004). Analyse factorielle de données mixtes. *Revue de Statistique Appliquée*, 52, 93–111.

Pearson, K. (1901). On lines and planes of closest fit to a system of points in space. *Philosophical Magazine*, 2, 559–572.

Peschar, J.L. (1975). *School, Milieu, Beroep*. Groningen, the Netherlands: TjeekWillink.

Post, W.J. and Snijders, T.A.B. (1993). A nonparametric probabilistic model for parallelogram analysis. *Kwantitatieve Methoden*, 42, 55–72.

R Development Core Team (2005). *R: A Language and Environment for Statistical Computing*. Vienna: R Foundation for Statistical Computing. http://www.R-project.org.

Rajadell Puiggros, N. (1990). Les Actitudes Envers la Lectura, un Model d'Anàlisi a l'Educació Primària. Unpublished Ph.D. thesis [in Catalan], Barcelona: Universitat de Barcelona.

Ramsay, J.O. and Silverman, B.W. (1997). *Functional Data Analysis*. New York: Springer.

Rao, C.R. (1964). The use and interpretation of principal component analysis. *Sankhya*, 26, 329–357.

Rao, V.R. (1977). Conjoint measurement in marketing analysis. In *Multivariate Methods for Market and Survey Research*, J. Sheth, Ed. Champaign-Urbana: University of Illinois, American Marketing Association, pp. 257–286.

Rasch, G. (1960). *Probabilistic Models for Some Intelligence and Attainment Tests*. Copenhagen: Nielsen and Lydiche.

Roberts, J.S. (1995). Item response theory approaches to attitude measurement, (doctoral dissertation, University of South Carolina, Columbia, 1995). *Dissertation Abstracts International*, 56, 7089B.

Roberts, J.S., Donoghue, J.R., and Laughlin, J.E. (2000). A general item response theory model for unfolding unidimensional polytomous responses. *Applied Psychological Measurement*, 24, 3–32.

Roberts, J.S. and Laughlin, J.E. (1996). A unidimensional item response model for unfolding responses from a graded disagree–agree response scale. *Applied Psychological Measurement*, 20, 231–255.

Rosenlund, L. (2000). Social structures and change; applying Pierre Bourdieu's approach and analytic framework. Working paper 85/2000, Stavanger University College, Stavanger, Norway.

Roskam, E.E.C.I. (1968). Metric Analysis of Ordinal Data in Psychology. Unpublished Ph.D. thesis. Leiden, the Netherlands: University of Leiden.

Rouanet, H., Ackermann, W., and Le Roux, B. (2000). The geometric analysis of questionnaires: the lesson of Bourdieu's La Distinction. *Bulletin de Méthodologie Sociologique*, 65, 5–18.

Rouanet, H. and Le Roux, B. (1993). *Analyse des Données Multidimensionnelles*. Paris: Dunod.

Rouanet, H., Lebaron, F., Le Hay, V., Ackermann, W., and Le Roux, B. (2002). Régression et analyse géométrique des données: réflexions et suggestions. *Mathématiques et Sciences Humaines*, 160, 13–45.

Rovan, J. (1994). Visualizing solutions in more than two dimensions. In *Correspondence Analysis in the Social Sciences*, M.J. Greenacre and J. Blasius, Eds. London: Academic Press, pp. 267–279.

Roy, S.N. and Kastenbaum, M.A. (1956). On the hypothesis of "no interaction" in a multi-way contingency table. *Annals of Mathematical Statistics*, 27, 749–757.

Ruìz-Gaze, A. (1996). A very simple robust estimator of a dispersion matrix. *Computational Statistics and Data Analysis*, 21, 149–162.

Samejima, F. (1969). Estimation of latent ability using a response pattern for graded scores. *Psychometrika*, Monograph 17.

Sanders, K. and Hoijtink, H. (1992). Androgynie bestaat. Vrouwelijkheid en mannelijkheid: twee onafhankelijke eigenschappen (Androgyny exists. Femininity and masculinity: two independent concepts). *Nederlands Tijdschrift voor Psychologie*, 47, 123–132.

Saporta, G. (1976). Discriminant analysis when all the variables are nominal, a stepwise method. Spring meeting of the Psychametric Society, Murray Hill, NJ.

Saporta, G. (1990a). *Probabilités, Analyse des Données et Statistique*. Paris: Technip.

Saporta, G. (1990b). Simultaneous analysis of qualitative and quantitative variables. *Società Italiana di Statistica, Atti della XXXV Riunione Scientifica*, 1, 63–72.

SAS (1992). SAS/STAT Software: Changes and Enhancements. Technical report P-229, Cary, NC: SAS Institute Inc.

Saville, D.J. (1990). Multiple comparison procedures: the practical solution. *American Statistician*, 44, 174–180.

Schein, A.I., Saul, L.K., and Ungar, L.H. (2003). A generalized linear model for principal component analysis of binary data. In *Proceedings of the Ninth International Workshop on Artificial Intelligence and Statistics*, C.M. Bishop and B.J. Frey, Eds., Jan. 3–6, 2003, Key West, FL, pp. 14–21.

Scheuch, E.K. (2000). The use of ISSP for comparative research. *ZUMA-Nachrichten*, 47, 64–74.

Schriever, B.F. (1985). Order Dependence. Unpublished doctoral dissertation, Amsterdam: Centre for Mathematics and Computer Science.
Schuman, H. and Converse, J.M. (1971). The effects of black and white interviewers on black responses in 1968. *Public Opinion Quarterly*, 35, 44–68.
Siciliano, R. and Mooijaart, A. (1997). Three-factor association models for three-way contingency tables. *Computational Statistics and Data Analysis*, 24, 337–356.
Smith, T. (2003). Developing comparable questions in cross-national surveys. In *Cross-Cultural Survey Methods*, J.A. Harkness, F.J.R. Van de Vijver, and P.P. Mohler, Eds. Hoboken, NJ: Wiley, pp. 69–91.
Smith, T.W. (1984). Nonattitudes: a review and evaluation. In *Surveying Subjective Phenomena*, C.F. Turner and E. Martin, Eds. New York: Russel Sage Foundation, pp. 215–255.
Statcorp (2003). *Stata Statistical Software: Release 8.* College Station, TX: Statcorp.
Stephenson, W. (1953). *The Study of Behavior.* Chicago: University Press.
Stone, M. (1974). Cross-validatory choice and assessment of statistical predictions. *Journal of the Royal Statistical Society B*, 36, 111–147.
Takane, Y. (2002). Relationships among various kinds of eigenvalue and singular value decompositions. In *New Developments in Psychometrics*, H. Yanai, A. Okada, K. Shigemasu, Y. Kano, and J. Meulman, Eds. Tokyo: Springer, pp. 45–56.
Takane, Y. and Hunter, M.A. (2001). Constrained principal component analysis: A comprehensive theory. *Applicable Algebra in Engineering, Communication, and Computing*, 12, 391–419.
Takane, Y. and Hwang, H. (2002). Generalized constrained canonical correlation analysis. *Multivariate Behavioral Research*, 37, 163–195.
Takane, Y. and Hwang, H. (2004). Regularized multiple-set canonical correlation analysis. Submitted for publication.
Takane, Y. and Yanai, H. (2003). A Simple Regularization Technique for Linear and Kernel Redundancy Analysis. Paper presented at the International Meeting of the Psychometric Society, July 2003, Sardinia, Italy.
Tateneni, K. and Browne, M. (2000). A noniterative method of joint correspondence analysis. *Psychometrika,* 65, 157–165.
Tenenhaus, M. (1998). *La Régression PLS.* Paris: Technip.
Ter Braak, C.J.F. (1986). Canonical correspondence analysis: a new eigenvector technique for multivariate direct gradient analysis. *Ecology*, 67, 1167–1179.
Ter Braak, C.J.F. (1990). *Update Notes: CANOCO Version 3.10.* Wageningen, the Netherlands: Agricultural Mathematics Group.
Thiessen, V. and Blasius, J. (1998). Using multiple correspondence analysis to distinguish between substantive and non-substantive responses. In *Visualization of Categorical Data*, J. Blasius and M.J. Greenacre, Eds. San Diego: Academic Press, pp. 239–252.
Thissen, D., Chen, W.H., and Bock, D. (2003). *Multilog 7: Analysis of Multiple-Category Response Data.* Chicago, IL: Scientific Software International.

Thomas, L.C., Edelman, D.B., and Crook, J.N. (2002). *Credit Scoring and its Applications*. Philadelphia, PA: SIAM Monographs on Mathematical Modelling and Computation.

Thurstone, L.L. (1927). A law of comparative judgment. *Psychological Review*, 34, 273–286.

Thurstone, L.L. (1947). *Multiple Factor Analysis*. Chicago: University of Chicago Press.

Thurstone, L.L. and Chave, E.J. (1929). *The Measurement of Attitude*. Chicago: University of Chicago Press.

Tock, K. and Suppes, P. (2002). The High Dimensionality of Students' Individual Differences in Performance in EPGY's K6 Computer-Based Mathematics Curriculum. http://epgy.stanford.edu/research.

Torres-Lacomba, A. (2001). Correspondence Analysis and Categorical Conjoint Measurement. Working paper 569, Departament d'Economia i Empresa, Universitat Pompeu Fabra, Barcelona. http://www.econ.upf.es/eng/research/onepaper.php?id=569.

Tucker, L.R. (1958). An inter-battery method of factor analysis. *Psychometrika*, 23, 111–136.

Tucker, L.R. (1960), Intra-individual and inter-individual multidimensionality. In *Psychological Scaling: Theory and Applications*, H. Gulliksen and S. Messick, Eds. New York: Wiley, pp. 155–167.

Tucker, L.R. (1966). Some mathematical notes to three-mode factor analysis. *Psychometrika*, 31, 279–311.

Underhill, L.G. (1990). The coefficient of variation biplot. *Journal of Classification*, 7, 41–56.

Van Blokland-Vogelesang, R. (1993). Unimodal social preference curves in unidimensional unfolding. *Kwantitatieve Methoden*, 42, 19–38.

Van den Wollenberg, A. (1977). Redundancy analysis: an alternative to canonical correlation analysis. *Psychometrika*, 42, 207–219.

Van der Burg, E. (1988). *Nonlinear Canonical Correlation and Some Related Techniques*. Leiden, the Netherlands: DSWO Press.

Van der Burg, E., de Leeuw, J., and Verdegaal, R. (1988). Homogeneity analysis with k sets of variables: an alternating least squares method with optimal scaling features. *Psychometrika*, 53, 177–197.

Van der Heijden, P.G.M. (1987). *Correspondence Analysis of Longitudinal Categorical Data*. Leiden, the Netherlands: DSWO Press.

Van der Heijden, P.G.M., de Falguerolles, A., and de Leeuw, J. (1989). A combined approach to contingency table analysis and log-linear analysis (with discussion). *Applied Statistics*, 38, 249–292.

Van der Heijden, P.G.M., Mooijaart, A., and Takane, Y. (1994). Correspondence analysis and contingency table models. In *Correspondence Analysis in the Social Sciences*, M.J. Greenacre and J. Blasius, Eds. London: Academic Press, pp. 79–111.

Van der Heijden, P.G.M. and Worsley, K.J. (1988). Comment on "correspondence analysis used complementary to loglinear analysis," *Psychometrika*, 53, 279–311.

Van de Vijver, F.J.R. (2003). Bias and equivalence: cross-cultural perspectives. In *Cross-Cultural Survey Methods*, J.A. Harkness, F.J.R. Van de Vijver, and P.P. Mohler, Eds. Hoboken, NJ: Wiley, pp. 143–155.

van Eeuwijk, F.A. (1995). Multiplicative interaction in generalized linear models. *Biometrics*, 51, 1017–1032.

Van Rijckevorsel, J. L. A. (1987). *The Application of Fuzzy Coding and Horseshoes in Multiple Correspondence Analysis*. Leiden, the Netherlands: DSWO Press.

Van Rijckevorsel, J.L.A. and de Leeuw, J. (1988). *Component and Correlation Analysis: Dimension Reduction by Functional Approximation*. Chichester: Wiley.

Van Schuur, W.H. (1993a). Masculinity and femininity B or androgyny? A comparison of three unfolding models. In *Advances in Longitudinal and Multivariate Analysis in the Behavioral Sciences: Proceedings of the SMABS Conference 1992*, J.H. Oud and R.A.W. Van Blokland-Vogelesang, Eds. Nijmegen, the Netherlands: Instituut voor Toegepaste Sociologie, pp. 219–233.

Van Schuur, W.H. (1993b). MUDFOLD. *Kwantitatieve Methoden*, 42, 39–54.

Van Schuur, W.H. (1993c). Nonparametric unfolding models for multicategory data. *Political Analysis*, 4, 41–74.

Van Schuur, W.H. (1998). From Mokken to mudfold and back. In *In Search of Structure: Essays in Social Science and Methodology*, M. Fennema, E. van der Eijk, and H. Schijf, Eds. Amsterdam: Het Spinhuis, pp. 45–62.

Van Schuur, W.H. (2003). Mokken scale analysis: between the Guttman scale and parametric item response theory. *Political Analysis*, 11, 139–163.

Vapnik, V. (1998). *Statistical Learning Theory*. New York: Wiley.

Venables, W.N. and Smith, D.M. (2003). An Introduction to R. www.r-project.org.

Verde, R. and Palumbo, F. (1996). Analisi fattoriale discriminante non-simmetrica su predittori qualitativi. Atti del Convegno della XXXVIII Riunione Scientifica della Società Italiana di Statistica.

Verdegaal, R. (1986). *OVERALS, Users Manual* (UG-86-1 ed.). Leiden, the Netherlands: Department of Data Theory.

Verhelst, N.D. and Verstralen, H.E.M. (1993). A stochastic unfolding model derived from the partial credit model. *Kwantitatieve Methoden*, 42, 73–92.

Vermunt, J.K. (1997). *Log-Linear Models for Event Histories*. Thousand Oaks, CA: Sage.

Vicente-Villardón, J.L. (2001). Biplot for binary data based on logistic response surfaces. Presented at Salamanca Statistics Seminar IV: Advances in Multivariate Analysis. Salamanca, Spain, December 2001.

Vinod, H.D. (1976). Canonical ridge and econometrics of joint production. *Journal of Econometrics*, 4, 47–166.

Vriens, M. (1995). *Conjoint Analysis in Marketing* Ph.D. thesis, University of Groningen, the Netherlands.

Walker, D.A. (1931). Answer pattern and score scatter in tests and examinations. *British Journal of Psychology*, 22, 73–86.

Webster, T., Gunst, R.F., and Mason, R.L. (1974). Latent root regression analysis. *Technometrics*, 16, 513–522.

Wermuth, N. and Lauritzen, S.L. (1990). Discussion of papers by Edwards, Wermuth and Lauritzen. *Journal of the Royal Statistical Society B*, 52, 51–72.

Whittaker, J. (1989). Discussion of paper by van der Heijden, de Falguerolles and de Leeuw. *Applied Statistics*, 38, 278–279.

Whittaker, J. (1990). *Graphical Models in Applied Mathematical Multivariate Statistics*. New York: Wiley.

Wickens, T.D. and Olzak, L.A. (1989). The statistical analysis of concurrent detection rating. *Perception Psychophysics*, 45, 514–528.

Wold, H. (1966). Estimation of principal components and related models by iterative least squares. In *Multivariate Analysis*, P.R. Krishnaiah, Ed. New York: Academic Press, pp. 391–420.

Wold, S., Martens, H., and Wold, H. (1983). The multivariate calibration problem in chemistry solved by the PLS method. In *Proceedings of the Conference on Matrix Pencils, Lecture Notes in Mathematics*, A. Ruhe and B. Kåstrøm, Eds. Heidelberg: Springer, pp. 286–293.

Wolff, M., Rouanet, H., and Grosgeorge, B. (1998). Analyse d'une expertise professionnelle: l'évaluation des jeunes talents au basket-ball de haut niveau. *Le Travail Humain*, 61, 281–303.

Wolters, M. (1982). Interspace Politics. Doctoral dissertation, Leiden, the Netherlands: University of Leijden.

Wong, R.S.K. (2001). Multidimensional association models: a multilinear approach. *Sociological Methods Research*, 30, 197–240.

Yang, Y., Dudoit, S., Luu, P., and Speed, T. (2000). Normalization for cDNA in microarray data. Technical report 589, University of California Berkeley: Department of Statistics.

Young, F.W. (1981). Quantitative analysis of qualitative data. *Psychometrika*, 46, 357–388.

Young, F.W., Takane, Y., and de Leeuw, J. (1978). The principal components of mixed measurement level multivariate data: an alternating least squares method with optimal scaling features. *Psychometrika*, 45, 279–281.

Zárraga, A. and Goitisolo, B. (2002). Méthode factorielle pour l' analyse simultanée de tableaux de contingence. *Revue de Statistique Appliquée*, 50, 47–70.

Zárraga, A. and Goitisolo, B. (2003). Étude de la structure inter-tableaux à travers l'analyse simultanée. *Revue de Statistique Appliquée*, 51, 39–60.

Index

A

B

C

D

E

F

G

H

I

N

O

P

R

S

T

U

V

W